Steven Eckhoff
Ag. Engineering
U of Illinois
217 244-4022

Sorghum and Millets: Chemistry and Technology

Edited by

David A. V. Dendy

Grains After Harvest
Upton, Oxford, United Kingdom

Published by the
American Association of Cereal Chemists, Inc.
St. Paul, Minnesota, USA

Cover photos courtesy Ciba-Geigy Corporation, Ciba Crop Protection, Greensboro, North Carolina.

Library of Congress Catalog Card Number: 94-72775
International Standard Book Number: 0-913250-84-8

Reference in this publication to a trademark, proprietary product, or company name by personnel of the U.S. Department of Agriculture or anyone else is intended for explicit description only and does not imply approval or recommendation of the product to the exclusion of others that may be suitable.

Printed in the United States of America on acid-free paper

American Association of Cereal Chemists, Inc.
3340 Pilot Knob Road
St. Paul, MN 55121-2097, USA

CONTRIBUTORS

D. J. Andrews, Department of Agronomy, University of Nebraska, Lincoln, Nebraska 68583-0915, U.S.A.

P. J. Bramel-Cox, Department of Agronomy, Kansas State University, Manhattan, Kansas 66506, U.S.A.

K. H. Daiber (*retired*), formerly with the Sorghum Beer Unit, Division of Food Science and Technology, Council for Scientific and Industrial Research (CSIR), Pretoria, South Africa

J. de Wet (*retired*), formerly with the International Crops Research Institute for the Semi-Arid Tropics (ICRISAT), Andhra Pradesh, India

D. A. V. Dendy, Grains After Harvest, Church Street, Upton, Didcot, Oxford. OX11 9JB, United Kingdom (formerly with the Natural Resources Institute, Chatham, U.K.)

M. Eustrom, U.S. Department of Agriculture, Federal Grain Inspection Service, Quality Assurance Research Division, FGIS Technical Center, 10383 North Executive Hills Blvd., Kansas City, Missouri 64153, U.S.A.

M. I. Gomez, formerly with the Southern Africa Development Council (SADC)/ICRISAT Sorghum and Millets Improvement Programme, Matopos, Zimbabwe

S. C. Gupta, SADC/ICRISAT Sorghum and Millets Improvement Programme, Matopos, Zimbabwe

L. Hallgren, Novo Nordisk A/S, Novo Allé, DK-2800 Bagsvaerd, Denmark (formerly with the Carlsberg Brewery A/S, Copenhagen, Denmark).

J. D. Hancock, Department of Animal Sciences and Industry, Kansas State University, Manhattan, Kansas 66506, U.S.A.

R. C. Hoseney, Department of Grain Science and Industry, Kansas State University, Manhattan, Kansas 66506, U.S.A.

L. R. House (*retired*), formerly with the SADC/ICRISAT Sorghum and Millets Improvement Program, Bulawayo, Zimbabwe

A. E. John, formerly with the Natural Resources Institute, Chatham, ME4 4TB, United Kingdom

C. F. Klopfenstein, Department of Grain Science and Industry, Kansas State University Manhattan, Kansas 66506, U.S.A.

K. A. Kumar, Pearl Millet Improvement Program, ICRISAT Sahelian Center, BP 12404, Niamey, Niger

J. A. McFarlane, formerly with the Natural Resources Institute, Chatham, ME4 4TB, United Kingdom

R. C. Marder, Natural Resources Institute, Chatham, ME4 4TB, United Kingdom

E. S. Monyo, SADC/ICRISAT Sorghum and Millets Improvement Programme, Matopos, Zimbabwe

L. Munck, Department of Food Technology, Royal Veterinary and Agricultural University, Thorvaldsensvej 40, DK-1871 Fredericksberg C, Denmark

D. S. Murty, ICRISAT West African Sorghum Improvement Program, PMB 3491, Kano, Nigeria

M. Osmanzai, SADC/ICRISAT Sorghum and Millets Improvement Programme, Matopos, Zimbabwe

L. W. Rooney, Department of Soil and Crop Science, Cereal Quality Laboratory, Texas A&M University, College Station, Texas 77843-2474, U.S.A.

R. E. Schaffert, Centro Nacional de Pesquisa de Milho e Sorgo (CNPMS)/ Empresa Brasileira de Pesquisa Agropecuaria (EMBRAPA), Caixa Postal 151, 35701-970 Sete Lagoas, Minas Gerais, Brazil

S. Serna-Saldivar, Instituto Tecnologico y de Estudios Superiores de Monterrey, Departamento de Tecnologia de Alimentos, Sucursal de Correos J, Monterrey, Nuevo León, Mexico

D. R. Shipman, U.S. Department of Agriculture, Federal Grain Inspection Service, Rm. 1094 South, P.O. Box 96454, Washington, D.C. 20090-6454, U.S.A.

J. R. N. Taylor, Cereal Foods Research Unit, Department of Food Science, University of Pretoria, Pretoria, Republic of South Africa

FOREWORD

The demand for cereals as food and feed is increasing. This is expected to continue well into the 21st century. The population explosion in developing countries is unprecedented in the history of the human species, and shortfalls in cereal production are common. In several developed countries, the demand for cereals as feed and industrial raw material also exceeds production capacity.

The total cereal production of the world in 1987 was estimated at 1.8 billion metric tons. Sorghum (62 million metric tons) and millets (32 million metric tons) accounted for about 5% of the cereal grain harvested. This is a minor contribution to the total word cereal production but belies the importance of sorghum and millets as human food; they are staples of some quarter million people in the semiarid tropics of Africa and Asia. Indeed, sorghum and the millets are the fifth and sixth most important cereals of the world, after wheat, rice, maize, and barley in declining order of importance.

The area planted to the four major cereals can no longer be greatly extended. Wheat and barley are cereals of temperate regions and already grown in habitats where production is marginally profitable. The cultivation of rice is limited by the availability of water for irrigation. This leaves maize, sorghum, and millets as the major cereals of the semiarid tropics.

Sorghum and millets cannot compete successfully with maize as a cereal in an agroecosystem with over 1,000 mm of annual rainfall; but it is inconceivable that maize will replace sorghum in areas with less than 900 mm of rainfall, or that any other cereal can compete successfully with pearl millet in the sahelo-sudanian ecological zone of Africa or the arid northwest of the Indian subcontinent with less than 600 mm of annual rainfall. Sorghum and pearl millet will remain the staples of people in a large part of the semiarid tropics.

There are other millets that play important roles in the diets of people in Africa and Asia. Important among these are finger millet in Africa and India, foxtail millet in south and west Asia, the fonios in West Africa, and little millet in central India. They supplement sorghum and pearl millet in agroecosystems where no other cereal will consistently produce a harvest. They also play an important part in the diet as specialty foods.

Sorghum and millets are used to prepare various kinds of porridges, breads, and beverages. Their grains are increasingly being used as feed for cattle and poultry, and their starches are finding increasing use in industry. They are used to produce malt and provide the starch in brewing and other industrial processes.

Substantial research on sorghum, pearl millet, finger millet, and foxtail millet is directed at increasing yield potential and stability and grain and fodder production. Little research is being done on improving grain or nutritional quality of these cereals or on their uses as industrial raw material. In India, as sorghum is becoming more abundantly available, it is already supplementing maize as a source of industrial starch. In Nigeria, sorghum malt is replacing barley malt, and the grain is replacing the more costly maize as a source of starch in the brewing industry. In southern Africa, the traditional sorghum beer is produced commercially; yet, we know very little about what is required in sorghum and millets to produce a good starch, a good malt, or a premium beer.

It is timely to have a compilation of the history, breeding, production, grain chemistry, nutritional quality, handling, and uses of sorghum and millet. This volume will be used not only by research scientists, but also by industrial users, other consumers, and by policy makers in determining future research strategy.

J. M. J. de Wet (*retired*)
International Crops Research Institute for the Semi-Arid Tropics (ICRISAT)
Andra Pradesh, India

CONTENTS

7. Traditional Uses of Sorghum and Millets.
D. S. MURTY and K. A. KUMAR, **185**

8. New Milling Technologies and Products: Whole Plant Utilization by Milling and Separation of the Botanical and Chemical Components.
L. MUNCK, **223**

9. Lager Beers from Sorghum. L. HALLGREN, 283

CHAPTER 1

SORGHUM AND MILLETS: HISTORY, TAXONOMY, AND DISTRIBUTION

L. R. House
Executive Director (retired)
SADC/ICRISAT Sorghum and Millet Improvement Program
Bulawayo, Zimbabwe

INTRODUCTION

The grasses (Poaceae) include some 8,000 species within 600 genera. It is estimated that historically some 300 species were harvested, but only 35 species in 20 genera have been domesticated (de Wet, 1986). Of greatest importance and mentioned in this chapter are sorghum (*Sorghum bicolor*), pearl millet (*Pennisetum glaucum*), finger millet (*Eleusine coracana*), proso millet (*Panicum miliaceum*), foxtail millet (*Setaria italica*), little millet (*Panicum sumatrense*, formerly *P. miliare*), barnyard (sawa) millet (*Echinochloa colona,* formerly *E. frumentacea*), kodo millet (*Paspalum scrobiculatum*), teff (*Eragrostis tef*), and fonio millet (*Digitaria exilis* and *D. iburua*) (Prasada Rao et al, 1991).

Excluding sorghum and pearl millet, these millets account for about 1% of the food grain used in the world but fit important niches that are generally marginal for other agriculture (de Wet, 1986).

SORGHUM
(*Sorghum bicolor* (L.) Moench)

History

As with most cereals, a major step in the process of sorghum domestication was the loss of the shattering habit (Mann et al, 1983). Vavilov considered the old Abyssinian (Ethiopian) area as the center of origin, but Harlan suggests that sorghum arose across a large area, where it was likely domesticated a number of times over a period of years. Snowden thought that sorghum arose in several separate centers and from different species: races durra and bicolor from *S. aethiopicum*, guinea from *S. arundinaceum*, and kafir from *S. verticilliflorum*. De Wet questions whether race virgatum, a desert grass, and arundinaceum, a forest grass, were involved with the direct origins of sorghum, since they grow outside the major habitat of the crop. He suggests that *S. verticilliflorum* was the first to be domesticated some 3,000 to 5,000 years ago (de Wet, 1986).

Taxonomy

Pliny (ca. 60 to 70 A.D.) was the first to give a written description of sorghum, and after that there was hardly any mention until the 16th century. During this period the genus name *Milium* was used, followed by several different species names. The words *sorghum* and *sorgo* were used in the early 1600s. Michelli in 1729 used sorghum as a generic name, and the word *dora* was used for a cultivated sorghum about 10 years later; this led to the word *durra*. Linnaeus in 1737 used the generic name *Holcus*. During the 1700s, several species were described: *H. dora, H. dochna*, and *H. cafer*. Moench in 1794 established the genus *Sorghum* and brought the sorghums under the name *S. bicolor*. *S. halepense* was described in 1848. Sorghum was also described under the genus *Andropogon,* which replaced the name *Holcus*. Alefold in 1866 and Koernicke in 1885 brought all cultivated sorghums in under *Andropogon sorghum*. Chiovenda in 1912 and Stapf in 1917 reworked the sorghums and, following Moench, brought them under the genus name *Sorghum*. (See Snowden [1936] for more detailed history.)

The most detailed classification of sorghum was made by Snowden (1936). While recognizing that sorghum could be subdivided into many races, he gave species status to 48 different taxa (31 cultivated species and 17 related wild species) to stress that they can individually be defined by a number of distinct characteristics. Snowden subdivided *Sorghum* into the following sections, subsections, and series:

Section *eu-sorghum* Stapf emend. Snowden
  Subsection *arundinacea* Snowden
    Series *spontanea* Snowden (10 wild species)
    Series *sativa* Snowden (31 cultivated species)
  Subsection *halepensia* Snowden (four wild grasses)
Section *para-sorghum* Snowden (eight to 10 annual and perennial grasses)

De Wet (1978) listed 28 of Snowden's cultivated taxa in subspecies *bicolor*, seven weedy taxa in subspecies *drummondii*, and 13 wild taxa in subspecies *arundinaceum*. Harlan and de Wet (1971) developed a simplified classification that is in common use. The cultivated taxa, including 28 (out of 31) species of Snowden's series *sativa,* are partitioned into the following races under *S. bicolor* subsp. *bicolor*:

Basic races—bicolor, guinea, caudatum, kafir, durra
Hybrid races—guinea-bicolor, caudatum-bicolor, kafir-bicolor, durra-bicolor, guinea-caudatum, guinea-kafir, guinea-durra, kafir-caudatum, durra-caudatum, kafir-durra

These 15 races can be identified by mature spikelet characteristics alone, although head type is frequently helpful (Harlan and de Wet, 1971).

Distribution

Sorghum is a tropical grass grown primarily in semiarid parts of the world. In Africa, a major growing area runs across West Africa south of the Sahara almost to the coast and eastward into Sudan, Ethiopia, and Somalia. It is grown in upper

Egypt but is minor along the north African coast. It is commonly grown in Uganda, Kenya, Tanzania, Rwanda, and Burundi. It is fairly important in Zambia, Tanzania, the Shiri Valley of Malawi, and drier areas of Mozambique. It is important in Botswana and Lesotho, common in South Africa, and minor in Namibia. It is an important crop in India and Pakistan, minor in Thailand, and important in central and northern China. It is important in Australia. It has been spreading in South America, particularly in the drier areas of Argentina, the Sao Paulo and northeast areas of Brazil, and drier areas of Venezuela and Colombia. It is important in Mexico and the southwestern United States, north into Nebraska, and in Europe is fairly common in France and Italy. The crop has spread over the drier areas of the world, in China being grown up to 45° north. Gradually, maize and sorghum are finding their places, with sorghum growing in areas too dry for maize. Sorghum does better when it is dry and cool, whereas pearl millet is better adapted to dry hot conditions.

PEARL MILLET
(*Pennisetum glaucum* (L.) R. Br.)

History

There are few substantiated early archaeological records of *Pennisetum.* Studies of the genus, including wild and cultivated types, have considered geographic distribution, morphological similarities, and genetic differences including isozymes and restriction fragment length polymorphism analysis (Andrews and Kumar, 1992; Tostain, 1992). Domestication probably occurred 5,000 or more years ago in Africa in the savanna south of the Sahara and west of the Nile (Andrews and Kumar, 1992). Domestication may have occurred among different isolated populations over time going back 7,500 to 7,000 years with indication of close coexistence with wild millet 3,000 years ago; however, intergression continues at the present time (Tostain, 1992). Genetic differences have been identified between wild and cultivated types, but domestication is considered to have involved relatively few genes (Andrews and Kumar, 1992).

Its introduction into India is uncertain, but is likely some 3,000 years ago. It was cultivated in the United States by 1873, while its introduction into Central and South America likely came from southern Europe and France. Its cultivation there more or less terminated toward the end of the 19th century. It has been grown some in Europe, but its use was more in the 19th than in the 20th century (Rachie and Majmudar, 1980).

Pearl millet is known by a number of names: bulrush, cattail, and spiked millet in English; *bajra* in Hindi; *dukhn* in Arabic; and *mil à chandelles* in much of West and North Africa (Rachie and Majmudar, 1980).

Taxonomy

Stapf, in 1934, divided the genus *Pennisetum* into five sections: *gymnothrix, eupennisetum, penicillaria, heterostachya,* and *brevivalvula.* The cultivated taxa belong to the section *penicillaria,* in which Stapf identified 32 species, of which six found in Africa are the likely progenitors of cultivated pearl millet. These are all annual species; there is a rhizomatous perennial, *P. purpureum.* Brunken et al

(1977) simplified the taxonomy of pearl millet by bringing the wild form *P. violaceum* (Lam.) L. Rich and the cultivated *P. glaucum* into the same biological species. In recent times, the cultivated pearl millet has been known taxonomically as *P. glaucum, P. typhoides,* and *P. americanum* (Rachie and Majmudar, 1980), but is currently known as *P. glaucum* (L.) R. Br. (Andrews and Kumar, 1992).

Wild and weedy pearl millet subspecies *monodii* and *stenostachyum* are most readily useful to cross with pearl millet for its improvement (Hanna, 1987).

Distribution

The most important countries growing pearl millet are India, Nigeria, Niger, Chad, Mali, Tanzania, Ethiopia, China (including Manchuria), and drier parts of the former U.S.S.R. Smaller producers include Uganda, Burkina Faso, Senegal, and the Sudan (Rachie and Majmudar, 1980). There are many other countries that grow the crop, particularly in Africa.

Early-maturing varieties are found from East Africa to Mali, with the greatest diversity found in Niger. Indian cultivars have cultural similarity to those found in southern Africa, and are more similar to the late-maturing cultivars of West Africa (Chowdhury and Smith, 1988).

Five geographically diverse groups have been identified among the wild types of northern Africa: 1) Darfur, Sudan, and eastern Chad; 2) western Chad and Nigeria; 3) the Air region of Niger; 4) eastern Mali and Niger; and 5) Senegal, Mauritania, and western Mali (Tostain, 1992). These groups are isolated from each other geographically.

Pearl millet is sown on about 15 million hectares in Africa and 12 million in Asia (Riley et al, 1993).

FOXTAIL MILLET
(*Setaria italica* (L.) P. Beauv.)

History

Seeds of foxtail millet have been found in Neolithic sites in eastern and central Asia, Europe, and the Middle East, suggesting that it is one of the most ancient cereals domesticated in Eurasia. *Setaria viridis* is a probable ancestor. Evidence suggests that domestication likely occurred from central Asia to Europe. Movement westward into Europe and eastward into east and southeast Asia and the Pacific Islands was gradual and accompanied by plant differentiation. Differentiation may have occurred in East Asia but not domestication. Findings from Neolithic Yang-shao suggest the possibility of a Chinese center of origin (Sakamoto, 1987). Seeds of foxtail have been found in early sites in Switzerland and Austria dating back 3,600 years, and it was widely used as food during the Bronze age. Remnants of foxtail have not been found in the Near East or ancient Egypt, nor in ancient sites in India (Prasada Rao et al, 1987).

Taxonomy

Cultivated foxtail was recognized by Linnaeus in 1753 as *Panicum italicum.* A number of taxa were recognized and brought into *Setaria italica* by Beauvois in 1812. The species is now classified as shown in Table 1.

TABLE 1
Taxonomic Classifications of the Minor Millets

Genus Species	Subspecies	Race	Subrace
Foxtail millet[a]			
Setaria			
pumila	*viridis* (wild)		
italica	*italica* (cultivated)	moharia	aristata, fusiformis, glabra
		maxima	compacta, spongiosa
		assamense	
		indica	ereeta, glabra, nana, profusa
Finger millet[b]			
Eleusine			
coracana	*africana*	africana	
		spontanea	
	coracana	elongata	laxa
		reclusa	
		sparsa	
		plana	seriata, confusa, grandigluma
		compacta	
		vulgaris	liliacea, stellata, incurvata, digitata
Proso millet[a]			
Panicum			
miliaceum	*miliaceum*	miliaceum	
		patentissimum	
		contractum	
		compactum	
		ovatum	
Little millet[a]			
Panicum			
sumatrense	*psilopodium*		
	sumatrense	nana	laxa, erecta
		robusta	laxa, compacta
Sawa millet[a]			
Echinochloa			
colona	*colona*		
	frumentacea	stolonifera	
		intermedia	
		robusta	
		laxa	
Japanese barnyard millet[a]			
Echinochloa			
crus-galli	*crus-galli*	crus-galli	
		macrocarpa	
	utilis	utilis	
		intermedia	
Kodo millet[c]			
Paspalum			
scrobiculatum		regularis	
		irregularis	
		variabilis	
orbiculare			
commersonii			
Teff			
Eragrostis			
tef			
Fonio			
Digitaria			
iburua			
exilis			

[a] Prasada Rao et al (1991).
[b] Prasada Rao et al (1987).
[c] Prasada Rao et al (1991) and Christopher (1987).

Distribution

Race moharia is found in Europe, southeastern Russia, Afghanistan, and Pakistan. Race maxima is grown in eastern China, Japan, and Korea and has been introduced into the United States. It also occurs in Nepal and Assam, the foothills of the Himalayas, northern India, and in Georgia of the former U.S.S.R. Race indica is found throughout India and Sri Lanka and likely arose from crosses between moharia and maxima (Prasada Rao et al, 1987).

FINGER MILLET
(*Eleusine coracana* L. Gaertn. subsp. *coracana*).

History

Finger millet is the second most important millet in Africa after pearl millet. It was domesticated in Africa, with archeological records going back some 3,000 years. In approximately the same period it was introduced into India. Five races are recognized and likely all arose in Africa; there has been little diversification in India (de Wet, 1978).

Taxonomy

The species *E. coracana* is divided as shown in Table 1.

Hybrids between subspecies *africana* and *coracana* form seeds that occur with the cultivated crop in Africa. These hybrids belong to race spontanea.

Distribution

Finger millet is most intensively grown around Lake Victoria in eastern Africa, in the southeastern parts of Karanataka, parts of Tamil Nadu, and Andhra Pradesh in southern India, and in Nepal. In eastern Africa, finger millet is most commonly found in Uganda, Tanzania, Rwanda, Burundi, eastern Zaire, Kenya, and to a lesser degree in Ethiopia, Sudan, and Somalia. It is also cultivated in Zimbabwe, Malawi, Zambia, Tanzania, Botswana, and Madagascar (Gupta et al, 1986). It is a minor crop scattered in provinces of south and southwest China (Jiaju, 1986). It is cultivated to a limited extent in Bangladesh (Majid et al, 1986) and Sri Lanka (Ponnuthurai, 1986).

PROSO
(*Panicum miliaceum* (L.))

History

The wild ancestor of proso has not yet been satisfactorily identified, but domestication is likely to have occurred in Manchuria (de Wet, 1986). Weedy forms are spread from the Aralo Caspian basin to Sinkiang and Mongolia. More recently, these wild types have spread into Europe and North America. The earliest remains of proso have been dated to the fifth millennium B.C. in eastern and central Europe (Zohary and Hopf, 1988). Sites were found in the Ukraine, Czechoslovakia, and Germany. Other finds dating to the same period have occurred in Georgia. Finds in the fourth millennium B.C. include Yugoslavia, and in the Neolithic Yang Shao villages of north China. Proso was probably introduced into Europe at least 3,000 years ago; remains of spikelet and florets have been found in early farming sites dated around 1600 B.C. (de Wet, 1986). It appeared after this period in the Near East and India (Zohary and Hopf, 1988).

Taxonomy

The species is subdivided as shown in Table 1. Race miliaceum resembles the wild *Panicum miliaceum* in panicle characteristics. The other races have different morphological characteristics.

Distribution

Proso is important in northwest China, being the main crop in some areas. It is used less in the northeast (Inner Mongolia, Shanxi, Gansu, Ningxia, and Heilongjing) (Jiaju, 1986).

Main growing areas in the former U.S.S.R. are in the Volga region and parts of Kazakhastan (Ilyin and Zolotukin, 1986). In India, proso is cultivated primarily in the central and southern states of Madhya Pradesh, Andhra Pradesh, Tamil Nadu, Karnataka, Gujarat, Uttar Pradesh, Maharashtra, and Orissa, and to a limited extent in hilly regions of Himachal Pradesh, Uttar Pradesh, and Jammu and Kashmir (Sampath et al, 1986). It is cultivated in Bangladesh (Majid et al, 1986), is of minor importance in the drier districts of Sri Lanka (Ponnuthurai, 1986), and is grown across temperate Eurasia from eastern Europe to Japan (de Wet, 1986).

LITTLE MILLET (*Panicum sumatrense* Roth. ex Roem. & Schult.)

Taxonomy

Races and subraces of *Panicum sumatrense* are identified in Table 1. The subspecies *psilopodium* includes the wild progenitor of *P. sumatrense* (sama). The two subspecies cross to produce fertile hybrids, which often are found as weeds in sama fields (Prasada Rao et al, 1991). Meiotic behavior in crosses suggests that the genomes of the two species are similar (Hiremath et al, 1980).

Distribution

Sama is cultivated in much of India, in Nepal, and western Burma. It is of particular importance in the Eastern Ghats of India. Race robusta is grown in northwestern Andhra Pradesh and part of Orissa where it crosses with race nana. Subspecies *psilopodium* is distributed from Sri Lanka to Pakistan and eastward to Indonesia.

BARNYARD MILLET (*Echinochloa crus-galli* (L.) P. Beauv.) SAWA (*Echinochloa colona* (L.) Link)

History

Echinochloa crus-galli, commonly known as Japanese millet, according to archaeological records was domesticated in Japan about 5,000 years ago and is found in the temperate regions of Eurasia (de Wet, 1986). *E. colona*, domesticated in India, spread through the tropics and subtropics of the Old World. It was harvested as a wild cereal in predynastic Egypt. While the two species are morphologically similar, hybrids between them are sterile; however, crosses of both species with their respective wild relatives are fertile (de Wet, 1986).

Taxonomy

Echinochloa has been classified as shown in Table 1. The species *colona* is weedy and invades cultivated fields. It has been divided into four races that do not have geographical, ecological, or ethnological similarities (Prasada Rao et al, 1991).

Distribution

E. crus-galli occurs as a common weed of the Old and New Worlds, being cultivated in China, Korea, and Japan (de Wet, 1986). It is widely distributed in the temperate parts of Eurasia, and weedy types found in the United States are known as barnyard grass (Prasada Rao et al, 1991).

KODO MILLET
(*Paspalum scrobiculatum* L.)

History

Kodo millet was domesticated in India almost 3,000 years ago. The species is found across the Old World in damp habitats of the tropics and subtropics. Crossing readily occurs between cultivated and weedy races, and seed from hybrids is harvested along with those of the sown crop; hence, racial differentiation is not distinct despite the years of cultivation in India (de Wet, 1986).

Taxonomy

The species has been described as shown in Table 1 primarily on the basis of characteristics of the florets (racemes).

Paspalum orbiculare and *Paspalum commersonii* are wild types. *Paspalum commersonii* is thought to be apomicitic and of hybrid origin (Christopher et al, 1987). The wild weedy and cultivated kinds merge in all characters evaluated for taxonomic purposes (Prasada Rao et al, 1991). However, irregularities in meiosis in crosses support the identification of the three species (Christopher et al, 1987).

Distribution

Paspalum scrobiculatum is widely distributed across the Old World tropics, particularly in damp areas. As a wild cereal, it is harvested in West Africa and India. It is widely spread in India from Tamil Nadu and Kerala in the south to Rajasthan and West Bengal in the north (de Wet, 1986). Kodo is grown to a limited extent in Bangladesh (Majid et al, 1986).

TEFF
(*Eragrostis tef* (Zucc.) Trotter)

History

Teff is an ancient crop believed to have been domesticated in the northern highlands of Ethiopia. Seeds were found in the pyramids of Dassur built around 3350 B.C. It is thought that teff was grown in Egypt before the eighth century B.C. (Seyfu, 1986).

Teff was likely derived from *Eragrostis pilosa* (de Wet, 1986).

Distribution

Teff is the most important cereal in Ethiopia, occupying a greater area than all other cereals combined (Seyfu, 1986). Teff is also grown in a small area in South Africa and Lesotho, but as a forage crop.

FONIO
(*Digitaria iburua* Stapf, *D. exilis* (Kippist) Stapf)

Black fonio, *Digitaria iburua* is grown primarily by the Hansa tribe of northern Nigeria but occurs in spots across much of semiarid West Africa. This drought-hardy crop is frequently grown between rows of pearl millet or sorghum. True fonio, *Digitaria exilis* is widely grown across the West African savanna (de Wet, 1986).

LITERATURE CITED

ANDREWS, D. J., and KUMAR, K. A. 1992. Pear millet for food, feed, and forage. Adv. Agron. 48:89-139.

BRUNKEN, J., DE WET, J. M. J., and HARLAN, J. R. 1977. The morphology and domestication of pearl millet. Econ. Bot. 31:163-174.

CHRISTOPHER, J., RAJ, P. S., and PILLAI, K. G. 1987. Cytomorphological studies of three species of *Paspalum* Linn. from South India. Cytologia 52(3):487-491.

CHOWDHURY, M. D. U., and SMITH, R. L., 1988. Mitochondrial DNA variation in pearl millet and related species. Theor. Appl. Genet. 76(1):25-32.

DE WET, J. M. J. 1978. Systematics and Evolution of Sorghum, Sect. Sorghum (Gramineae). Am. J. Bot. 654: 477-484.

DE WET, J. M. J. 1986. Origin, evolution and systematics of minor cereals. Pages 19-30 in: Small Millets in Global Agriculture. A. Seetharama, K. W. Riley, and G. Harinarayana, eds. Proc. Int. Small Millets Workshop, 1st, Bangalore, India. Oxford and IBH Publ. Co., New Delhi, India.

DE WET, J. M. J., HARLAN, J. R., and PRICE, E. G. 1970. Origin of the variability in the *Spontanea* complex of *Sorghum bicolor*. Am. J. Bot. 57:704-707.

DOGGETT, H. 1986. Small millets—A selective overview. Pages 3-17 in: Small Millets in Global Agriculture. A. Seetharama, K. W. Riley, and G. Harinarayana, eds. Proc. Int. Small Millets Workshop, 1st, Bangalore, India. Oxford and IBH Publ. Co., New Delhi, India.

GUPTA, S. C., RAO, S. A., and HOUSE, L. R. 1986. Finger millet research in the SADCC (southern African region). Pages 115-126 in: Small Millets in Global Agriculture. A. Seetharama, K. W. Riley, and G. Harinarayana, eds. Proc. Int. Small Millets Workshop, 1st, Bangalore, India. Oxford and IBH Publ. Co., New Delhi, India.

HANNA, W. W. 1987. Utilization of wild relatives of pearl millet. Pages 33-442 in: Proc. Int. Pearl Millet Workshop. International Crops Research Institute for the Semi-Arid Tropics (ICRISAT), Patancheru, A. P., India.

HARLAN, J. R., and DE WET, J. M. J. 1971. A Simplified classification of cultivated sorghum. Crop Sci. 12(2):172-176

HIREMATH, S. C., PATEL, G. N. V., and SALIMATH, S. S. 1980. Genome homology and origin of *Panicum sumatrense* (Gramineae). Cytologia 55(2):315-319.

ILYIN, V. A., and ZOLOTUKIN, E. N. 1986. Breeding of proso millet (*Panicum miliaceum* L.) in Volga. Pages 105-111 in: Small Millets in Global Agriculture. A. Seetharama, K. W. Riley, and G. Harinarayana, eds. Proc. Int. Small Millets Workshop, 1st, Bangalore, India. Oxford and IBH Publ. Co., New Delhi, India.

JIAJU, C. 1986. Importance and genetic resources of small millets with emphasis on foxtail millet (*Setaria italica*) in China. Pages 93-104 in: Small Millets in Global Agriculture. A. Seetharama, K. W. Riley, and G. Harinarayana, eds. Proc. Int. Small Millets Workshop, 1st, Bangalore, India. Oxford and IBH Publ. Co., New Delhi, India.

MAJID, M. A., HAMID, M. A., and MANNUJAN 1986. Importance, genetic resources and breeding of small millets in Bangladesh. Pages 71-76 in: Small Millets in Global Agriculture. A. Seetharama, K. W. Riley, and G. Harinarayana, eds. Proc. Int. Small Millets Workshop, 1st, Bangalore, India. Oxford and IBH Publ. Co., New Delhi, India.

MANN, J. A., KIMBER, C. T., and MILLER, F. R. 1983. The origin and early cultivation of sorghum in Africa. Texas Agric. Exp. Stn. Bull. 1454.

PONNUTHURAI, S. 1986. Importance, genetic resources and breeding of small millets in Sri Lanka. Pages 77-84 in: Small Millets in Global Agriculture. A. Seetharama, K. W. Riley, and G. Harinarayana, eds. Proc. Int. Small Millets Workshop, 1st, Bangalore, India. Oxford and IBH Publ. Co., New Delhi, India.

PRASADA RAO, K. E., DE WET, J. M. J., BRINK, D. E., and MENGESHA M. H. 1987. Intraspecific variation and systematics of cultivated *Setaria italica*. Foxtail Millet Poaecea. Econ. Bot. 4:108-116.

PRASADA RAO, K. E., DE WET, J. M. J., GOPAL REDDY, V., and MENGESHA, M. H. 1991. Genetic diversity in the small millets collection at ICRISAT. Int. Small Millets Workshop, 2nd, Bulawayo, Zimbabwe. Int. Crops. Res. Inst. Semi-Arid Tropics (ICRISAT), Patancheru, A. P., India.

RACHIE, K. O., and MAJMUDAR, J. V. 1980. Origin and history. Pages 1-9 in: Pearl Millet. Pennsylvania State University Press, University Park.

SAKAMOTO, S. 1987. Origin and dispersal of common millet and foxtail millet. JARQ (Jpn. Agric. Res. Q.) 21(2):84-89.

RILEY, K. W., GUPTA, S. C., SEETHARAMA, A., and MUSHONGA, V. N. 1993. Advances in Small Millets. Oxford and IBH Publ. Co., New Delhi, India.

SAMPATH, T. V., RAZVI, S. M., SINGH, D. N., and BONDALE, K. V. 1986. Millets in Indian agriculture. Pages 33-43 in: Small Millets in Global Agriculture. A. Seetharama, K. W. Riley, and G. Harinarayana, eds. Proc. Int. Small Millets Workshop, 1st, Bangalore, India. Oxford and IBH Publ. Co., New Delhi, India.

SEYFU, K. 1986. Production trends, germplasm resources, breeding and varietal improvement, with special emphasis on teff in Ethiopia. Pages 167-172 in: Small Millets in Global Agriculture. A. Seetharama, K. W. Riley, and G. Harinarayana, eds. Proc. Int. Small Millets Workshop, 1st, Bangalore, India. Oxford and IBH Publ. Co., New Delhi, India.

SNOWDEN, J. D. 1936. Cultivated races of sorghum. Adlard and Sons, London.

TOSTAIN, S. 1992. Enzyme diversity in pearl millet (*Pennisetum glaucum* L.). Theor. Appl. Genet. 83:733-742.

ZOHARY, D., and HOPF, M. 1988. Common millet (*Panicum miliaceum*). Pages 76-7 in: Domestication of Plants in The World. Clarendon Press, Oxford.

CHAPTER 2

SORGHUM AND THE MILLETS: PRODUCTION AND IMPORTANCE

David A. V. Dendy
Grains After Harvest
Didcot, Oxford
United Kingdom

INTRODUCTION

The map (Fig. 1) shows the current producing areas for sorghum and the millets, and it will be noted at a glance that these grains are of greatest importance in Africa, India, and central Asia.

Benjamin Disraeli once said that there are lies, damned lies, and statistics, referring, of course, to the almost habitual misinterpretation of statistics. Nowhere is this adage more true than with the production figures for the small grains, sorghum and the millets. Global figures provided by the Food and Agriculture Organization

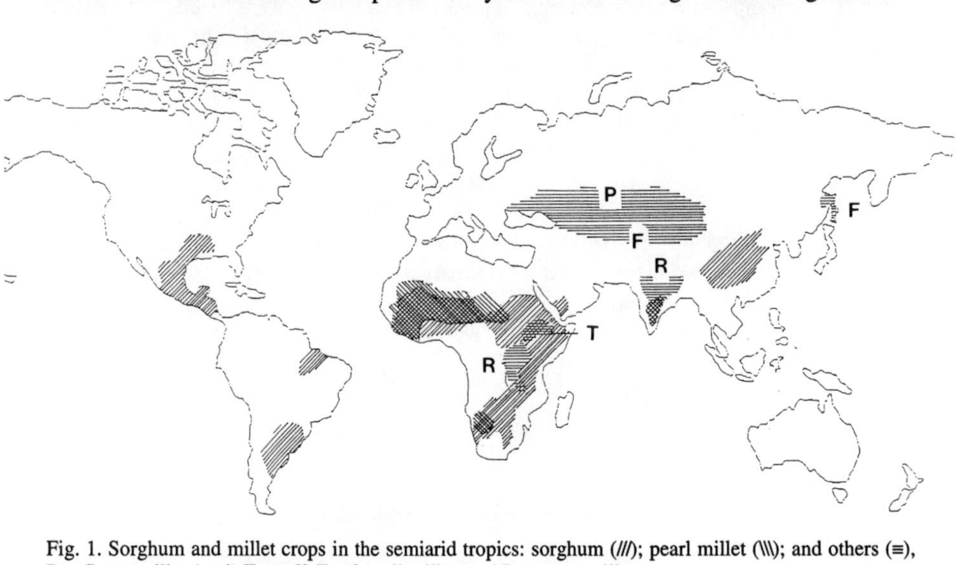

Fig. 1. Sorghum and millet crops in the semiarid tropics: sorghum (///); pearl millet (\\\); and others (≡), R = finger millet (ragi) T = teff, F = foxtail millet, and P = proso millet.

Formerly with the Natural Resources Institute, Chatham Maritime, United Kingdom.

of the United Nations (FAO) suggest that these grains are of little importance compared with wheat, rice, maize, and barley (Table 1) (FAO, 1962–1993) The truth is otherwise. Although they account for only 4.7% of world grain production—which represents, therefore, 4.4% of the world's total starchy staples on a food energy (Calorie) basis—they are extremely important in the semiarid zones, where, frequently, little else can be grown.

Thus, these small grains constitute a major source of energy and protein for millions of people in Asia and Africa. The combined figures are given in Table 2, listed regionally and by country, with the human population. The figure for staples was calculated by adding together the production of root crops, plantains, sago, and ensete, and converting this to cereal equivalents (25% of the wet weight).

Sorghum and the millets contribute annually some 100 million metric tons to the global food budget. In 13 countries, over half the cereal diet derives from sorghum and/or millets (Niger, Burkina Faso, Sudan, Mali, Senegal, Chad, Gambia, Togo, Namibia, Botswana, Uganda, Rwanda, and Yemen), and in four others they constitute over one third (Nigeria, Ghana, Ethiopia, and Somalia).

Table 2 takes FAO figures (FAO, 1962–1993) and compares production of sorghum and the millets (totaled from data in Tables 3 and 4) to the total cereals and staples production. The real value to the peoples of the less developed countries is expressed in terms of Calories per capita per diem. These figures are further developed in the four graphs (Figs. 2 and 3), which show sorghum and millets production for the years 1961–1989 for the world, the developed and developing countries, and for Africa, the continent where sorghum and several of the millets had their Vavilov center and are still very important subsistence crops.

These statistics, however, understate the importance of sorghum and the millets; it cannot be overemphasized that these grains are of immensely greater nutritional significance in the diets of poor people in the semiarid tropics than is wheat in the diets of contemporary Europeans and North Americans. Sorghum and the millets are the essential grains in regions where endemic drought causes the frequent failure of other crops. An example is Kenya, where in 1984 the maize crop was a total failure in Eastern Province, whereas the sorghum yield was only halved. In Namibia the people of the north (Ovamboland and Kavango) are almost entirely

TABLE 1
World Cereals Production[a]

	Million Hectares (ha)			Million Metric Tons (t)			Yield (t/ha)		
Cereal	**1979–1981**	**1985**	**1992**	**1979–1981**	**1985**	**1992**	**1979–1981**	**1985**	**1992**
Wheat	235	230	221	443	510	557	1.89	2.21	2.52
Paddy	144	145	148	396	466	521	2.76	3.22	3.52
Rice[b]				(264)	(310)	(347)	(1.84)	(2.15)	(2.35)
Maize	126	133	133	424	490	512	3.35	3.68	3.83
Barley	81	79	71	156	178	155	1.94	2.26	2.19
Sorghum	46	50	45	66	77	66	1.44	1.54	1.49
All millets	38	42	37	26	31	28	0.67	0.74	0.76
Other cereals	49	51	42	79	89	76	1.61	1.74	1.81
Total	719	730	697	1,590	1,841	1,915			
Calculated tons/hectare							2.21	2.52	2.75

[a] Data from production yearbooks of the Food and Agriculture Organization of the United Nations (FAO, 1962–1993).
[b] Paddy × 0.667.

TABLE 2
Production of Sorghum and Millets[a]

Region or Country	1992 Population (in 000s)	Range of Annual Rainfall (cm)	Total Cereals Production (000 t/annum)	Sorghum and Millets: 1992 Production (000 t/annum)	1992 Area (000 ha/annum)	Yield (t/ha)	Yield (Cal/capita per day)	As a Percentage of Cereals	As a Percentage of Total Staples
World	5,479,758	0–500	1,914,796	94,093	81,104	1.16	169	4.9	4.6
Developed	1,269,729	0–500	866,220	26,293	9,104	2.89	204	3.0	2.9
Developing	4,210,029	0–500	1,048,536	67,800	71,999	0.94	159	6.5	5.9
Africa total	681,674	0–500	79,771	23,498	33,538	0.70	340	29.5	21.9
West and North Africa									
Nigeria	115,664	20–300	12,523	7,300	7,900	0.92	622	58.3	33.3
Sudan	26,656	0–100	4,106	3,191	5,844	0.55	1,181	77.7	77.0
Ghana	15,959	20–200	1,436	250	420	0.60	155	17.4	8.6
Cameroon	12,198	60–500	1,003	463	560	0.83	374	46.2	31.1
Mali	9,818	0–60	2,232	1,523	1,850	0.82	1,530	68.2	67.1
Burkina Faso	9,513	0–60	2,234	1,964	2,396	0.82	2,036	87.9	86.8
Senegal	7,736	0–20	914	555	995	0.56	708	60.7	59.3
Niger	8,252	0–10	2,415	2,278	6,410	0.36	2,723	94.3	91.9
Chad	5,846	0–10	891	660	1,146	0.58	1,114	74.1	62.4
Togo	3,763	60–200	447	183	285	0.64	480	40.9	26.7
Gambia	908	20–40	108	61	68	0.90	663	56.5	55.7
East, central, and southern Africa									
Ethiopia	52,981	0–300	6,245	2,512	2,504	1.00	468	40.2	37.1
Tanzania	27,829	0–300	3,419	850	992	0.86	301	24.9	16.6
Uganda	18,674	20–200	1,652	850	600	1.42	449	51.5	17.2
Zimbabwe	10,583	10–60	505	59	100	0.59	55	11.7	11.0
Rwanda	7,526	60–300	330	172	182	0.95	225	52.1	11.3
Burundi	5,823	60–300	340	100	71	1.41	169	29.2	9.2
Namibia	1,534	0–150	114	23	70	0.33	148	20.2	12.6
Botswana	1,313	20–100	45	40	158	0.25	300	88.9	85.6
Asia total	3,233,333		876,719	34,822	35,347	0.99	106	4.0	3.7
China (People's Republic)	1,188,302	0–500	398,515	10,115	3,710	2.73	84	2.5	2.3
India	879,548	20–500	197,630	22,900	29,500	0.78	257	11.6	11.3
Nepal	20,577	40–150	5,505	225	197	1.14	108	4.1	3.9
Yemen	12,535	0–40	298	348	496	0.70	274	116.8	102.4
Other									
Mexico	88,153	10–400	23,511	4,467	1,273	3.51	500	19.0	18.8
Argentina	33,100	0–100	23,493	2,657	778	3.42	792	11.3	10.9
Colombia	33,424	20–500	3,939	750	265	2.83	221	19.0	14.8
U.S.S.R. (former)	293,092	0–100	179,975	2,250	3,130	0.72	76	1.3	1.1
United States	254,910	0–300	338,011	21,825	5,060	4.31	844	6.5	6.4

[a] Data from FAO (1993).

TABLE 3
Sorghum Production by Principal Producing Countries in 1992[a]

Region or Country	Production (000 t/annum)	Area (000 ha/annum)	Yield (t/ha)	Calories per Capita per Day	As Percentage of Cereals
World	66,333	44,452	1.49	119	3.5
Developed	23,946	5,885	4.07	186	2.8
Developing	42,387	38,566	1.10	99	4.0
Africa total	14,000	18,336	0.76	203	17.6
West and North Africa					
Burkina Faso	1,179	1,296	0.91	1,222	52.8
Sudan	2,941	4,694	0.63	1,088	71.6
Mali	729	750	0.97	732	32.7
Chad	365	496	0.74	616	41.0
Niger	378	2,010	0.19	452	15.7
Nigeria	4,100	4,000	1.03	350	32.7
Cameroon	400	500	0.80	323	39.9
Togo	106	165	0.64	278	23.7
Benin	110	150	0.73	221	20.6
Ghana	150	250	0.60	93	10.4
East, central, and southern Africa					
Botswana	38	153	0.25	285	84.4
Rwanda	170	180	0.94	223	51.5
Tanzania	587	692	0.85	208	17.2
Uganda	350	230	1.52	185	21.2
Ethiopia	900	880	1.02	168	14.4
Burundi	88	58	1.52	149	25.7
Kenya	143	146	0.98	56	5.0
Asia, total	18,962	17,144	1.11	58	2.2
Yemen	320	402	0.80	252	107.4
India	12,300	14,500	0.85	138	6.2
China (People's Republic)	5,615	1,480	3.79	47	1.4
Other					
Mexico	4,467	1,273	3.51	500	19.0
Argentina	2,592	728	3.56	772	11.0
Colombia	750	265	2.83	221	19.0
U.S.S.R. (former)	150	130	1.15	5	0.1
United States	21,675	4,935	4.39	839	6.4

[a] Data from FAO (1993).

TABLE 4
Sorghum: Feed Utilization in Selected Countries[a]

Country	Average (million tons) 1961–1965	1981–1985
United States	11.5	12.6
Mexico	0.5	8.0
Japan	0.7	3.4
Argentina	0.9	2.6
U.S.S.R. (former)	0.1	2.3
Venezuela	...	1.0
Others	3.5	6.5
World	17.2	36.4

[a] Source: FAO-IGG (1988); used by permission.

dependent on pearl millet, their subsistence crop. Yields, at 200 kg/ha, are low, but relatively dependable, even on the impoverished sandy soil and low rainfall. Food from other regions of Namibia and from imports must be brought over 600 km, at very high transport cost (for a discussion of pricing policy, currency valuation, and transport cost, see Dendy and Trotter, 1988).

In the semiarid tropics, sorghum and the millets cannot be replaced by wheat or maize except by importation; the effects of this on the economies of developing countries is open to debate. If sustainable income from tourism and exports of fossil fuels and other useful minerals, diamonds, and cash crops are available, then staples may well be imported; however, in the semiarid tropics this is rarely the case, and when it is, the real cost rather than the subsidized value should be used.

The long droughts of the 1970s and 1980s have so reduced harvests, grain reserves, and livestock populations that parts of Africa require massive food aid to

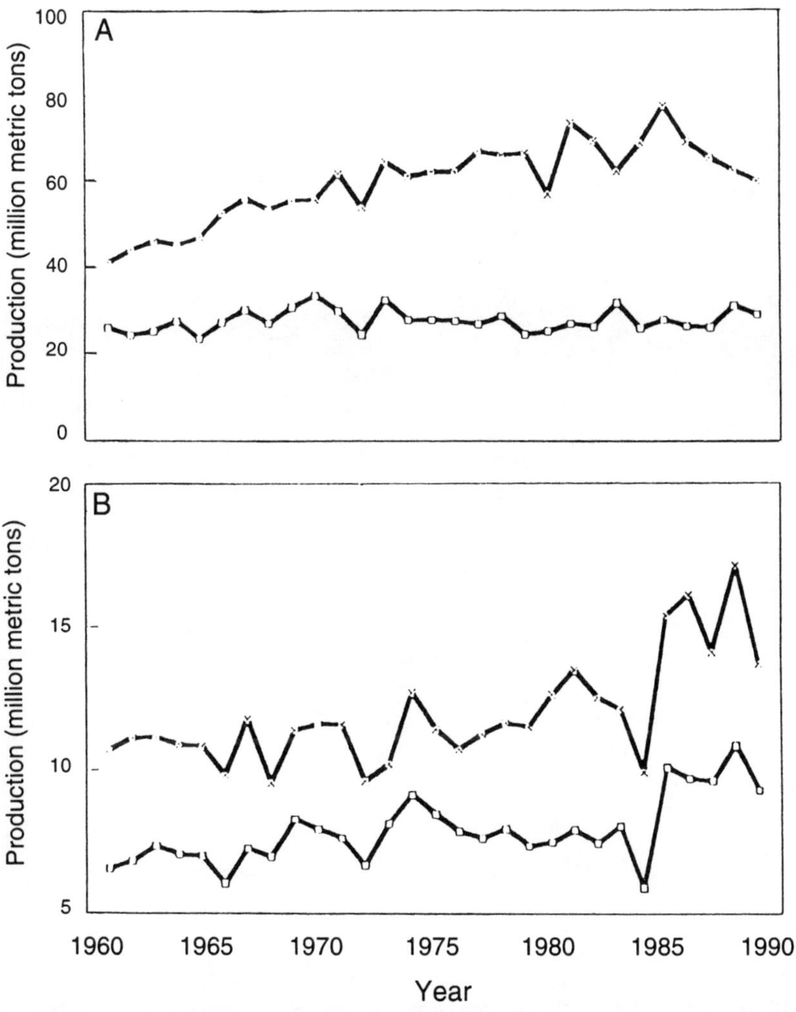

Fig. 2. Production of sorghum (top line) and millets (bottom line) in the world (A) and in Africa (B).

avert widespread starvation. These unfortunate circumstances brought before the whole world the depressed conditions that have long prevailed in these regions, where even at the best of times food supply has barely kept pace with the needs of increasing population. Nonetheless, as the graphs show, in the world as a whole, in the developing countries, in the developed countries, and in Africa, the production of both sorghum and the millets has been steadily increasing (Figs. 2 and 3).

Four countries in Sahelian Africa (Niger, Burkina Faso, Mali, and Sudan) with a combined population of 38 million, depend on sorghum and the millets (mainly pearl millet) for over 1,000 Calories a head daily. The two cereals account for between 72 and 90% of all Calories obtained from cereals. These grains are almost as important in Chad, Sudan, Nigeria, and the Gambia.

The figures are, of course, for nation states and give a misleading impression as to local importance—Kenya has already been mentioned. Nigeria is another ex-

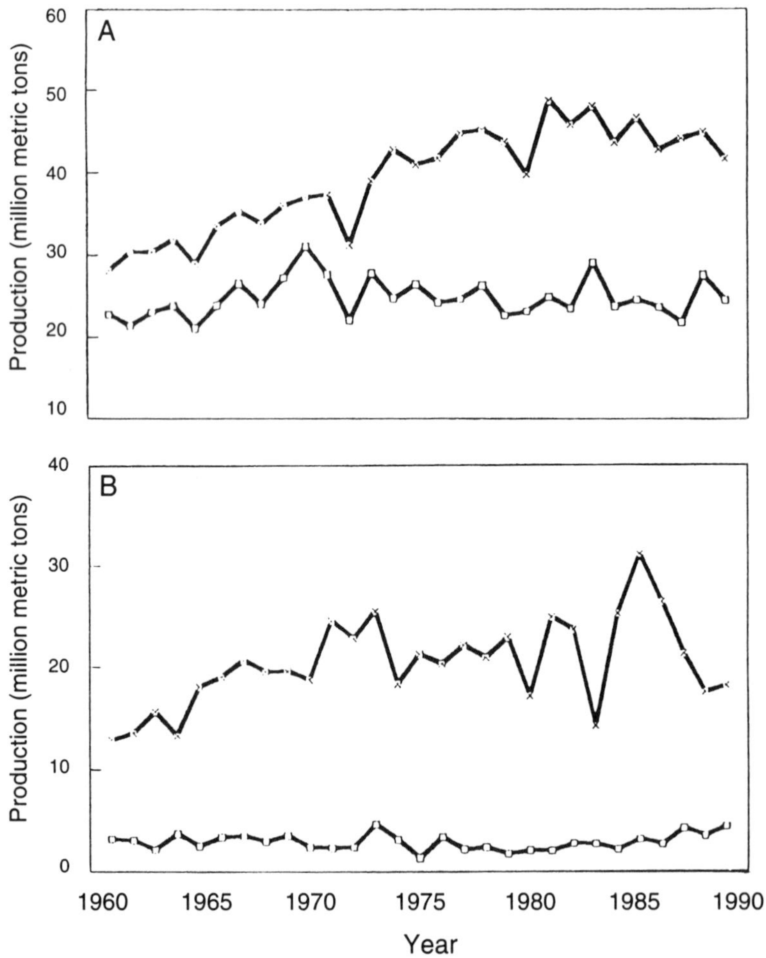

Fig. 3. Production of sorghum (top line) and millets (bottom line) in developing countries (A) and in developed countries (B).

ample, since these grains are the principal crops in the north, along the borders of the Sahara Desert, but are little used in the coastal rain forest belt of the south.

While it is now acknowledged that sorghum and the millets—especially pearl millet—are extremely important subsistence grains in the semiarid areas, international recognition of this fact was quite recent. The International Crops Research Institute for the Semi-Arid Tropics (ICRISAT) was founded as recently as 1972. By then research on sorghum as a feed grain was well in hand in the United States, and some programs of research on preharvest problems had been undertaken in some developing countries, but little work had been done on postharvest aspects of sorghum. The Study Group for Sorghum, Millets and Composite Flour of the International Association for Cereal Science and Technology (ICC) was formed in 1974.

From Table 2 it is clear that the order of importance of sorghum and millet is first in West Africa, second in East Africa, and third in Southern Africa, whether assessed in terms of production per capita or production as a percentage of all cereals. Using this last criterion, these two cereals together represent around 70% of total cereals production in West Africa, around 30% in East Africa, and around 10% in Southern Africa.

SORGHUM

FAO statistics for 1992 and earlier extracted from the FAO production yearbooks (FAO, 1962–1993) are incorporated into the tables in this chapter. For sorghum, the figures are quoted for major producers and are, within regions, in descending order of dietary importance, that is, in Calories per capita per diem.

Over 700 million hectares of the world's surface is laid down to cereals cultivation. Of this, only some 45 million hectares is sorghum, but over 80% of this is cultivated in developing countries (Tables 1 and 3). Sorghum is indigenous to Africa, where it is grown in the semiarid zone, which includes the large belt spreading from the Atlantic to Ethiopia and Somalia, bordering the Sahara desert in the north and the equatorial forest in the south and extending southwards through the drier parts of eastern and southern Africa, where rainfall is too low for other staples. Sorghum is, therefore, second in importance after maize in Africa south of the Sahara, and of disproportionate importance as a subsistence rather than a commercial crop.

In Asia, sorghum is an important staple in parts of China and of India and in the lowland parts of Yemen. It has recently become an important cash crop in Thailand, where it is grown in the dry season following maize, partly for export. Sorghum is widely grown as a staple in the drier parts of Central America in the Caribbean. In South America, sorghum production is concentrated in the dry areas of northern Colombia and Venezuela, northeastern Brazil, and in the states of Rio do Sul, Sao Paulo, and Paraná, as well as in the dry low and interior areas of Argentina, where grain sorghum is an important feed crop.

The graphs of sorghum and millet production are derived from FAO figures for 1961–1992 (FAO, 1962–1993) and show the increase in sorghum production worldwide, in the developing countries, in the developed countries (where the grain is used mainly for feed), and in Africa. World production of sorghum has increased from 40 million metric tons at the beginning of the sixties to 65 million tons at the middle of the seventies, after which growth fell back, mainly because of the African droughts. Thus, the annual increase in world sorghum production was relatively low (2.4% per annum) and erratic compared with the steady increase in production of the other major grains, e.g., wheat (3.4%), rice (3.1%), maize (3.6%), and barley (3.3%). Although it has been noted that "in the majority of countries where sorghum is regarded as a food item, mainly in Africa and Asia, per capita production declined as the increase in total output fell below population growth" (FAO-IGG, 1988), for developing countries as a whole this is, unfortunately, true of most staple production, even on fertile soils with good rainfall.

Over the past 25 years, world production of staple foods—of which cereals form over 90%—has increased by about 75%, although the cultivated area expanded by only 15%. This increase in yield has been very uneven between crops and between groups of countries; in developing countries as a whole, average yields by the mid-1980s, at roughly one metric ton per hectare, were about the same as 10 years before. In Africa, yields declined over an even longer period—by some 12% between 1961–1965 and 1981–1985 (FAO, 1962–1993).

As FAO publications have pointed out, there were several reasons for this development: the changing world climate with lower rainfall, which has affected the developing countries most; "the larger variation in total annual rainfall and the greater irregularity in its seasonal distribution than elsewhere; the fact that sorghum is traditionally not grown as a cash crop; the deteriorating nutritional status of the soil and the decline of the fallowing period in many areas; the squeeze of sorghum production in West Africa between the southward moving desertification and the expanding production of maize toward the north which entails that more marginal land has come under cultivation; and finally the absence until recently of large scale and locally oriented plant breeding programmes" (FAO-IGG, 1988).

In developed countries, led by the United States, the change to sorghum as a feed grain led to a rapid increase in research, with the development of hybrid seed, the use of chemical treatment against mildew and pests, the application of large quantities of fertilizers for the, by now "highly-tuned" varieties, and advanced land preparation techniques for maximum attainable yield. In developed countries, average yields are now over 3 metric tons per hectare, three times that for developing countries where sorghum is almost entirely a staple food.

Sorghum production in developed countries—mainly in United States, Argentina, and Mexico—is now a specialized commercial undertaking for feed grain and raw material for compound feeds. In Mexico, the grain was almost unknown until the mid-1950s, but is now nearly one-third of cereals production, with annual production at 6 million metric tons from 1.8 million hectares.

Grain use for animal feed has been a dynamic element in the stimulation of global sorghum consumption (FAO-IGG, 1988). Until the mid-1960s sorghum was a relatively minor feed crop, but a rapid expansion of production in North America led to feed utilization overtaking food use for the first time in 1966. To quote FAO (FAO-IGG, 1988):

> World feed use of sorghum has risen from 15 million tons to 40 million tons over the past 25 years, representing an average growth of 4% per annum. Its demand has been the main driving force in raising global production and international trade since the early sixties. However, demand is heavily concentrated in the developed countries where it accounts for 97% of total use and in some higher income developing countries, especially in Latin America, where over 80% of all sorghum estimated to be used as animal feed in developing countries is concentrated. The United States,

> Mexico and Japan are the main consuming countries, followed by Argentina, the former USSR, and Venezuela.... Together, these countries absorb over 80% of world use of sorghum as animal feed. All countries, with the exception of the United States, have shown very rapid growth over the 1961–65 to 1981–85 period.

In the developing countries, most sorghum is produced by subsistence farmers on small plots. Chapter 3 describes the various edaphic conditions under which sorghum is grown. It should also be noted that urbanization has had a profound influence on the traditional subsistence crops as farmers abandon traditional staples and attempt, at least for the market, to grow surpluses of maize on unsuitable soils and in uncertain climates. Thus, the falling per capita consumption levels in many countries have been caused not only by population increase but by changes in consumer habits brought about by this rapid rate of urbanization. The workers of the towns do not have the time or energy resources (fuel) to prepare traditional foods based on sorghum or millets. Furthermore, supplies are erratic, if available at all, and this, coupled with the lack of readily available sorghum products (meal) compared with other foodstuffs, has accelerated the changes in consumption habits in urban areas. By contrast, per capita food consumption of sorghum in rural producing areas remains stable in years of adequate rainfall and, in general, much higher than in the towns. As new processing techniques become better known, consumption of sorghum in the towns would be expected to rise, as the flavor of sorghum is favored as being "traditional."

With a lack of rural infrastructure, it may not yet be possible to assemble the large quantities of staples needed by even the present urban population. Medium-scale farming—above subsistence level but relying on sustainable and locally funded means of production and transportation—would seem a better option than consolidation of land into large holdings farmed by totally mechanical means. The problem of acceptability still has to be faced; the urban consumer purchases bread and maize meal from shops, cereal processing industries are already processing wheat and maize on a large scale, using mainly imported wheat and maize from large local farms. Sorghum is grown in Zimbabwe and South Africa on a large scale for the brewing of opaque beer *(chibuku)*, a subject dealt with in Chapter 10. The author has noted that in Zimbabwe, at least, government pricing policies are now giving encouragement to the growing of food varieties. However, in the Sudan, the expanded cultivation of sorghum on large mechanized farms led to export surpluses before the recent droughts and civil strife. Curiously, much of the feteritas imported to Europe for feeding racing pigeons came from Sudan. The species *Sorghum bicolor* covers a wide range of varieties, from white and yellow to brown, red, and almost black. The tannic pigment is polyphenolic, and its property of binding protein makes it an antinutritional factor (see chapter 4). The feteritas are unusual in that the pigment is contained in the inner layers of bran but not the outermost. The grains have a lustrous white color but become dark brown when the outer layers are removed by abrasion. The pigment is antinutritious, so that it is somewhat strange that feteritas should be chosen for pigeon feed rather than white or yellow varieties with a low polyphenol content.

Sorghum consumption as food, even in countries where it was a traditional staple, has also been negatively influenced by government policies. For example, wheat and rice were imported into Nigeria in extremely large quantities, and until recently this discouraged the consumption of locally produced cereals. In both

1983 and 1984, imports of wheat, rice, and cereal products such as barley malt accounted for 15% of the carbohydrate intake of the population of Nigeria. For economic reasons it became necessary to ban cereal imports in 1985, thus giving a great stimulus to local grain production, which leapt from 7.5 million metric tons from 7 million hectares in 1979–1981 to 11 million metric tons from 10 million hectares in 1989; few other countries have such areas available for planting. Composite flour and wheatless bread programs that had been proposed 20 years before but never implemented were at once put into effect, the market responding by producing the necessary local materials. Even clear (lager) beer and malted drinks are now made from sorghum rather than imported barley malt (see Chapter 9).

In India, the decline in the utilization of sorghum in urban areas may have been influenced by the discontinuation of sorghum sales by the Food Corporation of India. It has been noted by FAO (FAO-IGG, 1988), that in Yemen imports of cheap wheat and the limited availability of domestically produced sorghum led to sorghum prices in 1987 being almost three times higher than those of imported wheat. In many countries, subsidies or overvalued currencies kept prices of imports of wheat or rice at low levels, which reduced the competitiveness of domestically produced staples (Dendy and Trotter, 1988). With the recent restructuring policies, this situation should change, and a reversion to greater utilization of traditional staples such as sorghum may occur.

THE MILLETS

The most important millets are pearl millet (*Pennisetum glaucum*), finger millet (*Eleusine coracana*), proso millet (*Panicum miliaceum*), and foxtail millet (*Setaria italica*); other types of regional importance include fonio (*Digitaria exilis*) in West Africa, varagu or kodo (*Paspalum scrobiculatum*) in northern India, and teff (*Eragrostis tef*) in Ethiopia. The millets differ considerably from one another and cover a wide area of the tropics and subtropics (Fig. 1). They are mostly very small grains growing on slender grassy stems, and have been well described by Rachie (1975) and in Chapter 1. The exception is pearl millet, which has a strong stem resembling sorghum. Their agronomic characteristics are discussed in Chapter 3 and food preparations in Chapter 7.

The millets are common in much of the tropical and subtropical world, where they are cultivated on poor soils in areas of low or inconsistent rainfall. With a strong and deep rooting system and efficient use of available moisture, the millets can survive in areas with as little as 300 mm or less of seasonal rainfall. This compares with a minimum water requirement of 400 mm for sorghum and 500–600 mm for maize. As is to be expected under these conditions, yields are very low compared with other cereals grown elsewhere, and show great variation from one year to another. Mostly the millets are therefore grown as subsistence crops, although in highland Ethiopia the millet teff is the staple food of the whole population, without the familiar and recent differentiation of rural and urban diet. It is, therefore, an important cash crop for small farmers; curiously, the large-scale state farms concentrate on maize and wheat. Commercial farming is a risky undertaking in drought-prone areas, although in Namibia, for example, pearl millet is now being grown on a larger scale.

FAO statistics (FAO, 1962–1993) show that the average annual global production of the millets during the three-year period 1988–1990 was 30.3 million metric tons, from some 38.2 million hectares, well down from the figure of 38.1 million metric tons achieved in 1979–1981. Thus, world production of the millets is estimated to amount to only a very small share of total annual cereal output (Tables 1 and 2), including about 5% of the world's cereal area but only 1.6% of production. The detailed figures are given in Table 5.

FAO states that "national statistics on the millets are generally poor, especially data on utilization and international trade. Statistics are often fragmentary, based on rough estimates, frequently incomplete and usually do not identify the different types of the grain. Some countries do not separate production data on the millets and sorghum, or include the millets in their statistics under the heading 'other coarse grains', while in other countries official production figures are not collected annually but only become available periodically through agricultural censuses." (FAO-IGG, 1990). It is clear, therefore, that statistics on millets given in Tables 5, 6, and 7, should be treated with caution and regarded as approximations.

Table 6 was supplied by the FAO (FAO-IGG, 1990) and divides the figures between the species. Of this world total, almost half (13 million metric tons) is estimated to consist of pearl millet, followed by foxtail millet (6 million metric tons), proso millet (5 million metric tons), finger millet (4 million metric tons), teff (1 million metric tons), fonio, and others. Some 55% of the estimated world production of the millets is produced in Asia; Africa accounts for another 30%, and the

TABLE 5
Production of Millets by Principal Producing Countries[a]

Region or Country	Total Production (000 t/annum)	Area (000 ha/annum)	Yield (t/ha)	Calories per Capita per Day	As Percentage of Cereals
World	27,760	36,652	0.76	50	1.4
Developed	2,347	3,219	0.73	18	0.3
Developing	25,413	33,433	0.76	60	2.4
Africa, total	9,498	15,202	0.62	137	11.9
West and North Africa					
Niger	1,900	4,400	0.43	2,271	78.7
Burkina Faso	785	1,100	0.71	814	35.1
Mali	794	1,100	0.72	798	35.6
Senegal	470	900	0.52	599	51.4
Gambia	51	56	0.91	554	47.2
Chad	295	650	0.45	498	33.1
Nigeria	3,200	3,900	0.82	273	25.6
Togo	77	120	0.64	202	17.2
Sudan	250	1,150	0.22	93	6.1
East, central, and southern Africa					
Ethiopia	1,612	1,624	0.99	300	25.8
Uganda	500	370	1.35	264	30.3
Namibia	15	55	0.27	96	13.2
Tanzania	263	300	0.88	93	7.7
Angola	75	110	0.68	75	19.7
Zimbabwe	33	50	0.66	31	6.5
Asia, total	15,860	18,203	0.87	48	1.8
China (People's Republic)	4,500	2,230	2.02	37	1.1
India	10,600	15,000	0.71	119	5.4
Nepal	225	197	1.14	108	4.1
U.S.S.R. (former)	2,100	3,000	0.70	71	1.2

[a] Data from FAO (1993).

TABLE 6
Production of Millets by Species[a]

Region or Country	Total Millets Production (000 t/annum)		Minor Millets Production (000 t/annum) (1981–1985)					
	1991	1981–1985	Pearl (bullrush)	Finger (*Eleusine*)	Proso	Foxtail	Other (t/annum)	Species
World	28,970	29,716	13,351	3,763	5,061	5,780	1,760	
Developed	3,833	2,517	8	3	2,479	27	1	
Developing	25,137	27,199	13,343	3,760	2,582	5,753	1,759	
Africa, total	10,803	9,557	7,330	855	...	...	1,372	
West and North Africa								
Nigeria	4,200	3,191	3,133	...	...	...	58	fonio
Niger	1,853	1,229	1,227	...	...	...	2	fonio
Mali	792	731	695	...	...	...	36	fonio
Senegal	560	539	539	...	...	...	...	...
Burkina Faso	757	453	447	...	...	...	6	fonio
Sudan	308	362	355	7	...	...	...	...
Guinee	60	210	10	...	...	...	200	fonio
Chad	302	161	161	...	...	...	...	...
Cameroon	63	96	96	...	...	...	...	...
Ghana	112	92	92	...	...	...	...	...
Togo	77	59	59	...	...	...	...	...
Ivory Coast	53	39	33	...	...	...	6	fonio
Gambia	51	39	38	...	...	...	1	fonio
East, central, and southern Africa								
Ethiopia	1,612	1,266	5	198	...	...	1,063	teff
Uganda	600	451	40	451	...	...	...	...
Tanzania	270	239	144	95	...	...	...	...
Zimbabwe	122	136	90	46	...	...	...	...
Angola	63	50	50	...	...	...	...	...
Namibia	60	38	38	...	...	...	...	...
Kenya	63	36	14	22	...	...	...	...
Asia, total	14,285	17,469	6,013	2,905	2,409	5,753	386	varagu
India	9,000	9,624	5,625	2,613	500	500	386	varagu
China (People's Republic)	4,500	6,579	...	79	1,600	4,900	...	...
Korea (both)	64	487	...	...	150	337	...	...
Pakistan	220	258	252	2	2	2	...	...
Myanmar/Burma	138	167	10	5	150	2	...	...
Nepal	220	124	...	122	1	1	...	...
Yemen	25	108	108	...	...	...	...	...
Other								
Argentina	65	173	...	...	173	...	...	...
U.S.S.R. (former)	3,800	2,330	...	...	2,330	...	...	...

[a] Data from FAO-IGG (1990) and FAO (1992).

former U.S.S.R. for 14%—the world's most northerly production. In fact, about 70% of world output of the millets is produced in only six countries: China, Ethiopia, India, Niger, Nigeria, and the former USSR.

Per capita food consumption of the millets is highest in Africa, where production is distributed over a much larger number of countries than in other continents. Nigeria alone, the largest producer of the region, provides about one-third of the total. Cultivation is concentrated along the southern peripheries of the Sahara, which comprise the Sahel countries and the northern parts of the coastal countries in western Africa, and in the drier areas of eastern and southern Africa, including Ethiopia, Namibia, Tanzania, Uganda, and Zimbabwe. Referring again to Table 5, it will be seen that in Mali, Senegal, Burkina Faso, and Niger, millets account for over 600 Calories per diem per capita. Some six-tenths of the cereal food production of Niger is pearl millet, although production fell by 40% between 1983 and 1984, only to rise by over 70% from 1987 to 1988. Progress in raising yields here was principally in more assured rainfall areas rather than under marginal moisture conditions (FAO-IGG, 1990).

For many other countries, especially in eastern and Central Africa but also in the northern areas of the coastal countries of western Africa (notably Nigeria), the millets are an important food item for the population living in the drier parts of the countries. In Namibia, pearl millet is the principal crop, but mineral wealth ensures imports of wheat and maize.

Outside Africa, the millets are of local significance as food in regions of China, India, Myanmar (Burma), and North Korea. Although national per capita production is low in China and India, this is a reflection of the regional distribution of the millets and the high population of the countries; food use of the millets, including some of the more obscure species, is important in certain areas. "In China, although yields are among the highest in the world, production has declined over the years because of other crops offering higher returns and changes in consumer preferences" (FAO-IGG, 1990).

India is the largest producer of the millets in the world, with pearl millet making up 60% of the total millet production; it is mainly grown in the drier areas of the central and western regions. Finger millet is ubiquitous but is cultivated mostly in the south, the Deccan, and in Bihar in the north, and makes up a further 27% of the total. The other millets of India are interesting, including at least four species (little, proso, foxtail, and varagu [kodo]). These are highly regionalized. It is pleasing to note that the small millets, once even more neglected than sorghum and pearl millet, now warrant the occasional international workshop (Seetharam et al, 1989). Other substantial producers of the millets in Asia are Korea, Nepal, Pakistan, and Myanmar (Burma).

In the former U.S.S.R., the millets are cultivated mainly in the Russian Republic, the Ukraine, and Kazakhstan (Fig. 1). Production of the millets in the Americas, Australia, and Europe is low, although it is sometimes sown for grazing or fodder (see Chapter 11), and for game birds, when mixed grains including foxtail and proso millet are used. Some foxtail and barnyard millets are grown in Australia for export as pet food for small exotic cage birds.

There is a useful review of the trade prospects of proso millet in Wietgrefe (1989), with a comparison of U.S. and Argentine production and trade. This is mainly for cage birds and feed. Wietgrefe makes the point that proso millet is at least as good as maize as a constituent of feed, and its production for this should be encouraged.

Millet Consumption as Food

The consumption pattern for the millets is shown in Table 7 (from FAO-IGG, 1990). The graphs derived from FAO figures (FAO, 1962–1993) show that world production of millet and indeed regional production of millet has remained fairly static over the past 30 years, although there has been some increase of production in Africa. This compares unfavorably with the dramatic rise in production of all cereals worldwide. Between 1961 and 1990, cereals production increased from 970 to some 1,955 million metric tons, and the area grown increased in that period from 662 to 707 million metric tons, reflecting an increase of yield of from 1.46 to 2.76 metric tons/ha, that is almost a doubling of yield. Meanwhile the yield of millet barely increased, reflecting its marginality.

It is notable that millets are among the last cereals to enter the industrial system, so that whereas maize and wheat products are available in the shops of the developing countries, millet products rarely are. This however is not a reflection of the inadequacy of the millets for manufacturing foods but of the availability on a large scale of wheat and maize and lack of availability of millets to the processors in large towns. In the long term there is no reason why, with new technologies now available, the millets should not be used as raw materials for the manufacture of convenience foods for townspeople. This does assume, however, that there is a suitable infrastructure for assembling surpluses from, as yet, subsistence farmers. An exception to this is teff, which is the staple food of the highlands of Ethiopia discussed earlier in this chapter. It is grown by surplus-producing peasant farmers

TABLE 7
The Millets: Estimated Utilization by Type, Region, and Selected Countries[a]
(1981-82–1985-86 averages)

Region or Country	Food (000 t)	Feed (000 t)	Other (000 t)	Total (000 t)	Kilograms per Annum per Capita
World	22,335	3,144	3,642	29,121	4.8
Developing countries	21,535	1,878	3,231	26,644	6.1
Developed countries	800	1,266	411	2,477	0.7
Africa[b]	7,094	122	1,921	9,137	13.5
Burkina Faso	381	...	60	441	50.8
Ethiopia	1,020	...	196	1,216	24.9
Mali	516	1	88	605	67.7
Niger	977	21	215	1,213	168.9
Nigeria	2,365	86	700	3,151	26.5
Senegal	397	2	80	479	64.4
Uganda	259	47	150	456	17.8
Asia	14,441	1,665	1,305	17,411	5.3
China	4,857	1,120	480	6,457	4.7
India	8,794	150	710	9,664	11.9
Central America	...	...	...	...	...
South America	...	91	5	96	...
North America	...	104	6	110	...
Europe	...	42	3	45	...
U.S.S.R. (former)	800	1,107	400	2,307	2.9
Oceania	...	13	2	15	...

[a] Source: FAO-IGG (1990); used by permission.

and is traded to the towns for the preparation of traditional food, *injera,* a subject discussed in Chapter 7.

Millets for Feed

The FAO has attempted to estimate the total consumption of millets as animal feed, but the figures are very much estimates based upon assumptions made for China and the former USSR for which little reliable information is available (FAO-IGG, 1990). In Western Europe, North America, and Japan, utilization as feed is about 200,000 metric tons, excluding minor imports for the pet food market. This is a very small proportion of the total world trade in cereals of some 200 million metric tons, mainly wheat, maize, and rice. Very little millet is used for animal feed in India and Africa. Pearl millet and the hybrid of pearl millet with elephant grass are cultivated for grazing and silage and, of course, in the developing countries the stover and straw of the millets is valuable as fodder, bedding, and thatch. The use of millets for feed is dealt with in Chapter 11.

FORECAST FOR SORGHUM AND THE MILLETS

Sorghum and the millets are and will remain essential food grains for subsistence farmers in the semiarid zones of the tropics and subtropics. These zones are increasing rapidly in extent with the changes in world climate and overuse of marginal land by overgrazing and deforestation. These grains may, therefore, become of greater rather than lesser importance.

The FAO (FAO-IGG, 1990) pointed out that "experience has shown that increased productivity through higher yields can lead to a contraction in the area planted to the millets in favour of cash crops rather than to any significant expansion in output beyond overall subsistence requirements, given the limited domestic (with the exception of teff) and international commercial outlets for the grain in contrast for instance to competing crops such as cotton or groundnuts." This need not be so: for example, in Namibia, where pearl millet is the principal crop in the arid north growing on infertile sandy soil, there are now factories producing a packaged millet meal for sale to the northerners living in the industrial towns and the capital city. In Namibia, pearl millet is the preferred food, being more sustaining than maize, but has not until recently been available in the supermarkets. The development of new manufactured foods from sorghum in Kenya has also proved successful (author's observations; see also Chapter 8).

With the development of new processing techniques, described in Chapter 8, for sorghum and some of the millets, the prospects are good, and urban markets can now be developed so as to provide an incentive to farmers to grow a surplus.

As export crops, sorghum and the millets have been destined almost entirely for feed. With the developing world receiving wheat and maize as aid in many years, there is a strong case for the growing of varieties, especially of sorghum, for export as aid, so that indigenous eating habits may be sustained.

LITERATURE CITED

DENDY, D. A. V., and TROTTER, B. W. 1988. Wheatless and composite breads—Technologies awaiting adoption. Entwickl. Ländl. Raum 22(6):13-18.

FAO. 1962–1993. Production Yearbooks. Food and Agriculture Organization of the United

Nations, Rome.

FAO-IGG. 1988. Sorghum 1988: The Structure and Characteristics of the World Sorghum Economy. Working pap. Committee on Commodity Problems, Intergovernmental Group on Grains, 23rd session. FAO, Rome.

FAO-IGG. 1990. Millet 1990: The Structure and Characteristics of the World Millet Economy. Working pap. Committee on Commodity Problems, Intergovernmental Group on Grains, 24th session. FAO, Rome, 1990.

RACHIE, K. O. 1975. The Millets: Importance and Utilisation and Outlook. Int. Crops Res. Inst. Semi-Arid Tropics (ICRISAT), Patancheru, A. P., India.

SEETHARAM, A., RILEY, K. W., and HARINARAYANA, G. 1989. Small millets in global agriculture. Proc. Small Millets Workshop, 1st, Bangalore, 1986. Oxford and IBH Publ. Co., New Delhi.

WIETGREFE, G. W. 1989. Proso Millet: A Trade Summary. Author's publication, Pierre, SD.

CHAPTER 3

AGRONOMIC PRINCIPLES

L. R. House, M. Osmanzai, M. I. Gomez, E. S. Monyo, and S. C. Gupta
SADC/ICRISAT Sorghum and Millets Improvement Program
Matopos, Zimbabwe

INTRODUCTION

Sorghum and the millets are annual grasses that are extensively used in tropical areas of the world and have moved significantly into temperate regions. In tropical areas, these crops are wholly utilized, as they are often used for food, feed, building material, and fuel. The more important millets include pearl, finger, foxtail, and proso millets. There are other millets of more local significance, including teff, kodo, barnyard, fonio, and little millet. These crops have different areas of adaptation, growing periods, and grain characteristics. As a broad generalization, sorghum and millets are primarily used for food and beverage in tropical areas and as feed in temperate areas. Information on nutritional and quality characteristics is greatest for sorghum, less for pearl millet, and little to nonexistent for other millets. Both grain and stover traits have been studied, and in terms of genetic control and the influence of environmental factors, the greatest information is available for protein and amino acids and nutritional inhibitors, with little information available for carbohydrates and other constituents.

Germ Plasm Resources

Sorghum and millet breeders in a number of countries where these crops are important began collections of the cultivars and wild types growing in their countries some 50–70 years ago. Collections were developed for sorghum, pearl and finger millets, and foxtail, proso, barnyard, and small millets. During the mid-1960s, as part of the Indian Agricultural Program of The Rockefeller Foundation, with support from the U.S. National Academy of Sciences, existing collections were brought together, and cultivars growing in India were systematically collected.

These collections became the responsibility of the International Crops Research Institute for the Semi-Arid Tropics (ICRISAT) and continued to expand. Accessions with ICRISAT are presented in Table 1.

These collections provided a genetic base used by many breeders to diversify their breeding stock. Texas A&M University and the U.S. Department of Agriculture initiated a program to convert useful tropical sorghum cultivars to temperate adaptation. These converted cultivars have been widely used in sorghum improvement programs in both the temperate and tropical zones.

The collections also made it possible to breed for resistance to biotic and abiotic stresses (drought resistance being of particular interest). Variation in the sorghum collection increased the opportunity to breed for end use. The first effort was to match as closely as possible local preferences for foods and beverages made from sorghum, and to a somewhat lesser extent, from pearl and finger millets. Interest expanded into the preparation of specialty foods and to blending the flour of sorghum and pearl millet with that of other cereals (wheat and maize) and legumes.

Following the discovery of high lysine in maize, interest in nutritional quality grew and led to the discovery of high lysine in sorghum and to the antinutritional effect of high tannin. Traditional uses and techniques were documented, and these traits were extensively studied in a number of laboratories.

The study and use of resistance traits, nutritional characteristics, and quality of food, feed, and beverages has contributed to the development of a diverse array of useful breeding stock important to the strengthening of most sorghum and millet improvement programs around the world. A number of these traits are considered in this chapter.

The Structure of the Seed

The structures of sorghum and millet grains have much in common. A detailed discussion of the structure of the sorghum grain is presented by Rooney and Miller (1982). The major components of the seed are the pericarp or outer cover, the endosperm or storage organ, and the embryo, which germinates to reproduce a plant. There are four components of the pericarp; the outer layer is divided into the epidermis (frequently pigmented) and the hypodermis. The middle portion of the pericarp, the mesocarp, may vary from thin and translucent to thick, giving the pericarp a dull "chalky" appearance. The endocarp is the innermost layer of the pericarp, composed of narrow, long cells, and is the point of separation of the pericarp from the rest of the seed during the milling process.

The testa, which may or may not be present, is the layer between the pericarp and the endosperm. This is the tannin-containing layer, and therefore important to the nutritional characteristic of the grain.

TABLE 1
Sorghum and Millet Collections of ICRISAT

Crop	No. of Accessions
Sorghum[a]	33,108
Millets	
Pearl[a]	22,059
Finger[b]	3,220
Foxtail[b]	1,452
Proso[b]	831
Little[b]	423
Barnyard[b]	612
Kodo[b]	544

[a] Data from ICRISAT (1992).
[b] Data from Prasada Rao et al (1993).

The endosperm forms the bulk of the kernel, generally being corneous on the outer extremes and floury toward the center. Starch granules in the corneous outer portion are embedded in a protein matrix and are difficult to separate. Protein content in the floury endosperm is less than in the corneous types, and there can be voids in the structure contributing to a more opaque appearance of this portion of the endosperm. Starch is more easily recovered from the floury endosperm (Rooney and Miller, 1982).

The embryo appears at the lower portion on one side of the seed. Most of the oil content of the seed is in the embryo.

GENETICS OF SORGHUM GRAIN CHARACTERISTICS

Protein

Sorghum protein has several components based on solubility fractions. Of greater concern is the alcohol-soluble prolamin. Prolamin can be divided into fractions: fraction III (based on the Landry-Moureau fractionation) is highly cross-bonded and referred to as kafirin. It is this fraction that has a negative impact on sorghum protein quality. Fraction II is nutritionally better. Maize and pearl millet have some four to six times more fraction II than III, while sorghum has twice as much fraction III as fraction II (Doggett, 1988).

Nitrogen (N) is an important plant nutrient having significant impact on yield and nutritional quality of the grain. Varieties differ in the uptake and transport of nitrogen as well as its impact on plant performance.

Nitrogen uptake per plant and nitrogen transfer efficiency varied in one experiment, ranging from 0.22 to 1.14 g of N/plant and transfer efficiency from 57.8 to 86.6% at a given total N. A weak negative N content versus grain yield correlation was identified. A high positive correlation was observed between high nitrogen transfer efficiency and high harvest index (weight of grain as a proportion of total dry matter). Grain yield was positively correlated with total N in the plant, total N in the grain, and a high nitrogen transfer efficiency, suggesting the possibility of simultaneously selecting for these traits (Hulse et al, 1980).

Generally, the correlation between grain yield and protein is found to be negative but variable. Hybrids almost always produce higher grain yield and frequently have lower protein content than the parents; but, because of the higher grain yield, protein produced per hectare is higher than from the parents or lower yielding varieties.

Inheritance of protein is polygenic. From an experiment at two locations in Kansas both general and specific combining ability (GCA and SCA) were found significant for protein content and heritability was estimated to be 43%. In a number of experiments, heritabilities have ranged from 41 to 85% depending on varieties, populations used, and the method of calculation (Liang et al, 1969).

Although variable, additive gene action for protein inheritance has been found to be less important than the nonadditive component.

Heterosis, calculated as the mean of the F_1s compared with the average of the means of the parents, was found to be significant and negative. Genotypic correlations between percent protein and yield were found to be significant and negative. The same was found for the correlation between protein and lysine as a percentage of protein. In general, genotypic effects on protein content were greater than environmental effects (Hulse et al, 1980).

Amino Acids

"If one looks at the essential amino acid composition of sorghum grain, in comparison with monogastric nutritional requirements, it is obvious that lysine is deficient and that there is great excess of leucine. The methionine content of sorghum is low, but considering the cystine content of 1.5%, the overall sulfur amino acid content approaches the weanling rat requirement. The tryptophan content of sorghum seems to be adequate in contrast to the low tryptophan content in normal maize" (Axtell et al, 1974).

In an experiment conducted in India, Nanda and Rao (1975) found significant differences among hybrids for the traits that they studied; protein content, yield, and a number of amino acids: threonine, methionine, isoleucine, leucine, phenylalanine, cystine, tyrosine, and valine. The parent versus hybrid comparison was significant for yield attributes, lysine, leucine, phenylalanine, glutamic acid, proline, and cystine. Additive gene action contributed more to all attributes except methionine, isoleucine, leucine, phenylalanine, tyrosine, glutamic acid, proline, and alanine, where SCA was significant or predominated over GCA. For protein, lysine, and threonine, GCA was significant.

Significant negative correlations were observed between: protein and yield, seed size, and grain hardness; lysine with isoleucine, leucine, and cystine; leucine with methionine; carotene with protein, methionine, and cystine; methionine, with carotine, threonine, and yield; seed size with threonine; and yield with cystine.

Significant positive correlations were observed between yield with seed size and grain hardness; grain hardness with seed size and lysine; methionine with lysine; isoleucine with threonine; leucine with protein, threonine, and isoleucine; and beta-carotene with lysine, threonine, isoleucine, and leucine (Hulse et al, 1980).

Nanda and Rao (1975) found that the magnitude of additive variance was smaller than nonadditive variance for most amino acids.

The identification of high lysine content in some floury-opaque maize seeds spurred interest in improving protein quality in other cereals. Initially high lysine was found in two sorghum accessions from Ethiopia, IS 11167 and IS 11758. In these accessions, lysine content as percent of sample was approximately twice that of normal. Over 50% of the endosperm protein in sorghum is present in the kafirin fraction which is poor in lysine. It was found in these accessions that there was a shift from 55% kafirin found in normal sorghums to 39%, accounting for the increase in lysine; i.e., the relative concentration of protein components high in lysine increased (Axtell, 1974; Ejeta and Axtell, 1989). Inheritance studies indicated that high lysine was controlled by a single recessive gene designated *hl* (Axtell, 1974).

Seeds from these accessions were shrunken at maturity. It was found that the rate of starch synthesis and level of total sugars at various stages of grain development differed between normal and high-lysine grains. Between 21 and 31 days after pollination, starch synthesis was delayed, and there was an increase of sugars in the high-lysine grains (Ejeta and Axtell, 1987).

The *hl* gene, when homozygous in the endosperm, altered the amino acid pattern. Beside the increase in lysine, there was also an increase in arginine, aspartic acid, glycine, and tryptophan, and a decrease in the concentration of glutamic acid, proline, alanine, and leucine (Singh and Axtell, 1973).

A second approach was to use a mutagenic agent, diethyl sulfate. From 23,000 M_3 heads, 475 were found segregating for opaque. Thirty-three were found with at least a 50% increase in lysine, and P721 had an increase of 60% (Axtell, 1974).

Opaque seeds from P721 were 11 to 14% lower in seed weight than normals. Dry matter accumulation in the seeds of P721 stopped at 31 days, seven days earlier than in normals. Also associated with the floury opaque endosperm were slower rates of drying at harvest, greater mold and insect susceptibility, and more grain cracking following threshing than found in normals (Ejeta and Axtell, 1990). This resulted in problems of acceptance as it did for *opaque-2* maize. To attempt to resolve this problem, hundreds of crosses were made between P721 and an array of other parents to diversify the genetic background. One hundred and fifty-eight lines with high lysine were tested internationally, and both opaque and vitreous types exceeded eight tons of grain per hectare, indicating that opaque seeds could yield well. Several high-lysine lines with vitreous endosperm were identified. Vitreous high-lysine types were also identified by treating P721 with diethyl sulfate. Generally, the vitreous seeds had higher kernel weight with lower protein and lysine contents than opaque seeds (Axtell and Ejeta, 1990). While vitreous high-lysine types were found, it was disturbing that the modification to vitreousness was unstable and would vary from generation to generation even when sowings were made from well-developed vitreous seeds (Ejeta and Axtell, 1990). Additional effort is required to realize commercial application of high-lysine sorghum.

Carbohydrates

The information available on seed carbohydrates in sorghum relates to waxy and sugary endosperm types. Endosperm texture, an important consideration for the preparation of foods, ranges from completely corneous to very floury types, reflecting the degree of protein matrix holding carbohydrate particles. Some corneous types pop well and are used in the preparation of snacks.

The starch in normal sorghum is approximately 25% amylose and 75% amylopectin. The starch in waxy kernels is almost all amylopectin. Normal endosperm stains blue with iodine, whereas that of waxy starch stains red. Waxy is inherited as a single recessive gene; the gene must be recessive at all three loci in the endosperm for the trait to be expressed. Xenia is expressed in the individual arising from fertilization with pollen from a normal plant; i.e., pollen from a normal plant fertilizing ovules of a waxy plant will result in normal seeds.

Waxy sorghum has been used for feed but the advantage compared with normal grain is slight. Waxy grain was used during World War II to make a tapioca substitute (Rooney and Miller, 1982). Waxy sorghum has been used to make specialty foods in China but generally does not make good traditional type foods.

The sugary trait is inherited as a simple recessive gene. At maturity the kernel is crystalline and shrunken in appearance. Sugary seeds are about 25% smaller than normal but contain about twice as much or more sugars than normal. Reducing sugars ranged from 2.52 to 2.67% compared with 0.34% in normal (Hulse et al, 1980). Starch content in sugary lines is much reduced, ranging from 22.2% to 34.0% compared with normal sorghum (60.8%). Amylose content of the starch in sugary lines ranged from 22 to 62% higher than in normal. Water-soluble protein in sugary is higher than that in normal.

There are several sugary genes (dented *R*, *W*, and *F*) and defective traits that influence kernel carbohydrate, but they are of no economic interest.

In some countries, sugary grains are harvested in the dough stage and parched to use as a snack.

Tannins

Tannins, a group of high molecular weight polyphenols, occur in the testa. The presence of a testa is determined by a pair of complementary genes, B_1 and B_2. Fragments of testa may be present in some sorghums if recessive b_1 or b_2 is present. In the presence of B_1 and B_2 and the recessive *tptp*, the testa is purple, and in the presence of *Tp*—it is brown (Doggett, 1988). *S* is a spreader gene and in the presence of B_1 and B_2 results in a brown color in the pericarp. If the testa is fragmented, brown spots occur in the pericarp above the fragments.

High tannin, not lysine, is the primary nutrient-limiting component in sorghum grain.

Initially, it was thought that tannins would bind with endosperm proteins, reducing their biological value; i.e., the complexed proteins were not easily digested. More recent observation indicates that the primary antinutritional effect is the inhibition of the metabolism of digested and absorbed nutrients, especially protein (Ejeta et al, 1989). Polyphenols are in high concentration in the grains with a brown pericarp and pigmented testa, but all polyphenols involved may not act as tannins. It was demonstrated that rat weight gain was significantly reduced as tannin content increased (Ejeta et al, 1989). Cummins and Axtell identified a polyphenol group (type II) occurring in seeds that have a colored testa, usually purple, with vanillin-hydrochloride values less than one and catachin equivalents of no more than two. In rat feeding tests, the protein efficiency ratio was as good as that of white grain (0.63 and 0.58). These types may not repel birds or hold down molds anymore than white types (without testa) (Doggett, 1988). However, once it became possible to highly purify tannin, it was not possible to distinguish the tannins between group II and group III sorghums (Ejeta et al, 1989).

In both group II and group III sorghums, the polymerization that occurs to form large molecules of tannin takes place following wounding of the tissue or grain drying at maturity. Proteins differ significantly in their affinity for tannins; the proline-rich proteins generally have a strong affinity for tannin (Ejeta et al, 1989).

High-tannin sorghums have been cultivated in a number of places in the world where birds are a serious problem. Such grains were usually processed to reduce their negative impact on nutrient value. Dehulling improves the biological value substantially. Traditionally the negative impact has been reduced by treating with wood ash and by malting (Mukuru, 1990). It has been found that treatment with a base such as NaOH reduces toxic effects. Beside the antinutritional aspect, color of testa and pericarp add a color to food that does not occur in food from pearly white sorghum.

Oil

Oil is found primarily in the embryo. Oil content has been found to be positively correlated with protein content, so both traits can be selected for simultaneously. Additive gene action was found to be significant for the inheritance of oil content in some experiments and dominance variance in others. The magnitude of additive variance ranged from quite small to large (Hulse et al, 1980).

Other Grain Traits

Grain color can occur in the pericarp, testa, and endosperm. Color of the testa can be seen if the pericarp has no color and little starch. For endosperm color to be visible, it is necessary to have both a colorless pericarp and no testa.

There are a number of genes affecting grain color. The B_1, B_2, and *Tp* genes were mentioned in the discussion of testa and tannin. A number of genes influence pericarp color. The presence of *Y* results in a yellow pericarp. The presence of *R* results in a red pericarp but only in the presence of *Y* (Doggett, 1988). Colorless or white pericarp occurs when the genetic composition is *R—yy* or *rryy* (Rooney and Miller, 1982). The gene *I* augments the intensity of expression of *Y—* and *Y—R—*. Kernels that are *R—Y—I—* appear bright red compared with *R—Y—ii*. The gene *I* intensifies the lemon yellow color of the pericarp. The gene *M* also modifies the expression of *Y* and *YR*. Bw_1 and Bw_2 result in a brown wash in the pericarp if the spreader gene is recessive (*ss*). As indicated in the previous section, when dominant *S* (spreader) is present with B_1—B_2—, the pericarp color is brown. Rs_1— sunred causes the pericarp to turn red when exposed to the sun at the time the grain matures. The presence of *Pb* results in purple blotches and *Pt* a purple tip on the grain.

A thin pericarp occurs in the presence of the *Z*-gene. Thin pericarp is translucent, permitting testa and endosperm color to show. If there is no testa and the endosperm is corneous, the grain has a pearly white appearance. Grain of this type, particularly in a plant with tan plant color, is preferred by many sorghum consumers, particularly in India, for the preparation of food. If the plant color is red or purple, color may appear in food prepared from pearly white grain. The pericarp varies greatly in thickness, particularly the mesocarp, which can range from 12 to 120 μm. The thicker cells usually contain starch granules. The thick pericarp is associated with a chalky appearance of the grain and can mask or modify color expression. In some crosses *Z* and *S* may interact such that *ZZSS* or *ZZss* results in pearly grain, *zzss* in a chalky mesocarp and *zzSS* in a thin chalky pericarp (Doggett, 1988). Traditionally, in India and much of Africa, round grain with a thick pericarp is liked because it pearls easily by traditional methods.

Damage to the grain by insects or diseases results in spots or blotches of color depending on the presence of *P* or *Q* genes. In continuous wet weather, pigment can penetrate from the testa into the endosperm or from the glumes into the pericarp. These colors result in discoloration of food products. Extensive mold attack can result in loss of grain weight, increase in floury endosperm, and finally in loss of usable grain. Sprouting of the grain in the head can also result if the period of grain maturation coincides with warm temperature and heavy rain or excessive moisture.

Endosperm color is either white or yellow; yellow is controlled by the gene *Ye*. A kernel with a red colored pericarp over a yellow endosperm has a bronze appearance (Rooney and Miller, 1982).

Endosperm hardness (texture) varies with variety; for example, the seeds of the guineas grown in West Africa are completely corneous, while the grain of varieties Safra and Mugud, grown in eastern Sudan, are very floury. These are highly heritable traits.

In some parts of Africa, sorghum is used in the brewing of beer, more frequently as adjunct, but also for malt. In southern Africa, the traditional opaque beer is produced commercially. There is a tremendous range in diastatic power among sorghum types, from less than 10 to more than 60 diastatic units. The desired level for opaque beer is around 30 units. The trait is heritable and selection is possible, but there is an important environmental component. Sorghum diastatic units per gram of malt relates to the conversion of starch to sugars. It is used in the calculation of diastatic power (Daiber, 1971).

FORAGE CHARACTERISTICS OF SORGHUM

Sorghum, sudangrass, and the cross between them are commonly used for animal feed: pasture, hay, and silage. The crop is sown thick and usually cut around the boot stage; three or four cuts or more are commonly taken during a growing season. There are a number of genetic traits contributing to forage uses.

There is a cyanogenic glycoside called dhurrin found in sorghum. It varies with variety and is in highest concentration during the seedling stage. Some 45 days after emergence, the danger from dhurrin is past.

When dhurrin is hydrolyzed, it breaks down into HCN (prussic acid) and *p*-hydroxybenzaldehyde. A cow can be killed by as little as 0.5 g of HCN. Environmental factors such as drought or frost, particularly in the seedling stage, that break down tissue result in the formation of HCN, and the crop can be dangerous to feed. The making of hay or silage results in the loss of HCN, and the feed is safe. Small amounts of glucose and maltose fed to animals reduce the amount of hydrolysis and release of HCN. The feeding of concentrates prior to putting animals to pasture also reduces the danger from poisoning.

There have been a number of studies on the inheritance of high dhurrin content. The inheritance has been reported to be recessive, partially dominant, and dominant. Inheritance appears to be multigenic. Selection for low prussic acid has been effective and varieties with low prussic acid content have been identified (Doggett, 1988.

Brown Midrib

A number of brown midrib genes (*bmr*) have been identified in sorghum, and some (*bmr* 6, 12, and 18) reduce the lignin content and modify the lignin molecule improving animal digestibility by 20–30%. Lignin is not digested and limits the breakdown of cell wall polysaccharides. The phenotypic expression of this trait is a reddish brown color in the midribs of leaves. Expression begins in the four- to six-leaf stage of a developing plant. The color may fade as the plant matures, being difficult to see at maturity.

Inheritance of the *bmr* trait is by simple recessive genes. Allelism tests indicate that *bmr* 12 and 18 may be linked but *bmr* 6 is independent. Brown midrib genes have been induced in sorghum and pearl millet using mutagens, and they have been found naturally in maize and pearl millet and sorghum. Chemically induced mutants in sorghum have reduced lignin content in the leaves from 5 to 25% and in the stem from 5 to 51%. Increased digestibility of some *bmr* lines compared with their near isogenic counterpart was as much as 33% (Cherney et al, 1991).

Other Forage Traits

The stem of sorghum can be insipid or sweet, dry or juicy. Traditionally, sweet juicy stems are used for chewing. There have been indications that sweet juicy stems result in a more palatable forage. The sugar content of sorghum stems can be near to that of sugarcane, and the juice has been expressed commercially to make syrup, refined sugar, and traditionally in some parts of the world (India, for example) to make a caramelized product.

Sorghum plants may senesce, dry, and become brown as grain matures, or leaves and stems may stay green long after the grain matures and dries. The stay-green trait contributes to drought tolerance and resistance to the stress-related disease, charcoal rot.

There is a varietal type known as broomcorn. The panicle has a very short central rachis and very long panicle branches. Broomcorn is grown commercially as a source of material to manufacture brooms.

GENETICS OF MILLET GRAIN CHARACTERISTICS

Pearl Millet

GENERAL

"Pearl millet kernels are about one-third the size of sorghum kernels. The average weight of each kernel is about 9 mg compared with 20–30 mg for sorghum. Sorghum kernels are generally spherical in shape; however, pearl millet kernels are generally tear shaped. The most common color is a slate gray, but colors from creamy white through yellow to black are known. The germ of pearl millet is a much larger percentage of the total kernel than is the germ of sorghum (17.4% compared with 9.8% in sorghum). This difference explains in part the higher protein and oil contents of pearl millet as compared to sorghum" (Sullivan et al, 1990).

The protein content of pearl millet ranges from 8 to 19%. Pearl millet is low in lysine, tryptophan, threonine, and the sulfur-containing amino acids. In an evaluation of several cereals, methionine content was found to be highest in proso, followed by sorghum, pearl millet, and then maize. The level of lysine content of pearl millet grain on a dry matter basis was 0.357%, 21% greater than corn and 36% greater than low-tannin sorghum (Sullivan et al, 1990). With increasing protein content, lysine as a percent of protein decreases, but as yields go up, the total lysine per hectare will increase. Generally the amino acid profile of pearl millet compares favorably with that of wheat, barley, and rice (Hulse et al, 1980).

Starch represents about 56 to 65% of the kernel and is about 20 to 22% amylose; free sugars range from 2.6 to 2.8% of the grain, with sucrose representing about 66% of the total sugars (Sullivan et al, 1990).

Ether-extractable lipids range from 3.0 to 7.4%. They are mostly in the germ and similar in composition to sorghum. There is a "mousy" odor associated with whole milled grain that was thought to be due to rancidity of the oil, but it has been found to be associated with a flavinoid, apigenin, found primarily in grains with a slate gray color. Oxidation of lipids is still considered to be a contributing factor (Hulse et al, 1980).

The carotene content of yellow endosperm pearl millet is similar to that of sorghum and lower than for maize.

Pearl millet is generally not troubled with antinutritional factors, although phytic acid and two trypsin inhibitors have been found (Sullivan et al, 1990). A thionamide-like substance has been identified that interferes with the formation of a thyroid hormone contributing to undesirable gastrogenic effects.

PROTEIN AND AMINO ACIDS

The relationship of yield to protein and lysine content was carefully evaluated at ICRISAT, and results were presented in their annual reports from 1975–1976 to 1983. A negative but low correlation (–0.15 to –0.50) was found between yield and protein content (1977–1978). The variability for protein (5.1 to 18.2%) and for lysine (1.6 to 4.7 g/100 g of protein) contents were found to be high (1975–1976) and, coupled with the weak correlation coefficient, suggested the possibility of breeding for high-yielding, high-protein types (1983). Fractionation of the protein of a number of varieties indicated that albumins and globulins ranged from 22 to 28% of total N, prolamin and prolamin-like fractions from 22 to 35%, and glutelin and glutelin-like fractions from 28 to 32%, indicating that the prolamin fraction in pearl millet is smaller than in sorghum. A hybrid, ICH 105, consistently had a low prolamin content in all locations, which resulted in a high-lysine content (1977–1978). After six generations of pedigree selection, 440 progenies ranged from 14.0 to 23.3% grain protein (1983). After four cycles of selection, 22 high-yielding, full sib progenies with above average protein content (mean, 12.9%) and an estimated lysine content of 2.9% (of protein) were selected to be used for the fifth cycle (1979–1980). From crosses between lines selected for high protein, 18 F_1s showed 14.1 to 15.8% grain protein, and five yielded as well as the high-yielding variety BJ104, which had a protein level of 11.7% (1983). It appears possible with pearl millet to select for yield and protein content and keep a reasonable level of lysine.

Considerable environmental variation for protein content exists. There has been no consistent expression of heterosis. As in other cereals, it appears that the inheritance of protein involves many genes, most of them recessive. In different experiments, additive variance was found to be small, while dominance and epistatic effects appeared to be greater. The predominance of nonadditive effects indicates that diversity of alleles rather than heterozygosity contributed to adaptive changes (Hulse et al, 1980).

Nonadditive variance was also found to be of greater importance in the inheritance of tryptophan, methionine, and total sulfur. Although of less importance, additive effects were also found to be significant.

Godawat and Kaul (1981) evaluated protein, tryptophan, lysine, and threonine content and 1,000-grain weight. The highest genotypic coefficient of variation was observed for protein content. Estimates of broad sense heritabilities were highest for 1,000-grain weight and lowest for lysine content. Protein content was positively correlated with tryptophan content and 1,000-grain weight.

Burton et al (1980) studied the effect of male and female gametes on 1,000-grain weight, nitrogen content, proline content, the nitrogen-proline ratio, and dye-binding capacity, and found that the genetic effects for female gametes were significant for all traits, but for male gametes, significant effects were found for 1,000-grain weight, proline content, and nitrogen-proline ratio.

Gupta et al (1976) evaluated protein content, mineral content, and grain yield of 46 lines from India, Africa, and America grown in India. Tiller number and grain size had the greatest effect on yield. Earing span (time of flowering from tip to base of panicle) had a direct effect on protein content. Plant mineral content has a direct effect on plant height, length of the vegetative period, grain size, and yield.

Gartan et al (1987), from an incomplete diallel analysis, found a predominance of nonadditive variance for the inheritance of protein content, although the heritability estimate was low. Tyagi and Paroda (1982), in a study involving six environments, found a preponderance of nonadditive variance for protein content.

Kumar et al (1983) found no significant correlation between grain yield and protein content. Likewise dye-binding capacity (estimate of basic amino acids in protein) and protein content were not significantly correlated.

CARBOHYDRATES

Gupta and Indoo (1983, *personal communication*) studied the changes in activity of peroxidase and amylase and contents of soluble proteins and soluble sugars at 14 stages of plant development of six pearl millet inbreds and their crosses. They observed that plasticity of plant response in its environment was related to biochemical versatility as plants were able to quickly utilize soluble sugars and proteins in early stages of germination. Both amylase and peroxidase activities were highest during the seedling phase and lowest in the reproductive phase. At different developmental stages, the relative level of recessive and dominant gene action was found to vary based on changes in biochemical constituents.

Champingy et al (1982) studied the growth and accumulation of starch by the parents and F_1 hybrids obtained by crossing one ecotype of *Pennisetum mollissimum* and a cultivar of *Pennisetum glaucum*. Leaves of 25-day-old seedlings of *P. mollissimum* were twice as long as those of *P. glaucum*. Studies of reciprocal F_1 hybrids revealed that each hybrid exhibited heterosis for leaf length in relation to its male parent. Many starch grains were observed in the bundle sheath chloroplasts of *P. mollissimum* and of the hybrid *P. glaucum* × *P. mollissimum*. Quantitative analysis of leaf starch during seedling growth confirmed the observation. Fast growth was associated with starch accumulation, and each hybrid behaved like its male parent.

Khangura et al (1980) analyzed the combining ability of beta-carotene, total carotenoids and other grain characteristics in pearl millet. They observed that specific combining ability was more important than general combining ability for all traits. In a similar study, Bhardwaj et al (1987) showed the absence of non-allelic interaction in the inheritance of grain hardness and beta-carotene. Additive and nonadditive gene effects were important with nonadditive predominating. Grain hardness was controlled by dominant genes and beta-carotene by recessive genes. High heritability was observed for total carotenoids and beta-carotene.

Pande et al (1985) investigated grain sink activity in pearl millet inbreds and the hybrid from crossing them. They observed heterosis for grain starch and protein content in the grain of hybrids.

LIPIDS, FATTY ACIDS, AND FAT CONTENT

Singh and Gupta (1987) observed significant genetic variability for crude protein, soluble protein, total lipids, and fatty acids in a 12 × 12 diallel cross. Additive gene effects were predominant for all traits, and maternal effects were observed for fatty acids only. Estimates of specific combining ability were low, indicating lesser scope for exploitation of heterosis. In a study on genetics of yield and some quality traits in pearl millet, Sachdeva (1981) observed moderate narrow sense heritability for fat content.

FORAGE CHARACTERISTICS OF PEARL MILLET

Pearl millet is widely used as forage. Much work has been done in the United States at Tifton, Georgia, to develop pearl millet varieties and hybrids for feed. Much has also been learned about the management of pearl millet as a feed crop.

Brown Midrib

Brown midrib (*bmr*) genes have been found naturally and as a consequence of exposure to a mutagenic agent (diethyl sulfate). These genes appear to be inherited as simple recessive traits. As with sorghum and maize, the concentration of lignin was reduced (40 g/kg in *bmr* compared with 50 g/kg in the normal counterpart). The reduction of alkalilabile phenolics is not as much as it is in sorghum. At anthesis, digestibility of *bmr* was found to be 726 g/kg versus 659 g/kg for normal. Experimental results support the theory that a 10 g/kg reduction in lignin results in a 40 g/kg increase in digestibility (Cherney et al, 1991).

Other Forage Traits

Forage pearl millet is free of HCN.

The genes *d* (dwarf) and *tr* (trichomeless) have been useful for the improvement of forage quality. The dwarfing gene d_2 results in as much as a 50% reduction in plant height, but leaf number remains the same; the greatest feed value is in the leaves. Dwarf plants, although with lower dry matter yield, have 15% more protein and 17% less lignin, and animal gains are higher than from taller plants.

The trichomeless trait is inherited as a single recessive gene. It improves palatability and contributes to drought tolerance. These positive aspects are balanced by slower digestion contributing to reduced feed intake.

The orange node trait (*on*) resembles the *bmr* trait in improving forage digestibility. Crosses between *on* and *bmr* indicate that these traits are controlled by genes at the same locus.

Heterosis has been found in hybrid pearl millet forage. It has been found that hybrids produced with a pollinator that does not restore male fertility in the hybrid improves forage quality, provides longer grazing, and yields better under stress (Andrews and Kumar, 1992).

CROSSES WITH NAPIER GRASS

The cross between pearl millet and napier grass provides a useful forage plant. With adequate moisture and fertility, dry matter yields of 80 t/ha have been realized, falling to 10 t/ha or less as environmental stress increases. In southern Africa this hybrid is referred to as banagrass. Breeding to improve banagrass was undertaken as part of the SADC/ICRISAT Sorghum and Millets Improvement Program. Yield increases of 50% of the control and protein content of 14% were achieved. Banagrass has withstood severe moisture stress and survived winter temperatures of 0–5°C. Its rapid regrowth following cutting has been attractive to farmers in Zimbabwe.

OTHER MILLETS

Kostantinov in Russia crossed high-protein lines of foxtail millet with low protein lines of common millet. Protein content in the F_1 hybrids was similar to mid-parent values. In the F_2, the range of protein content was large—14 to 20%. It was noted that red seeds had higher protein content than creamy or yellow ones.

He treated common millet with a mutagenic agents and found three mutants with improved yield, equivalent milling quality, but lower protein quality. Other mutants had 16% protein compared with 14.9% for the untreated control and 12.9% for the standard line (Hulse et al, 1980).

AGRONOMY: NUTRITIONAL CONSIDERATIONS

The section on agronomy is divided; the first part relates to the interaction of agronomic practices on protein and amino acid composition of sorghum and then the millets. A later section outlines interactions of plant nutrients and moisture on plant performance. This division of the agronomic section keeps both heritable and crop management considerations relative to nutritional factors together and places management of crop performance with other crop considerations.

Sorghum

PROTEIN AND AMINO ACIDS

The percent protein in sorghum grain is influenced by variety and by the environment. Significant interactions have been found for variety, fertilization (in amount, in kind of fertilizer, and in methods of application), location, and season. A number of experiments indicate a greater effect of variety than environment on protein content.

Response to application of N fertilizer has been variable: significant in some cases, significant in one or more seasons in a multiseason experiment, or not significant. Applications of phosphorous and potassium fertilizers usually do not influence the protein content of the grain; however, application of fertilizer that resulted in increased grain yield resulted in a greater protein yield per hectare.

There is an effect of location on amino acids. One study in India involving three locations and eight sorghum varieties evaluated the effect on protein and several amino acids: lysine, isoleucine, leucine, methionine, and tryptophan. Protein content varied from 8.1 to 10.5% across varieties and from 6.4 to 11.2% (mean of the eight varieties) across locations. In the location with low protein content, the concentration of the five amino acids as a percentage of protein was higher. In the experiment as a whole, there were negative correlations between protein and lysine, methionine, and tryptophan; lysine and leucine; protein and leucine; and protein and isoleucine (Hulse et al, 1980).

Several experiments have indicated that the change in percent protein is generally due to an increase in the prolamin fraction, with little effect on the albumin and globulin fractions. This indicates that an increase in percent protein is not associated with an improvement in protein quality. In an experiment comparing soil and foliar applications, it was found that foliar application increased protein yield.

The prolamin content increased with nitrogen application (up to 180 kg/ha of N). The glutelin fraction did not increase much beyond 60 kg/ha application, and foliar application was the same as for soil. At 180 kg/ha of N application, protein content increased from 8.9 to 11.6%, associated with an increase in glutamic acid, proline, alanine, aspartic acid, and cystine but a decrease in lysine, histidine, arginine, and phenylalanine (mg/g of N). The leucine-isoleucine ratio increased from 2.5 to 3 as nitrogen application increased from 0 to 180 kg/ha (Warsi and Wright, 1973).

In an experiment in Australia involving six varieties of sorghum and four levels of nitrogen fertilization (0, 68, 136, and 173 kg/ha), it was found that the three basic amino acids, lysine, histidine, and arginine, were negatively correlated with protein content, and other amino acids showed a positive relationship. Only leucine and threonine were significantly affected by nitrogen fertilization and variety (McKenzie and Wallace, 1954).

Results of an experiment conducted in Kansas indicated that the content of protein and amino acids increases in the grain, but the increase in amino acids is not proportional to the increase in protein with increasing N fertilization. There were significant increases (mg/g of N) in glutamic acid, proline, alanine, isoleucine, leucine, and phenylalanine, and a decrease in lysine, histidine, arginine, threonine, and glycine. Changes for aspartic acid, serine, half-cystine, valine, methionine, and tyrosine were not significant. Only glutamic acid, alanine, and leucine were significantly changed due to location (mg/100 g of grain). Expressed as mg/100 g of total N, lysine, histidine, arginine, glutamic acid, glycine, alanine, half-cystine, methionine, and leucine were significantly affected (Waggle et al, 1967).

Different levels of application of P_2O_5 and K_2O fertilizer have not been found to influence levels of protein or lysine and leucine. While K_2O had no effect, P_2O_5 resulted in a significant increase in the concentrations of Ca, Mg, P, Fe, and Mn in the grain (Deosthale et al, 1972).

It has been found that as protein content increased with increased fertilization that the starch content decreased.

In central Sudan, in a three-year study, the crude protein content increased when sowing was delayed from the end of June to the end of August, but yield declined with sowings after the fourth week of July. The decrease in yield was associated with an increase in N and P content, but the increase was not proportional to the loss in yield (Rai, 1964).

Plant population was found to have no significant effect on protein, lysine, or leucine contents.

OTHER TRAITS

Location was found to have a significant effect on the contents of Ca, P, Fe, Mg, Mn, Zn, and Mo in sorghum. The effect on Cu was not significant. Varietal differences for these elements were found to be not significant. Fe was the only element found to be significantly associated with protein (Deosthale et al, 1972).

An experiment in India indicated a high level of variation in carotene content with location. Beta-carotene content continues to decline while grain is in storage.

The above information has been summarized from (Hulse et al, 1980).

Pearl Millet

A comparison of soil and foliar (urea) application of N was studied. Urea application had a significant impact on increasing grain weight. Protein content was also increased significantly with greatest increase coming at 45 kg N/ha. Effects of N fertilization on fat, fiber, ash, and P_2O_5 were not significant. The Ca content increased significantly at low levels of fertilization (11 kg N/ha foliar and 22 kg N/ha soil application). As fertilization increased, carbohydrate content decreased (Khanna, 1966).

In another experiment with application of N fertilizer from 0 to 200 kg N/ha, yield increased from 1,100 to 1,500 kg/ha and protein 6.7 to 8.1%. Again, with N ranging from 0 to 50 + 50 kg/ha (split application), yield ranged from 781 to 2,178 kg/ha and protein from 7.3 to 13.6% (Blondel, 1970).

An experiment in Senegal with nitrogen fertilization ranging from 0 to 150 kg/ha resulted in a marked increase in protein content, the increase being linear with N application. Increased N fertilization resulted in increased glutamic acid and leucine in the protein and decreased lysine, glycine, histidine, arginine, and to a small degree threonine. As the range of N fertilizer increased from 0 to 50 kg/ha, the average decrease in essential amino acids was less than 2% of the protein. There was no significant change in moisture, ash, lipid, cellulose, Fe, Ca, and P with increasing N fertilization (Ganry and Bideau, 1974).

An experiment was conducted in India using $(NH_4)_2SO_4$ as a source of fertilizer. Application was made at 0, 44, 88, and 176 kg N/ha. Both protein and sulfur content increased significantly at 44 kg/ha of N, and there was a significant increase in the concentration of tryptophan. Above this rate of fertilization, there was an increase of lysine in the grain but a reduction as percentage of protein. The increase in protein content resulted from a 40% increase in the prolamin fraction, with decrease in the albumin and globulin fractions. This resulted in a redistribution of sulfur in the protein fractions; sulfur in the prolamin fraction increased from 20 to 30%, stayed the same in the albumin fraction (20%), and decreased in the globulin fraction (from 28 to 19%).

It was found in pearl millet that the amount of tryptophan in the prolamin fraction is higher than in sorghum and maize (Sawhney and Naik, 1969).

A comparison of seed rate of 2.5 versus 5.0 kg/ha at sowing had no effect on protein content (Tomer, 1970).

Iron and manganese in the seed increased with increasing N fertilization. The concentration of Ca and Mn were found to be inversely related (Shukla and Bhatia, 1971). Thiamin content in the grain was also found to increase with increasing N fertilization.

Finger Millet

As with sorghum and pearl millet, the concentration of protein increased with increasing N fertilization, but there was no consistent pattern with P_2O_5 or K_2O_5 (Venkataramana and Krishna Rao, 1961). The level of Ca fell with increasing application of nitrogen fertilizer (Krishnamurthy, 1968).

As the content of protein increased, the levels of lysine and the sulfur amino acids methionine and cystine as a percentage of total N decreased (Stabursvik and Heide, 1974).

Finger millet has a relatively high content of sulfur-containing amino acids. With fertilization, methionine and cystine increased to about 8% of protein then leveled off. Fertilization, in a ratio of 1:10 sulfur-nitrogen has been found to be appropriate for finger millet (Stabursvik and Heide, 1974).

AGRONOMY: MANAGEMENT ASPECTS

Crop Fertility

From a practical point of view, agronomy involves the management of adapted plants in sole or mixed crop situations; nutrient and moisture availability; cultural practices with respect to sowing and weed management; and to some degree, the expression of pests. Traditionally, sorghum and the millets frequently are not fertilized or are fertilized with animal manure. With some traditional varieties, selected under these conditions for long periods of time, response to fertilizer is reduced compared with varieties and hybrids selected to significantly benefit from good levels of fertility. Several experiments performed in India in the 1960s and later at ICRISAT demonstrated that there is little or no interaction between sorghum variety and response to fertilization. Differences between lines to be selected are greater at high levels of fertility than at low levels. Greater differences facilitate selection. In the absence of interaction between yield and fertility, entries selected at high fertility will also perform better at lower levels of fertility justifying the use of fertilizer in breeding nurseries.

Nutrients may not be limiting. This needs to be determined for each environment. If commercial fertilizers are used by traditional farmers, 30 kg/ha of N, with half to the same amount of P_2O_5, and some K_2O is common, and there is plant response. Unless it is very dry, application of 80 kg of N/ha is economic, and application above 120 kg of N/ha is not as common in countries where most of the agriculture remains traditional.

In situations where rainfall is in the 250 to 500 mm range, distribution of rainfall can be as important as amount. With reasonable distribution and no fertility, yields in the 400 to 800 kg/ha range can be expected, but, with fertility yields can be 1.0 to 2.0 t/ha. With good moisture and fertility, yields of varieties can be 3,000 to 4,000 kg/ha and 5,000 to 6,000 kg/ha for hybrids. Highest yields can reach 8,000 to 12,000 kg/ha depending on length of growing season and good crop management. It is important, but commonly not realized, that hybrids almost always have higher yields than varieties, and while the yields of both drop with poor growing conditions, the magnitude of hybrid yield over varietal yield is the greatest in stressed environments.

R. L. Vanderlip (1979) and his colleagues have studied nutrient uptake by sorghum. They find that curves for nutrient uptake are above those for dry matter accumulation. They point out that at 60 days after emergence, about one-half of the plant weight has been produced, but "nearly 60% of the phosphorous, 70% of the nitrogen, and 80% of the potassium have already been taken up."

Of the three major nutrients, potassium is taken up most rapidly, followed by nitrogen, then phosphorus. Translocation of nutrients to the grain is largely nitrogen and phosphorus with smaller amounts of potassium. A crop with grain yield of 8,500 kg/ha uses approximately 210 kg of nitrogen, 40 kg of phosphorus, and 245 kg of potassium per hectare (Vanderlip, 1979).

Plant growth is slow during the first 20 to 25 days as leaf area (photosynthetic capacity) increases. Leaf growth continues during the next 50 to 80 days, when the upper 8 to 10 leaves develop, providing the important leaf area to produce nutrients for grain production. Following floral initiation, there is rapid stem elongation, followed by flowering and grain formation. Later-maturing varieties and hybrids tend to have more weight at each growth stage. A two-year average of three hybrids indicated a dry matter accumulation of 193 kg/ha per day. Later-maturing hybrids yield more than earlier types—for normally adapted cultivars, yield tends to increase linearly with length of growing period.

Micronutrients are important to plant growth and must be available if high yields are to be realized. Iron concentration was initially high in stems and gradually decreased as the plants developed. Grain and panicle parts remained relatively constant at 75 ppm.

Concentrations of zinc were highest in early stages of growth, particularly in the stems. All plant parts had roughly the same concentration after vegetative growth ceased. About 50% of total zinc uptake was in the grain at physiological maturity (Jacques et al, 1985).

The highest concentration of copper during plant growth was in the stem. Concentration gradually decreased with plant development including grain formation. There was a greater uptake of copper during the vegetative stage than during panicle development and seed formation.

Manganese concentrations differ substantially among plant parts, with higher concentrations in leaf blade and leaf sheaths compared with the stem and panicle. Concentration is higher in younger plants than when they mature. Manganese concentration changes little once maximum dry weight is reached. Leaf sheaths have the highest concentration. At physiological maturity, 50 to 60% of the manganese is in leaf blades and sheath, 5% in culms, and most of the rest in the grain.

Concentrations of magnesium are high in leaf sheaths and remain so throughout the growing season. Concentration in the stems decreases as the season progresses, while it remains fairly constant in leaves. Manganese concentration in the heads is relatively low and decreases as grains reach physiological maturity. Manganese was translocated from the stems into the panicles.

Calcium was found to be lower in the panicles than other plant parts. The concentration of calcium increases in leaves following their attaining maximum weight and during grain formation. Calcium remains in higher concentration in leaf blades than in other plant parts. By contrast, calcium concentration in stems and leaves increases during grain development but remains the same in the sheaths (Jacques et al, 1975).

As with major nutrients, uptake of minor elements preceded dry matter production. Generally, nutrient content changed among plant parts with development, was highest following emergence, and then decreased during the rapid growth stage when nutrient uptake could not keep up with growth and elongation. It is apparent that nutrients should be plentiful during early plant growth. This can be provided by soil application or by spray.

Moisture

Sorghum and millets are crops of semiarid areas, and moisture availability for plant growth is critically important. A characteristic of the semiarid tropics is high variability in amount of rainfall per season and distribution within the season. With an average rainfall of 700 mm, seasonal variation can range from 300 to 1,200 mm. During the period 1980 to 1992 in southwestern Zimbabwe, where long-term average rainfall is around 600 mm per year, rainfall was below expectation every year but two. Even with low rainfall (300 to 350 mm) but with good distribution, plot yields exceeding 3 t/ha of grain have been realized. If there are prolonged breaks in rainfall, particularly at the time of floral initiation or flowering, yield of grain would drop to a ton per hectare or less.

Management of moisture becomes important, particularly when it is generally in short supply but can be excessive in a season with few but heavy rains. Various practices have been developed to manage such rainfall while minimizing soil erosion, such as leaving crop residue, alternating vegetative strips with cultivated strips, drainage systems to carry off water with minimum erosion, furrowing fields so that each furrow carries water onto or off the soil surface. The furrows may be tied; i.e., small barriers are built periodically in the furrow to hold water where it falls. If sufficient power is available, fields may be deep-chiseled periodically to break hard pan and improve water penetration. The important thing is that plans are made to retain adequate water on the land and to drain excess water off the land with minimal soil loss.

Fertility is important to effective moisture utilization. An experiment conducted at Matopos, Zimbabwe, with stubble and 50 kg/ha nitrogen resulted in a yield increase of 8.8 kg/ha per millimeter of rainfall for a total yield of 3,878 kg/ha. Without the stubble and nitrogen treatment, yield increase dropped to 1.6 kg/ha of grain per millimeter of rainfall with a yield of 705 kg/ha. It is difficult to indicate how much fertilizer to use, but approximately 30 to 50 kg of N/ha for rainfall of 250 to 400 mm; 60 to 80 kg of N/ha for rainfall of 400 to 650 mm; and 120 kg N/ha for rainfall of 650 mm and above. Large-scale commercial farmers, especially with irrigation, may use in excess of 150 kg/ha of N. Generally phosphorus is 50 to 100% that of nitrogen, and the need for potassium should be determined.

Traditional farmers, particularly in West Africa, frequently use photoperiod-sensitive varieties. At times, first and second sowings fail, and resowing is necessary. With photoperiod-sensitive types, flowering and grain maturity occur during the same calendar days regardless of planting date, so that even with delayed sowing, plants mature when chances are good that there is moisture to finish the crop.

There are frequently advantages in sowing into dry soil early in the season, as yields are higher with a longer growing period. Deep sowing is advantageous because more moisture is accumulated in the soil before germination. Many sorghum varieties emerge from 5 cm and some from 10 cm. The greater the depth, the more moisture accumulates in the soil before germination and the longer seedlings can survive if rain following emergence is delayed.

Control of weeds, particularly during the seedling stage, is very important; it can be even more important than fertilization. An experiment conducted at Matopos in Zimbabwe is revealing. With no treatment, yield came to 814 kg/ha; with 50 kg of N/ha the yield was 920 kg/ha; with two weedings the yield came to 2,630 kg/ha; and with two weedings and 50 kg of N/ha the yield came to 3,134 kg/ha. Nitrogen fertilization was most effective with weeding.

Plant stand should be lower when moisture is limiting; with nontillering traditional sorghum varieties, stands may be as low as 60,000 plants/ha. With 700 mm of expected rainfall and good management, a plant stand of 180,000 plants/ha is good; with poorer management or fertility a stand of 120,000 would be better. Sorghum and millets have a great capacity to compensate for stand.

Cropping Systems

Different cropping systems, well adapted to different regions where sorghum and millets are grown, have been developed and improved upon by farmers and scientists involved in the production of these crops. The differences in cropping systems are dictated by the technology available to the farmers and the agroclimatic conditions that prevail in a particular region.

INTERCROPPING

The main consideration for mixing crops together is to reduce the risk of total crop failure. Food crops are usually mixed with cash crops to help ensure both sustenance and disposable income. Cereals and legumes are often intercropped more for dietary reasons than for any beneficial effect the nitrogen-fixing powers of the legumes convey to the associated cereal crop or to a subsequent one (Swindale, 1979).

Small-scale farmers are often burdened with the twin problem of low crop productivity and limited land resources. Intercropping provides these farmers with a possible means of increasing crop productivity. Intercropping aims to increase total productivity per unit land area and to equitably and judiciously utilize land resources and farming inputs including labor. One obvious limitation is the reduced population of both crops and the effect on their productivity.

Intercropping is a risk-reducing method of agriculture that is particularly common in traditional agriculture where amount and distribution of rainfall are uncertain. The variety of crops is great, frequently involving a cereal and a legume. Many other combinations of cereals with cereals, with oil crops, or with vegetables are undertaken. The most effective intercrops and their management require evaluation. The two crops should not make peak demand on the environment at the same time. One plant that heavily shades another seriously affects the yield of the shorter crop. Sowing the crops at different times, with different row combinations (two-to-one or three-to-one ratio, etc.) and fertilization are important management considerations. Insect and disease problems may be different in an intercrop than a sole crop, influencing pest management.

Results from intercropping experiments have had variable results. An intercropped pigeon pea in sorghum had higher yield than a sole crop. However, the yield of mung beans in sorghum declined (1.03 to 0.42 t/ha), but the monetary value of the intercrop was greater. The conclusion was that 50% sorghum and 50% mung beans was most profitable (Chaudhry et al, 1987).

When sorghum was intercropped with pearl millet, land equivalent ratios showed no advantage from intercropping. Grain yields were consistently highest from pure stand sorghum. When soil moisture was adequate, yield tended to increase with the later-maturing sorghum varieties in the intercropped system; however, with inadequate moisture, millet was more competitive for water than sorghum. The adverse effects on sorghum outweighed the favorable ones on pearl millet. Sorghum intercropped with finger millet reduced the light available to finger millet, and dry matter produced by the finger millet was reduced (Stutzel and Vanderlip, 1988).

In an intercrop of sorghum and soybeans, the sorghum performance was good, but the soybean performance was much reduced.

In comparative performances of sorghum–pigeon peas, cotton–black gram, cotton-soybeans, pearl millet–pigeon peas, and pigeon peas–pearl millet intercropping systems under dryland conditions in India, it was found that highest net returns were realized with sorghum–pigeon peas followed by the pigeon peas–pearl millet system. The pearl millet–pigeon peas systems however showed the highest water use efficiency on a yield basis, whereas on a net returns basis the highest water use efficiency was for the sorghum–pigeon peas system (Dhoble et al, 1990).

An experiment was carried out in India involving pearl millet with soybeans or cowpeas. Sowing involved pure stands, two rows of legume to one of pearl millet and a mixture of legume and pearl millet sown in the same row. Green fodder and dry matter yields were generally higher for intercrops and seed mixtures than for pure stands. Dry matter yields ranged from 3.65 t/ha for the pure stand of pearl millet to 5.93 t/ha for the mix of soybean and pearl millet sown in the same row (Narwal et al, 1988).

At the ICRISAT Center, groundnuts were intercropped with pearl millet in the ratio of three to one. There were several interesting observations. Wind speed above the groundnuts decreased, and the radiation use efficiency of the groundnuts increased by 21–35%. The uptake of nitrogen by pearl millet increased in the intercrop. The best arrangement was to sow the groundnuts three weeks after the pearl millet.

It was found that indeterminate-spreading cultivars of cowpeas produced greater seed and fodder yield than erect types and caused the greatest millet yield reduction. For best yield results, it was suggested that cowpeas be selected in the intercropped situation.

Finger millet is included in intercrops. At Coimbatore, India, it was intercropped with sorghum, sunflower, okra, onions, and cluster bean. Total biomass production ranged from 11.73 t/ha in the finger millet–onion intercrop to 15.10 t/ha in the finger millet–sunflower intercrop. Finger millet grain yield ranged from 2.72 t/ha when grown in association with sunflower to 4.57 t/ha in pure stand (Siddeswaran et al, 1989). Finger millet production was reduced because of competition with sorghum and sunflowers; light availability was reduced to cluster beans, onions, and okra because of shading by finger millet.

SORGHUM AND MILLETS IN CROP ROTATION

Crop rotation is a sequence of different crops grown on the same land. Generally, the strategy is to improve soil and at the same time gain an income from it. Cropping systems tend to be relatively location specific.

Crop rotation usually involves two or three crops; most frequently one or more cereals alternate with legumes. Rotations can involve summer-winter crops; sorghum or pearl millet with chickpeas or sorghum and millet with wheat or mustard are examples. Sorghum and millet can rotate with summer legumes, pigeon peas, mung beans, cluster beans, etc., and can be rotated with a green manure crop such as sunnhemp, which is plowed in before maturity. Both foxtail and proso millets and sorghum have been used in rotation with transplanted rice.

Incorporation of crop residue and green manure crops and crop rotation improve soil tilth and can reduce nematode and other soilborne problems.

There are examples of application of fertilizer on one crop carrying over to the next. However, residual nitrogen from previous legume crops on a cereal is limited, and addition of commercial fertilizer is important to increasing yield.

When farmers grow cash crops such as chilies, onions, potato, tobacco, etc., they are sometimes followed by finger, pearl, or foxtail millet. Finger millet is also grown (transplanted finger millet) in relay cropping with amaranths.

In China, foxtail millet is grown after corn, sorghum, soybean, or spring wheat. However, repeated cropping of foxtail millet may cause a yield decrease in the succeeding crop; hence, it may be beneficial to have one crop of foxtail millet at an interval of two or three years.

Finger millet is also included in the cropping system of a number of southern and eastern African countries. Some of these cropping systems include finger millet as a mixed crop in rotation or as a sole crop. A millet rotation system might follow the sequence, cotton, millet, cotton, millet, cowpea, groundnut. Finger millet is also preceded by sorghum, maize, peas, beans, or sweet potatoes. A mixture of these crops is also quite common. Sorghum and millets can fit into many cropping patterns.

PLANT DEVELOPMENT AND STRUCTURE

Sorghum

When seeds are sown into moist soil they take up moisture and germination occurs. Emergence of sorghum and millets is usually about three days from sowing in warm climates and around 10 days in temperate areas where soil temperatures are cold at the time of sowing. As the seed swells, the seed coat breaks, and the coleoptile and radicle (primary root) emerge. The coleoptile emerges from the ground, and the first leaf breaks through the tip. The young plant continues to grow, and the coleoptile remains as a sheath at the base of the plant. Frequently, one or more buds are formed at the base of the coleoptile from which tillers arise. Some tiller profusely, especially sudangrass and forage sorghums.

Secondary roots begin to develop when the plant is three to seven days from emergence. About the time that secondary roots begin to develop, the primary root begins to die, and the plant then continues to develop from secondary or adventitious roots.

Cultivated sorghums are annuals or weakly perennial and essentially nonrhizomatous. The roots will continue to live and support ratoon crops (second and third crops from the same roots).

The stem consists of a sequence of nodes and internodes that ranges from 0.5 to 4 m in length and from 0.5 to 5 cm in diameter at the base.

Leaves arise alternately at nodes along the stem. The leaves can be as much as a meter long and 10 to 15 cm wide, although they can be much smaller. In well-adapted plants, the leaf number ranges from 14 to 17, but in less well-adapted types and some very long duration types the number may reach 30. Generally, the embryo has five to seven embryonic leaves, the remainder being formed by the vegetative bud at the tip of the stem. At maturity, a well-adapted plant generally has seven to eight functional leaves after lower leaves have died and dropped.

The vegetative bud becomes a floral bud, usually 30 to 40 days after emergence, although it can range from 17 to 90 days. At this time, all leaves have been formed, and the plant is knee to waist high. This is a period of high moisture demand. Moisture stress during or just after floral initiation can reduce panicle size and floret number. The plant enters the grand period of growth where stem elongation is primarily by cell enlargement. After floral initiation, a panicle develops, and about six to 10 days before flowering, a boot or swelling forms at the top of the plant. Sorghum usually flowers in 55 to 70 days in warm climates, which can be 10–15 days longer in cooler, temperate climates.

Flowering begins at the tip of the panicle and proceeds toward the base, taking four to five days. Sorghum is primarily a self-pollinated crop with 2 to 20% cross-pollination, the percentage being higher on more open-headed panicles. Flowering is another period of high moisture demand; severe moisture stress at this time will reduce yield.

About 10 days after flowering, the seed becomes conspicuous as a green spherical body. It takes about 30–40 days for the seed to reach maximum dry weight (physiological maturity). There are three stages of seed development: milk dough, early dough, and late dough. The seed is green during the milk stage and takes

Fig. 1. Sorghum.

TABLE 2
Stages of Sorghum Development[a]

Growth Stage	Days from Emergence	Identifying Characteristic
0	0	Emergence; coleoptile visible at soil surface
1	10	Collar of third leaf visible
2	20	Collar of fifth leaf visible
3	30	Growing point differentiation; approximate eight-leaf stage
4	40	Final leaf visible in whorl
5	50	Boot; head extended in flag leaf sheath
6	60	Half bloom; half of plants at some stage of bloom
7	70	Soft dough
8	85	Hard dough
9	95	Physiological maturity; maximum dry weight

[a]Source: Vanderlip (1979).

color at soft dough, when the seed is still soft, and juice can be expressed by squeezing. During the hard dough stage, the seed has colored and is firm. At physiological maturity the seed has about 25 to 30% moisture. During the following 20 to 25 days, the seed dries to around 12% moisture, the upper limit for safe storage (House, 1978).

The development of the plant is frequently considered to have three stages; emergence to floral initiation, floral initiation to flowering, and flowering to physiological maturity. Vanderlip (1979) refined this to include more stages of development, which has proved useful for agronomic, pathological, and entomological studies. His stages of development are shown in Table 2.

The first 30 to 35 days after emergence, growth is primarily in the leaves. The stem then begins rapid elongation. Leaves and stalks continue growth until about 60 days to maximum leaf weight and 65 days until maximum stalk weight is reached. Panicle weight begins to increase rapidly after 50 days. Following pollination, seed weight rapidly increases, sometimes faster than dry weight accumulation. This results in a decrease in stalk weight as materials move from the stalk to the panicle. The time may vary with maturity—the time from emergence to half bloom is about two-thirds of the time from planting to physiological maturity. For a detailed discussion of stages of sorghum development see Vanderlip (1979).

Pearl Millet

The development and structure of pearl millet is similar to that for sorghum with some exceptions. Pearl millet, an annual grass, generally tillers more profusely than does sorghum, with the exception of sudangrass. Tillering is primarily from basal nodes. Plant heights range generally from 1.2 to about 3.5 m depending on dwarfing genes. Leaves are generally smaller than for sorghum, 90 to 100 cm in length and 5 to 8 cm in width. The stem, on the average, is thinner than for sorghum (1 to 3 cm in diameter). The panicles are long and cylindrical, ranging from 15 cm to a meter in length and 2 to 5 cm in width. At maturity, the stem tends to be more woody than that of sorghum.

Pearl millet is earlier than sorghum, maturing a crop in 80 to 110 days; some varieties are even earlier. It is primarily cross-pollinating, as stigmas emerge one to three days before the anthers (protogynous) and therefore have opportunity to accept foreign pollen. Seeds are one-third to one-half the size of sorghum and tend to be more elongated and tear shaped. Spikelets can be bristled (awned), but in most varieties there are no awns.

Fig. 2. Pearl millet.

Fig. 3. Finger millet.

Finger Millet

Finger millet is a highly tillering annual grass. Average height is a little over 1.0 m but can reach 1.6 m. Tillers come from the base of the plant and axillary buds along the stem. Each tiller or branch produces a panicle. Leaves are generally 30 to 40 cm in length, but can reach 70 cm, and are narrow (1.5 to 3.0 cm). Panicle branches commonly come from the same place giving a fingerlike appearance. The number of branches ranges from four to 19; they can be straight (3 to 10 cm long), or curved like a hand with fingers partially closed. Seeds are formed in florets generally arranged in two rows along the panicle branch. Seeds are generally dark brown, red brown, or purple, although light brown and cream colored seeds are found. Seeds are very small, up to 2-mm in diameter and hard. Seeds of finger millet store well. Some of the brown seeded types have high levels of tannin.

Finger millet is generally grown in higher rainfall areas (600 to 1,200 mm) and is one of the better crops for acid soils. Maturity is reached in from 100 to 130 days.

Foxtail Millet

Foxtail millet is an annual grass that is variable in its morphology. Plants range from single-stemmed to highly tillered. The highly tillering types range from 30 cm to almost a meter, and are grasslike in appearance—each tiller having a panicle pretty well described by its name, foxtail. Panicles are frequently small, 10 to 15

Fig. 4. Foxtail millet.

cm in length and 1.5 to 3 cm in diameter. The central rachis is frequently lax, so the panicle droops. By contrast, some varieties, particularly common in China, are single-stemmed and range from 60–70 cm up to 150 cm tall. Leaves are 30 to 35 cm long and 1.5 to 3.0 cm wide. The panicle has short branches; the central rachis is frequently stiff, but heads on some varieties droop. Panicles can range 12 to 15 cm long and 4 to 6 cm wide at the lower part of the panicle, tapering slightly toward the tip. Seeds are small (2 to 3 mm) and generally light cream colored.

Foxtail millet is adapted to temperate regions although found in the tropics. It has a broad range of maturity, from 70 to 120 days.

Proso

Proso is an annual grass adapted to temperate parts of the world. It is highly variable in its morphology, ranging from as short as 25 cm to 230 cm. Many tillers are formed, each with an open panicle with drooping branches. Panicle branches range from 4–5 cm up to 15 cm in length. Individual seeds weigh from 5 to 9 mg.

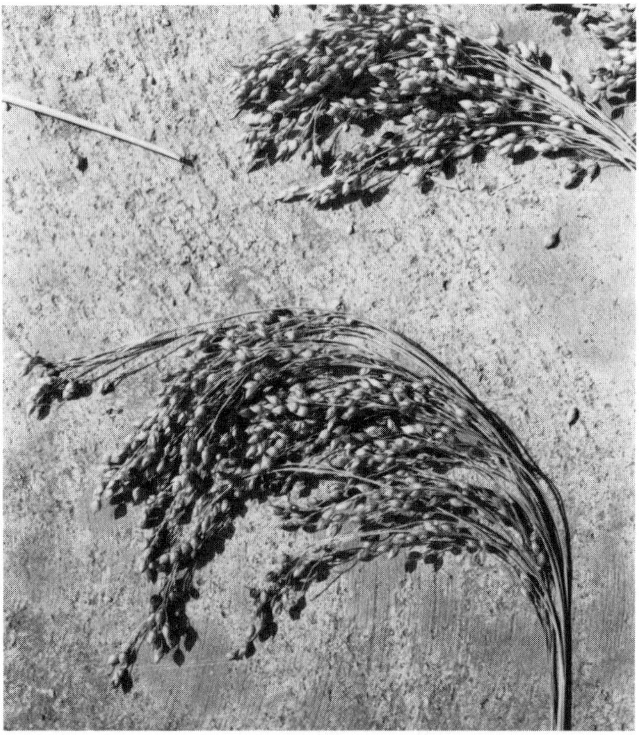

Fig. 5. Proso millet.

Kodo Millet

As a commercial crop, kodo millet is grown only in India. Kodo is an annual millet that varies in height from 30 to 90 cm and has many basal tillers, 10 to 48. The inflorescence is small, 2 to 12 cm. It matures late compared with other small millets. It is highly self-pollinating; florets generally remain closed during the flowering period. The grain occurs in a hard husk, making debranning difficult. The crop is drought hardy and frequently grown on poor soils.

Little Millet

There are two races of little millet, nana and robusta. Plants in race nana vary from 60 to 170 cm in height. The inflorescence is 14 to 15 cm long, erect, open, and highly branched. These branches sometime droop at maturity. Plants in race robusta are 120 to 190 cm tall. The inflorescence is 20 to 45 cm long, open or compact, and highly branched. It is a primarily self-pollinated crop with up to 3.5% cross pollination.

SORGHUM DISEASES

Sorghum suffers from a broad range of pathogens, in part, because of the wide diversity of environments in which it is cultivated. "In the areas where sorghum is traditionally grown, plants may be attacked by as many as five or six foliar pathogens: an array of soilborne organisms including beneficial mycorrhizae, one or more viruses, a mycoplasmalike organism, and at least two systemic fungal diseases. Overlap of symptoms is common in diseases caused by foliar pathogens" (Frederiksen, 1986). Disease identification is further complicated by variation in plant colors and some other morphological characteristics. Visual field identification is frequently difficult and uncertain.

The range of sorghum diseases can be appreciated from the table of contents of the *Compendium of Sorghum Diseases* (Frederiksen, 1986):

- Diseases caused by bacteria: bacterial leaf stripe, bacterial leaf streak, bacterial leaf spot
- Diseases caused by fungi. Foliar diseases: leaf blight, leaf anthracnose, gray leaf spot, zonate leaf spot, sooty stripe, rough leaf spot, oval leaf spot, tar spot, target leaf spot. Smuts and rust: head smut, loose kernel smut, covered kernel smut, long smut, rust. Downy mildews: crazy top, sorghum downy mildew. Root and stalk diseases: anthracnose stalk rot, Fusarium root and stalk rot, charcoal rot, pokkah boeng (twisted top), milo disease, Pythium root rot, Acremonium wilt. Leaf sheath diseases, banded leaf and sheath blight, zonate leaf spot on leaf sheaths, southern sclerotial rot. Panicle and seed diseases: grain mold, head blight, ergot, panicle and grain anthracnose, storage molds
- Diseases caused by viruses and viruslike organisms: maize dwarf mosaic, sugarcane mosaic, brome mosaic, maize stripe, peanut clump, sorghum stunt mosaic, sugarcane Fiji disease, yellow sorghum stunt
- Nematodes: stunt nematodes, root-knot nematodes, cereal root-knot nematodes, *Pratylenchus* spp.
- Parasitic plants: witchweed (*Striga* spp.)

Bacterial Diseases

Three bacterial diseases are important and there are eight more of minor importance. The important ones are bacterial leaf stripe (*Pseudomonas andropogonis*), found most frequently in humid tropical and subtropical regions, and bacterial leaf streak (*Xanthomonas campestris* pv. *holcicola*) and bacterial leaf spot (*P. syringae* pv. *syringae*), which are more common in humid temperate areas.

Generally, yield loss due to bacterial diseases is minor. Bacterial leaf spot occurs on pearl and foxtail millet in addition to sorghum. Symptoms are interveinal lesions on the leaves. The color of the lesion relates to the red, purple, or tan plant color. A water-soaked appearance occurs with leaf streak and leaf spot but not with leaf stripe. Lesions can increase in size and coalesce to form blotches or diseased areas.

Control of these diseases includes use of resistant cultivars, crop rotation, and destruction of crop residue.

Fungal Diseases

Fungal diseases are the most common. "Fungi, perhaps the most abundant plant pathogens, are small organisms that lack chlorophyll and usually grow as filamentous threads called hyphae. They usually reproduce by forming spores, which can be spread by wind, rain, machinery, and other means. Fungi can penetrate plants directly or via natural openings, such as stomata. They survive as spores or resting structures (e.g., sclerotia and teliospores) in soil or plant debris or as hyphae (i.e., mycelium) within living plants or plant debris" (Frederiksen, 1986).

As indicated in the listing of diseases, fungal diseases attack at any stage of development and all plant parts. Several different fungi may be involved. Fungi affecting seedling emergence and early growth are favored by cool damp soils. Generally, humid wet conditions and warm temperatures favor fungal diseases. Some fungal diseases occur after flowering and have minor impact on yield, but if conditions favor earlier disease expression (before flowering), there can be loss of both grain and forage yield.

The first expression of a fungal problem is failure of seeds to germinate. This problem is frequently associated with poor seed quality. After emergence, seedlings may show symptoms varying from leaf necrosis, to stunting, to death (damping off). The root system also suffers, with restricted growth to discoloration and death. *Pythium* spp. are the most important causal organisms.

Fungicides such as captan and thiram have been used with limited effect. Systemic fungicides, such as metalaxyl and fosetyl Al, used with fungicides such as captan and thiram have proven to be effective. The use of good quality seed, and seeds of hybrids rather than varieties (because of increased vigor) help reduce disease damage.

Among the fungi causing foliar disease is leaf blight (*Exserohilum turcicum*). This disease is spread over much of the world where sorghum is grown, particularly where it is humid. Initial expression is small lessons on the leaves that enlarge and can coalesce resulting in dead areas. Characteristic lesions are elliptical, 2.5 to 15 cm long and 1.2 cm wide; older lesions have yellow to gray centers with reddish margins. Control is primarily by use of resistant cultivars; rotation with nonsusceptible cultivars can be advantageous.

Leaf anthracnose (*Colletotrichum graminicola*), gray leaf spot (*Cercospora* sp), and zonate leaf spot (*Gloeocercospora sorghi*) are found around the world where sorghum is grown. If anthracnose becomes severe before flowering, yield reduction follows. Gray leaf spot and zonate leaf spot seldom have much impact on yield. Use of resistant cultivars is an effective means of control, except for zonate leaf spot, where crop rotation and clean cultivation help reduce disease severity.

Sooty stripe (*Ramulispora sorghi*) occurs on sorghum primarily in the central states of the United States and in West Africa. It occurs on sorghum at all stages of development. Initial spots enlarge to form long elliptical or spindle-shaped lesions. As the lesions mature, abundant conidia give a grayish color in the center, which later becomes black. Control by resistance is useful. Destruction of infected leaves will reduce spread; crop rotation will help.

There are several other leaf diseases; the interested reader is referred to the *Compendium of Sorghum Diseases* (Frederiksen, 1986).

SMUTS

There are several head smuts, each different, resulting in different methods of control. The important smuts are head smut (*Sporisorium reilianum*), loose kernel smut (*Sphacelotheca cruenta*), covered kernel smut (*Sporisorium sorghi*), and long smut (*Tolyposporium ehrenbergii*). The panicle is altered by the expression of head smut, with panicle branches being distorted. As the disease matures, panicle structure becomes black with the rupture and release of teliospores. Besides the complete alteration of the panicle, cultivars may be dwarfed, tillering may be more than normal, and the root system may weaken, predisposing infected plants to other diseases.

Infection by loose kernel smut may also reduce plant height and increase tillering. Usually all spikelets are infected. The disease develops on the pistils and stamens and grows to form a black conical body protruding from the glumes. This body ruptures, releasing spores. Control is possible with a fungicidal seed dressing, and highly susceptible cultivars should be avoided.

Covered kernel smut results in infection of the ovules and the eventual production of an oval to cylindrical body (sorus) protruding from the glumes. At maturity, this body ruptures and releases spores. The use of fungicidal seed treatment controls the disease.

Long smut also produces cylindrical, slightly curved but long (2 to 4 cm) sacs that extend from the glumes. Spores develop in these structures that are released at maturity. The best control is the use of resistant varieties. Destruction of newly infected heads and of alternate hosts help provide control.

Head smut is most common in some sorghum-growing areas of Africa, Asia, Australia, Europe, and North America. Loose kernel smut commonly occurs in most sorghum areas of the world except Australia, Oceania, and the Malaysian–Indonesian archipelago. Covered kernel smut is found wherever the crop is grown, and long smut is found in a number of African and Asian countries. Usually, smuts, when they occur, reduce yields.

RUST

Rust (*Puccinia purpurea*) is widespread where sorghum is grown (except in the United States). If the disease occurs before flowering, it can reduce both grain and forage yields. With severe infection, grains shrivel. Rust appearing after grain begins to form generally has little impact on yield but can reduce forage quality.

Rust appears as scattered small (2.0-mm) lesions on both surfaces of the leaves, and is red, purple, or tan depending on plant color. As the disease matures, lesions darken as spores develop, rupture covering tissue, and are dispensed. The use of resistant varieties in areas where the disease is severe offers the best control.

DOWNY MILDEW

Downy mildew (*Peronosclerospora sorghi*) occurs wherever sorghum is grown and can be an important reducer of yield. Infection can be systemic or from lesions on the leaves. Systemic infection can occur in the seedling stage, stunting and killing the plant. Leaves have chlorotic streaks, and in cool climates a white downy growth develops generally on the lower leaf surface, consisting of conidia and conidiophores. Eventually, the chlorotic tissue dies, and the leaves have a shredded appearance. Conidial infection of the leaves results in rough-looking, irregular lesions. The systemic infection is much more common.

A level of control can be achieved by treating seeds with the fungicide, metalaxyl, use of resistant varieties, and crop rotation. In southern Africa, plants from early sowings were less infected than those from later ones.

ROOT AND STALK ROTS

There are a number of root and stalk rots; the more important are Fusarium rots caused by *Fusarium moniliforme*, charcoal rot (*Macrophomina phaseolina*), and milo disease (*Periconia circinata*).

Fusarium moniliforme occurs in tropical and temperate soils, where it causes rots of seed, seedlings, roots, and stalks on a wide range of crops including maize, millet, sudangrass, and sorghum. Crop losses frequently range from 5 to 10% but can destroy a crop. Control of the disease is best achieved by reducing crop stress; i.e., adequate moisture during the growing season, weed and insect control, good fertility, reasonable population, and use of cultivars with good stalk strength.

Plants are conditioned to charcoal rot by stress, frequently from drought. The roots are first attacked, where lesions turn brown or black. As the disease progresses, soft pith in the stem is destroyed, and vascular tissue turns black. Diagnostic symptoms are lodging, black vascular tissue, and the ease with which stems at the base of the plant can be crushed between the fingers. Control is achieved (as for *Fusarium*) by maintaining good growing conditions. Use of drought-tolerant, lodging-resistant, nonsenescing cultivars of sorghum contributes to control.

Milo disease, a rot of root and crown in some genotypes of milos, has been severe in the United States. Susceptibility is determined by a single semidominant gene that mutates to resistance at a fairly high rate. This resistance has been used to provide control.

PANICLE AND SEED DISEASES

Grain molds are an important constraint to sorghum production. Seeds also discolor in wet weather from pigments in the plant, and sprouting of seed in the panicle can occur. There are a number of fungal species that cause molding: *Fusarium moniliforme*, *F. semitectum*, *Curvularia lunata*, *Olpitrichum* spp., *Alternaria* spp., *Helminthosporium* spp., *Phoma sorghina*, and *Colletotrichum graminicola*. *Phoma* appears as small black spots, *Fusarium* as a pink powdery mass that can envelop the seed, and *Curvularia* as a stringy, black, fluffy growth on the seed. Symptoms vary with the pathogenic organism. Control is largely by avoidance; i.e., use of cultivars and/or sowing dates so that grain matures in dry weather. Some cultivars are more resistant to molding than others; some tannins contribute resistance, and flavin-4-ol has been found to contribute. Corneous grains are generally more resistant than floury types. Control of molding is important to grain quality and subsequent uses of the crop. Discolored grain adds unwanted color to products, and severely molded grain does not mill well, resulting in a low starch yield.

Ergot (*Sphacelia sorghi*) is an important disease of sorghum, particularly on male-sterile seed parents in hybrid seed production fields. The disease affects individual florets, interfering with or preventing seed formation. The first sign of ergot is the presence of a sugary, pinkish, syrupy substance that appears in drops coming from infected florets. Severely infected panicles can be almost covered with this sticky syrup. With age the syrup hardens into a white, hard mass, which usually becomes black from saprophytic fungal growth. Under some conditions, a hard irregular cylindrical body protrudes from infected florets. Control is best achieved, particularly for hybrid seed production, by sowing in seasons and locations where infection does not occur, and use of seeds from uninfected crops is beneficial.

Viruses

Eighteen viruses have been reported to infect sorghum—nine only in experimental situations. Viruses cause a number of symptoms: mottling; discoloration (reddening or yellowing); and necrosis of leaves, stems, and peduncles. Stunting, a rosette formation where numerous tillers lie prostrate on or almost on the ground, excessive tillering, and sterility follow infection.

Maize dwarf mosaic virus (MDMV) is found wherever sorghum and johnsongrass occur. Epidemics have occurred in the United States, Europe, Australia, and South America. Maize, broomcorn, millet, and sudangrass are also susceptible. Symptoms include mottling. Symptoms begin with young leaves and spread to older ones. The mosaic appears as chlorotic streaks or dark green islands on a yellow background. A reddish discoloration may appear on leaves, sheaths, and the peduncle. These symptoms tend to disappear if plants grow to reach maturity. Infected plants may be dwarfed. Tillering, head size, and grain number may also be reduced.

The best control is by use of resistant or tolerant cultivars. Destruction of susceptible host plants, johnsongrass for example, contributes.

Sugarcane mosaic virus (SCMV) is a potential problem only where sorghum is grown in the same areas as sugarcane. Many symptoms are similar to MDMV. The disease is aphid-transmitted, and control of aphids offers control of the disease.

Other viruses of potential significance are maize chlorotic dwarf, brome mosaic, maize stripe, sorghum stunt mosaic, sugarcane Fiji disease, and yellow stunt virus.

Nematodes

Nematodes are minute multicellular animals that live in the soil. They are in virtually all of the world's soils. "In general, they are spindle-shaped, colorless, nonsegmented roundworms with complete sensory, digestive, excretory, and reproductive systems" (Frederiksen, 1986).

Nematodes range from 0.4 to 4.0 mm long, averaging about 1 mm, and are 0.01 to 0.05 mm in diameter. Plant nematodes possess a stylet used to pierce host tissue, inject digestive fluids, and withdraw nutrients.

Damage due to nematodes can be difficult to determine. Yield losses ranging up to 10% can go unnoticed. Initially, there can be a decrease in growth and apparent wilted appearance. In more severe situations, areas or patches of a field may show considerably reduced height and vigor than surrounding areas. Another expression is for individual plants scattered at random to be stunted compared with their neighbors.

Control is not easy. It is worthwhile knowing nematode levels in fields of interest, particularly if there have been such problems in the area. If nematodes are below or near a threshold level, crop rotation is recommended, particularly with a crop not susceptible to the same nematode. If nematodes are a problem, avoid transport of soil and plant remains that can infect other areas. Mulching may help reduce nematode populations. There are resistant cultivars, and insecticide-nematicides can be used. Experience of the author indicates that while nematodes are usually not a problem, it is useful to have nematodes checked if plant growth problems are detected. Nematode problems can be anticipated in fields sown repeatedly to sorghum over extended periods of time.

The nematode that has been of greatest concern is *Pratylenchus* spp. (a root-lesion nematode). The two species *P. hexincisus* and *P. zeae* are parasitic on sorghum. Other relevant nematodes are stunt nematodes (*Tylenchorhynchus*, *Quinisulcius*, and *Merlinius*), root-knot nematodes (*Meloidogyne incognita* and *M. acronea*) and the cereal root-knot nematode (*Meloidogyne naasi*).

Witchweed (*Striga* spp.)

Striga is a parasitic higher plant. The extremely small seed germinates in the presence of a stimulus from the roots of sorghum, pearl millet, or maize. There are a number of species of importance; *S. hermonthica* is found in West Africa eastward into the Sudan and Ethiopia and south around Lake Victoria and northern Namibia. There are two races of *S. asiatica*, a white-flowered type found on the Indian subcontinent and a red-flowered race found from Tanzania southward into South Africa. These are the two most important species, *hermonthica* being more robust, up to 1 m tall, while *asiatica* seldom exceeds 50 cm. *S. forbesii* has been found parasitizing sorghum in relatively restricted areas of Tanzania and Zimbabwe, and *S. densiflora,* found in India, resembles *S. asiatica.*

Striga has already damaged plants before it emerges from the soil. The haustoria of *Striga* penetrate the root and extract nutrients from it. It may also inject substances into the roots toxic to its host. *Striga*-infested plants have symptoms similar to those of drought; leaves wilt and may scorch. *S. hermonthica* may induce chlorotic lesions on the leaves. Infested plants are reduced in height, lighter in stature, have reduced yield, and in cases of heavy infestation, may not produce a head.

This is a serious pest and can be the most limiting factor to yield where it is severe. It is a well-adapted parasite, each plant producing between one and four hundred thousand dustlike seeds that have a 15- to 20-year viability. During underground development, the parasite is completely dependent on the host for nutrients. When it emerges above ground, it is green and there is photosynthesis. *Striga* flowers about 25 to 30 days after emergence. Flowers are lined along stems in leaf axils, being small and white in the Indian species, small and red in the southern African *S. asiatica*, and larger (12 to 15 mm in length) and reddish purple in *S. hermonthica* . Seeds are formed in pods.

Control of *Striga* is difficult. Highly resistant varieties have been developed for the white-flowered *S. asiatica* in India and are being used by farmers. Development of resistance has been promising against the red-flowered species in southern Africa and shown promise in reducing attack of *S. forbesii*. Useful levels of resistance to *S. hermonthica* have also been found, and resistant varieties have found their way onto farmers fields, particularly in the Sudan.

Artificial stimulants to germination, which cause *Striga* to germinate when there are no host plants, have been identified but not found their way to on-farm use. Ethylene was used to control a *Striga* out break in the United States.

There are several trap crops, cotton for example, that germinate the seed but are not infested. If there is a heavy seed load in the soil, the effectiveness of this method of control is limited.

It is beneficial to hand pull *Striga* plants before seed is formed or to control with herbicide. As this is a problem of many traditional farmers, hand pulling is practiced most. *Striga* is generally more severe on fields of low fertility.

On badly infested fields, it may prove necessary for a few years to plant crops that are unaffected, such as cotton, groundnut, and sunflower.

The above discussion of diseases, nematodes, and *Striga* has been taken primarily from the *Compendium of Sorghum Diseases* (Frederiksen, 1986).

INSECT PESTS OF SORGHUM

There are a large number of insects that attack sorghum. The *Sorghum Insect Identification Handbook* (Teetes et al, 1983) lists the following:

- Soil insects: wireworms, white grubs, cutworms, southern corn rootworm, foliage feeders, greenbugs, corn leaf aphid, sugarcane aphid, yellow sugarcane aphid, shoot bug, spittle bug, chinch bug, fall armyworm, oriental armyworm, African (nutgrass) armyworm, red-headed hairy caterpillar, flea beetle, leaf weevil, grasshoppers, migratory locust, banks grass mite
- Stem feeders: shoot fly, spotted stem borer, maize stalk borer, pink borer (Africa), sugarcane borer (Africa), sugarcane borer (Americas), lesser cornstalk borer, sugarcane rootstock weevil
- Head feeders: sorghum midge, bollworm, corn earworm, sorghum webworm, earhead webworm, earhead (Christmas berry) webworm, hairy caterpillar, blister beetle, earhead bug, sap-sucking bug, panicle-feeding bugs, iridescent blue-green cotton bug
- Predators and parasites of sorghum insect pests: ladybird beetle, syrphid fly, common green lacewing, predacious bugs, predacious beetles (ground beetles)
- Insect pests of stored sorghum grains: maize weevil, rice weevil, flat grain beetle, confused flour beetle, lesser grain borer, rice moth, Angoumois grain borer, Indian meal moth

The following comments on sorghum insect pests are summarized from the *Sorghum Insect Identification Handbook* (Teetes et al, 1983). Some of these insects are more important than others and to a degree have broad geographic areas of adaptation. The more important insect pests will be mentioned.

Soil insects attack seeds and seedlings. Wireworms, (several species), white grubs (*Phyllophaga crinita*), and cutworms (several species) are distributed worldwide, whereas the southern corn rootworm is a pest of the Americas. Wireworms feed on seed, destroying the embryo resulting in patches in fields with no plants. White grubs and wireworms feed on young plants destroying roots and cutting the young stem just below the soil surface (cutworms). If plants do not die, growth is stunted and yield reduced. Southern corn rootworm bores into the roots or the base of the stem, destroys the crown, and plants die. Damage from these insects is occasional.

Wireworms can be controlled with seed treatment (heptachlor, lindane, diazinon). Control of the other pests includes crop rotation, plowing six weeks or more before sowing, control of weeds that can be alternate hosts, and use of insecticides.

A number of aphids attack sorghum. The greenbug (*Schizaphis graminum*) is the most destructive. Other aphids include the corn leaf aphid (*Rhopalosiphum maidis*), sugarcane aphid (*Melanaphis sacchari*), and the yellow sugarcane aphid (*Sipha flava*). These are sucking insects; greenbugs attack all leaves, while the sugarcane and yellow sugarcane aphids prefer older leaves. Corn leaf aphids are found in the whorls. Discoloration of leaves occurs where aphids are feeding, and this discoloration expands as the numbers of aphids increase. Toxins are injected into the plants, leaves die, and plants can be stunted, even die. Aphids produce a sticky honeydew that may turn black with time.

Generally, aphid populations are held down by predators. If aphid numbers increase to damaging levels, insecticides can be used; organophosphates are generally effective. In the United States, hybrids with resistance to greenbugs are available. Other than the greenbug, aphids are occasionally a problem, but generally yield is not much reduced.

There are several armyworms, among them the fall armyworm (*Spodoptera frugiperda*), oriental armyworm (*Mythimna separata*), and the African (nutgrass) armyworm (*Spodoptera exempta*). The fall armyworm is an important pest in the southeastern United States and the tropical Americas, where it is frequently an important factor in reducing yield. The oriental army worm occurs throughout Asia, Australia, and New Zealand. This armyworm feeds on leaves, leaving only the midrib, and also attacks panicles. It is a periodic pest, but when insect populations are high, it can devastate the crop. The African armyworm, occurring in East and West Africa, is only occasionally a pest.

There are a number of predators that attack fall and oriental armyworms. Fungus infects the fall armyworm. Insecticides are effective in the control of all three armyworms. Clean cultivation contributes to control.

The chinch bug (*Blissus leucopterus*) is found in North America, where it attacks young sorghum plants. It is a sucking insect, drawing sap from stems and roots. Attacked plant parts redden, and plants become weak and stunted and frequently lodge.

Use of insecticides and barriers against migrating nymphs are the usual control methods. Early planting with a high seed rate into a good seed bed to stimulate rapid growth contributes control.

Shoot fly (*Atherigona soccata*) is an important pest on sorghum in India and Pakistan. It occurs in Africa but at lower levels of severity. It is not found in the Americas or Australia.

Eggs are laid on the lower side of seedling leaves. The maggot hatches and migrates to the growing point, which it kills, and then feeds on decaying tissue. The central leaf wilts and dies and can be pulled from the whorl. Attack occurs in young plants from about one week to one month of age.

Control can be achieved with early sowing. The problem becomes severe with late-sown crops or when it is necessary to resow because first sowings fail due to drought. The use of organophosphates applied in the soil close to the seed, but not in direct contact, will provide reasonable control.

Stem borers are found wherever sorghum and millets are grown. On a worldwide basis, they are the most devastating insect pest. There are a number of stem borers; the spotted stem borer (*Chilo partellus*) is found on the Indian subcontinent and the Far East, and in eastern and southern Africa. This is the most important stem borer in these regions. The maize stalk borer (*Busseola fusca*) is an important stem borer in Africa. There are two species of the pink borer (*Sesamia inferens* and *S. calamistis*), the former widespread in India and Asia, while the latter is found in Africa. The sugarcane borer (*Eldana saccharina*), found in Africa south of the Sahara, is generally not as damaging as the previously mentioned borers. Another sugarcane borer, primarily *Diatraea saccharalis* but including other species, is distributed throughout North, Central, and South America. These borers generally infest sugarcane and maize more than they do sorghum. The lesser cornstalk borer (*Elasmopalpus lignosellus*) is a sporadic pest on sorghum in the Americas.

There are a number of similarities in the damage caused by stem borers. With the exception of the lesser cornstalk borer, which will feed in the roots, feeding occurs in all aboveground parts. There are a number of symptoms: an early symptom is leaf scraping, leaving a transparent window on the lower surface. This tissue dies, and leaves break, giving leaves a ragged appearance. Borers may cut through leaves rolled in the whorl resulting in a row of holes when the leaf grows and unrolls. The growing point may be cut, resulting in a dead heart. Borers tunnel in the stem and the peduncle, which leads to lodging and peduncle breakage. Plant growth is stunted, and when severe, seeds become chaffy.

Destruction of stubble and crop residue is an important factor in stem borer control. A level of resistance has been found for several of the borers, but when attack is heavy substantial damage is done to the crop. Early sowing at a high seed rate followed by destruction of infected plants helps in small fields. A number of insecticides provide control, including organophosphates. Parasites and predators offer some control.

The sorghum midge (*Contarinia sorghicola*) is found in various levels of severity throughout the world where sorghum is grown except Southeast Asia. The midge, a member of the fly family, is 1.3 to 1.6 mm long with an orange red thorax and abdomen. Eggs are laid in flowering spikelets; the maggots feed on and destroy developing seeds, leaving empty florets at maturity. Each female lays eggs in up to 75 florets, and with a large number of flies damage to the crop can be extensive. The fly has an 11-day life cycle and can build up on early flowering varieties or early sown crops and inflict extensive damage on later-flowering sorghum crops.

An effective control can be achieved if the community will sow, preferably an early-maturing cultivar, at the same time. The midge population will then not build up on the first flowering fields and attack later flowering ones. Midge resistant sorghums have been developed in the United States and India. If insecticides are used, repeated application is generally required.

Earhead bugs are serious in parts of India and West Africa. The important head bug in India is *Calocoris angustatus*. Head bugs suck the juice from developing seeds; they then shrivel and discolor, and yield is reduced.

Reasonable control can be realized using an insecticide such as carboryl, first applying about the time of flowering. Open-headed varieties are less attacked than compact headed types. There are predators, but control is modest.

There are several grain-feeding insects including weevils, beetles, and moths. These insects infest a variety of stored grains and flour. The maize and rice weevils (*Sitophilus zeamais* and *S. oryzae*) feed on the seed, frequently leaving pieces of grain. The small flat grain beetle (*Cryptolestes pusillus*) follows the attack of other grain feeding insects feeding on damaged grain. The confused flour beetles (*Tribolium confusum* and *T. castaneum*) are found all over the world, where they attack both grain and stored grain products. The lesser grain borer (*Rhyzopertha dominica*) attacks most cereal grains and flours including those from sorghum and the millets.

The rice (*Corcyra cephalonica*) and Angoumois grain (*Sitotroga cerealella*) moths bore into the grain until they emerge as adults, leaving a characteristic webbing.

Grain should be dry and kept in clean conditions, preferably in insect-proof containers. Fumigation provides control and is available to small farmers in many countries.

Both the adult and nymph of the ladybird beetle (several species of Coccinellidae) effectively feed on aphids and spider mites. The larva of the syrphid fly (several species of Syrphidae) and the common green lacewing (several species of Chrysopidae) are voracious feeders of aphids, mites, and other soft-bodied insects from which they can suck juice. There are several bugs and beetles that feed on sorghum insects and several wasps that develop into adults within the host insect.

DISEASES OF PEARL MILLET

Downy mildew (*Sclerospora graminicola*) is the most important disease of pearl millet. Of second importance is *Striga*, particularly *S. hermonthica* in West Africa. *S. asiatica* also attacks pearl millet. Smut (*Tolyposporium penicillariae*), ergot (*Claviceps fusiformis*), and rust (*Puccinia penniseti*) are widespread on the crop but of less importance. Bacterial and viral diseases occur, but their identity is frequently not known; they are of minor importance. The same is true of nematodes for which not much is known (King, 1992). An estimate of the relative importance of these diseases follows: downy mildew 45%, *Striga* spp. 32%, smut 9%, ergot 7%, rust 3%, viruses, 7% and other diseases 3% (King, 1992).

Downy mildew is widespread in Africa and Asia where the crop is grown, it is somewhat more important in Asia. Virulence differences exist in India and Africa. Systemic infection, streaking of leaves as with sorghum, can occur at any stage of plant growth, but is less likely to appear in the seedling stage. The pearl millet panicle is frequently deformed to various degrees, producing a leafy structure (green ear) (Werder and Manzo, 1992). A large number of sporangia form at temperatures between 15 and 25°C resulting in the white downy appearance on the lower surface of leaf lesions (Balusubramaniam, 1992).

The best method for control is the use of resistant cultivars. Such cultivars have been developed and are in wide use. As for sorghum, the fungicide metalaxyl has been found effective as a seed treatment and to provide relief as a foliar spray. Destruction of infected crop residue and destruction of infected plants in a standing crop contribute to control (King, 1992).

The same species of *Striga* that attack sorghum also infest pearl millet. Contrary to the situation in sorghum, there is no known resistance in pearl millet.

There are four rust pathogens: *Puccinia penniseti, P. substriata* var *indica, P. substriata* var. *penicillariae*, and *P. stenotaphri.* Rust is widespread in Asia and Africa, being moderately important in Asia and of less importance in Sahelian and sub-Sahelian Africa. When the disease occurs, it is generally most severe late in the plants' development so has little impact on yield (King, 1992). Resistance has been identified but has not been transferred to high-yielding cultivars. Inheritance of hypersensitivity appears to be controlled by dominant genes (King, 1992).

Ergot infection first occurs from windblown ascospores arising from sclerotia in the soil. Stigmas are infected, leading to infection of the ovule and the production of a viscous, sticky, sugary liquid. Infected florets do not produce grain but sclerotia that matures with the seed. These sclerotia contain several alkaloids (of the agroclavine group) that are toxic to man and livestock (King, 1992).

In pearl millet, stigmas appear before pollen is shed, the time difference ranging from near simultaneous to three days. One method of control is to have varieties where this difference is small to reduce the time opportunity for infection. An abundance of pollen also contributes. Early sowing may help avoid infection. Other techniques include sowing of sclerotia-free seed and deep plowing to bury sclerotia.

Ergot is a particular problem in the production of hybrid seed because male-sterility in the seed parent increases the time from stigma appearance to pollination. Within a region, it may be possible to find places where infection is minimum or does not occur and use these for seed production.

In the case of smut, soilborne teliospores germinate and produce windborne spores that infect florets. Secondary spread occurs on late-maturing varieties from infected early-maturing types. Infected florets produce top-shaped sori that are somewhat larger than the grain. They are green at first, but turn brown when mature and rupture easily, releasing spores.

Many resistant lines have been bred, resistance appears to be controlled by one or a few dominant genes. Varieties with resistance to ergot are also resistant to smut, but the reverse is not true (King, 1992).

In Zimbabwe, false mildew (*Beniowskia sphaeroidea*) was first observed in the 1985–86 cropping season. Not much is known about the disease, but it does not appear to influence yield much. Resistance to the disease has been found (De Milliano, 1992).

Charcoal rot (*Macrophomina phaseolina*) has been found in lodged plants, but the disease is not severe (De Milliano, 1992).

There are two characteristic viral symptoms, streak and mosaic. These diseases have been found to be severe only occasionally in India, Mali, Nigeria, Niger, and Zambia. Streak viral infection is characterized by intermittent streaking of the leaves, dwarfing, and sometimes excessive tillering. Mosaic symptoms have been associated with dwarfing (King, 1992).

There are some 25 nematodes found on pearl millet, but only three (*Meloidogyne, Pratylenchus*, and *Tylenchorhynchus*) appear to cause any damage. Nematodes in pearl millet are not well understood.

DISEASES OF FINGER MILLET

The most important disease of finger millet is blast, a fungal disease caused by *Pyricularia* spp. Blast is especially severe on the panicles and peduncles, but it also infects other aboveground plant parts. Some resistance is available. This disease is important in India and in East and southern Africa.

Other diseases are Cylindrosporium leaf spot, tar spot, and bacterial blight. Of minor importance are damping off (*Pythium* spp.), foot rot (*Fusarium* sp.), Pyrenophora seedling blight, leaf streak, and some viruslike infections (Adipala, 1992).

INSECT PESTS OF PEARL AND FINGER MILLETS

The insect problems on pearl millet are of less importance than on sorghum. The major problem is from stem borers; this is also true for finger millet. There is a head girdler that produces a spiral-like effect moving lengthwise up the pearl millet head. In some situations, armyworms and aphids can be a problem. The species are essentially the same as on sorghum, and controls are similar (Teetes and Gilstrap, 1988).

Stored grain insects are less severe on pearl and finger millet than on sorghum. Soft grain pearl millet is more subject to attack than hard grain.

CONCLUSION

This chapter was designed to familiarize the reader with sorghum and millet crops. Emphasis has been on quality characteristics, but crop management, cropping systems, and pest problems are also indicated in sufficient detail to provide background information.

ACKNOWLEDGMENTS

The writing of this chapter took place while the container of the senior author including books was on the high seas. I am indebted to a number of colleagues who generously provided me material: J. Axtell and G. Ejeta (Purdue University); R. Frederiksen, L. Rooney, and G. Teetes (Texas A&M University); D. J. Andrews (University of Nebraska); and R. L. Vanderlip and P. Bramel Cox (Kansas State University).

LITERATURE CITED

ADIPALA, E. 1992. Diseases of finger millet in Uganda. In: Sorghum and Millets Diseases: A Second World Review. W. A. De Milliano, R. A. Frederiksen, and G. D. Bengston, eds. Int. Crops Res. Inst. Semi-Arid Tropics (ICRISAT), Patancheru, A. P., India.

ANDREWS, D. J., and KUMAR, K. A. 1992. Pearl millet for food, feed and forage. Advances in Agronomy, Vol. 48, Academic Press.

ANDREWS, D. J., and BRAMEL-COX, P. 1993. Breeding varieties for sustainable crop production in low input dryland agriculture in the tropics. Pages 211-224 in: International Crop Science I. D. R. Buxton, R. Shibles, R. A. Forsberg, B. L. Blad, K. H. Asay, G. M. Paulsen, and R. F. Wilson, eds. Crop Science Society of America, Madison, WI.

AXTELL, J. D., and EJETA, G. 1990. Improving sorghum grain protein quality by breeding. Pages 117-125 in: Proc. Int. Conf. on Sorghum Nutritional Quality. G. Ejeta, E. T. Mertz, L. Rooney, R. Schaffert, J. Yohe, eds. Purdue University, W. Lafayette, Indiana.

AXTELL, J. D., MOHAN, D. P., and CUMMINGS, D. P. 1974. Genetic improvement of biological efficiency and protein quality in sorghum. Proc. Annu. Corn Sorghum Res. Conf. 29.

BALUSUBRAMANIAM, K. A. 1992. Pearl millet diseases in India. In: Sorghum and Millets Diseases: A Second World Review. W. A. J. De Milliano, R. A. Frederiksen, and G. D. Bengston, eds. Int. Crops Res. Inst. Semi-Arid Tropics (ICRISAT), Patancheru, A. P., India.

BHARDWAJ, D. S., SINGH, M., and SATIJA, D. R. 1987. Genetic architecture of quality traits including beta-carotene in pearl millet. India J. Genet. Plant Breed. 47(1):94-98.

BLONDEL, D. 1970. New results on increasing the protein content of the grain of millet

(*Pennisetum typhoides*) and sorghum (*Sorghum vulgare*) by nitrogen fertilization in Senegal. Rep. Centre National de Recherches Agronomiques, Bambey, Senegal.

BURTON, G. W., RABSON, R., and AXMANN, H. 1980. Immediate effect on male and female gametes and N on five pearl millet grain characteristics. Crop Sci. 20(1):7-10.

CHAMPIGNY, M. D., BISMUTH, E., and SARDA, C. 1982. Genetic analysis of the relationships between spontaneous and cultivated forms of pearl millet. IV Study of growth and accumulation of starch by the parents and the F1 hybrids obtained by crossing one ecotype of *Pennisetum mollissimum* Hochst (wild species) and a cultivar of *Pennisetum americanum* (L) Leeke. Agronomy 2(4):65-372.

CHAUDHRY, G. A., SORWAR, G. M., and CHEENA, N. M. 1987. Legume-cereal mixture for rainfed areas. Pakistan J. Agric. Res. 8(3):266-269

CHERNEY, J. H., CHERNEY, D. J. R., AKINS, D. E., and AXTELL, J. D. 1991. Potential of Brown-midrib, low lignin mutants for improving forage quality. Adv. Agron. 46:157-198.

DAIBER, R. H. 1971. Grain and malting quality of sorghum cultivars II. 1970 harvest. CSIR Spec. Rep. CHEM 180. Natl. Food Res. Inst. Pretoria, South Africa.

DE MILLIANO, W. A. J. 1992. Pearl millet diseases in southern Africa. In: Sorghum and Millets Diseases: A Second World Review. W. A. J. De Milliano, R. A. Frederiksen, and G. D. Bengston, eds. Int. Crops Res. Inst. Semi-Arid Tropics (ICRISAT), Patancheru, A. P., India.

DEOSTHALE, Y. G., NAGARAJAN, V., and VISWESWAR RAO, K. 1972. Some factors influencing the nutrient composition of sorghum grain. Indian J. Agric. Sci. 42:100-108.

DHOBLE, M. V., BHOSLE, B. R., and KHATING, E. A. 1990. Studies on comparative performance of different intercropping systems under dryland conditions. India J. Agron. 35(4):346-352

DOGGETT, H. 1988. Sorghum, 2nd ed. Longman Scientific and Technical, Harlow, Essex, England

EJETA, G., and AXTELL, J. D. 1987. Dry matter accumulation and carbohydrate composition in developing normal and high lysine sorghums. J. Agric. Food Chem. 35:981-985.

EJETA, G., and AXTELL, J. D. 1990. Development of hard endosperm high lysine sorghum lines. Pages 126-140 in: Proc. Int. Conf. on Sorghum Nutritional Quality. G. Ejeta, E. T. Mertz, L. Rooney, R. Schaffert, J. Yohe, eds. Purdue University, W. Lafayette, Indiana.

EJETA, G., AXTELL, J. D., BUTLER, L. G., and KIRLEIS, A. W. 1989. Proc. Bienn. Grain Sorghum Res. Utilization Conf.,16th, Lubbock, TX.

FREDERIKSEN, R. A., ed. 1986. Compendium of Sorghum Diseases. The American Phytopathological Society, St. Paul, MN.

GANRY, F., and BIDEAU, J. 1974. Effect of nitrogenous fertilization on the yields and nutritional value of a millet, Souma III. Agron. Trop. Nogent. 29:1006-1015.

GARTAN, S. L., TOMER, Y. S., TOMER, R. P. S., and SINGH, M. 1987. Genetics of protein content in pearl millet (*Pennisetum typhoides*) (b). S & H Agric. Sci. Digest. Haryana Agric. Univ. Hissar, India 7(4):188-190.

GODAWAT, S. L., and KAUL, B. K. 1981. Genetic variability in pearl millet. Indian J. Agric. Sci. 51(3):147-149.

GUPTA, V. P., DHILLON, B. S., and SETHI, A. S. 1976. The path analysis of protein, mineral content, and grain yield of pearl millet. Pages 80-85 in: Crop Improvement. Punjab Agric. Univ. Ludhiana, India.

HOUSE L. R. 1978. A Guide to Sorghum Breeding. Int. Crops Res. Inst. Semi-Arid Tropics (ICRISAT), Patancheru, A. P., India.

HULSE, J. H., LAING, E. M., and PEARSON, O. E. 1980. Ch. 3 in: Sorghum and the Millets, Their Composition and Nutritive Value. Academic Press, New York.

ICRISAT. 1992. Annual Report 1991. International Crops Research Institute for the Semi-Arid Tropics (ICRISAT), Patancheru, A. P., India.

ICRISAT. 1975–1983. Annual Reports. International Crops Research Institute for the Semi-Arid Tropics (ICRISAT), Patancheru, A. P., India.

JACQUES, G. L., VANDERLIP, R. L., and WHITNEY, D. A. 1975. Growth and nutrient accumulation and distribution in grain sorghum I. Dry matter production and Ca and Mg uptake and distribution. Agron. J. 67:607-610.

JACQUES, G. L., VANDERLIP, R. L., and ELLIS, R. 1985. Growth and nutrient accumulation and distribution in grain sorghum II. Zn, Ca, Fe, and Mn uptake and distribution. Agron. J. 67:611-616.

KHANGURA, B. S., GILL, K. S., and PHUL, P. S. 1980. Combining ability analysis beta-carotine, total carotenoids and other grain characteristics in pearl millet. Theor. Appl. Genet. 56:91-96.

KHANNA, M. L. 1966. Composition of bajra (*Pennisetum typhoides*) grain as influenced by fertilization. Indian J. Agron. 11:47-49.

KING, S. B. 1992. World review of pearl millet diseases: Knowledge of future research needs in sorghum and millets diseases. In: Sorghum and Millets Diseases: A Second World Review W. A. J. De Milliano, R. A. Frederiksen, and G. D. Bengston, eds. International Crops Research Institute for the Semi-Arid Tropics (ICRISAT), Patancheru, A. P., India.

KRISHNAMURTHY, K. 1968. Nutritive content of ragi varieties in relation to fertilizer levels. J. Nutr. Diet. 5:10-12.

KUMAR, K. A., GUPTA, S. C., and ANDREWS, D. V. 1983. Relationship between nutritional quality characters and grain yield in pearl millet. Crop Sci. 23:232-234.

LIANG, G. H. L., OVERLEY, C. B., and CASADY, A. V. 1969. Interrelations among agronomic characters in grain sorghum (*Sorghum bicolor* Moench). Crop Sci. 9:299-302.

MCKENZIE, H. A., and WALLACE, H. S. 1974. The Kjeldahl determination of nitrogen—A ritical study of digestion conditions—temperature: Catalyst and oxidising agent. Aust. Chem. 7:55-70.

MERTZ, E. T. 1990. Improvement of cereal protein quality through mutations: Historical review. In: Proc. Int. Conf. on Sorghum Nutritional Quality. G. Ejeta, E. T. Mertz, L. Rooney, R. Schaffert, J. Yohe, eds. Purdue University, W. Lafayette, Indiana.

MUKURU, S. Z. 1990. Traditional food grain processing methods in Africa. Pages 216-221 in: Proc. Int. Conf. on Sorghum Nutritional Quality. G. Ejeta, E. T. Mertz, L. Rooney, R. Schaffert, J. Yohe, eds. Purdue University, W. Lafayette, Indiana.

NANDA, G. S., and RAO, N. G. P. 1975. Genetic analysis of some exotic × Indian crosses in sorghum. Nutritional quality and its association with yield. Indian J. Genet. Plant Breed. 35:131-135.

NARWAL, S. S., SINGH, S., MALIK, D. S., and GUPTA, P. C. 1988. Performance and quality of summer cowpea (*Vigna unguiculata* subsp. *cylindrica*), soybean (*Glycine max*) and pearl millet (*Pennisetum glaucum*) grown in mixture and intercropping under different moisture regimes. Indian J. Agric. Sci. 58(11):817-822.

PANDE, P. C., POKHRIYAL, S. C., and SINGH, B. 1985. Heterosis in grain sink activity in pearl millet. Indian J. Genet. Plant Breed. 43(3):409-413.

PRASADA RAO, K. E., DE WET, J. M. J., REDDY, G. V., and MENGESHA, M. H. 1993. Diversity in the samll millets collections at ICRISAT. In: Advances in Small Millets. Oxford and IBH Publ. Co. Pvt. Ltd., New Delhi.

RAI, K. D. 1964. Study of rain-grown sorghum and maize in the central rainlands of the Sudan I. Effect of date of sowing, varieties and spacing on crude protein content and nitrogen accumulation. Indian J. Agron. 9:175-183.

ROONEY, L. W., and MILLER, F. R. 1982. Variation in the structure and kernel characteristic of sorghum. Pages 145-162 in: Proc. Int. Symp. on Sorghum Grain Quality. L. W. Rooney and D. S. Murty, eds. Int. Crops Res. Inst. Semi-Arid Tropics (ICRISAT), Patancheru, A. P., India.

SACHDEVA, A. K. 1981. Genetics of yield and some quality traits in pearl millet (*Pennisetum typhoides* (Burm.) Stapf and Hubb.) (Abstr.) Thesis Abstr. 7(1):77-78. Harayana Agric Univ., Hissar, India.

SAWHNEY, S. K., and NAIK, M. S. 1969. Amino acid composition of protein fractions of pearl millet and the effect of nitrogen fertilization on its proteins. Indian J. Genet. Plant Breed. 29:395-406.

SHUKLA, U. C., and BHATIA, K. N. 1971. Effect of nitrogen levels on the Fe and Mn concentrations and uptake in hybrid and local varieties of pearl millet (*Pennisetum typhoides* (Burm. F.) Stapf & C. E. Hubb.) Indian J. Agric. Sci. 41:790-794.

SIDDESWARAN, K, RAMASWAMY, S., and MORACHARAN, Y. B. 1989. Nutrient uptake of finger millet as influenced by intercrops, border crops and N fertilization. Madras Agric. J. 76(7):361-365.

SINGH, P., and GUPTA, V. P. 1987. Combining ability for grain quality traits in pearl millet. Indian J. Genet. Plant Breed 47(2):141-144.

SINGH, R., and AXTELL, J. D. 1973. High lysine mutant gene (hl) that improves protein quality and biological value of grain sorghum. Crop Sci. 13:535-539.

STABURSVIK, A., and HEIDE, O. M. 1974. Protein content and amino acid spectrum of finger millet (*Eleusine corocana* (L.) Gaertn) as influenced by nitrogen and sulfur fertilizers. Plant Soil 41:549- 571.

STUTZEL, H., and VANDERLIP, R. L. 1988. Grain yield of intercropped sorghum and pearl millet as influenced by sorghum genotype and cropping pattern. J. Agron. Crop Sci. 160(3):191-197.

SULLIVAN, T. W., DOUGLASS, J. H., ANDREWS, D. J., BOND, P. L., HEMCOCK, J. D., BRAMEL-COX, P. J., STEGMEIR, W. D., and BRETHOUR, J. R. 1990. Nutritional value of pearl millet for food and feed. Pages 83-94 in: Proc. Int. Conf. on Sorghum Nutritional Quality. G. Ejeta, E. T. Mertz, L. Rooney, R. Schaffert, J. Yohe, eds. Purdue University, W. Lafayette, Indiana.

SWINDALE, L. D. 1979. Proc. Int. Workshop on Intercropping, Hyderabad, India. Int. Crops Res. Inst. Semi-Arid Tropics (ICRISAT), Patancheru, A. P., India.

TEETES, G. L., and GILSTRAP, F. E. 1988. Development and evaluation of systems for controlling insect pests of sorghum/millet by integration of resistant varieties, cultural manipulation and biological control. Annu. Rep. INTSORMIL, University of Nebraska, Lincoln.

TEETES, G. L., REDDY, SESHU, K. V., LEUSCHNER, K., and HOUSE, L. R. 1983. Sorghum Insect Identification Handbook. Inf. Bull. 12. Int. Crops Res. Inst. Semi-Arid Tropics (ICRISAT), Patancheru, A. P., India.

TOMER, P. S. 1970. Test weights and protein content in bajra in relation to methods of sowing, manurial treatments and seed rates. Ann. Arid Zone 9:159-162.

TYAGI, C. S., and PARODA, R. S. 1982. Genetic analysis of protein content in pearl millet. Indian J. Agric. Sci. 52(11):718-722.

VANDERLIP, R. L. 1979. How a sorghum plant develops. Kans. Agric. Exp. Stn. Contr. 1203.

VENKATARAMANA, R. S., and KRISHNA RAO, D. V. 1961. Chemical composition and nutrient uptake of ragi J. Indian Soc. Soil Sci. 9:245-252.

WAGGLE, D. H., DEYOE, C. W., and SMITH, F. W. 1967. Effect of nitrogen fertilization on the amino acid composition and distribution in sorghum grain. Crop Sci. 7:367-368.

WARSI, A. S., and WRIGHT, B. C. 1973. Effects of rates and methods of nitrogen application on the quality of sorghum grain. Indian J. Agric. Sci. 43:722-726.

WERDER, J., and MANZO, S. K. 1992. Pearl millet diseases in western Africa. In: Sorghum and Millets Diseases: A Second World Review. W. A. De Milliano, R. A. Frederiksen, and G. D. Bengston, eds. Int. Crops Res. Inst. Semi-Arid Tropics (ICRISAT), Patancheru, A. P., India.

CHAPTER 4

STRUCTURE AND CHEMISTRY OF SORGHUM AND MILLETS

Sergio Serna-Saldivar
Instituto Tecnologico y de Estudios Superiores de Monterrey
Departamento de Tecnologia de Alimentos
Monterrey, N.L., Mexico

Lloyd W. Rooney
Department of Soil and Crop Science
Cereal Quality Laboratory
Texas A&M University
College Station, Texas

INTRODUCTION

Sorghum and millets constitute a major source of calories and protein for millions of people in Africa and Asia. These cereals are mainly considered as subsistence crops because of their unique tolerance to drought and adaptation to dry tropical and subtropical ecosystems throughout the world. Their nutritional quality is dictated mainly by their chemical composition and the presence of antinutritional factors.

The grasses known collectively as millets are a set of highly variable small-seeded plants indigenous to many areas of the world (Table 1). Millets are of special value in semiarid and tropical regions because of their hardiness, short growing season, and productivity under heat and drought conditions. Pearl millet is the most widely grown on an estimated 27 million hectares (Singh et al, 1987) followed by finger millet. World production of millets has been stable during the last decade because it is mainly cultivated as a subsistence crop. In 1991, 37.1 million hectares were sown with an average yield of 781 kg/ha totaling 28.9 million metric tons (FAO, 1992); 30% of the production was harvested in Africa, 56% in Asia, and 13% in the former U.S.S.R. Major producers were India (31.1%), China (15.8%), and the former U.S.S.R. (12.4%).

It is our purpose to review the structure and chemistry of sorghum and millets and to discuss the variation in composition due to genetics, agronomic practices, and processing. Significant reviews on the chemistry and technology of sorghum and millets have been published (Hoseney et al, 1981, 1987; Hulse et al, 1980; Rachie and Majmudar, 1980; Rooney et al, 1980, 1982, 1986; Rooney and Serna-Saldivar, 1991; Serna-Saldivar et al, 1991a).

TABLE 1
Probable Cytogenetic Origin and Scientific and Common Names of Sorghum and Major Types of Millets[a]

Scientific Name	Common Names[b]	Cytogenetic Origin	Location Grown
Sorghum bicolor	Sorghum, milo, jowar, kafir, kaffir-corn, Guinea corn, cholam	Equatorial Africa	Worldwide, Africa, India, China, United States
Pennisetum glaucum *P. americanum* *P. typhoides*	Pearl, bajra, cattail, bulrush, candlestick, sanyo, munga, seno	West Africa	Africa, India
Eleusine coracana	Finger, ragi, African, bird's foot, rapoko, Hansa	Originated in Africa and domesticated in India	East and central Africa, India, China
Setaria italica	Foxtail, Italian, kangni, navane, German, Siberian, Hungarian	Eastern Asia	Asia (U.S.S.R., China, India, Japan), North Africa, southeast Europe, Near East
Panicum miliaceum	Proso, common, Hershey, panivarigu, broomcorn, hog, samai, Russian	China	Eastern Asia, U.S.S.R., China Mongolia, Middle East Main millet in United States
Echinochloa frumentacea *E. crus-galli* *E. utilis*	Japanese, barnyard, sanwa, kweichou, kudiraivali, sawan, Korean	Java/Malaysia	East Asia, India, Egypt
Paspalum scrobiculatum *P. commersoni*	Kodo, varagu, bastard, ditch, naraka	Africa or India	India
Eragrostis tef	Teff	Ethiopia	East Africa (Ethiopia)
Digitaria exilis *D. iburua*	Fonio, fundi, hungry rice, acha, crabgrass, raishan	Domesticated in Nigeria	West Africa (savanna)

[a] Data from Brunken et al (1977), Hulse et al (1980), Porteres (1976), Rachie and Majmudar (1980), Serna-Saldivar et al (1991a).
[b] Underlined common names are used throughout the manuscript.

CLASSIFICATION

Sorghum

Sorghum is a cereal of remarkable genetic variability. More than 30,000 selections are present in the world sorghum collections in India and elsewhere. These banks are used by plant breeders to improve the crop. Sorghum is difficult to classify due to its wide diversity. The plant originated in equatorial Africa and is distributed throughout the tropical, semitropical, and arid regions of the world (Table 1).

From the utilization viewpoint, sorghum can be classed as grain sorghum, forage sorghum, grass or Sudan sorghum, and broomcorn. The latter is grown for its long, fibrous panicle branches that are manufactured into brooms. According to U.S. Grain Standards, grain is classed into sorghum, tannin sorghum, white sorghum, and mixed sorghum. The sorghum class does not contain more than 3% sorghum with a pigmented testa or undercoat. The white class contains sorghum with a white pericarp without a pigmented testa. Mixed sorghum contains mixtures of sorghum with and without pigmented testa. The kernel characteristics of sorghum have been documented by Rooney and Miller (1982). Appearance and quality of sorghum are affected significantly by these genetically controlled characters.

Millets

PEARL MILLET

There are more than 15,000 pearl millet lines in the World Germplasm Collection. Pearl millet cultivars are often classified into late (130–150 days) and early (60–95 days) maturing groups. The early cultivars, called Souna or Gero millets, dominate the area under production. They are insensitive to photoperiod and produce smaller plants (1.5–3 m tall) than the late types called Sanio or Naiwa. The late millets are photoperiod sensitive and produce tall plants (3 to 6 m) (Adrian and Sayerse, 1956; Kassam, 1976; Perten et al, 1983). The Souna millets provide an early maturing grain crop for consumption during the hungry season when grain is in short supply. The Souna millets have poor dehulling properties because they are usually lanceolate, while the Sanio are globular and dehull more completely with high yields.

FINGER MILLET

Two groups of finger millet are recognized: African highland types and Afro-Asian types. The African types are cultivated at 1,000–2,000 m with average rainfall of 900–1,250 mm and have grains enclosed within the floret. The Afro-Asian types are cultivated at altitudes of 500–1,000 m with lower rainfall; the mature grain protrudes from the florets, is globular to oval, and has a diameter that ranges from 1.0 to 1.5 mm. The grain varies widely in color and has a 1,000-kernel weight of 2.5 g (Hilu and deWet, 1967; Hulse et al, 1980). Some varieties of finger millet contain significant levels of condensed tannins.

FOXTAIL MILLET

Foxtail millet is a short plant that grows and matures rapidly; it is ideally suited as a catch crop. The inflorescence is cylindrical and has the appearance of a false spike (Narayanaswami, 1956). The spikelets are persistent and yellowish when mature and consist of two florets subtended by a pair of glumes. The lower floret is sterile and contains only a lemma and a rudimental axis. The upper floret is fertile and bisexual (Narayanaswami, 1956). The seeds are smaller (2 mm length) than those of proso millet and are enclosed in thin outer glumes, which are removed during threshing. The color of the caryopses can be white, red, yellow, brown, or black. Foxtail millet has good adaptation to low rainfall from sea level to 2,000 m and can tolerate a wide range of soils. Various types differ in time to maturity (80–120 days). Foxtail is capable of producing two types of grains based on starch properties: the normal or nonglutinous and the waxy or glutinous type (Tomita et al, 1981).

PROSO MILLET

The proso or common millet is a short plant that yields flattened kernels in a short time (60–90 days). Proso is the only millet of economic importance in the United States, where it is grown as an emergency crop and for bird seed (Yanez and Walker, 1986). It grows under hot conditions when the rainy season is too short and the soil too poor for other cereals. The inflorescence is a slender panicle up to 45 cm long, which may be open or compact. The caryopses (2-mm long and 2-mm wide) are covered by smooth, hard and shiny glumes (lemma and palea). The kernels are generally white, oval, and smooth with a 1,000-kernel weight of approximately 5 g (Angold, 1979; Hulse et al, 1980). Waxy cultivars of proso millet exist (Tomita et al, 1981).

JAPANESE MILLET

Japanese millet is a short plant (50–100 cm) frequently sown as a reclamation crop in Egypt on land too saline for rice or as a fast-growing crop in Southeast Asia, China, and Japan (Table 1; Hulse et al, 1980). It is the earliest maturing of all millets (45 days).

FONIO

Fonio (*Digitaria exilis*) is a short plant (45 cm) usually grown on poor, shallow, rocky soils in the savanna zone of West Africa, where rainfall commonly does not exceed 400 mm. The period to maturity varies from 90 to 130 days. In some areas of Guinea and Nigeria, fonio is the staple crop. The plant produces extremely small kernels (0.5 g/1,000 kernels). Black fonio (*D. iburua* Stapf) is a white-seeded millet with brown spikelets (Hulse et al, 1980; Porteres, 1976).

TEFF

Teff is one of the major crops of Ethiopia, where it may represent half of the total grain production. It is cultivated at high altitudes (1,700–2,800 m), maturing in 2 to 6 months depending upon variety. Two major types of teff are recognized: plants that produce white or colored (brown or reddish) seeds. White-seeded cultivars are preferred for food production (i.e., *injera*). The grain is extremely small (1–1.5 mm long and 0.75–1 mm wide; 0.3–0.4 g/1,000 seeds) (Hulse et al, 1980; Porteres, 1976).

APPEARANCE AND GENETICS OF SORGHUM GRAIN

Several interacting factors affect the color and overall appearance of sorghum caryopses. Appearance is mainly affected by pericarp color and thickness, presence of pigmented testa, and endosperm color (Rooney and Miller, 1982). Plant and glume color also influence appearance of the kernel. This is especially true in weathered sorghums. Thus, environment and insect and mold damage play a major role in the condition and appearance of the grain.

Pericarp color is genetically controlled by the *R* and *Y* genes. The combination of these genes can produce white (*R—yy* or *rryy*), lemon yellow (*rrY—*), or red (*R—Y—*) color. The presence of the testa is controlled by the B_1—B_2 genes. Sorghums with a pigmented testa (B_1—B_2—) that contain a dominant spreader (*S—*) gene contain condensed tannins and are classified as type III (also called brown, bird-resistant sorghum). The *Z* gene controls pericarp thickness. Sorghums with homozygous recessive (*zz*) genes possess a thick mesocarp. Thick pericarps usually mask the color of the testa and endosperm, and kernels look chalky due to the presence of small starch granules in the pericarp (Earp and Rooney, 1982; Rooney and Miller, 1982). The pigmented seed coat is absent when these genes are homozygous recessive ($b_1b_1b_2b_2$). The *tp tp* genes control testa color which can be brown or purple.

Endosperm color is mostly affected by the maternal parent. Yellow endosperm cultivars contain high levels of carotenoid pigments. Endosperm color affects grain appearance especially in kernels with a thin pericarp without a pigmented testa. When the pericarp is thin and colorless and the testa is absent, the kernel appears yellow. When the kernel has a thick, colorless pericarp, the kernel appears white. Bronze sorghums contain a thin red pericarp with yellow endosperm. Heteroyellow endosperm sorghums are hybrids from crossing a yellow endosperm with a nonyellow endosperm parent (Rooney and Miller, 1982).

Waxy endosperm cultivars contain three (*wx*) genes in the recessive form. Heterowaxy genotypes contain one or two of these genes in the dominant form, whereas nonwaxy endosperm cultivars contain all three genes in the dominant form. Waxy cultivars contain nearly 100% amylopectin and an endosperm that appears waxy to the naked eye (Rooney and Miller, 1982).

The environmental conditions during grain maturation greatly affect the appearance of the grain, because the sorghum head is readily exposed to biotic and abiotic agents. An environment that is hot and humid during maturation negatively affects grain quality due to weathering and deterioration, which results from insects and molds. Insect-attacked kernels secrete phenolic compounds, which produce discolored spots in the affected areas. Molds discolor the grain, break down the kernel, and significantly affect processing qualities. Sorghums from open panicles, with thin pericarp, condensed tannins, corneous endosperm, and large, tight glumes are generally considered more resistant to weathering (Waniska et al, 1989).

GRAIN MORPHOLOGY

Sorghum

The sorghum kernel is considered a naked caryopsis, although some African types retain their glumes after threshing. Kernels differ widely in size or weight (3–80 mg), test weight (55–61 lb/bu), and density (1.15–1.38 g/cm^3). Commercial U.S. sorghums are generally 4-mm long, 2-mm wide, and 2.5-mm thick, a kernel weight of 25–35 mg, test weight of 58–60 lb/bu, and densities ranging from 1.28 to

1.36 g/cm^3. The caryopsis consists of three distinctive anatomical components (Fig. 1): pericarp (outer layer), germ (embryo), and endosperm (storage tissue). The relative proportion of these components varies depending upon variety and environment (Table 2; Haikerwal and Mathieson, 1971; Hubbard et al, 1950; Jambunathan and Mertz, 1973; Rooney and Clark, 1968; Taylor and Schussler, 1986).

McDonough and Rooney (1990) monitored grain development of six sorghum varieties using scanning electron microscopy. The pericarp began to differentiate into distinctive layers before days 3–6. By six to nine days postanthesis, the pericarp was fully developed and began to compress. The testa in a brown cultivar was apparent at three to six days, whereas the aleurone took longer to develop (six to

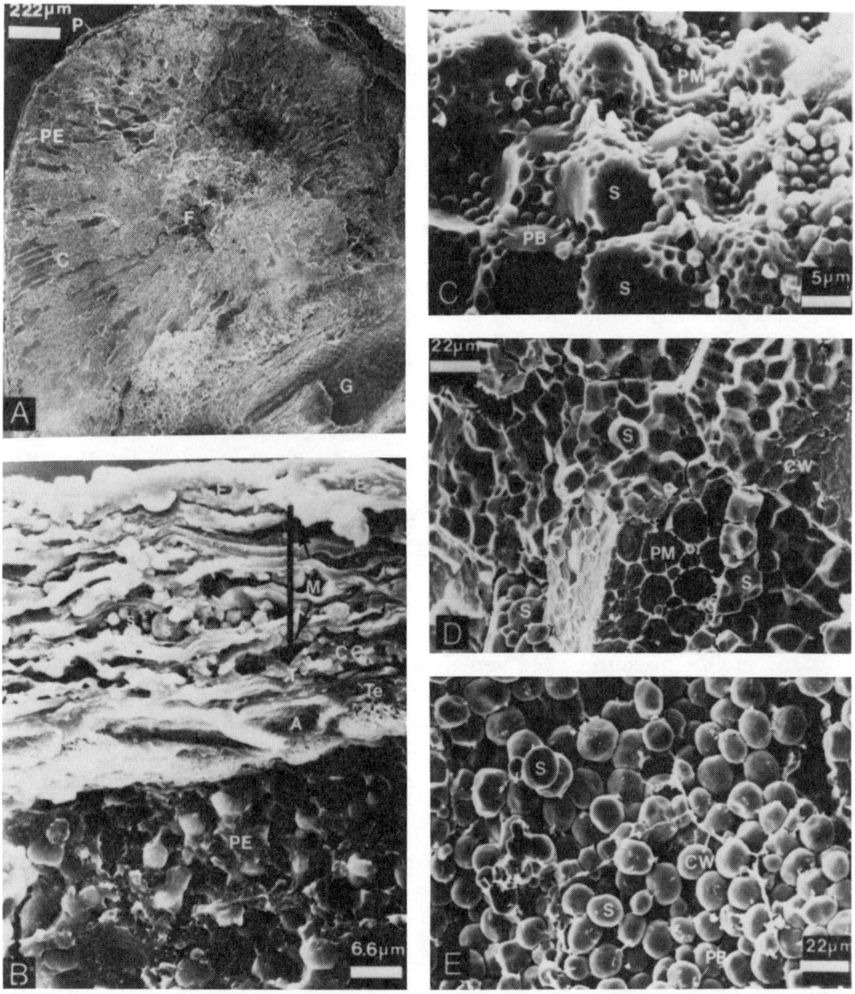

Fig. 1. Structure of the mature sorghum kernel. Cross section (A), pericarp (B), peripheral endosperm (C), corneous endosperm (D), floury endosperm (E). P = pericarp, PE = peripheral endosperm, C = corneous endosperm, F = floury endosperm, G = germ, E = epicarp, M = mesocarp, CC = cross cells, T = tube cells, Te = testa, A = aleurone, S = starch granule, PB = protein body, PM = protein matrix, Cw = cell wall.

12 days). Simple and compound starch granules were in the ovary walls of all varieties at anthesis (Glennie et al, 1984; McDonough and Rooney, 1990). Simple granules and protein bodies started to develop in the endosperm three to six days postanthesis. At this stage, protein bodies were covered by a thin filamentous webbing, which later developed into a distinctive protein matrix. Glennie et al (1984) indicated that by the soft dough stage the endosperm had gained most of its storage material; and, thereafter there was a considerable loss of moisture. During the milk stage, the endosperm cell walls contained a large number of pits to facilitate translocation of storage and chemical components. The pits gradually disappeared by the hard dough stage.

The outer layer or pericarp originates from the ovary wall (Glennie et al, 1984; Saunders, 1955) and is divided into three histological tissues: epicarp, mesocarp, and endocarp (Earp and Rooney, 1982). The outermost layer or epicarp is generally covered with a thin layer of wax. The epicarp is two or three layers thick and consists of rectangular cells often containing pigmented material. Unlike other cereals, the sorghum mesocarp may contain starch granules. Sorghum with thick pericarp usually contains three or four mesocarp cell layers filled with small starch granules (Fig. 1). The inner pericarp tissue, endocarp, is composed of cross and tube cells. Pericarp thickness (determined by the *Z* gene) ranges from 8 to 160 μm (Earp and Rooney, 1982) and varies within an individual kernel. The areas below the style and near the hilum are the thickest and the sides of the kernel are thinnest. The inner tube cells conduct water during grain germination, whereas the outer cross cells impede moisture loss by forming a blocky seal on top of the tube cells. A stylar area is located on the tip of the caryopsis, opposite the germ. The black layer or hilum is the colored scar tissue that develops on the tip of the germ when it reaches physiological maturity.

The seed coat or testa is derived from the ovule integuments and is pigmented only in sorghums with dominant B_1 and B_2 genes. These are often referred to as type II or type III sorghums. Type II sorghums have a pigmented testa and contain condensed tannins. Type III sorghums contain the greatest amount of tannins be-

TABLE 2
Chemical Composition of Whole Sorghum and Its Anatomical Parts[a]

Component	Whole Grain (%)	Endosperm (%)	Germ (%)	Pericarp (%)
Whole kernel	100	84.2	9.4	6.5
		81.7–86.5	8.0–10.9	4.3–8.7
Protein	12.3	10.5	18.4	6.0
Range	11.5–12.3	8.7–13.0	17.8–19.2	5.2–7.6
Total protein	100	80.9	14.9	4.0
Fat	3.6	0.6	28.1	4.9
Range	. . .	0.4–0.8	26.9–30.6	3.7–6.0
Total fat	100	13.2	76.2	10.6
Ash	1.6	0.4	10.4	2.0
Range	1.6–1.7	0.3–0.4	. . .	. . .
Total ash	100	20.6	68.6	10.8
Starch	73.8	82.5	13.4	34.6
Range	72.3–75.1	81.3–83.0	. . .	. . .
Total starch	100	94.4	1.8	3.8

[a]Data from Haikerwal and Mathieson (1971), Hubbard et al (1950), Jambunathan and Mertz (1973), and Taylor and Schussler (1986).

cause they have a pigmented seed coat in the presence of a dominant spreader (*S*) gene. The type III sorghums have the greatest bird resistance, and affect nutritional value the most. According to Earp and Rooney (1982), the thickness of the testa ranges from 8 to 40 μm. Differences in thickness of the testa within individual kernels occur. The thickest area is usually observed below the style and the thinnest on the side of the kernel (Blakely et al, 1979).

The endosperm tissue is triploid, resulting from the fusion of a male gamete with two female polar cells. It is composed of the aleurone layer, peripheral, corneous, and floury areas (Fig. 1). The aleurone is the outer cover and consists of a single layer of rectangular cells adjacent to the testa or tube cells. The cells possess a thick cell wall, large amounts of proteins (protein bodies, enzymes), ash (phytin bodies), and oil (spherosomes). The peripheral endosperm tissue is composed of several layers of dense cells containing large quantities of protein and small starch granules. This part of the endosperm affects processing and nutrient digestibilities. Waxy sorghums have the smallest proportion of peripheral endosperm and contain less dense protein matrices with larger starch granules than regular endosperm sorghums (Sullins and Rooney, 1974, 1975).

The corneous and floury endosperm cells are composed of starch granules, protein matrix, protein bodies, and cell walls rich in β-glucans and hemicellulose (Fig. 1). Sorghum exhibits a blue autofluorescence in the cell walls of the pericarp, aleurone, and endosperm, which is mainly due to ferulic acid. In the corneous endosperm, the protein matrix has a continuous interphase with the starch granules and protein bodies embedded in the matrix (Hoseney et al, 1974; Seckinger and Wolf, 1973). According to Taylor et al (1984), the protein bodies are largely circular and vary from 0.4- to 2.0-μm across. High-lysine cultivars contain fewer and smaller protein bodies than regular sorghums; thus, they contain significantly lower amounts of alcohol-soluble kafirins (Paulis and Wall, 1979; Seckinger and Wolf, 1973). The starch granules are polygonal and often contain dents from the protein bodies. Their size varies from 4 to 25 μm (average 15 μm). The appearance of the corneous endosperm is translucent or vitreous.

The opaque, floury endosperm is located around the center of the caryopsis. It has a discontinuous protein phase, air voids, and loosely packaged, round-lenticular starch granules (Hoseney et al, 1974). The presence of air voids diffracts incoming light, giving an opaque or chalky appearance.

The germ is diploid due to the sexual union of one male and one female gamete. It consists of two major parts: the embryonic axis and the scutellum (Fig. 1). The embryonic axis contains the new plant and is divided into a radicle and plumulae. Upon germination and development, the radicle forms primary roots, whereas the plumulae form leaves and stems. The scutellum is the single cotyledon of the sorghum caryopsis. It is the germ reserve tissue with large amounts of oil, protein, enzymes, and minerals and serves as the bridge or connection between the endosperm and germ.

Millets

PEARL MILLET

The structure of mature caryopses of pearl millet is very similar to the structure of sorghum but with several important differences (Fig. 2, Table 3). In comparison with sorghum, pearl millet contains smaller seeds (3–15 g/1,000 kernels vs. 5–80 g/1,000 kernels) and proportionally a larger germ and consequently a smaller endosperm (Abdelrahman et al, 1984). Abdelraham et al (1982, 1984) reported that pearl millet consists of 71–76.2% endosperm, 15.5–21% germ, and 7.2–10.6% pericarp (Table 4).

On the surface of the pericarp there is a thin, waxy cutin layer that could decrease the effects of weathering (Rooney and McDonough, 1987). The pericarp consists of the epicarp, mesocarp, and endocarp and varies in thickness (Fig. 2). However, thick pericarp cultivars do not have starch granules in the mesocarp. The epicarp contains 1–2 or 2–4 layers with very thick cell walls (McDonough and

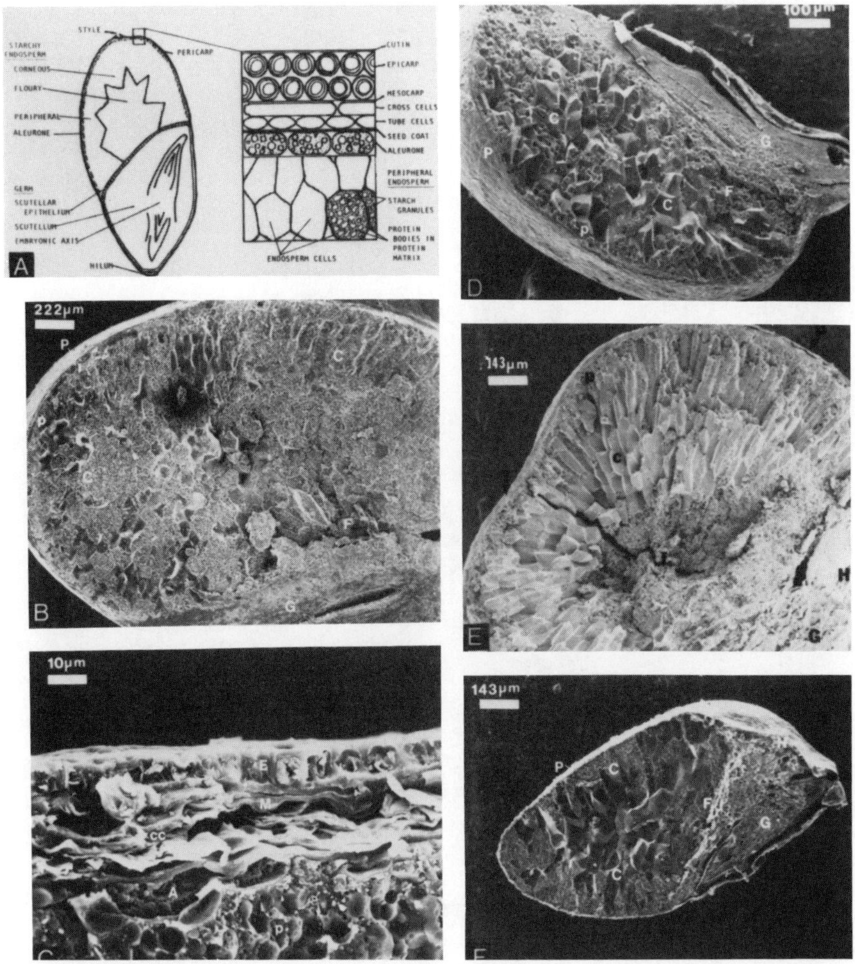

Fig. 2. Scanning electron micrographs of the structure of some millets. A, Internal structure of *Pennisitum americanum* (pearl millet) showing details of the pericarp. B–C, Scanning electron micrographs of pearl millet. D, *Eragrostis tef* (red teff). E, *Eleusine coracana* (finger millet or ragi). F, *Digitaria exilis* (fonio). P = pericarp, p = peripheral endosperm, C = corneous endosperm, F = floury endosperm, G = germ, A = aleurone, M = mesocarp, cc = cross cells, s = seed coat, H = hilum.

TABLE 3
Structural Comparisons of Some Millets[a]

	Pearl	Finger	Foxtail	Proso	Japanese	Fonio	Kodo	Teff
Seed type	Caryopsis	Utricle	Caryopsis	Utricle	. . .	Caryopsis	. . .	Caryopsis
Kernel wt.	7.8	2.3	2.0	6.1	. . .	0.5	. . .	0.8
1,000 kernels	2.5–14.7	2.0–2.5	. . .	5.0–7.1	. . .	. . .	. . .	0.3–1.9
Testa (layers)	1	5	Remnant	1	Yes	1	. . .	1
Pigmented	Sometimes	Yes	. . .	No	. . .	No	. . .	No
Thickness	0.4	10.8 × 24.2	. . .	0.2–0.4	. . .	0.5–1.5	. . .	1.5–2.0
Aleurone (layers)	1	1	1	1	1	1	. . .	1
Starch granules	6.4–12.0	8.0–21.0	5.0–25.0	1.3–17.0	2.5–20.0	6.5–7.8	7.0–15.0	5.6–22.4
Starch type granule	Simple	Compound simple	Simple	Simple	. . .	Simple	Simple	Compound simple
Waxy type	No	No	Yes	Yes	No	No	No	No
Protein body size (μm)	0.6–1.2	2.0	1.0–2.0	0.5–2.5	. . .	1.2–1.4	. . .	1.2–1.4
Germ size (L × W)	1,420 × 620	980 × 270	. . .	1,100 × 977	. . .	517 × 977	. . .	160 × 725
Endosperm-germ ratio	2.5:1	11:1	. . .	12:1	. . .	6:1	. . .	3:1

[a] Data from McDonough and Rooney (1989), McDonough et al (1986), Nuralikrishna et al (1982), Paramahans et al (1980), Rost and Lesten (1973), Serna-Saldivar et al (1991a), Sullins and Rooney (1977), Tomita et al (1981), Zee and O'Brien (1971).

Rooney, 1989; Rooney and McDonough, 1987; Sullins and Rooney, 1977). The mesocarp layer varies in thickness and is composed of several tiers of collapsed cells. The endocarp is composed of cross and tube cells: the first arranged longitudinally across the kernel, and the second lying perpendicular to the cross cells.

A partial or thin (0.4-μm) seed coat is present in all varieties (Table 3); the partial seed coat predominates in the slate color varieties (Rooney and McDonough, 1987). According to Narayanaswami (1953), during grain maturation of the young caryopsis there is progressive disintegration of the testa. The testa is thin and contributes little to the overall color perceived in kernels with a thin, colorless pericarp (Rachie and Majmudar, 1980). The aleurone layer is a one-cell thick, single layer of blocky cells ranging from 16- to 30-μm long and 14- to 33-μm wide. Aleurone cell walls are thick and fluoresce a deep royal blue under fluorescence microscopy, which also reveals that aleurone cells contain large amounts of protein and spherosomes (McDonough and Rooney, 1989; Rooney and McDonough, 1987).

The rest of the endosperm is subdivided into three distinctive areas: peripheral, corneous, and floury endosperm (Fig. 2). The peripheral endosperm has a very dense protein matrix with only small starch granules in the first one or two cell layers. The corneous area contains large, uniformly sized polygonal starch granules embedded in protein matrix with small amounts of protein bodies. The respective average sizes of starch granules and protein bodies are 6.4–7.6 and 0.6–0.7 μm (Fig. 2, Table 3) (McDonough and Rooney, 1989). Badi et al (1976a,b) reported

TABLE 4
Chemical Composition of Pearl Millet and Its Anatomical Parts[a]

Component	Whole Grain (%)	Endosperm (%)	Germ (%)	Pericarp (%)
Whole kernel	100	75.1	16.5	8.4
		73.9–76.2	15.5–17.4	7.2–10.6
Protein	13.3	10.9	24.5	17.1
Percent of total protein	100	59.5	31.2	9.4
Fat	6.3	0.5	32.2	5.0
Percent of total fat	100	6.2	87.8	5.9
Ash	1.7	0.3	7.2	3.2
Percent of total ash	100	13.9	72.2	13.9
Essential				
Phe	5.5	5.5	4.4	4.3
Ile	3.1	3.1	2.1	2.4
Leu	9.6	11.2	6.5	7.1
Lys	3.0	1.4	5.2	4.4
Met	2.3	2.6	1.8	1.6
Thr	3.6	3.3	3.9	4.7
Val	3.9	3.9	3.6	3.5
Nonessential				
Asp	8.8	8.1	11.4	11.4
Glu	19.6	22.8	15.1	13.5
Ala	8.2	9.4	8.0	8.7
Arg	4.8	2.0	9.0	5.0
Gly	3.3	1.9	6.1	7.3
Pro	6.4	8.0	5.0	6.8
Ser	5.2	5.2	5.7	5.6
Tyr	3.8	3.6	4.0	4.7

[a] Adapted from Abdelrahman et al (1984).

that the starch granule and protein body sizes in pearl millet examined by scanning electron microscopy were 10 and 1.5 μm, respectively. The floury endosperm contains large, round starch granules, loosely packaged in a discontinuous protein matrix with a small number of protein bodies (Rooney and McDonough, 1987). The proportion of floury to corneous endosperm varies among cultivars and among kernels within the same cultivar and is affected significantly by environment (Rooney and McDonough, 1987).

The pearl millet germ is larger than most other cereals (Fig. 2). It contains an embryo and scutellum. Its scutellar epidermal or epithelial cells are very similar to those of sorghum (Zeleznack and Varriano-Marston, 1982). A dark pigmented material (black layer) is deposited in the basal adgerminal (chalazal cells) surface of the grain during seed development (Fussell and Dwarte, 1980).

McDonough and Rooney (1989) found great variation in kernel shape, endosperm color, and external appearance in 96 varieties of pearl millet. Kernel color is due to the combined effects of pigmentation in the pericarp, aleurone, and endosperm in addition to pericarp thickness. White kernels have few pigmented areas, yellow colored kernels have pigments primarily in the pericarp and endosperm, and brown grains contain pigments in the epicarp, aleurone, and endosperm. The slate gray color of pearl millet kernels is primarily due to endosperm pigmentation, which can be affected by pH.

FINGER MILLET

The finger millet grain is small, round, or globose and 1.2–1.8 mm in diameter. Finger millet has a distinctive morphological property that is not found in the other major cereals. The kernel is not a true caryopsis but a utricle (Angold, 1979; McDonough et al, 1986). The pericarp in utricles is not completely fused to the seed coat or testa. Thus, the pericarp is easily removed by rubbing or soaking in water and often detaches during harvesting. Finger millet possesses a five-layered testa that varies from red to purple (McDonough et al, 1986; Table 3). The aleurone is located beneath the testa and is single-layered. The cells are small (18 × 7.6 μm) and are rich with aleurone bodies ranging from 0.9 to 2.2 μm in diameter. The endosperm of finger millet has distinctive peripheral, floury, and corneous areas with polygonal and/or round starch granules of varied sizes (8–21 μm) (McDonough et al, 1986; Paramahans et al, 1980; Wankhede et al, 1979a,b). The endosperm cell walls fluoresce strongly, indicating the presence of phenolic compounds. The starch granules are primarily compound (McDonough et al, 1986; Paramahans et al, 1980), with some simple granules in the corneous section. The endosperm protein bodies are about 2 μm in diameter.

Finger millet has a small germ (270 × 980 μm) located in a depression surrounded by a characteristic ridge (Table 3). The scutellum is separated from the floury endosperm by the scutellar epithelium.

FOXTAIL MILLET

The surface of the glumes of foxtail millet is covered with small peaks arranged symmetrically (Lorenz, 1977). The pericarp is thin with two characteristic epidermal layers firmly compressed to each other. According to Narayanaswami (1956), the immature pericarp of foxtail contains large numbers of starch granules that disappear after maturation. The testa collapses and is reabsorbed during grain maturation. Foxtail kernels have a one-layered aleurone (Hodson and Parry, 1982;

Lorenz, 1977), which accommodates a dense cytoplasm. The aleurone layer completely surrounds the endosperm and germ, and it is composed of cells that are 25–50 μm in length with protein bodies and distinctive spherosomes (Table 3). Similar to other millets, the starch granules of the corneous endosperm are angular, whereas the ones located in the floury area are spherical (Lorenz, 1977). The size of the starch granules ranges from 2 to 10 μm (Table 3). The protein bodies are concentrated in the peripheral cells of the endosperm, becoming more scattered and less frequent towards the inside. The protein bodies are spherical, 1–2 μm in diameter, with a subtle internal structure consisting of alternating light and dark concentric layers (Rost, 1971). The germ also contains protein bodies but does not have any inclusions. The scutellum consists of irregularly shaped cells and appears in the form of two winglike expansions. Upon maturation, the chalazal region of the embryo becomes a part of the scutellum. Unlike other cereals, the embryo lacks a distinctive epiblast.

PROSO MILLET

Unlike foxtail millet, the glumes of proso millet have a smooth and even surface. However, the internal structure of proso and foxtail millet kernels is essentially the same (Lorenz, 1977). Proso is also classified as an utricle instead of a true caryopsis (Table 3). The pericarp of utricles is loosely attached to the kernel; therefore, it can be easily removed. However, the glumes firmly adhere to the utricle and are very difficult to remove by conventional milling.

JAPANESE MILLET

Little information is available about the microstructure of Japanese millet. The inflorescence produces one fertile and one infertile floret subtended by two unequal glumes. The glumes totally enclose the kernel. The mature pericarp consists of two epidermal layers. The cells of the inner epidermis are closely compressed against those of the outer part. Japanese millet has an aleurone layer thought to contain strongly cutinized cell walls (Narayanaswami, 1955; Zee and O'Brien, 1971). The testa is multilayered in the developing seed. However, the only persistent portion in the mature caryopsis is the rim of the inner integument (elongated cells tangentially compressed). The endosperm structure is similar to other members of the *Paniceae* family.

CHEMICAL COMPOSITION

Proximate Analysis

SORGHUM

Sorghum proximate composition varies significantly due to genetics and environment. The protein content is usually the most variable (Table 5). Sorghum is similar in composition to corn (*Zea mays*). In general, sorghum in the United States contains slightly more protein and has less oil than yellow dent corn. The lysine content of sorghum is slightly lower than that for yellow corn, but sorghum contains more tryptophan.

The pericarp is rich in fiber, whereas the germ is high in crude protein, fat, and ash (Table 2). The endosperm contains mostly starch and protein with small amounts of fat and fiber. In general, yellow dent corn and yellow sorghum are

TABLE 5
Nutrient Composition of Sorghum and Millets[a,b]

	Proximate Composition					
Cereal	Protein (%)	Ether Extract (%)	Crude Fiber (%)	Ash (%)	NFE[c] (%)	Starch (%)
Sorghum	11.0 7.3–15.6	3.2 0.5–5.2	2.7 1.2–6.6	1.8 1.1–4.5	81.3 68.1–89.9	70.8 55.6–75.2
Millets						
Pearl	14.5 8.6–19.4	5.1 1.5–6.8	2.0 1.4–7.3	2.0 1.6–3.6	76.4 62.9–86.9	71.6 63.1–78.5
Finger	8.0 6.9–10.9	1.5 1.0–4.6	3.0 2.0–6.8	3.0 2.3–3.9	84.5 73.8–88.7	59.0 ...
Foxtail	11.7 6.0–14.0	3.9 1.2–5.2	7.0 2.6–8.6	3.0 1.5–3.6	74.2 68.3–88.7	55.1 51.0–59.1
Proso	13.4 6.4–15.9	6.7 2.9–11.1	6.3 4.6–10.7	4.2 1.4–8.8	69.4 53.5–84.7	57.1 56.1–58.0
Japanese	11.8 11.2–12.7	4.9 2.5–6.3	14.3 13.9–14.7	4.9 4.7–5.0	64.1 61.3–68.0	60.3 58.6–62.0
Fonio	8.7 5.1–10.4	2.8 2.1–5.2	8.0 4.6–11.3	3.8 1.8–6.0	76.7 67.1–86.4	... ...
Kodo	10.4 6.2–13.1	3.7 3.2–4.9	9.7 8.4–11.0	3.6 3.0–4.1	72.6 66.9–79.2	72.0 ...
Teff	10.9 7.9–12.6	2.4 2.3–2.5	2.4 ...	2.2 ...	82.1 ...	... ...

[a] Data from Jambunathan et al (1984), Hulse et al (1980), Lorenz (1983), Parvathy and Sadasivam (1982), Pore and Magar (1979).
[b] All values are expressed on dry matter basis. Protein conversion factor = N × 6.25.
[c] Nitrogen-free extract.

nearly equivalent in feeding value, with corn slightly higher in gross and metabolizable energy. Some sorghums contain condensed tannins, but most sorghums do not.

MILLETS

Pearl millet contains high protein and oil (Tables 4 and 5), which is due to the large proportion of germ to endosperm (Tables 3 and 4). Proso millet also contains high levels of protein and fat. Fonio, finger, and teff millets are relatively low in protein and fat; consequently, they contain high amounts of nitrogen-free extract. Japanese and kodo millets contain higher crude fiber contents than the other millets, likely due to the higher proportion of bran present in the caryopses (Table 5).

Carbohydrates

The carbohydrates of sorghum and millets are composed of starch, soluble sugar, pentosans, cellulose, and hemicellulose. Starch is the most abundant chemical component, while soluble sugars and crude fiber are low.

STARCH

From one-half to three-fourths of grain weight is starch. Starches exist in highly organized granules in which amylose and amylopectin molecules are held together by hydrogen bonding. In the native form, they are considered pseudo crystals that have crystalline and amorphous areas (Rooney and Pflugfelder, 1986). The true density of starch varies from 1.4 to .6 g/cm^3. The native granules are insoluble in cold water, swell reversibly, rotate the plane of polarized light (exhibit birefringence), and are relatively inaccessible to enzyme attack.

Starch is the main source of energy required for germination. It is composed of linear chains of glucose units joined by α-1,4 glycosidic bonds called amylose and branched chains of glucose units joined by α-1,4 and α-1,6 glycosidic bonds called amylopectin. Amylose exists as a helix that forms a blue or purple clathrate with iodine. Amylopectin is a much larger molecule. The linear chains in amylopectin are classed as A, B (branched), and C (the single central chain containing the reducing end) (Rooney and Pflugfelder, 1986). Among sorghum and millets, only sorghum, finger millet, and proso millet have varieties that are waxy or glutinous (Tomita et al, 1981). These caryopses contain starch without amylose. For sorghum, a cross between waxy and regular parental lines produces heterowaxy endosperm, which contains around 12% amylose in the bulk seed sample.

Sorghum. Sorghum starch has properties and uses similar to maize starch for production of derived foods or industrial products. The procedure for sorghum wet milling is similar to the one used for maize (Rooney and Serna-Saldivar, 1991; Watson, 1984). The pigments of the sorghum pericarp may sometimes discolor the starch, yielding a light pink color. Bleaching with $NaClO_2$ produces acceptable coloration (Freeman and Watson, 1971).

Regular endosperm sorghum types contain from 23 to 30% amylose (Horan and Heider, 1946; Ring et al, 1982). Waxy varieties contain up to 5% amylose. Sullins and Rooney (1977) estimated that the starch granules range from 2 to 30 μm in diameter. Granules present in the corneous endosperm are smaller and angular, whereas those in the floury endosperm are larger and round. Isolated starch from regular and waxy sorghum exhibits a type A X-diffraction pattern, which corresponds to the ones observed in other cereals (Gomez, 1988). Amylose and amylopectin have molecular weights of $1–3 \times 10^5$ and $8–10 \times 10^6$, respectively (Gomez, 1988). Corneous endosperm starches have lower iodine-binding capacity, higher gelatinization temperature, and intrinsic viscosity compared with starch isolated form the floury endosperm (Cagampang and Kirleis, 1985). The gelatinization temperature range as determined by differential scanning calorimetry for sorghum starch is 71–80°C (Sweat et al, 1984).

The pasting behavior and thermal properties of waxy and nonwaxy sorghum are presented in Tables 6 and 7. Starch from waxy sorghum is notable for rapid cooking, high peak viscosity, poor stability during cooking, paste clarity, high water-binding capacity, and resistance to gel formation and retrogradation. The pastes are rather stringy and cohesive (Watson, 1984; Akingbala et al, 1982). The water-binding capacity of sorghum starch is significantly lower than that of yellow maize. Sorghum starch has a higher swelling power at 90°C but lower solubility than maize starch. In comparison with maize starch, sorghum starches have similar amylograph initial gelatinization and lower temperature at maximum viscosity but higher peak and cold viscosities (Abd Allah et al, 1987).

According to Hibberd et al (1982), Rooney and Pflugfelder (1986), and Waniska et al (1990), isolated waxy starch or waxy grains have higher in vitro starch digestibility than normal endosperm types.

Millets. According to Beleia et al (1980), pearl millet starches have an intrinsic viscosity of 1.47–1.71, iodine affinity of 4.13–4.38, and an amylose content of 20–21.5% (Table 8). The viscosity values are lower than intrinsic viscosities reported for sorghum and maize and may reflect smaller molecular weight in starch molecules. X-ray diffraction patterns of millet starches show an A-type pattern seen in other cereals (Beleia et al, 1980). The initial and end-point temperatures of starch gelatinization are 59–63°C and 68–70°C. The amylograph pasting characteristics of pearl millet starch and other important related properties are in Table 9. Interestingly, pearl millet appears to require a higher temperature to initiate pasting or to develop viscosity than other cereal starches (Abd Allah et al, 1987). The water-holding capacity of pearl millet starch was higher than that of sorghum but lower than that of maize. Pearl millet starch appears to have a higher swelling power and solubility than other starches (Abd Allah et al, 1987).

Starches from finger and foxtail millets exhibit single-stage swelling and low solubility. The respective molecular weights of amyloses from finger and foxtail millets ranged from 20×10^4 to 26×10^4 and 19.8×10^4 to 25.5×10^4. Amylopectin molecular weights were above 20×10^6. These molecular weights are similar to the

TABLE 6
Endosperm and Isolated Starch Properties of Different Sorghum Types[a]

Sorghum Type	Amylose (%)	Differential Scanning Calorimetry: Endosperm T_o	T_p	T_e	Starch T_o	T_p	T_e	Gelatinization Energy $-\Delta$ Hg (cal/g)
Normal	23.6	72.6	75.1	77.8	71.7	75.3	79.7	3.25
WxWxWx	23.3–23.9	72.0–72.9	74.3–75.7	76.3–79.2	71.0–73.0	74.0–76.5	77.0–81.1	3.11–3.48
Waxy	3.5	73.2	77.6	81.9	74.6	77.9	82.0	3.76
wxwxwx	. . .	. . .	. . .	. . .	73.5–75.6	77.6–78.2	81.1–82.8	3.49–4.03
Heterowaxy *Wxwxwx*	19.6	74.3	78.0	80.9	75.8	78.4	81.6	4.07

[a] Adapted from Akingbala et al (1982).

TABLE 7
Viscoamylograph Properties of Sorghum Flours and Isolated Starches[a]

Sorghum	Pasting: Temperature (°C)	Pasting: Viscosity (BU)	Viscosity (BU): 1 hr at 95°C	50°C	1 hr at 50°C
Normal					
Starch[b]	73.3	873	459	874	1,109
Range	68.5–77.5	750–1,170	320–675	610–1,250	700–1,915
Flour[c]	82.1	504	351	939	743
Waxy					
Starch[b]	69.8	1,130	370	490	425
Flour[c]	72.7	223	125	240	220

[a] Data from Akingbala et al (1981). Values for normal sorghum starch and flour are means of 24 cultivars.

[b] A 10% (db) starch slurry (4.5 g in 42 ml of distilled water) cooked at 75 rpm using the 125-mcg sensitive cartridge.

[c] Flour was obtained from decorticated/degermed grain. A 12% (db) sorghum flour (5.44 g in 42 ml of distilled water) was cooked similar to isolated starch.

ones reported for other cereal starches (Wankhede et al, 1979a,b). Viscoamylograph properties of their isolated starches are presented in Table 9.

Fujita et al (1989) studied the starch properties of 13 foxtail cultivars. The maximum absorbance of the iodine-treated starch ranged from 535 to 606 nm. Waxy types contained from 1.3 to 2.7% amylose while regular endosperm types contained from 17.4 to 24%. Waxy starches required a higher temperature for

TABLE 8
Properties of Millet Starches[a]

Property	Millet: Pearl	Finger	Proso	Foxtail	Japanese	Kodo
Gelatinization temp. range[b] (°C)	61–69	65–69	68–76	55–63 56–83	50–70	57–68
Amylose (%)	17.0–21.5	15–16	0.0–1.0[c] 26.3–28.4[d]	0.8–4.2[c] 17.4–24.0[d]	19.7–25.6	24
Amylopectin (%)	78.9–83.0	84–85	99–100[c] 72–73.7[d]	95.8–99.2[c] 82.6–76.0[d]	74.6–80.3	72
Starch						
Granule size (μm)	4.0–12.0	10–16	1.3–17.5	7.0–25.0	2.5–40	7–15
Form	Polygonal, round	Polygonal, round	Hexagonal, round	Polygonal, round	Hexagonal, round	Polygonal, round
Hilum	Large	Centric	Centric	Faint	Centric	Centric

[a]From Beleia et al (1980), Fujita et al (1989), Hoseney et al (1981), Nuralikrishna et al (1982), Paramahans et al (1980), Tharanathan et al (1980), Tomita et al (1981), Wankhede et al (1979a,b), Yañes et al (1986).
[b]Initial and final gelatinization.
[c]Waxy.
[d]Regular.

TABLE 9
Viscoamylograph Properties of Isolated Millet Starches

Millet	Amylograph Pasting: Temp. (C)	Amylograph Pasting: Viscosity (BU)	Amylograph Viscosity[a] (BU): A	B	C	D	Starch: Swelling Power (90°C)	Starch: Solubility (90°C)
Pearl[b]	...	...	460	396	568	536	13.1	9.14
Pearl[c]	90.7	185	185	185	230	...	19.8	19.20
Finger[d]	...	...	1,633	1,286	1,796	...	11.4	6.50
Proso[e]	...	...	688	520	826	1,203	12.0	6.89
Foxtail[e]	...	...	840	620	1,100	1,220	11.2	4.65
Foxtail[d]	...	...	1,780	1,540	2,000	...	9.8	4.80
Kodo[f]	83.5	700	300	270	390	...	12.0	5.50

[a]A = Viscosity at 93–95°C; B = viscosity after holding sample at 93–95°C, C = viscosity cooling to 35 or 50°C; D = viscosity after holding sample at 35 or 50°C.
[b]From Beleia et al (1980). The amylograph slurry contained 9% starch (14% mb) and was suspended in disodium phosphate/citric acid buffer. First holding period (B) = 1 hr; second holding period (D) = 1 hr. The slurry was cooled to 50°C.
[c]From Abd Allah et al (1987).
[d]From Wankhede et al (1979). The amylograph slurry contained 9.5% parts of starch solids per 100 g of water. First holding period (B) = 30 min. The slurry was cooled to 50°C.
[e]From Lorenz and Hinze (1976). The amylograph slurry contained nine parts of starch solids per 100 parts water. First holding period (B) = 30 min. The slurry was cooled to 50°C.
[f]From Paramahans et al (1980). Values were extrapolated from amylograph curve.

gelatinization than regular counterparts. The heat of gelatinization for the waxy and regular starches averaged 3.7 and 3.1 cal/g (Fujita et al, 1989).

Proso and foxtail millet starches had higher water-binding capacities and gelatinization temperatures than wheat starch. The swelling power at 90°C was comparable but lower than that of rye starch. Amylograph viscosities were higher than those of wheat at all reference points (Lorenz and Hinze, 1976).

X-ray diffraction patterns of isolated starch from proso, foxtail, and Japanese millets had the A-type crystalline structure, which is typical of cereal starches (Tomita et al, 1981). The iodine absorption spectra of these starches revealed that the maximum absorption was at 584–605 nm, showing that they are of normal type. The waxy starches of proso and foxtail millets showed an absorption maxima at 521–529 nm. When subjected to α-amylase hydrolysis, waxy starches of proso and foxtail millets degraded faster than nonwaxy counterparts and nonwaxy maize starch (Tomita et al, 1981).

SOLUBLE SUGARS

Sorghum. Murty et al (1985) determined the soluble sugar content of sorghum grain throughout maturation. Sorghum at 22 days postanthesis and at physiological maturity contained 5.2 and 2.3% soluble sugars, respectively. High-lysine (*hl*) and sugary cultivars contained more soluble sugars than regular endosperm counterparts (Table 10). Jambunathan et al (1984) reported that the average soluble sugar content of 100 sorghum samples was 1.3% (0.8–4.2%). The soluble sugars of sorghum consist of 75% sucrose (Subramanian et al, 1980; Table 10).

The total sorghum soluble sugars and reducing sugars gradually decreased during grain maturation (90–120 days after sowing). Mature kernels contained 2.2–3.8% soluble sugars, 0.9–2.5% free reducing sugars, and 1.3–1.4% nonreducing sugars. Free glucose and fructose ranged from 0.6–1.8% and 0.3–0.7%, respectively. Bound glucose and fructose contents were 0.7–1.2 and 0.1–0.6%, respectively (Bhatia et al, 1972).

During sorghum germination there is sugar accumulation in the endosperm after the second day, and maximum concentration after eight days' germination (Newton et al, 1980). Sucrose is the major soluble carbohydrate in the dry seed. After two days' germination, glucose and fructose were located both in the endosperm and scutellum; however, by the fourth day sucrose localization was concentrated in the scutellum (Newton et al, 1980). Pflugfelder et al (1986) also found that the total soluble sugar content and free glucose and maltose increased in sorghum kernels subjected to germination.

TABLE 10
Sorghum Soluble Sugar Content and Composition[a]

Sorghum Type	Total Soluble Sugars (%)	Stachyose (%)	Raffinose (%)	Sucrose (%)	Glucose + Fructose (%)
Normal	2.25	0.10	0.23	1.68	0.25
Range	1.3–5.2	0.04–0.21	0.10–0.39	0.93–3.90	0.06–0.74
Sugary	2.21	0.06	0.39	0.81	0.95
High lysine	2.58	0.11	0.39	0.94	1.13
Range	2.5–2.7	0.10–0.12	0.36–0.42	0.90–0.99	1.04–1.22

[a]From Murty et al (1985) and Subramanian et al (1980). Values are expressed on dry weight basis.

TABLE 11
Soluble Sugar Content and Composition of Millets[a]

Millet	Total Soluble Sugars (%)	Stachyose (%)	Raffinose (%)	Sucrose (%)	Glucose + Fructose (%)
Pearl[b]	2.56	0.09	0.71	1.64	0.11
Range	2.16–2.78	0.06–0.10	0.65–0.86	1.32–1.82	0.08–0.16
Proso[c]	...	...	0.08	0.66	...
Range	...	...	0.04–0.12	0.48–0.90	...
Foxtail[c]	...	...	0.10	1.01	...
Range	...	...	0.08–0.12	0.90–1.12	...
Foxtail[d]	0.46	...	0.04	0.15	0.10
Finger[d]	0.65	...	0.07	0.21	0.18

[a] All values are expressed on dry weight basis.
[b] Subramanian et al (1981).
[c] Becker and Lorenz (1978).
[d] Wankhede et al (1979a).

Millets. The total soluble sugar content of kernels of nine pearl millet cultivars and 36 samples ranged from 2.16 to 2.78% and 1.4 to 2.6%, respectively (Jambunathan et al, 1984; Subramanian et al, 1981; Table 11). Fussel and Pearson (1978) reported an average value of 2.4%. Sucrose and raffinose were the predominant sugars. Compared to sorghum, the proportion of sucrose in pearl millet was low (Subramanian et al, 1980). Other sugars present in considerable amounts are stachyose and the monosaccharides glucose and fructose (Subramanian et al, 1981). Opoku et al (1983) reported that after 96 hr of germination, the soluble sugars of pearl millet increased from 2.5 to 15%. Raffinose is quickly broken down during malting with the corresponding increase in glucose and sucrose.

Two pearl millet varieties harvested 110 days after sowing contained 4.5–5.4% soluble sugars, 2.2–3.4% free reducing sugars, and 2.1–2.3% nonreducing sugars. Free glucose and fructose ranged from 1.4 to 2.5% and 0.8 to 1.0%, respectively. Bound glucose and fructose contents were 0.7 and 1.4–1.7%, respectively (Bhatia et al, 1972).

The main sugar in pearl, foxtail, finger, and proso millets is sucrose (Table 11). According to Pore and Magar (1979), the reducing sugar content of finger millet ranges from 1.2 to 1.8%. The effect of germination on reducing sugar content of foxtail, Japanese, and proso millets was determined by Parvathy and Sadasivam (1982); reducing sugars in ungerminated kernels were 0.30, 0.15 and 0.16%, respectively, and germination for eight days increased contents to 10.1, 12.3, and 11.3%. Bhatia et al (1972) determined changes in sugar composition of foxtail millet throughout grain maturation (45–77 days after sowing). The total soluble sugar and reducing sugar contents diminished from 4.1 to 2.9% and 2.4 to 1.2%, respectively. The amount of nonreducing sugars stayed about the same during progressive grain germination.

FIBER

Cereal grains are a rich source of fiber. Fiber consists of endogenous components of plant materials (i.e., cereal grains) that are resistant to digestion by enzymes in the monogastric stomach and upper gastrointestinal tract. The major individual components are cellulose, hemicellulose, lignin, pectins, and gums mainly located in the pericarp and endosperm cell walls. Dietary fiber components are not created equally; each different source of fiber has its unique physicochemical properties. However, in general, dietary fiber is divided into soluble and insoluble water categories. Insoluble fiber components are present primarily in the pericarp (Tables 2 and 4) where they have structural and protective functions. Therefore, the fiber content in cereal products depends to a large extent on the degree of milling.

The fiber contents of sorghum and millets are reported in Table 12. Most of the dietary fiber of millets and sorghum is insoluble; therefore, sorghum and pearl millet fiber may decrease transit time and prevent gastrointestinal problems. However, pearl millet and sorghum are low in soluble dietary fiber; therefore, they probably do not reduce blood cholesterol and arteriosclerosis as oat fiber does. No information is available about the dietary fiber content and composition of the other millets.

TABLE 12
Fiber Composition of Sorghum and Pearl Millet as Determined by Various Methods[a]

Component	Sorghum	Pearl Millet
Fiber method		
Crude fiber, %	2.7	1.7
Range	1.2–6.6	1.3–6.2
Detergent		
Neutral	...	2.7
Range	...	1.8–3.5
Acid	3.6	11.3
Range	2.9–4.0	9.4–13.1
Dietary fiber 1		
Total	9.8	8.5
Range	9.5–10.1	8.0–9.0
Dietary fiber 2		
Total	11.8	...
Range	7.5–16.1	...
Insoluble	10.8	...
Range	6.5–15.1	...
Soluble	0.9	...
Range	0.5–1.2	...
Dietary fiber 3		
Total	6.7	6.9
Range	5.8–7.6	...
Insoluble	5.7	6.3
Range	5.4–6.0	...
Soluble	1.0	0.6
Range	0.4–1.6	...
Fiber components		
Nonstarch polysaccharides	5.9	...
Cellulose	2.7	1.9
Range	...	1.8–2.0
Noncellulosic polysaccharides	3.2	...
β-D-Glucans	1.0	...
Range	0.6–1.4	...
Pentosans	4.1	2.5
Range	2.5–5.6	2.0–3.0

[a]From Bach Knudsen and Munck (1985), Bach Knudsen et al (1988), Bailey and Sumrell (1979), Goswami et al (1969), Nyman et al (1984), Prentice et al (1980), San Buenaventura et al (1987), Serna-Saldivar et al (1987a,b, 1991a), Singh et al (1987).

According to Bach Knudsen and Munck (1985) the majority of the fiber in sorghum is insoluble (86.2%). Sorghum contains 6.5–7.9% insoluble and only 1.1–1.23% of soluble β-D-glucans which, comprise most of the soluble fiber. The acid detergent fiber ranges from 2.92 to 3.63% (Table 12). Bach Knudsen et al (1988) compared the fiber composition of a white decorticated sorghum (28% removal of grain weight) with that of whole red sorghum. Raw and decorticated flours contained, respectively, 5.9 and 2.1% nonstarch polysaccharides, 2.7 and 0.8% cellulose, 3.2 and 1.3% noncellulosic polysaccharides (β-glucans), 9.5 and 2.1% dietary fiber, 9.3 and 1.7% total dietary fiber, and 4.0 and 1.3% acid detergent fiber.

Cooking sorghum into porridges and fermented *kisra* (Bach Knudsen and Munck, 1985) and tortillas (Serna-Saldivar et al, 1987a) increases insoluble dietary fiber due to increased levels of bound protein (mainly kafirins) and enzyme-resistant starch. In brown sorghum, cooking may increase the amount of dietary fiber present due to formation of polyphenol-protein complexes (Bach Knudsen et al, 1988). Among major cereals, sorghum is thought to contain the most protein associated with the dietary fiber fraction (Bach Knudsen and Munck, 1985).

Woolard et al (1977) identified cold-water-soluble gums in the endosperm of sorghum grain. The α- and β-glucans were 0.14 and 0.06% of the endosperm weight. They had both 1-4- and 1-6-linked glucopyranose residues in the ratio of 2:1 and 6:1, respectively. The α-glucan had a preponderance of α-D glucosidic linkages, whereas the β-glucan showed the presence of α and β linkages in the ratio of 4:1. A hemicellulose, β-D-glucan, was also identified in the sorghum endosperm. The water-soluble glucan accounted for 4.8% of the total hemicellulose and 0.015% of the whole grain. The hemicellulose was linear and had both β-linked 1-3- and 1-4-linked D-glucopyranose residues in the ratio of 3:2 (Woolard et al, 1976). Karim and Rooney (1972) fractionated and characterized sorghum pentosans. The main simple sugar present in the pentosan polymers was glucose (Table 13).

Feeding of sorghum bran to human subjects significantly increased stool weight, decreased intestinal transit time, and increased frequency of evacuation (Fedail et al, 1984). Klopfenstein et al (1981) compared several sources of cereal fibers and found that sorghum bran was effective in lowering serum and liver cholesterol levels in guinea pigs. Rooney et al (1992) decorticated pearl millet and white and brown sorghum to remove approximately 7.5% of the grain. The result-

TABLE 13
Sorghum Pentosan Content and Composition[a]

Pentosan	Yield (%)	Protein (%)	Total Carbohydrate (%)	Galactose[b] (% TC)	Glucose[b] (% TC)	Arabinose[b] (% TC)	Xylose[b] (% TC)
Water soluble							
Whole grain	0.90	10.5	76.3	9.4	68.7	16.7	5.2
Endosperm	0.16	3.2	69.6	5.0	85.1	6.5	3.4
Pericarp	0.62	6.8	68.7	13.7	30.2	37.8	18.4
Alkali soluble							
Whole grain	0.42	1.3	74.0	7.6	55.4	22.5	14.4
Endosperm	0.09	3.7	85.7	3.9	73.5	14.5	8.2
Pericarp	20.00	5.2	76.2	5.2	37.2	32.4	25.3

[a]Source: Karim and Rooney (1972); used by permission.
[b]As a percentage of total carbohydrates.

ing pericarp-rich (bran) fractions of white and brown sorghum contained 47.8 and 35.1% insoluble dietary fiber and 1.6 and 1.0% soluble fiber. Pearl millet bran contained 37.9% insoluble fiber and 1.1% soluble fiber. In rats, the sorghum brans did not lower blood cholesterol levels but were excellent bulking agents. Brown sorghum bran (high in tannins) exhibited better bulking ability than white sorghum bran. Pearl millet bran slightly decreased serum cholesterol levels. Pearl millet bran also had excellent bulking capacity.

Sorghum and pearl millet brans obtained from raw and parboiled grains decorticated to remove 17.5% of the total weight were analyzed according to the Englyst technique. Brans from parboiled kernels were richer in fiber and oil than corre-

TABLE 14
Pearl Millet Sugar Proportions in Pentosan Fractions[a]

	Extraction			
		Water		
Sugar	80% EtOH	Room Temperature	50°C	Alkali
Rhamnose	14.2	4.4	12.5	4.0
Fucose	14.6	1.5	13.6	9.6
Ribose	2.3	...	...	...
Arabinose	9.4	36.3	16.0	35.6
Xylose	10.5	24.2	11.4	27.1
Mannose	2.2	3.8	4.8	2.3
Galactose	18.5	22.2	28.8	15.7
Glucose	28.4	7.7	13.0	5.7

[a]Data from Bailey et al (1979).

TABLE 15
Landry-Moureaux Protein Fractions of Sorghums[a]

	Fraction[c] (% of total protein)				
Sorghum[b]	I	II	III	IV	V
Regular	12.7	21.1	28.9	4.4	25.9
Range	10.0–15.3	15.7–26.4	26.5–31.3	4.3–4.5	22.5–29.3
Regular	15.3	26.4	26.5	4.3	22.5
High lysine	24.6	13.1	19.7	5.1	30.1
Range	22.4–26.0	10.3–15.2	19.3–20.2	4.3–6.5	27.2–33.5
High lysine	25.9	10.3	17.2	4.5	42.3
Range	23.1–28.6	9.9–10.7	15.3–19.0	4.1–4.8	42.1–42.4
High lysine	22.4	13.7	20.2	4.3	33.5
Low tannin	16.1	14.6	17.3	5.6	34.9
Range	15.4–16.8	10.6–18.4	14.9–18.9	4.4–6.4	29.9–40.5
High tannin	3.7	1.9	21.1	13.6	47.0
High tannin	5.1	3.2	10.4	15.1	54.6
Range	4.1–6.0	2.5–3.8	6.9–13.8	12.8–17.3	49.3–59.9

[a]Data from Guiragossian et al (1978), Jambunathan and Mertz (1973), Jambunathan et al (1975), and Mertz (1974).
[b]Repeated terms indicate data from different sources.
[c]Landry-Moureaux fractionation scheme. Fraction I (albumins and globulins) extracted with 0.5*M* NaCl solution; fraction II (prolamins or kafirins) extracted with 70% isopropanol; fraction III (alcohol-soluble glutelins) with 70% isopropanol and 0.6% 2-mercaptoethanol; fraction IV, borate buffer with 2-mercaptoethanol; and fraction V, borate buffer + 2-mercaptoethanol and sodium lauryl sulfate.

sponding brans from raw grains. The insoluble and soluble dietary fiber values for sorghum bran ranged from 15 to 24% and 2.7 to 3.7%, respectively. Pearl millet bran contained from 21.1 to 27.6% insoluble and 1.5 to 4.7% soluble dietary fiber (Serna-Saldivar et al, 1993).

The pearl millet fiber content as estimated by various techniques is summarized in Table 12. Bailey et al (1979) fractionated and identified the sugars of pearl millet pentosans. The most important sugars were arabinose, xylose, glucose, and galactose (Table 14).

Proso and Japanese millets contain 0.38 and 0.44% hemicellulose and 3.79 and 2.69% cellulose, respectively. The hemicellulose is mainly composed of glucose, arabinose, uronic acid, and xylose units.

Proteins

Protein content and composition varies due to genotype water availability, soil fertility, temperatures, and environmental conditions during grain development.

TABLE 16
Osborne Protein Fraction Distribution of Sorghum and Millets[a]

	Protein Fraction[c]			
Grain[b]	**Albumin (%)**	**Globulins (%)**	**Prolamins (%)**	**Glutelins (%)**
Sorghum	5.3	14.8	33.2	32.8
	2.7–9.0	12.9–16.3	27.0–43.1	26.2–39.5
Sorghum	8.0	13.4	31.2	30.5
	6.9–9.6	12.0–14.8	24.2–36.9	28.3–31.8
Sorghum	10.2	9.3	28.3	31.9
	8.8–11.5	7.1–10.7	23.2–34.9	27.8–41.8
Pearl	16.6	12.9	26.6	20.9
	16.0–17.1	11.6–14.2	26.2–26.9	16.0–25.8
Pearl	8.0	14.8	39.1	37.6
	6.4–9.6	11.6–16.8	33.1–49.5	30.6–45.3
Pearl	15.1	8.7	30.2	30.3
	6.1–26.5	3.5–14.7	21.3–38.0	23.8–37.7
Pearl	19.2	10.4	39.2	18.9
	15.8–23.5	8.6–12.0	35.4–44.5	16.9–21.5
Finger	12.7	17.9	21.5	12.2
	10.2–15.7	14.3–20.4	15.0–24.8	7.4–17.0
Finger	14.4	10.7	7.6	29.6
	13.0–15.8	8.0–13.3	6.0–9.1	26.2–33.0
Foxtail	4.4	18.0	38.5	22.7
	3.2–5.6	12.2–23.5	37.8–39.7	21.0–23.7
Proso	9.0	10.9	31.0	8.0
	8.3–9.7	10.5–11.3	25.1–36.9	7.7–8.3
Japanese	2.8	3.0	5.7	7.5
	2.0–3.2	2.6–3.4	2.6–8.1	7.1–8.1
Kodo	22.0	15.3	25.3	31.3
	16.0–28.5	14.4–18.9	17.5–35.6	4.6–7.9

[a] Data from Chandna and Matta (1990), Gupta and Gupta (1972), Naik (1968), Indira and Naik (1971), Sawney and Naik (1969), Singh et al (1987), and Warsi and Wright (1973).
[b] Repeated terms indicate data from different sources.
[c] Data are means and ranges as a percentage of total protein.

Prolamins constitute the major protein fraction in sorghum, pearl, finger, foxtail, and proso millets (Tables 15–17) followed by glutelins. These fractions are mainly located within the protein bodies and protein matrix of the starchy endosperm. Nitrogen fertilization significantly increases amounts of protein due to accumulation of prolamins (Warsi and Wright, 1973).

TABLE 17
Landry-Moureaux Protein Fractions of Millets

Millet	Fraction[a] (% of total protein) I	II	III	IV	V
Pearl[b]	26.0	39.6	4.0	7.7	16.0
	22.0–29.0	38.0–41.0	0.0–7.0	5.0–9.0	11.0–21.0
Pearl[c]	25.0	28.4	2.7	5.5	18.4
	22.6–26.6	22.8–31.7	1.8–3.4	4.7–7.2	16.4–19.2
Finger[c]	12.1	10.3	32.3	2.8	21.2
	10.3–13.0	7.0–14.6	24.6–36.2	2.5–3.0	12.4–28.2

[a] Landry-Moureaux fractionation scheme. Fraction I (albumins and globulins) extracted with 0.5*M* NaCl solution; fraction II (prolamins) extracted with 70% isopropanol; fraction III (alcohol-soluble glutelins) with 70% isopropanol and 0.6% 2-mercaptoethanol; fraction IV, borate buffer with 2-mercaptoethanol; and fraction V, borate buffer + 2-mercaptoethanol and sodium lauryl sulfate.
[b] From Ejeta et al (1987) and Nwasike et al (1979). Average of three cultivars.
[c] From Virupaksha et al (1975). Average of six cultivars.

TABLE 18
Amino Acid Composition of Landry-Moureaux Protein Fractions of Normal and High-Lysine Sorghums[a]

Amino Acid	I Normal	I High Lysine	II Normal	II High Lysine	III Normal	III High Lysine	IV Normal	IV High Lysine	V Normal	V High Lysine
Essential										
Phe	3.2	3.3	5.9	5.4	5.5	5.0	2.3	2.4	5.3	5.3
His	2.1	2.2	0.9	0.8	1.0	0.7	5.8	3.0	2.2	2.1
Ile	2.6	2.8	4.1	3.9	3.6	3.6	2.6	2.6	4.8	3.1
Leu	5.0	5.0	17.0	16.0	15.9	15.9	7.7	7.2	12.9	12.7
Lys	4.5	4.7	0.1	0.3	0.2	0.1	1.3	1.7	2.2	2.8
Met	1.8	1.2	0.6	0.5	1.3	1.2	1.1	1.4	1.5	2.4
Thr	3.3	3.3	1.9	2.2	2.3	1.7	3.9	3.1	3.1	3.6
Trp	...	...	...	...	...	...	...	...	...	...
Val	4.6	4.7	4.0	4.0	3.9	3.8	5.3	5.0	6.1	6.8
Nonessential										
Asp acid	8.4	8.8	6.2	6.4	5.2	4.5	3.2	3.2	6.5	6.8
Glu acid	11.5	11.6	24.8	23.5	23.3	22.4	16.2	15.4	25.7	25.7
Ala	5.1	5.3	10.0	9.8	9.6	8.7	4.0	4.0	10.4	10.4
Arg	8.5	8.6	1.1	1.4	1.7	1.3	4.0	3.8	3.5	4.1
Cys	1.9	1.6	0.4	0.5	1.3	0.7	1.0	0.4	0.4	0.4
Gly	4.9	5.2	1.2	1.4	1.5	1.3	5.3	5.2	3.2	3.5
Pro	3.6	3.7	8.5	8.5	9.9	8.5	14.0	11.5	9.3	8.0
Ser	4.0	4.0	3.4	3.6	4.0	2.6	3.7	3.3	3.9	4.6
Tyr	2.6	2.8	4.2	4.4	4.6	4.0	2.8	2.4	4.8	5.2

[a] Values are expressed as grams per 100 g of protein. Adapted from Guiragossian et al (1978).

Grain protein is notoriously deficient in the essential amino acid lysine. The albumin-globulin and glutelins fractions are rich in lysine and other essential amino acids (Table 18). Cultivars with improved protein quality usually contain higher amounts of albumins, glutelins, and globulins and correspondingly lower proportions of prolamins (Table 18).

Protein quality is critically important in developing countries where human diets consist mainly of cereals. Supplementation of sorghum- and millet-based diets with relatively small amounts of legumes (i.e., cowpeas, beans, chickpeas) can overcome the protein deficit by improving the overall quality of the diet. This is especially valid for weanling children and infants who have a high essential amino acid requirement per kilogram of body weight.

TABLE 19
Amino Acid Composition from Two Sources of Sorghum Anatomical Parts[a]

Amino Acid	Whole Grain		Endosperm		Germ		Pericarp	
Essential								
Phe	3.8	4.4	4.4	5.5	3.9	5.2	4.0	5.4
His	1.7	2.6	1.7	2.5	2.6	3.4	1.7	2.8
Ile	2.8	4.0	4.0	4.4	3.5	3.8	3.9	4.1
Leu	12.2	13.1	14.2	13.8	7.8	7.5	9.0	8.1
Lys	1.6	2.6	1.2	1.7	3.8	6.7	3.0	5.0
Met	1.5	1.7	1.3	1.5	1.0	1.3	1.1	1.1
Thr	3.7	3.5	3.4	3.0	4.5	4.1	4.9	5.0
Val	4.3	5.6	5.3	5.1	6.6	6.3	6.4	6.4
Nonessential								
Asp	6.6	7.4	5.6	6.9	8.0	9.2	8.3	9.9
Glu	18.6	19.7	19.5	20.8	14.0	14.2	12.4	11.9
Ala	14.4	8.8	12.8	9.2	10.5	6.6	11.3	6.8
Arg	2.6	3.6	2.2	2.9	6.9	8.9	4.1	6.8
Cys	1.1	1.4	0.8	1.5	1.1	1.7	1.6	2.0
Gly	1.6	4.1	1.2	2.7	3.8	6.6	3.0	7.8
pro	9.7	8.8	9.9	9.3	5.2	5.1	6.4	6.0
Ser	6.5	3.9	5.5	3.9	7.0	5.1	7.2	5.6
Tyr	2.6	3.3	2.1	3.7	1.8	3.5	2.1	3.7

[a] Values are expressed as grams per 100 g of protein. Data from Haikerwal and Mathieson (1971) and Taylor and Shussler (1986).

TABLE 20
Lysine and Tryptophan Content of Osborne Protein Fractions of Sorghum and Millets[a]

	Albumin		Globulins		Prolamins		Glutelins	
Grain[b]	Lys	Trp	Lys	Trp	Lys	Trp	Lys	Trp
Sorghum	4.74	2.35	3.53	2.09	0.22	0.22	3.12	0.65
Sorghum	4.64	2.43	3.50	1.76	0.26	0.18	3.23	1.00
Pearl	1.51	1.62	0.27	0.33	0.39	0.56	0.33	0.78
Pearl	3.92	1.52	3.78	0.54	1.66	2.83	2.14	0.59
Finger	1.81	2.51	0.99	0.97	0.64	1.38	8.35	3.76
Foxtail	3.39	1.45	1.74	1.48	0.41	0.56	5.25	1.76
Proso	3.31	1.79	8.40	1.99	0.23	0.31	11.31	3.57
Japanese	7.30	2.51	5.99	1.48	0.41	1.10	6.09	1.69
Kodo	3.80	0.82	6.66	0.89	0.54	0.41	5.48	7.09

[a] Values are expressed as grams per 100 g protein. From Ahuja et al (1970), Chandna and Matta (1990), Indira and Naik (1971), Naik (1968), Sawhney and Naik (1969), and Taira (1962).
[b] Repeated terms indicate data from different sources.

TABLE 21
Amino Acid Composition of Sorghums[a]

Amino Acid	Normal			High Lysine	
(g/16 g N)	**Regular**	**Brown**	**P721 N**	**P721 O**	**IS111758**
Protein	12.0	9.9	11.3	12.6	12.8
	8.5–15.6	9.4–10.8	10.5–12.0	10.6–14.6	11.9–12.8
Essential					
Phe	5.0	5.1	5.6	5.0	5.1
	2.1–7.6	4.8–5.3	5.4–6.1	4.9–5.0	4.5–5.6
His	2.2	2.2	2.2	2.3	2.3
	0.6–3.9	2.1–2.3	2.2–2.3	2.1–2.5	2.2–2.4
Ile	4.1	3.9	4.0	4.0	4.3
	1.9–6.2	3.8–4.0	3.8–4.4	3.9–4.1	4.0–4.5
Leu	14.6	12.9	13.9	12.3	11.9
	7.8–24.9	12.4–13.4	13.6–14.2	11.0–12.3	10.3–13.1
Lys	2.2	2.2	2.1	3.0	3.5
	0.7–3.9	2.0–2.4	2.0–2.1	2.9–3.1	3.2–3.7
Met	1.4	1.4	1.8	1.7	1.6
	0.2–3.5	0.8–1.9	1.7–1.8	1.6–1.9	1.3–2.0
Thr	3.3	3.1	3.4	3.3	3.6
	1.6–5.0	3.0–3.1	3.1–3.7	3.3–3.4	3.5–3.7
Trp	1.1	0.8	0.8	0.9	...
	0.1–2.3	...	...	...	...
Val	5.4	6.8	5.2	5.2	5.4
	2.9–8.1	4.7–7.9	5.1–5.4	5.1–5.4	5.4–5.5
Nonessential					
Asp	6.4	6.4	6.8	7.5	8.2
	2.8–10.1	5.8–6.9	6.7–7.4	7.4–7.7	7.9–8.5
Glu	22.6	20.8	21.1	19.9	18.3
	13.3–36.1	20.1–21.8	19.2–22.1	18.9–20.5	17.8–19.8
Ala	9.1	8.7	9.1	8.1	8.6
	5.9–13.9	8.3–9.3	8.7–9.3	7.0–8.7	8.0–9.2
Arg	4.3	3.9	3.7	4.5	5.5
	1.5–9.0	3.6–4.2	3.5–3.9	4.5–5.2	3.5–3.9
Cys	1.7	1.1	1.5	1.5	1.0
	0.5–3.0	...	1.4–1.6	1.5	1.0–1.1
Gly	3.2	2.7	3.2	3.5	4.3
	1.9–5.2	2.4–2.9	2.9–3.4	3.5	3.9–4.7
Pro	7.6	6.1	7.4	7.0	7.9
	3.0–11.4	5.3–7.4	5.2–8.6	5.0–7.8	7.0–8.5
Ser	4.2	4.2	4.5	4.2	4.8
	1.2–6.7	4.2–4.3	4.3–4.6	4.0–4.3	4.7–4.9
Tyr	3.2	4.3	4.2	4.2	3.7
	0.8–5.4	4.1–4.5	3.5–4.6	4.0–4.3	3.2–4.5
Amino acid score, %	40.4	40.4	38.6	55.1	64.3

[a] Amino acid data is expressed in grams per 100 g protein. From Cousins et al (1981), Ejeta et al (1987), Featherston et al (1975), Guiragossian et al (1978), Haikerwal and Mathieson (1971), Hulse et al (1980), Mohan and Axtell (1975), and Serna-Saldivar et al (1987, 1991b).

[b] The essential amino acid requirements for infants (g/100 g protein) are: lysine 5.44 methionine + cysteine 3.52, threonine 4.0, isoleucine 4.0, leucine 7.04, phenylalanine + tyrosine 6.08, histidine 1.4; and tryptophan 0.96. Tyrosine and cysteine are not essential amino acids but they can reduce the requirement for phenylalanine and methionine, respectively.

TABLE 22
Amino Acid Composition of Millets[a]

Amino Acid (g/16g N)	Pearl	Finger	Proso	Japanese	Foxtail	Kodo	Teff
Essential[b]							
Phe	5.2	6.2	5.2	5.9	5.3	5.8	5.1
	4.4–5.6	4.4–8.4	4.3–5.6	5.5–6.3	4.3–6.7	5.6–6.0	4.5–5.9
His	2.2	2.6	2.2	1.9	2.3	1.8	2.6
	1.8–2.6	1.5–4.0	1.8–2.9	1.8–2.0	1.7–4.0	1.5–2.2	1.6–3.7
Ile	4.4	5.1	4.5	4.5	5.0	5.4	4.2
	3.6–5.9	3.8–8.5	3.1–6.5	4.5–4.6	3.9–7.6	3.0–7.7	3.6–4.8
Leu	12.2	13.5	12.9	11.5	13.3	10.2	7.9
	8.0–25.1	9.2–16.2	10.6–15.4	11.4–11.7	11.4–16.7	6.7–12.7	6.5–8.8
Lys	3.3	3.7	2.2	1.7	2.1	3.3	3.1
	1.7–6.5	2.6–5.5	1.4–4.3	1.6–1.8	1.5–2.8	3.0–3.5	2.0–4.0
Met	2.2	2.6	2.0	1.8	2.6	1.7	3.7
	1.5–2.9	1.3–4.3	1.3–2.6	1.6–2.0	2.3–2.8	1.5–1.8	2.4–4.6
Thr	3.9	5.1	3.4	2.7	3.9	2.9	3.8
	1.2–4.8	3.5–5.8	2.3–4.5	3.6–3.7	2.8–4.4	2.7–3.1	3.1–4.4
Trp	1.6	1.3	0.9	1.0	1.5	0.8	1.4
	1.1–2.8	1.0–1.7	0.6–1.7	1.0	0.4–1.9	0.5–1.0	1.2–1.7
Val	5.7	7.9	5.1	6.1	5.2	5.6	5.4
	4.8–7.0	5.8–10.4	4.0–6.5	6.1–6.2	3.8–6.9	3.8–7.2	5.0–5.9
Nonessential							
Asp	8.7	7.9	5.5	6.3	6.9	6.3	6.7
	4.9–10.3	6.5–10.0	3.7–6.3	6.0–6.9	6.4–7.3	6.1–6.4	5.6–7.2
Glu	21.2	27.1	20.5	20.7	18.8	23.1	24.8
	12.3–25.4	20.3–37.8	14.9–22.3	15.2–25.0	17.6–19.9	12.2–33.9	23.1–27.1
Ala	8.5	8.0	9.3	9.2	8.9	5.7	5.7
	7.5–10.5	5.9–8.9	3.9–12.2	8.8–9.9	8.6–9.1	5.3–6.0	5.0–6.5
Arg	4.8	5.2	4.4	3.2	6.1	4.2	4.0
	3.2–8.1	3.8–8.2	2.7–9.1	2.5–3.8	2.7–9.5	3.6–5.0	2.6–6.2
Cys	1.5	1.6	1.7	1.5	1.4	0.9	1.2
	0.7–2.8	0.7–2.9	0.5–2.8	1.4–1.6	1.4–1.5	0.7–1.0	0.7–1.8
Gly	3.6	4.8	2.2	2.7	2.9	3.8	3.8
	2.8–5.8	3.6–5.9	1.7–2.5	2.2–3.9	2.8–3.1	3.0–4.7	3.1–4.4
Pro	7.2	6.7	7.2	10.3	10.6	7.2	5.2
	5.9–14.2	4.2–10.1	5.3–10.4	8.9–12.4	10.5–10.7	5.3–9.2	4.5–6.3
Ser	4.9	6.9	6.3	5.8	5.8	4.1	4.4
	3.7–5.6	5.1–8.7	4.8–6.9	5.4–6.2	5.7–5.9	4.0–4.2	3.9–5.6
Tyr	3.2	3.6	3.9	2.7	2.7	3.8	2.9
	1.7–4.8	2.0–5.6	1.8–4.0	2.3–3.4	2.2–3.2	3.4–4.2	1.7–4.0
Amino acid score, %	60.6	68.0	40.4	31.2	42.3	60.7	57.0

[a] Amino acid data is expressed as grams per 100 g of protein. From Almeida-Dominguez et al (1993), Badi et al (1976), Ejeta et al (1987), Goswami et al (1969 a,b, 1970), Hulse et al (1980), Jambunathan et al 1984, Jansen et al (1962), Jones et al (1970), Lester and Bekele (1981), Lorenz and Dilsaver (1980), Sawhney and Naik (1969), Serna-Saldivar et al (1991a), Singh et al (1987), Taira (1968), Vincent Monteiro et al (1982), Virupaksha et al (1975).

[b] Essential amino acid requirement for infants (g/100 g protein); lysine 5.44; methionine + cysteine 3.52; threonine 4.0; isoleucine 4.0; leucine 7.04; phenylalanine + tyrosine 6.08; histidine 1.4; and tryptophan 0.96. Tyrosine and cysteine are not essential amino acids, but they can spare the requirement for phenylalanine and methionine, respectively.

SORGHUM

Approximately 80, 16, and 3% of the sorghum protein is located in the endosperm, germ, and pericarp, respectively (Taylor and Schussler, 1986; Table 2). The germ is rich in albumins and globulins, whereas the endosperm contains the kafirins and glutelins. Kafirins are mainly located within the protein bodies and increase with increased levels of protein in the kernel (Warsi and Wright, 1973).

The sorghum prolamin fraction has been divided into three subfractions: α-, β-, and τ-kafirins. These fractions are similar to the zein subfractions of corn (Shull et al, 1991). α-Kafirins can be extracted with 40 to 80% alcohol, are rich in glutamic acid, and have molecular weights ranging from 22 to 25 kDa. β-kafirins have lower molecular weights (16–20 kDa), are richer in sulfur-containing amino acids than α-kafirins, and are soluble in 10–60% alcohol solutions. The τ-kafirins were extracted with alcohol and reducing agents (mercaptoethanol) because they are highly cross-bonded with disulfide bonds. They had the highest molecular weight (28–29 kDa) prolamins and were rich in proline and sulfur-containing amino acids. The τ-kafirins were distributed within the dark-staining areas in the core and periphery of the protein bodies. They had many disulfide bonds and resisted digestion by proteases. Importantly, τ-kafirins become less digestible after cooking due to additional intramolecular cross bonding. Protein digestibility measured with a pepsin in vitro assay was greatly enhanced when reducing agents (i.e., metabisulfite, mercaptoethanol) that break disulfide bonds were used (Hamaker et al, 1987; Kirleis, 1990; Watterson et al, 1990).

The amino acid composition of sorghum anatomical parts is summarized in Table 19 (Haikerwal and Mathieson, 1971; Taylor and Schussler, 1986). Glutelins are the second most predominant protein fraction and are mainly located within the protein matrix. Albumins and globulins predominate in the germ and have the highest amount of lysine and tryptophan (Table 19).

The protein fractions of regular and high-lysine sorghums determined by the Landry-Moureaux and Osborne fractionation schemes are summarized (Tables 15 and 16). Their lysine and tryptophan contents are in Table 20. The endosperm (Table 19) and prolamin fraction (Table 20) are poor in both lysine and tryptophan so high-lysine cultivars were bred to contain a higher germ-to-endosperm ratio or genetically modified to produce higher amounts of glutelins and or albumins and globulins.

The amino acid composition of regular, high-lysine, and brown sorghums is summarized in Table 21. In all types, sorghum protein is first limited in lysine followed by threonine. Sorghum lysine meets approximately 40% of the recommended level for infants. Brown and regular sorghums have similar amino acid composition; however, the high-lysine cultivars contain approximately 50% more lysine and consequently a better amino acid score than regular counterparts. Nutritionally, brown sorghums have lower protein quality than regular counterparts due to reduced protein digestibility. In monogastrics, high-tannin cultivars consistently have 5–10% reduced protein digestibility, which lowers nitrogen retention and growth.

MILLETS

Among millets, pearl is known to contain a higher protein content (Table 2) and better amino acid balance (Tables 21 and 22) than sorghum. The higher ratio of germ to endosperm is responsible for the higher protein (Table 5), albumin, and globulin (Tables 6, 16, and 17) contents and improved amino acid composition (Table 22). Finger, teff, and kodo millets contain similar amounts of lysine as pearl millet, whereas the lysine levels of Japanese, proso, and foxtail millets are similar to that of regular sorghum (Tables 21 and 22). The protein fractions of millets are summarized in Tables 16 and 17. As in other cereals, the albumins and globulins are rich in lysine and tryptophan, whereas the prolamins are low in these essential amino acids (Table 22).

Lipids

Lipids are relatively minor constituents in cereal grains (Table 5). Most of the lipids of sorghum and millets are located in the scutellum (Tables 2 and 4). Thus, lipid contents are significantly reduced when kernels are decorticated and/or degermed. The whole-grain lipids can be subdivided into polar, nonpolar, and nonsaponifiable lipids. The most abundant are by far the nonpolar lipids consisting of triglycerides. They serve as a reserve material for germination. The less abundant polar lipids (i.e., phospholipids, glycolipids) have important biochemical functions in the grain. The nonsaponifiable compounds include phytosterols, carotenoids, and tocopherols.

SORGHUM

Osagie (1987) reported that sorghum lipids extracted with hexane-ether (4:1) consist mainly of nonpolar or neutral lipids (93.2%). The polar glycolipids and phospholipids totaled 5.9 and 0.9%, respectively. The composition of the nonpolar lipids was clearly dominated by triglycerides (85%) followed by diglycerides. The main glycolipids found were esterified sterol glycoside (38%), digalactosyldiacylglycerol (20.9%), and sterol glycoside (10.2%). Phospholipids were dominated by lysophosphatidylcholine (41.8%), phosphatidylcholine, (36.3%) and lysophosphatidylethanolamine (13.6%). The typical fatty acid composition of sorghum oil is similar to that of maize and is dominated by linoleic and oleic acids (Table 23).

TABLE 23
Fatty Acid Composition of Oil Extracted from Sorghum Grain

Fatty Acid	Source 1[a] (%)	Sources 2 and 3[b] (%)
14:0	0.5	0.4
	0.5–0.5	. . .
16:0	12.9	13.8
	11.6–13.4	13.2–14.3
16:1	0.9	1.2
	0.4–1.5	1.0–1.3
18:0	3.2	2.1
	2.9–4.0	2.0–2.1
18:1	37.7	30.8
	33.2–40.9	30.5–31.0
18:2	38.4	49.4
	33.2–43.7	49.0–49.7
18:3	2.9	2.4
	1.8–4.0	2.0–2.7

[a] Data from Neucere and Sumrell (1980).
[b] Data from Rooney (1978) and Wall and Blessin (1970).

Sorghum grain contains about 1.5 ppm of total carotenoids. Some yellow endosperm types contain up to 8 to 9 ppm (Blessin et al, 1958). Barnes (1983) and Chung (1991) report that regular and yellow sorghums contain 1.5 and 25–30 mg of carotenes per kilogram, respectively. The major identified carotenoids in a yellow endosperm sorghum grain containing a total of 7.7 mg/kg were: zeaxanthin (36.3%), lutein (28.6%), xanthophylls (24.7%), and β-carotenes (10.4%). Weathered sorghum lost about 50% of the carotenoids compared with protected counterpart grains. The presence of red pigments in the pericarp did not inhibit loss of carotenoids (Blessin et al, 1962).

Heupel et al (1986) determined that the major sterols in sorghum were β-sitosterol followed by stigmasterol and dihidrobrassicasterol. Barnes (1983) and Chung (1991) reported that sorghum contains 0.08 and 1.15 mg/100 g α- and τ-tocol derivatives.

MILLETS

The free and bound lipid contents of pearl millet ranged from 5.55 to 6.08 and 0.57 to 0.90%, respectively (Lai and Varriano-Marston, 1980b). Pruthi and Bhatia (1970), studying two strains of pearl millet, found that the free and bound lipid contents were 5.0 and 0.5%, respectively. Triglycerides constituted from 83.5 to 86.9% of the total nonpolar lipid fraction. Lecithin or phosphatidylcholine was the major component of the polar fraction. The pearl millet contained other phospholipids, cerebrosides, and sterol-containing glycolipids. Osagie and Kates (1984) reported that the lipids of pearl millet accounted for 7.2% and consisted of 85% neutral lipids, 12% phospholipids, and 3% glycolipids. Neutral lipids contained approximately 85% triglycerides and small amounts of mono- and diglycerides, sterols, and free fatty acids. Sterols consisted of camesterol, stigmasterol, lysophosphatidylcholine (42%), phosphatidylcholine (24%), and lysophosphatidylethanolamine (21%) were the major phospholipids. The major glycolipids were esterified sterol glycoside, sterol glucoside and mono and digalactosyldiacylglycerol.

Pearl millet is the cereal grain that most rapidly develops off-odors and flavors after milling. The reasons for the rapid deterioration are 1) high lipid content (Kaced et al, 1984), 2) higher amounts of unsaturated fatty acids than other cereals, 3) insufficient naturally occurring antioxidants, and 4) high enzymatic-hydrolytic activity. Hexanal is a major product of oxidative degradation of lipids. Whole pearl millet flour becomes rancid after six to 10 days' storage and inedible after 11–14 days. Moisture, sugars, free fatty acids, fat acidity, and peroxide levels increased throughout storage (Chaudhany and Kapoor, 1984; Lai and Varriano-Marston, 1980a). Reddy et al (1986) indicated that the mousy, acidic odor generation in ground pearl millet is not due to oxidative rancidity. The generation of off-odors requires high moisture content in the flour or meal, suggesting enzymatic deterioration. The off-odor precursor was methanol soluble and had characteristics similar to apigenin, the aglycone of the major C-glycosylflavone present in pearl millet. Sietz et al (1993) found that the "mousy" odor from raw pearl millet was 2-acetyl-1-pyrroline. Parboiling of pearl millet before milling enhanced the shelf life of milled products.

The carotene content of six samples of pearl millet varied from 0.9 to 2.0 ppm. In contrast, a yellow corn hybrid and a yellow endosperm sorghum contained 4.0 and 0.6 ppm, respectively (Curtis et al, 1966).

TABLE 24
Fatty Acid Composition of Oil Extracted from Millets

	Fatty Acid[a] (% of total fatty acids)					
Millet	**16:0**	**16:1**	**18:0**	**18:1**	**18:2**	**18:3**
Pearl[b]	20.5	0.6	3.9	25.6	45.1	3.5
	16.7–23.7	0.4–0.8	1.8–8.0	20.2–30.6	40.3–51.7	2.6–6.8
Pearl[c]	20.2	0.7	5.1	26.7	42.6	3.3
	17.9–22.2	0.3–1.1	2.0–7.9	23.4–29.1	36.7–54.4	1.9–4.8
Finger[d]	24.7	...	...	49.8	24.2	1.3
Proso[e]	8.3	0.2	2.0	20.8	65.8	1.7
	7.2–10.0	0.1–0.3	1.5–3.4	18.1–24.2	62.2–69.5	1.3–2.2
Proso	11.3	...	...	21.4	64.9	2.5
Foxtail[f]	8.6	0.2	3.3	14.4	69.0	2.7
	6.9–9.9	0.1–0.3	1.1–77.2	11.1–16.4	64.9–72.2	2.1–4.1
Kodo	19.2	...	1.0	38.2	42.3	...
	18.4–19.9	...	...	37.7–38.7	41.5–43.0	...

[a] 16:0 palmitic, 16:1 palmitoleic, 18:0 oleic, 18:2 linoleic, 18:3 linolenic.
[b] From Jellum and Powell (1971). Average of 65 lines grown at Tifton, Georgia.
[c] From Chadhary and Kapoor (1984), Lai and Varriano-Marston (1980), and Opoku et al (1983). Average of eight varieties.
[d] From Kahadevappa and Raina (1978).
[e] From Lorenz and Hwang (1986). Average composition of five different flours (extraction rates 79.2–87.5%).
[f] From Taira (1984).
[g] From Paramahans and Tharanathan (1980).

Pearl millet contains 0.13, 5.5, 0.5, 0.1, and 0.2 mg/100 g of α-, τ-, δ-, and δ-3-tocol derivatives, respectively. In general, the amounts reported are comparable to those in other cereal grains (Barnes, 1983; Chung, 1991).

Pore and Magar (1979) determined that of the total finger millet lipids (1.7–2.1%), 19.0 and 1.5% are unsaponifiables and sterols, respectively.

Lorenz and Hwang (1986) determined the lipid composition of flours and brans of nine proso millets. Free lipids in flours ranged from 3.2 to 4.1% and in bran from 3.5 to 6.8%; bound lipids ranged from 0.5 to 0.9% and from 0.3 to 0.7%, respectively. As in other cereals, the most predominant fatty acids were linoleic and oleic. Major components of the free lipids were triglycerides. The bound lipid fraction contained important components such as lysophosphatidylcholine and digalactosyl diglycerides.

The major fatty acids of the oil extracted from millets were linoleic (18:2) and oleic (18:1) acids (Table 24).

The major tocol derivative in foxtail and finger millets is τ-tocol (2.78 and 1.76 mg/100 g, respectively) followed by α-tocol (0.19 and 0.32 mg/100 g) (Barnes, 1983; Chung, 1991).

Minerals

Sorghum and millets are important sources of minerals. The pericarp, aleurone layer, and germ are rich sources of ash (Tables 2 and 4); therefore, refined sorghum and millet products lose part of these important nutrients. All cereals except finger millet and teff are poor sources of dietary calcium. Phosphorus is the mineral found in greatest amounts (Tables 25 and 26); its availability is negatively related to the amount bound by phytates. Malting and fermentation significantly increase

TABLE 25
Mineral and Vitamin Composition of Sorghum[a]

Minerals		Vitamins	
Ca, %	0.04	Thiamin, mg/g	0.46
P, %	0.35	Riboflavin, mg/g	0.15
Phytic acid, %	0.77	Niacin, mg/g	4.84
K, %	0.38	Pyridoxine, mg/g	0.59
Na, %	0.05	Pantothenic acid, mg/g	1.25
Mg, %	0.19	Biotin, mg/g	0.02
Fe, ppm	50.0	Folacin, mg/g	0.02
Co, ppm	3.1	Carotenes, mg/kg	29.0
Cu, ppm	10.8	Vitamin E, mg/kg	12.0
Mn, ppm	16.3		
Zn, ppm	15.4		

[a] Data from Adrian and Sayerse (1956), Chavan and Kadam (1989a,b), Chung (1991), Gazzaz et al (1989), Hulse et al (1980), Rooney et al (1982), Rooney and Serna-Saldivar (1991), Tanner et al (1947), Serna-Saldivar et al (1991b, 1993).

TABLE 26
Mineral and Vitamin Composition of Millets[a]

	Millet						
Nutrient	Pearl	Finger	Proso	Foxtail	Kodo	Teff	Fonio
Minerals							
Ca, %	0.01	0.33	0.02	0.01	0.01	0.17	0.03
P, %	0.35	0.24	0.23	0.31	0.32	0.45	0.18
K, %	0.44	0.43	0.32	0.27	0.17	0.31	0.16
Na, %	0.01	0.02	0.01	0.01	0.01	0.02	0.02
Mg, %	0.13	0.11	0.14	0.13	0.13	0.18	0.40
Fe, ppm	74.9	46.0	52.0	32.6	7.0	14.9	36.0
Co, ppm	0.50	0.10	...	...	...	0.06	3.30
Cu, ppm	6.2	0.3	8.3	9.2	...	4.4	15.0
Mn, ppm	18.0	7.5	18.1	21.9	...	2.5	30.0
Zn, ppm	29.5	15.0	17.2	21.4	...	6.7	30.0
Vitamins					0.32	0.45	0.30
Thiamin, mg/g	0.38	0.48	0.63	0.48	0.05	0.10	0.10
Riboflavin, mg/g	0.22	0.12	0.22	0.12	0.70	2.000	3.00
Niacin, mg/g	2.70	1.30	1.82	3.70	...	...	...
Pyridoxine, mg/g	...	...	...	...	...	...	...
Pantothenic acid, mg/g	1.09	...	1.10	0.82	...	...	...
Biotin, mg/g	...	...	...	...	...	...	...
Folacin, mg/g	...	...	...	0.02	...	...	...
Carotenes, mg/kg	5.4	...	...	...	...	...	...
Vitamin E, mg/kg	19.0	22.0	...	31.0			

[a] From Adrian and Sayerse (1956), Chavan et al (1989 a,b), Chung (1991), Hulse et al (1980), Lorenz et al (1976), Pore and Magar (1979), Serna-Saldivar et al (1991a, 1993), Simwemba et al (1984), Varriano-Marston and Hoseney (1980).

phosphorus availability due to increased phytase activity. These processes considerably improve bioavailability of other minerals as well. Sorghum and millets are considered a good source of potassium and are practically devoid of sodium (Tables 25 and 26). Whole grains are considered an adequate source of magnesium, iron, zinc, and copper that are reduced by degermination and decortication.

Vitamins

Sorghum and millets are an important source of B vitamins (Tables 25 and 26), except for vitamin B-12 (Gazzaz et al, 1989). Dried, matured kernels do not contain vitamin C. The B vitamins are concentrated in the aleurone layer and germ. Removal of the these tissues by decortication reduces amounts. Niacin in cereals is found in free and bound forms and can be synthesized from tryptophan. The alkali treatment to produce tortillas or *tô* improves bioavailability because the glycosidic bond that renders niacin unavailable is alkali labile. Malting and fermentation increase the amount of B vitamins and their availability.

Liposoluble vitamins are mainly located in the germ. Cereals are good sources of tocopherols (vitamin E); only kernels with yellow endosperm contain some provitamin A activity.

Enzymes

SORGHUM

Unlike barley, the sorghum aleurone layer does not produce endosperm-degrading enzymes such as α-amylase, proteases, pentosanases, and endo-β-(1,3)(1,4)-glucanases. Sorghum α-amylase is secreted from the scutellum. Sorghum contains moderate to very limited β-amylase activity (Dyer and Novellie, 1966). α-Amylase activity in sorghum starts 36 hr after germination (Aisien and Palmer, 1983; Aisien et al, 1983). A rapid increase in dextrinase activity is observed 24 hr after germination. Limit dextrinases and proteases are found mainly in the endosperm, whereas carboxypeptidases are located primarily in the germ. Sorghum also has pentosanase and β-D-glucanase activities (Palmer, 1989).

MILLETS

According to Jain and Date (1975) and Sheroain and Wagle (1973) the α-amylase activity of pearl millet is higher than that of barley.

The amylase activity of finger millet malt is approximately twice that of ungerminated grain (Chandrasekhara and Swaminathan, 1953). Optimum pH for β-amylase was between 4.6 and 5.0.

Parvathy and Sadasivam (1982) determined that the α-amylase activity of proso, foxtail, and Japanese millets was very low at initial stages of germination. Maximum α- and β-amylase activities were reached after five to seven days' germination. Foxtail millet malt showed lower levels of α-amylase activity than other millets.

The analysis of eight cultivars of proso millet indicated that all had β-amylase, protease, and cellulase activities. Optimum pH values for β-amylase and protease were around 5.0 and between 3 and 5, respectively (Skovron and Lorenz, 1979).

EFFECT OF PROCESSING ON CHEMICAL COMPOSITION AND NUTRITIONAL VALUE

Sorghum and millets are processed into a wide array of foods. For production of most foods, the grains are decorticated using mortars and pestles or mechanical dehullers. Germination and/or fermentation enhance the nutritional value of sorghum and millets by causing significant changes in chemical composition and elimination of antinutritional factors (Chavan and Kadam, 1989a,b).

TABLE 27
Effect of Decortication on the Chemical Composition of Sorghum

	Regular[a] (90% yield)		Brown[b] (63% yield)		High Lysine[c] (82.5% yield)		Dabar[d] (79% yield)	
Component	Whole	Dehulled	Whole	Dehulled	Whole	Dehulled	Whole	Dehulled
Protein, %	10.6	10.0	9.4	8.2	14.6	14.5	11.6	11.3
Fat, %	3.0	2.5	...	...	4.1	3.9	4.0	3.3
Crude fiber, %	...	...	...	...	...	...	2.0	1.0
Dietary fiber (%)								
Insoluble	15.1	10.3	...	...	6.0	4.4	...	...
Soluble	0.9	1.1	...	...	1.5	1.1	...	...
Starch, %	77.9	81.9	...	...	...	...	73.4	79.9
Ash, %	1.4	1.1	...	...	2.1	1.4	1.7	1.4
Amino acids, g/100 g protein								
His	2.1	2.1	2.2	1.3	2.1	2.1	...	...
Ile	3.8	3.5	3.6	3.3	4.1	4.1	...	...
Leu	12.6	12.2	12.0	11.8	12.3	12.6	...	...
Lys	2.0	1.8	2.2	1.3	2.9	2.8	2.1	1.9
Met + Cys	...	...	4.2	2.8	3.7	3.2	3.6	3.5
Phe + Tyr	8.6	8.6	8.8	7.6	9.0	9.0	...	...
Thr	2.8	2.8	3.1	2.4	3.4	3.3	3.1	3.1
Val	4.7	4.7	4.6	3.8	5.2	5.2	...	...
Digestibility, %								
Protein	81.7	87.0	...	...	88.9	90.3	95.4	100.0
Energy	90.4	94.7	...	...	...	...	...	...
Tannin content, catechin/g	...	...	4.5	0.2	...	...	...	...

[a] From Serna-Saldivar et al (1987a,b).
[b] From Chibber et al (1978).
[c] From Serna-Saldivar et al (1994).
[d] From Eggum et al (1993). The value for starch included soluble sugars. True protein digestibilities are reported.

TABLE 28
Effect of Decortication on the Chemical Composition of Millets

Component	Pearl Millet[a] (82.5% yield)		Pearl Millet[b] (76% yield)		Hamsa Finger[c] (undetermined)		Proso[d] (78.5% yield)	
	Whole	Dehulled	Whole	Dehulled	Whole	Dehulled	Whole	Dehulled
Protein, %	18.0	17.4	15.4	13.9	8.2	5.3	14.7	15.7
Fat, %	6.8	6.1	5.2	3.9	...	...	3.6	2.1
Crude fiber, %	...	...	...	...	...	...	4.9	1.0
Dietary fiber (%)								
Insoluble	6.3	3.5	...	...	...	...	...	...
Soluble	0.6	2.1	...	...	...	...	...	...
Starch, %	...	...	70.8	78.7	...	...	...	...
Ash, %	1.8	1.4	1.7	1.2	...	...	...	...
Amino acids, g/100 g protein								
His	2.3	2.2	2.4	2.3	2.7	1.3	1.9	1.8
Ile	4.6	4.7	5.2	5.5	4.5	5.2	2.6	2.6
Leu	11.0	11.5	11.9	13.1	14.1	14.4	11.9	12.6
Lys	2.9	2.5	3.1	2.4	3.8	3.2	1.6	0.8
Met + Cys	4.2	4.3	3.8	3.9	3.2	3.2	2.3	2.3
Phe + Tyr	9.0	8.9	8.9	9.2	8.9	9.0	7.7	7.7
Thr	3.9	3.7	4.1	4.1	5.0	4.8	2.9	2.5
Trp	1.5	1.6	1.7	2.0	...	...	...	...
Val	6.1	5.9	6.4	6.6	7.9	8.8	3.6	3.3
Digestibility, %								
Protein	89.0	91.7	71.7	77.2	85.1	91.0	...	...

[a] From Serna-Saldivar et al (1993).
[b] From Almeida-Dominguez (1990) and Almeida-Dominguez etal (1993). Protein digestibility was determined in vitro.
[c] From Ramachandra et al (1978) and Virupaksha et al (1975). A high-tannin finger millet lost 88% of tannins upon decortication.
[d] From Lorenz and Dilsaver (1980). Proso millet flour was obtained by roller milling. Values for whole seeds were calculated from the composition of flour and respective bran.

TABLE 29
Effect of Decortication on the Chemical Composition of Minor Millets[a]

Component	Kodo		Fonio	
	Whole	Decorticated	Whole	Decorticated
Protein, %	12.0	13.1	8.7	8.3
Fat, %	4.7	1.5	3.5	2.0
Crude fiber, %	11.3	0.4	8.5	1.8
Ash, %	5.6	4.0	3.8	1.8

[a] Adapted from Hulse et al (1980).

Decortication

Decortication by mortar and pestle or with abrasive dehullers removes the grains' outer layers; therefore, decorticated or "pearled" kernels have reduced levels of fiber, ash, and fat (Tables 27–29). Sorghum and millets are usually decorticated to remove from 12 to 30% of the grain. Increased decortication causes increased losses of fiber, ash, and fat. Decortication slightly reduces protein and lysine content due to partial degermination (Tables 27 and 28). Amino acid analysis of decorticated grains showed a progressive decrease in the content of lysine, histidine, and arginine. Eggum et al (1982) reported that the removal of 20 to 25% of the sorghum weight by traditional manual decortication decreased lysine levels by 40%. Serna-Saldivar et al (1987a, 1988b) reported losses of 14% lysine in sorghum decorticated to remove 10% of its initial grain weight.

Nutrient digestibilities of decorticated sorghums were slightly higher than those of the parent whole grains. However, nitrogen retention and protein efficiency ratios were lower due to removal of germ (Serna-Saldivar et al, 1988a,b).

Decortication of brown sorghums significantly reduces the amount of condensed tannins and presumably can overcome their deleterious effect. Decortication to remove 37% of the original grain weight reduced tannins from 4.5 to 0.2 catechin equivalents (Chibber et al, 1978; Nwasaru et al, 1988). Stepwise removal of tannins gradually increased the percentage of protein solubilized by pepsin and by trypsin and chymotrypsin (Chibber et al, 1980).

Cooking and Parboiling

In general, cooking of sorghum greatly reduces the amount of protein solubilized by the pepsin assay (Hamaker et al, 1987; Kirleis, 1990) due to insolubilization of kafirins and greater cross-bonding with disulfide bonds.

Sorghum is cooked with alkali (i.e., ashes and lime) for the production of foods such as *tô* and tortillas. These cooking conditions have only a slight detrimental effect on protein availability. Lime-cooking for sorghum tortilla production slightly decreases in vivo protein and lysine digestibilities as estimated with ileum-cannulated swine and with rats (Serna-Saldivar et al, 1987a, 1988b). During tortilla chip preparation from maize and sorghum, there is a progressive decrease in extractable albumins-globulins and prolamins, with a consequent increase in the amount of residual protein (Vivas et al, 1987; Vivas-Rodriguez et al, 1990). Pepsin digestibility of tortilla chips was less than one half of the raw grain. However, in vivo protein digestibilities of tortillas were slightly lower than corresponding raw grains (Serna-Saldivar et al, 1987b, 1988b).

Bach Knudsen et al (1988) also found reduced protein digestibilities when decorticated or whole sorghum flour was processed into acid, neutral, or alkaline *tô*. As a consequence of cooking *tô*, the digestible energy was slightly reduced (3.4–4.3% absolute units) due to formation of resistant starch, and slightly reduced true protein digestibility. Lysine true digestibility was slightly lower in alkaline *tô* than that observed for neutral or acidic *tô*. This phenomenon was also observed when sorghum was lime-cooked into tortillas (Serna-Saldivar et al, 1987a).

The lime process for tortilla production considerably upgrades its calcium (Ca) content. Serna-Saldivar et al (1991b, 1992) determined that sorghum tortillas contain about 150 mg of Ca/100 g, whereas the raw grain only contains 18 mg of Ca/100 g. The tortilla calcium was highly bioavailable to growing rats. Rats fed tortillas absorbed and retained more Ca and had heavier, stronger, and denser femurs than those of counterparts fed raw grains. Lime-cooking also increases the bioavailability of niacin in maize (Koetz and Neukom, 1977); however, no studies have been conducted on sorghum or millets.

Parboiling of sorghum and pearl millet is aimed towards the more efficient removal of pericarp in grains that are subjected to decortication (Young et al, 1990). The process firms the endosperm structure and increases efficiency of decortication, especially in soft textured grains (i.e., brown sorghums). Serna-Saldivar et al (1994) determined that parboiled-decorticated sorghum and pearl millet have slight differences in chemical composition but essentially the same nutritional value as raw grains decorticated to the same extent. The outer layers of parboiled kernels were decorticated more efficiently, as estimated by losses of fat, ash, and insoluble dietary fiber. Protein digestibilities of decorticated, parboiled, and raw sorghum and pearl millet were around 89%. Protein efficiency ratios for pearl millet and for regular and P721 (high-lysine) parboiled, decorticated sorghums were 1.3, 1.1, and 1.7, respectively.

The cooking or treatment of high-tannin sorghum grain with CaO, K_2CO_3, NH_4OH, $NaHCO_3$, wood ashes, and Magadi soda reduces assayable tannin content and improves the nutritional value of the grain to a point that it is equivalent to that of sorghum without tannins (Blessin et al, 1971; Butler, 1982, 1990; Mukuru et al, 1988; Price et al, 1979).

Malting

Malting is a common practice in sorghum and millet producing areas (Palmer, 1989). The effect of sprouting on the chemical composition of sorghum and millets was recently reviewed by Chavan and Kadam (1989b). Grains are malted for production of opaque beers, weaning foods, and other traditional dishes. Malting causes 8–30% dry matter loss (Chavan and Kadam, 1989b; Jayatissa et al, 1980); decreased levels of prolamins, fat, tannins, and starch; and increased levels of free amino acids, albumins, lysine, reducing sugars, and most vitamins including synthesis of vitamins B-12 and C (Table 30; Almeida-Dominguez, 1990; Almeida-Dominguez et al, 1993; Chavan et al, 1981; Malleshi and Desikachar, 1986; Okoh et al, 1989; Opoku et al, 1981, 1983; Osuntogun et al, 1989; Taylor, 1983; Wang and Fields, 1978; Wu and Wall, 1980). The activation of intrinsic amylases, proteases, phytases, and fiber-degrading enzymes disrupts protein bodies (Taylor et al, 1985) and protein matrix and breaks down polymers into simpler compounds, thereby increasing nutrient accessibility (Aisien et al, 1983, 1986;

TABLE 30
Effect of Germination on Chemical Composition of Sorghum and Millets

	Regular Sorghum[a]		Regular Sorghum[b]		Regular Sorghum[c]		High-Lysine Sorghum[c]		Pearl Millet[d]	
Component	**Grain**	**Malt**	**Grain**	**Malt**	**Grain**	**Malt**	**Grain**	**Malt**	**Grain**	**Malt**
Protein, %	10.5	10.5	10.8	13.1	10.1	11.9	15.9	21.3	8.6	11.8
Fat, %	3.4	2.8	...	...	3.5	3.3	4.9	7.0	7.5	2.5
Crude fiber, %	...	...	...	...	...	...	...	...	10.4	18.6
Starch, %	79.1	71.9	...	...	75.0	54.0	...	...	...	...
Dry matter loss, %	...	...	...	18.2	...	...	...	...	...	...
Ash, %	1.6	1.5	...	...	1.7	1.8	1.8	2.3	4.1	3.2
Protein fractions,										
% of total fraction	...	...	...	...	15.0	43.0	18.0	48.0	...	...
Albumin	...	...	...	...	1.0	3.0	3.0	4.0	...	...
Globulin	...	...	...	...	17.0	10.0	19.0	5.0	...	...
Prolamins	...	...	...	...	31.0	6.0	22.0	2.0	...	...
Cross-linked prolamins	...	...	...	...	25.0	19.0	23.0	23.0	...	...
Glutelins										
Amino acids,	2.4	2.5	3.3	3.9	...	...	...	...	...	...
g/100 g protein	4.9	5.1	3.8	3.3	3.9	4.2	4.2	4.4	...	...
His	15.9	15.6	11.9	7.4	13.2	8.8	13.7	8.2	...	...
Ile	2.5	3.0	2.7	3.6	2.2	3.2	3.0	6.7	...	...
Leu	3.4	3.4	4.5	4.1	2.6	2.5	2.6	2.8	...	...
Lys	10.9	10.9	8.7	7.0	10.6	7.3	9.8	7.6	...	...
Met + Cys	3.6	3.7	3.1	3.2	3.4	3.7	3.4	3.9	...	...
Phe + Tyr	0.9	1.1	...	...	...	...	...	...	...	...
Thr	6.0	6.3	5.6	5.6	5.4	6.1	4.9	5.3	...	...
Trp										
Val	72.6	78.7	...	...		...	...	...	51.0	77.2
Digestibility, %										
Protein										

[a] Data from Almeida-Dominguez (1990) and Almeida-Dominguez et al (1993). Protein digestibility was determined in vitro.
[b] Data from Okoh et al (1989). Protein and dry matter loss values were extrapolated from figures.
[c] Data from Wu and Wall (1980). Normal and high-lysine sorghum were germinated for 10 and nine days, respectively.
[d] Data from Opoku et al (1981). The carbohydrate contents of grain and malt were 53.9 and 48.5%, respectively. In vitro protein digestibility values are from Khetarpaul and Chauhan (1990).

Dyer and Novellie, 1966; Watson and Novellie, 1975). According to Taylor (1983), Okoh et al (1989), and Wu and Wall (1980), the kafirins and glutelin fractions decrease during malting, while soluble nitrogen and albumins-globulins increase.

The lysine content of germinated normal and high-lysine sorghums increased from 2.2 to 3.2 and 3.0 to 7.8 g/100 g of protein, respectively (Table 30). This is due to a large increase in albumins and a large decrease in kafirins and cross-linked kafirins (Wu and Wall, 1980). Germinated sorghum grain has better nutritional value due to the chemical changes mentioned above. Wang and Fields (1978) determined that sorghum malt had higher available lysine, tryptophan, and methionine, relative nutritional value, and protein efficiency ratio (PER) compared with ungerminated grain.

Rate of in vitro starch digestibility of pearl millet increased by more than threefold when kernels were germinated for 24 hr (Dhanker and Chauhan, 1987a,b; Khetarpaul and Chauhan, 1990). The PER values of pearl, finger, and foxtail millets were slightly improved after germination due to improved protein quality (Malleshi and Desikachar, 1986). Germination for 48 hr significantly reduced phytate and phytate phosphorus contents of pearl, finger, and foxtail millets (Malleshi and Desikachar, 1986) due to activation of phytases (Khetarpaul, 1988).

Hemanalini et al (1980) determined that sprouted finger millet had better calcium and B vitamin availability with a PER equal to the whole grain.

Fermentation

Fermented cereal products are widely consumed in India and many countries of central and southern Africa. Fermentation usually involves malting and souring by mixed cultures of yeast and lactobacilli. Fermentation causes degradation of grain components, especially starch and soluble sugars, by both intrinsic grain enzymes and enzymes of the fermenting media (Chavan and Kadam, 1989a,b). Fermented cereals have enhanced nutrient digestibilities. Natural lactic acid fermentation of sorghum meal for five days slightly increased protein content due to the degradation of carbohydrates (Chavan and Kadam, 1989a; Chavan et al, 1988; Kazanas and Fields, 1981). Protein quality is enhanced by improved digestibility and better essential amino acid composition resulting from the de novo production of some important amino acids. Chavan et al (1988) observed a significant increase in water-soluble proteins and free amino acids in sorghum subjected to 24 hr of fermentation. Hamad and Fields (1979) determined that pearl millet subjected to fermentation at 37°C had increased levels of available lysine (3 vs. 45 mg/g of nitrogen). Fermented products are also richer in vitamins synthesized by fermenting microorganisms; minerals are usually more bioavailable. Gazzaz et al (1989) observed synthesis of vitamin B-12 during fermentation of sorghum.

Graham et al (1986) determined in malnourished children that *nasha* (a fermented weaning food from Sudan) had more digestible energy and protein than unfermented whole grain. Eggum et al (1983) compared the nutritional value of *kisra, aceda, ugali,* and *marissa* with the original raw and decorticated grains. Cooking and fermentation for production of these foods slightly decreased digestible energies, true protein digestibilities, and biological values in low-tannin grains. Fermentation of a sorghum with testa (feterita) for production of *ugali* and *kisra* slightly improved biological and net protein utilization values. Axtell et al (1981) determined that *kisra* and *abrey* were more digestible than the unfermented grain. Aliya and Geervani (1981) concluded that rats fed fermented sorghum had better protein digestibilities, biological and net protein utilization values, and PER values than counterparts fed the unfermented grain.

The true protein digestibility values of pearl and finger millets did not change upon fermentation. However, biological and net protein utilization values of finger millet significantly increased when fermented, whereas pearl millet values remained unchanged. Thiamin and riboflavin content of pearl and finger millet increased as fermentation time increased.

Dhankher and Chauhan (1987b) and Khetarpaul and Chauhan (1990) observed improved in vitro protein and starch digestibilities of pearl millet when subjected to germination and even better when further fermented. Dhankher and Chauhan (1987a) determined that fermentation of pearl millet for 9 hr for production of Indian *rabadi* decreased levels of phytic acid (27–30%) and polyphenols (10–12%) significantly.

El-Tinay et al (1979) observed a slight increase in thiamin, a greater increase in niacin, and no appreciable change in riboflavin in sorghum fermented for production of *kisra*.

ANTINUTRITIONAL FACTORS

Tannins

SORGHUM

Sorghum is unique among major cereals because some cultivars produce polymeric polyphenols known as tannins (Butler, 1990). All sorghums contain phenols, and most contain flavonoids; however, only cultivars with a pigmented testa (B_1— B_2— genes) produce condensed tannins. Most cultivated sorghums do not contain any condensed tannins even though the literature often reports phenols as tannins.

Phenolic compounds have been divided into three major categories: phenolic acids, flavonoids, and tannins. Phenolic acids are derivatives of benzoic or cinnamic acid. Flavonoids consist of two units: a C6–C3 fragment from cinnamic acid and a C6 fragment from malonyl-coenzyme A. The major groups of flavonoids in sorghum are the flavans. Flavan 3-en 3-ols with double bond between C3 and C4, hydroxylated at the C3 are anthocyanidins. Tannins are polymers of five to seven flavan-3-ol units (catechin) linked through acid-labile carbon-carbon bonds (Butler, 1990; Hahn et al, 1984; Mehansho et al, 1987a,b). Sorghum tannins apparently occur only in the pericarp and testa layers (Hahn and Rooney, 1986) and in some glumes (Doherty et al, 1987).

Tannins protect the grain against insects, birds, and weathering (Waniska et al, 1989). Rate of preharvest germination or early sprouting is significantly lower in most high-tannin cultivars. These beneficial effects ensure that brown sorghums will continue to be produced in certain pest-ridden areas of the world (Butler, 1990). The agronomic advantages are accompanied by nutritional disadvantages and reduced food qualities. The antinutritional effects of high-tannin sorghums have been reviewed by Butler (1989a,b, 1990) and Salunkhe et al (1982, 1990).

The color of sorghum grain and flour plays an important role in its acceptance. In general, white sorghums produce the most acceptable food products. However, in some countries, brown sorghum products are preferred, e.g., for opaque sorghum

TABLE 31
Assays Commonly Used to Determine Sorghum Phenols and Tannins[a]

Method	Standard	Reagents	Extraction	Time	Assayed Compounds
Vanillin	Catechin	4% HCl, 1% vanillin in methanol	Methanol, 1% HCl in methanol	24 hr	Leucoanthocyanidins
Prussian blue	Catechin	$FeCl_3$ in HCl	Methanol	1 min	Total phenols
Folin-Denis	Tannic acid	Folin-Denis reagent	Water	5 hr	All reducing compounds
Protein precipitation	Tannic acid	$FeCl_3$, alkaline detergent	Methanol	20 min	Proanthocyanidins (tannins)
Folin-Ciocalteu	Catechin or gallic acid	Folin-Ciocalteu reagent	Methanol; 1% HCl in methanol, dimethylformaide	1 hr 2 hr	Total phenols
Relative degree of polymerization	. . .	4% HCl, 0.5% vanillin in acetic acid, 4% HCl in acetic acid	Methanol	20 min	Leucoanthocyanidins, anthocyanidins, proanthocyanidins

[a] Source: Hahn et al (1984).

TABLE 32
Polyphenol Levels of Three Types of Sorghums[a]

		Vanillin-HCl		Modified Vanillin-HCl			
Sorghum Type[b]	Prussian Blue	With Blank	Without Blank	With Blank	Without Blank	24 hr	α-Amylase Inhibition (%)
Type I	0.07	0.03	0.12	0.02	0.37	0.55	3.7
	0.04–0.09	0.01–0.04	0.05–0.24	0.01–0.03	0.20–0.58	0.33–0.96	1.8–15.4
Type II	0.07	0.03	0.19	0.07	0.52	0.98	6.0
	0.05–0.09	0.02–0.04	0.09–0.52	0.02–0.19	0.34–0.88	0.68–0.98	2.3–15.8
Type III	0.39	2.39	2.68	1.10	1.75	2.14	31.7
	0.24–0.99	0.88–6.67	1.12–7.45	0.36–3.48	1.00–4.85	1.49–4.85	13.9–77.0

[a] From Earp et al (1981). Data are means and ranges of catechin (mg) per 100 g of sample.
[b] Type I sorghums do not have pigmented testa. Type II are sorghums with pigmented testa without a dominant spreader gene. Type III or brown sorghums contain pigmented testa and dominant spreader gene.

beer (see chapter 10). Food color is the result of factors such as grain color and type (pericarp color, pigmented testa, endosperm color), degree of milling, and pH of the food system. Pigmentation in the pericarp and testa is primarily due to phenolic compounds. The color intensity greatly depends on pH. Anthocyanins are very unstable in acid medium and are readily converted to the corresponding anthocyanidin under slightly acid conditions (Hahn, 1984). Pericarp color of sorghum appears to be due to a combination of anthocyanin and anthocyanidin pigments as well as other flavonoid compounds (Hahn, 1984). Kambal and Bate-Smith (1976) found no flavonoids in the pericarp of a white sorghum, whereas Nip and Burns (1971) found four major anthocyanins in the pericarp of six white sorghums. Luteforol was identified as the major pigment in red pericarp sorghums (Nip and Burns, 1969). Cooking in alkali (i.e., alkaline *tô*, tortillas) usually promotes off-colors, especially in colored grains. The glume or plant color also affects the color of the grain and food products. Grains with straw-colored glumes produce better products than similar grains with colored glumes. This is especially true when fabricating lime-cooked products such as tortillas and chips.

Several techniques have been used to characterize phenolic compounds in sorghum grain (Butler, 1982; Hagerman and Butler, 1981; Hahn et al, 1984; Maxson and Rooney, 1972). Most tannin assays measure the level of phenols, which may or may not be condensed tannins (Table 31). In general, these assays usually correlate well with time-consuming measurements of antinutritional effects determined in vivo. The catechin equivalents for different types of sorghums measured by different techniques show that significant levels of tannins are often erroneously reported, even though it is known that condensed tannins are not present in type I sorghums (Table 32). The absolute amount of tannin present in the sorghum kernel is virtually impossible to determine, because a significant proportion cannot be extracted and assayed (Butler, 1990; Hahn et al, 1984; Hahn and Rooney, 1986), and a suitable standard for sorghum tannins is unavailable. The highly polymerized or condensed tannin molecules are the most difficult to extract, which may account for the decreased amounts of extractable tannin as grain matures. Different assays are likely to yield different tannin values because they respond to different chemical parts of the tannin molecule (Hagerman and Butler, 1981).

Hahn et al (1983) separated, via reverse-phase high-performance liquid chromatography, free and bound phenolic acids of sorghum. Eight main phenolic acids with different polarity were identified in extracts (Table 33). A typical brown sorghum contained the highest amount of free phenolic acids. Sorghum cultivars resistant to fungal attack contained both a greater variety and larger amounts of phenolic acids in the free form. Waniska et al (1989) also partitioned sorghum phenolic acids and concluded that white cultivars without pigmented testa contained the lowest amounts of phenolic acids. Brown cultivars were more resistant to weathering and contained higher levels of free phenolic acids.

The major antinutritional effects of tannins found in rats, hamsters, swine, poultry and ruminants are diminished growth rates, protein digestibility, and feed efficiency (Cousins et al, 1981; Jambunathan and Mertz, 1973; Knabe, 1990; Maxson et al, 1973a,b; Mehansho et al, 1987a,b; Muindi and Thomke, 1981; Sell et al, 1983).

According to Hagerman and Butler (1981), tannins bind strongly and specifically to certain proteins and inhibit virtually all enzymes in in vitro assays. Tannin-binding, salivary, proline-rich proteins share some biochemical characteristics with

TABLE 33
Free and Bound Phenolic Acid Composition of Several Sorghum Types[a]

	White Sorghum (CS3541)		Red Sorghum (SC0630)		Brown Sorghum (SC0719)	
Phenolic Acid	**Free (μg/g)**	**Bound (μg/g)**	**Free (μg/g)**	**Bound (μg/g)**	**Free (μg/g)**	**Bound (μg/g)**
Gallic	...	19.7	...	46.0	...	26.1
Protocatechuic	7.4	133.9	13.0	83.0	8.0	15.8
p-Hydroxybenzoic	4.0	11.4	6.7	16.0	9.3	24.2
Vanillic	8.3	...	7.7	19.2	23.3	27.4
Caffeic	3.4	22.2	4.1	48.0	8.7	26.8
p-Coumaric	45.7	138.5	13.5	72.5	6.4	79.9
Ferulic	45.4	297.2	8.9	95.7	26.0	91.9
Cinnamic	9.4	...	10.7	...	...	19.7

[a] Source: Hahn et al (1983). All data are on a dry weight basis.

the tannins associated with sorghum protein (Asquith et al, 1987). Recent information indicates that tannins increase saliva secretion and endogenous protein excretion and inhibit protein metabolism once they have been digested and absorbed. C^{14}-labeled tannin studies indicate that low molecular weight tannins (dimers and others) are absorbed and cause metabolic effects (Mehansho et al, 1987a,b). Phenolic compounds (monomers, polyphenols, and tannins) are considered to interfere with iron absorption by complex formation with iron in the gastrointestinal lumen. Iron absorption was strongly inhibited by tannic and gallic acid followed by chlorogenic acid, whereas no inhibition was observed when catechins or condensed tannins were added to the diet (Brune et al, 1989).

Several different approaches for detoxification of brown sorghums have been proposed. The most common and practical ways are by decortication or malting. Decortication removes the pericarp and testa and, therefore, most tannins. However, decortication is sometimes inefficient, because most brown sorghums have soft-floury endosperm (Chibber et al, 1980; Nwasaru et al, 1988; Reichert et al, 1988). Parboiling of brown sorghum prior to decortication can be a better alternative for tannin removal (Young et al, 1990). Chemical decortication with alkali solutions followed by neutralization with acid diminishes the levels of assayable tannins (Blessin et al, 1971; Butler, 1990; Kock et al, 1985). Malting effectively lowers (up to 43%) the assayable levels of sorghum tannins (Osuntogun et al, 1989).

Detoxification of high-tannin sorghum with alkaline materials is effective. Kernels treated with moisture and wood ashes have reduced extractable tannins (up to 97%) and improved protein digestibility and rat weight gains (Mukuru et al, 1988). Similar effects are achieved when brown sorghums are treated with Magadi soda. Butler (1990) and Price et al (1979) proposed several experimental alkaline treatments to detoxify brown sorghums.

MILLETS

Polyphenol and tannin levels of several millets are presented in Table 34. Millet changes color reversibly from gray to yellow-green at alkaline pH and partially reversibly from gray to creamy white under acidic conditions due to the presence of glucosylvitexin, glucosylorientin, and vitexin (Reichert, 1979). These phenolic

TABLE 34
Phenolic Acid Composition of Some Millets[a]

Phenolic Assay	Pearl	Finger	Foxtail[b]	Proso	Teff	Fonio
Folin-Ciocalteu[c]	0.19–0.33	0.55–0.59	0.12	0.05–0.10	0.09–0.15	0.14
Vanillin-HCL[c]	0.05	0.17–0.32	0.0	0.0	0.0	0.0
Folin-Denis[d]	516–528	. . .	. . .	. . .	. . .	. . .
HPLC phenolic acids, μg/mg of sample						
Protocatechuic	11.8	23.1	. . .	. . .	25.5	. . .
Gentisic	96.3	61.5	. . .	. . .	15.0	. . .
p-OH Benzoic	22.0	8.9	14.6	. . .	. . .	. . .
Vanillic	16.3	15.2	87.1	. . .	54.8	. . .
Caffeic	21.3	16.6	10.6	. . .	3.9	. . .
Syringic	17.3	7.7	93.6	. . .	14.9	. . .
Coumaric	268.2	56.9	2,133.7	. . .	36.9	. . .
Ferulic	679.7	387.0	765.8	. . .	285.9	. . .
Cinnamic	345.3	35.1	781.7	. . .	46.0	. . .

[a] Data from McDonough and Rooney (1985).
[b] With lemma and palea (glumes) attached.
[c] Expressed as milligrams per 100 mg of catechin equivalents, dry weight basis.
[d] Data from Dhankher and Chauhan (1987). Extraction with 1% HCL methanol of defatted samples and estimated as tannic acid equivalents according to the Folin-Denis procedure.

compounds are concentrated in the bran. Kernels bleach from gray to creamy white when grains are soaked in acidic solutions (sour milk, tamarind juice). Decortication of the grains expedites the bleaching process due to partial removal of the phenolic compounds and easier access to the grain (Reichert and Youngs, 1979).

Many of the millets test positive in the vanillin assay for trace amounts of tannins (Lorenz, 1983); however, only finger millet contains condensed tannins. Certain varieties of finger millet contain tannin levels ranging from trace amounts to levels comparable to some brown sorghums (Ramachandra et al, 1978). A strong relationship exists between the tannin levels and the in vitro protein digestibilities among finger millet varieties. Decortication significantly decreases the amount of tannins, with a corresponding increase in protein digestibility.

Phytic Acid

Phytic acid and/or phytates complex with essential dietary minerals such as calcium, zinc, iron, and magnesium to make them biologically unavailable for absorption (Lasztity and Lasztity, 1990). The primary role of phytates may be to store phosphorus and inositol, which are gradually utilized during germination. Germination or malting, therefore, significantly reduces the amount of phytates due to production of phytases (Lasztity and Lasztity, 1990). The phytic acid in monocotyledons is mainly stored in the aleurone layer (phytic bodies or aleurone grains) and to a lesser extent in the germ. Thus, milling and/or decortication greatly reduces the amount of phytates.

SORGHUM

Levels of phytate-phosphorus in thirty varieties of sorghum ranged from 0.17 to 0.38%, accounting for 80 to 87% of the total phosphorus present in the kernel (Doherty et al, 1981, 1982). Sorghum bran contained the highest levels of phytates. Forty to fifty percent of phytate and total phosphorus can be removed by abrasive dehulling (Doherty et al, 1981, 1982). Levels ranging from 0.16 to 4.3% have been reported for whole sorghum (Wang et al, 1959). Levels of phytate phosphorus in sorghum are no higher than those reported for wheat, barley, and corn but are significantly lower than soybeans and other oilseeds.

MILLETS

Hulse et al (1980) reported from 208 to 246 mg of phytic acid/100 g in finger millet, whereas phytate content values for proso millet ranged from 0.17 to 0.47%. Values are higher than in rice but lower than in corn and durum wheat. Dehulling of proso millets caused a 27 to 53% reduction in phytate content. Hulls of proso millet contained two to six times as much phytate as the dehulled grains (Lorenz, 1983).

Cyanogenic Glycosides

Cyanogenic glycosides occur in most sorghum varieties, although the quantity depends upon variety and environment. The main cyanogenic glycoside, dhurrin, when hydrolyzed yields hydrocyanic acid (prussic acid) and *p*-hydroxybenzaldehyde. The occurrence of this toxin is mainly in the leaves of developing sorghum plants. According to Panasiuk and Bills (1984) caryopses of grain and sweet sorghums are low in cyanogenic glycosides (1–29 ppm). However, when sorghums were germinated for three days at 30°C, the resulting seedlings contained 258–1,030 ppm. Hulse et al (1980) reports that sorghum malt might contain as much as 15 μmol of dhurrin/g fresh weight (Hulse et al, 1980). Osuntogun et al (1989) reported that sorghum grains and malt contain 8–38 μg/g and 82–150 μg/g, respectively. Newton et al (1980) determined that the precursors of dhurrin are glucose and fructose and that it is concentrated in the coleoptile-shoot primordium region (90 μg/mg) and the upper growing portion of the mesocotyl (28 μ g/mg) of the germinating seed. The maximum concentration of dhurrin was reached 10 days postgermination; it was 3 to 4% of the total dry seedling weight. Panasiuk and Bills (1984) reported a mean prussic acid value of 61.3 mg/100 g of sprouted kernels and reported that neither drying nor milling reduced the prussic acid content. Saunders and Conn (1978) demonstrated that vacuoles present in the protoplast were the major site for the accumulation of dhurrin.

In traditional African Culture (see chapter 7, on traditional foods), sorghum malt is used to make opaque beer (chapter 10) and other foods (chapter 8). The dried malted grain is pounded and, usually, the shoots and roots are removed. The fact that no one seems ever to have suffered cyanide poisoning by ingestion of opaque beer suggested that the cyanogenic glycosides are destroyed during the drying or brewing processes. Dada and Dendy (1988) therefore measured the residual cyanide in opaque beers and weaning foods made from malted sorghum by a variety of traditional processes and showed that during these processes the potential cyanide level was lowered to zero or to well below that considered toxic.

Pearl Millet Goitrogens

Epidemiological studies have suggested that pearl millet might be at least partly responsible for the higher goiter incidence in the Sudan (Klopfenstein et al, 1983a,b; Osman, 1981, 1983). The goitrogen in pearl millet is a thioamide and/or other compounds derived from flavonoids such as C-glucosylflavones, vitexin, glucosylvitexin, and glucosylorientin (Birzer and Klopfenstein, 1988). Klopfenstein et al (1983a,b, 1985) reported that the goitrogen is mainly found in the bran (Klopfenstein et al, 1991) and that it produces goiter due to the inhibition of the normal conversion of thyroxine (T4) to triiodothyronine (T3). The T4 level was higher and T3 lower in rats fed pearl millet than counterparts fed sorghum. Feeding pearl millet enlarged thyroid colloid follicles and flattened cells. A semi-wet-milling process successfully removed the bran and antithyroid properties as demonstrated by serum thyroid hormone levels and histopathology (Klopfenstein et al, 1991). Grain fermentation did not reduce the levels or activity of the antinutritional factor. However, autoclaving appears to diminish the negative effect of the goitrogen (Klopfenstein et al, 1983).

Enzyme Inhibitors

SORGHUM

Harish Kumar et al (1978) found proteinase inhibitors in sorghum that specifically inhibited serine proteinases (i.e., trypsin and chymotrypsin). Three inhibitors were isolated. One had a molecular weight of 25 kDa and consisted of a linear peptide chain with isoleucine as the N-terminal group.

MILLETS

Nagaraj and Pattabiraman (1985) isolated and identified an amylase inhibitor of approximately 24,000 kDa from foxtail millet. The inhibitor was made up of two dissimilar polypeptides. The inhibitor acted in human salivary amylase and human and porcine pancreatic amylase. It was labile to heat and extreme acidic and alkaline conditions.

According to Chandrasekher and Pattabiraman (1981), pearl millet contains two trypsin inhibitors with molecular weights around 11,000 Da. The trypsin inhibitors were extracted with 0.1 N HCl, fractionated with ammonium sulfate, and identified with gel chromatography.

A trypsin-chymotrypsin inhibitor was purified from Japanese millet by ammonium sulfate fractionation. The size of the inhibitor was found to be approximately 14,000 Da (Udupa and Pattabiraman, 1985). Heat treatment at 100°C for 1 hr at pH 7 reduced the antitryptic and antichymotryptic activities by 35 and 50%, respectively. The protease inhibitor from Japanese millet resembles the counterparts from sorghum (Harish et al, 1978), pearl millet (Chandrasekher and Pattabiraman, 1981), finger (Shivaraj and Pattabiraman, 1981) and foxtail (Udupa and Pattabiraman, 1985) millets in molecular size, heat stability, and pH stability.

LITERATURE CITED

ABD ALLAH, M. A., MAHMOUD, R. M., EL-KALYOUBI, M. H., and ABOU ARAB, A. A. 1987. Physical properties of starches isolated from pearl millet, yellow corn, sorghum, sordan and pearl millet. Starch/Staerke 39:9-12.

ABDELRAHMAN, A., HOSENEY, R. C., and VARRIANO-MARSTON, E. 1982. A simple technique for hand dissecting of pearl millet kernels. Cereal Foods World 27:452.

ABDELRAHMAN, A., HOSENEY, R. C., and VARRIANO-MARSTON, E. 1984. The proportions and chemical compositions of hand-dissected anatomical parts of pearl millet. J. Cereal Sci. 2:127-133.

ADRIAN, J., and SAYERSE, C. 1956. Composition of Senegal millets and sorghums. Br. J. Nutr. 11:99-105.

AHUJA, V. P., SINGH, J., and NAIK, M. S. 1970. Amino acid balance of proteins of maize and sorghum. Indian J. Genet. Plant Breed. 30:727-731.

AISIEN, A. O., and PALMER, G. H. 1983. The sorghum embryo in relation to hydrolysis of the endosperm during germination and seedling growth. J. Sci. Food Agric. 34:113-121.

AISIEN, A. O., PALMER, G. H., and STARK, J. R. 1983. The development of enzymes during germination and seedling growth in Nigerian sorghum. Starch/Staerke 35:316-320.

AISIEN, A. O., PALMER, G. H., and STARK, J. R. 1986. The ultrastructure of germinating sorghum and millet grains. J. Inst. Brew. 92:162-167.

AKINGBALA, J. O., ROONEY, L. W., PALACIOS, L. G., and SWEAT, V. E. 1982. Thermal properties of sorghum starches. Pages 251-262 in: Proc. Int. Symp. on Sorghum Grain Quality. L. W. Rooney and D. S. Murty, eds. Int. Crops Res. Inst. Semi-Arid Tropics (ICRISAT), Patancheru, A. P., India.

ALIYA, S., and GEERVANI, P. 1981. An assessment of the protein quality and vitamin B content of commonly used fermented products of legumes and millets. J. Sci. Food. Agric. 32:837.

ALMEIDA-DOMINGUEZ, H. D. 1990. Production of weaning foods from composites of pearl millet and cowpea. Ph.D. diss. Texas A&M University, College Station, TX.

ALMEIDA-DOMINGUEZ, H. D., SERNA-SALDIVAR, S. O., GOMEZ-MACHADO, M. H., and ROONEY, L. W. 1993. Production and nutritional value of weaning foods from mixtures of pearl millet and cowpeas. Cereal Chem. 70:14-18.

ANGOLD, R. E. 1979. Cereals and Bakery Products. Pages 75-138 in: Food Microscopy. R. G. Vaughan, ed. Academic Press, London.

ASQUITH, T. N., UHLIG, J., MEHANSHO, H., PUTMAN, L., CARLSON, D. M., and BUTLER, L. 1987. Binding of condensed tannins to salivary proline-rich glycoproteins: The role of carbohydrate. J. Agric. Food Chem. 35:331-334.

AXTELL, J. D., KIRLEIS, A. W., HASSEN, H. M., MANSON, N. O., MERTZ, E. T., and MUNCK, L. 1981. Digestibility of sorghum proteins. Proc. Natl. Acad. Sci. 78:1333-1335.

BACH KNUDSEN, K. F., and MUNCK, L. 1985. Dietary fiber content and composition of sorghum and sorghum based foods. J. Cereal Sci. 3:153-164.

BACH KNUDSEN, K. F., KIRLEIS, A. W., EGGUM, B. O., and MUNCK, L. 1988. Carbohydrate composition and nutritional quality for rats of sorghum t prepared from decorticated white and whole grain red flour. J. Nutr. 118:588-597.

BADI, S. M., HOSENEY, R. C., and CASADY, A. J. 1976a. Pearl millet. I. Characterization by SEM, amino acid analysis, lipid composition and prolamine solubility. Cereal Chem. 53:478-487.

BADI, S. M., HOSENEY, R. C., and FINNEY, P. L. 1976b. Pearl millet. II. Partial characterization of starch and use of millet flour in bread making. Cereal Chem. 53:733.

BAILEY, A. V., SUMRELL, G., and BURTON, G. W. 1979. Pentosans in pearl millet. Cereal Chem. 56:295-298.

BARNES, P. J. 1983. Non-saponifiable lipids in cereals. Pages 35-55 in: Lipids in Cereal Technology, P. J. Barnes, ed. Academic Press, New York.

BECKER, R., and LORENZ, K. 1978. Saccharides in proso and foxtail millets. J. Food Sci. 43:1412-1414.

BELEIA, A., VARRIANO-MARSTON, E., and HOSENEY, R. C. 1980. Characterization of starch from pearl millets. Cereal Chem. 57:300-303.

BHATIA, I. S., SINGH, R., and DUA, S. 1972. Changes in carbohydrates during growth and development of Bajra (*Pennisetum typhoides*), Jowar (*Sorghum vulgare*) and Kangni (*Setaria italica*). J. Sci. Food Agric. 23:429-440.

BIRZER, D. M., and KLOPFENSTEIN, C. F. 1988. The pearl millet goitrogens. Cereal Foods World 33:229-231.

BLAKELY, M. E., ROONEY, L. W., SULLINS, R. D., and MILLER, F. R. 1979. Microscopy of the pericarp and the testa of different genotypes of sorghum. Crop Sci. 19:837-842.

BLESSIN, C. W., VANETTEN, C. H., and WIEBE, R. 1958. Carotenoid content of the grain from yellow endosperm type sorghums. Cereal Chem. 35:359-364.

BLESSIN, C. W., DIMLER, R. J., and WEBSTER, O. J. 1962. Carotenoids of corn and sorghum. II. Carotenoid loss in yellow endosperm sorghum grain during weathering. Cereal Chem. 39:389- 392.

BLESSIN, C. W., ANDERSON, R. A., DEATHERAGE, W. L., and INGLETT, G. E. 1971. Effect of alkali dehulling on composition and wet milling characteristics of sorghum grain. Cereal Chem. 48:528-532.

BRUNE, M., ROSSANDER, L., and HALLABERG, L. 1989. Iron absorption and phenolic compounds: Importance of different phenolic structures. Eur. J. Clin. Nutr. 43:547-558.

BRUNKEN, J., DE WET, J. M. J., and HARLAM, J. R. 1977. The morphology and domestication of pearl millet. Econ. Bot. 31:163-174.

BUTLER, L. G. 1982. Polyphenols and their effects on sorghum quality. Page 294 in: Proc.

Int. Symp. on Sorghum Grain Quality. L. W. Rooney and D. S. Murty, eds. Int. Crops Res. Inst. Semi-Arid Tropics (ICRISAT), Patancheru, A. P., India.

BUTLER, L. G. 1989a. Effects of condensed tannins on animal nutrition. Pages 391-402 in: Chemistry and Significance of Condensed Tannins. R. W. Hemingway and J. J. Karchesy eds., Plenum Press, New York.

BUTLER, L. G. 1989b. New perspectives on the antinutritional effects of tannins. Pages 402-409 in: Food Proteins. J. E. Kinsella and W. G. Soucie, eds. Am. Oil Chem. Soc., Champaign, IL.

BUTLER, L. G. 1990. The nature and amelioration of the antinutritional effects of tannins in sorghum grain. Pages, 191-205 in: Proc. Int. Conf. on Sorghum Nutritional Quality. G. Ejeta, E. T. Mertz, L. Rooney, R. Schaffert, J. Yohe, eds. Purdue University, W. Lafayette, Indiana.

CAGAMPANG, G. B., and KIRLEIS, A. W. 1984. Relationship of sorghum grain hardness to selected physical and chemical measurements of grain quality. Cereal Chem. 61:100-105.

CAGAMPANG, G. B., and KIRLEIS, A. W. 1985. Properties of starches isolated from sorghum floury and corneous endosperm. Starch/Staerke 37:253-257.

CASEY, P., and LORENZ, K. 1977. Millet functional and nutritional properties. Bakers Dig. 51:45-51.

CHANDNA, M., and MATTA, N. K. 1990. Characterization of pearl millet protein fractions. Phytochemistry 29:3395-3399.

CHANDRASEKHARA, M. R., and SWAMINATHAN, M. 1953. Enzymes of ragi (*Eleusine coracana*) and ragi malt. I. Amylases. J. Sci. Indian Res. 12B:51-56.

CHANDRASEKHER, G., and PATTABIRAMAN, T. 1981. Natural plant enzyme inhibitors: Isolation and characterization of two trypsin inhibitors from Bajra (*Pennisetum thyphoideum*). Ind. J. Biochem. Biophys. 19:1.

CHAUDHANY, P., and KAPOOR, A. C. 1984. Changes in the nutritional value of pearl millet flour during storage. J. Sci. Food Agric. 35:1219-1224.

CHAVAN, J. K., and KADAM, S. S. 1989a. Nutritional improvement of cereals by fermentation. Crit. Rev. Food Sci. Nutr. 28(5):349-400.

CHAVAN, J. K., and KADAM, S. S. 1989b. Nutritional Improvement of Cereals by Sprouting. Crit. Rev. Food Sci. Nutr. 28(5):401-437.

CHAVAN, J. K., KADAM, S. S., and SALUNKHE, D. K. 1981. Changes in tannin, free amino acids, reducing sugars, and starch during seed germination of low and high tannin cultivars of sorghum. J. Food Sci. 46:638-639.

CHAVAN, U. D., CHAVAN, J. K., and KADAM, S. S. 1988. Effect of fermentation on soluble proteins and in vitro protein digestibility of sorghum, green gram and sorghum-green gram blends. J. Food Sci. 53:1574-1575.

CHIBBER, B. A. K., MERTZ, E. T., and AXTELL, J. D. 1978. Efects of dehulling on tannin content, protein distribution and quality of high and low tannin sorghum. J. Agric. Food Chem. 26:679-683.

CHIBBER, B. A. K., MERTZ, E. T., and AXTELL, J. D. 1980. In vitro digestibility of high tannin sorghum at different stages of dehulling. J. Agric. Food Chem. 28:160-161.

CHUNG, O. K. 1991. Cereal lipids. Ch. 13 in: Handbook of Cereal Science and Technology. K. J. Lorenz and K. Kulp, eds. Marcel Dekker, New York.

COUSINS, B. W., TANKSLEY, T. O., KNABE, D. A., and ZEBROWSKA, T. 1981. Nutrient digestibility and performance of pigs fed sorghums varying in tannin concentration. J. Anim. Sci. 53:1524-1537.

CURTIS, D. L., BURTON, G. W., and WEBSTER, O. J. 1966. Carotenoids in pearl millet seed. Crop Sci. 6:300-301.

DADA, L. O., and DENDY, D. A. V. 1988. A study of various processing techniques on the cyanide content o germinated sorghum. In: Proc. Int. Workshop Household Technologies for Improving Young Child Feeding, Nairobi. International Development Research Centre, Ottawa.

DHANKHER, N., and CHAUHAN, B. M. 1987a. Effect of temperature and fermentation time on phytic acid and polyphenol content of rabadi—A fermented pearl millet food. J. Food Sci. 52:828-829.

DHANKHER, N., and CHAUHAN, B. M. 1987b. Effect of temperature and fermentation time on protein and starch digestibility (in vitro) of rabadi—A fermented pearl millet food. J. Food Sci. 52:489-491.

DOHERTY, C. A., ROONEY, L. W., and FAUBION, J. M. 1981. Phytin content of sorghum and sorghum products. Pages 328-333 in: Proc. Int. Symp. on Sorghum Grain Quality, L. W. Rooney and D. S. Murty, eds. Int. Crops Res. Inst. Semi-Arid Tropics (ICRISAT), Patancheru, A. P., India.

DOHERTY, C., FAUBION, J. M., and ROONEY, L. W. 1982. Semiautomated determination of phytate in sorghum and sorghum products. Cereal Chem. 59:373-377.

DOHERTY, C. A., WANISKA, R. D., ROONEY, L. W., EARP, C. F., and POE, J. H. 1987. Free phenolic compounds and tannins in sorghum caryopsis and glumes during development. Cereal Chem 64:42-46.

DYER, T. A., and NOVELLIE, L. 1966. Kaffircorn malting and brewing studies. XVI. The distribution and activity of α and β amylases in germinating kaffircorn. J. Sci. Food Agric. 17:449-456.

EARP, C. F., and ROONEY, L. W. 1982. Scanning electron microscopy of the pericarp and testa of several sorghum varieties. Food Microstruct. 1:125-134.

EARP, C. F., AKINGBALA, J. O., RING, S. H., and ROONEY, L. W. 1981. Evaluation of several methods to determine tannins in sorghums with varying kernel characteristics. Cereal Chem. 58:234-238.

EGGUM, B. O., BACH KNUDSEN, K. E., MUNCK, L., AXTELL, J. D., and MUKURU, S. Z. 1982. Milling and nutritional value of sorghum in Tanzania. Pages 211-225 in: Proc. Int. Symp. on Sorghum Grain Quality. L. W. Rooney and D. S. Murty, eds. Int. Crops Res. Inst. Semi-Arid Tropics (ICRISAT), Patancheru, A. P., India.

EGGUM, B. O., MONOWAR, L., BACH KNUDSEN, K. E., MUNCK, L., and AXTELL, J. 1983. Nutritional quality of sorghum and sorghum foods from Sudan. J. Cereal Sci. 1:127-137.

EJETA, G., HASSEN, M. M., and MERTZ, E. T. 1987. In vitro digestibility and amino acid composition of pearl millet (*Pennisetum typhoides*) and other cereals. Proc. Natl. Acad. Sci. USA. 84:6016-6019.

EL-TINAY, A. H., ABDEL GADIR, A. M., and EL-HIDAI, M. 1979. Sorghum fermented kisra bread. I. Nutritive value of kisra. J. Sci. Food Agric. 30:859-863.

FEATHERSTON, W. R., ROGLER, J. C., AXTELL, J. D., and OSWALT, D. L. 1975. Nutritional value of high lysine sorghum grain for the chick. Poultry Sci. 54:1220-1225.

FEDAIL, S. S., BADI, S. E. M., and MUSA, A. R. M. 1984. The effects of sorghum and wheat bran on the colonic functions of healthy Sudanese subjects. Am. J. Clin. Nutr. 40:776-779.

FAO. 1992. Production Yearbook. FAO Statistics No. 145, Vol. 45. Food and Agriculture Organization of the United Nations, Rome.

FREEMAN, J. E., and WATSON, J. E. 1971. Influence of sorghum endosperm pigments on starch quality. Cereal Sci. Today. 16:378-381.

FUJITA, S., DONGHUI, L., SUGIMOTO, Y., INOUCHI, N., and FUWA, H. 1989. Some properties of starch from foxtail millet (*Setaria italica* Beauv.). Agric. Biol. Chem. 53:1163-1165.

FUSSELL, L. K., and DWARTE, D. M. 1980. Structural changes of the grain associated with black region formation in *Pennisetum americanum*. J. Exp. Bot. 31:645-654.

FUSSELL, L. K., and PEARSON, C. J. 1978. Course of grain development and its relationship to black region appearance in *Pennisetum americanum*. Field Crops Res. 1:21-31.

GAZZAZ, S. S., RASCO, B. A., DONG, F. M., and BORHAN, M. 1989. Effects of processing on the thiamine, riboflavin, and vitamin B12 content of fermented whole grain cereal products. J. Food Process. Preserv. 13:321-334.

GLENNIE, C. W., LIEBENBERG, N. W., and VAN TONDER, H. J. 1984. Morphological development in sorghum grain. Food Microstruct. 3:141-148.

GOMEZ, M. H. 1988. Physicochemical characteristics of fresh masa from alkaline processed corn and sorghum and of corn dry masa flour. Ph.D. diss.. Texas A&M University, College Station, TX.

GOSWAMI, A. K., SHARMA, K. P., and GUPTA, B. K. 1969a. Chemical composition of bajra grains. 2. American entries. J. Nutr. Diet. 6:291-294.

GOSWAMI, A. K., SHARMA, K. P., and SEHGAL, K. L. 1969b. Nutritive value of proteins of pearl millet of high yielding varieties and hybrids. Br. J. Nutr. 23:913-916.

GOSWAMI, A. K., SEHGAL, K. L., and GUPTA, B. K. 1969c. Chemical composition of bajra grains. 1. African entries. J. Nutr. Diet. 6:287-290.

GOSWAMI, A. K., SHARMA, K. P., and GUPTA, B. K. 1970. Chemical composition of bajra grains. 3. Indian inbreds. J. Nutr. Diet. 7:5-9.

GRAHAM, G. G., MACLEAN, W. C., MORALES, E., HAMAKER, B. R., KIRLEIS, A. W., MERTZ, E. T., and AXTELL, J. D. 1986. Digestibility and utilization of protein and energy from nasha, a traditional Sudanese fermented sorghum weaning food. J. Nutr. 116:978-984.

GUIRAGOSSIAN, V., CHIBBER, B. A. K., VAN SCOYOC, S., JAMBUNATHAN, R., MERTZ, E. T., and AXTELL, J. D. 1978. Characteristics of proteins from normal, high lysine and high tannin sorghums. J. Agric. Food Chem. 26:219-223.

GUPTA, R. K., and GUPTA, Y. P. 1972. Effect of nitrogen application on the protein quality of *Sorghum vulgare* Pers. Indian J. Agric. Res. 6:191-195.

HAGERMAN, A. E., and BUTLER, L. G. 1981. Choosing appropriate methods and standards for assaying tannin. J. Chem. Ecol. 15:1795-1810.

HAHN, D. H. 1984. Phenols of sorghum and maize: The effect of genotype and alkali processing. Ph.D. diss. Texas A&M University, College Station, TX.

HAHN, D. H., and ROONEY, L. W. 1986. Effect of genotype on tannins and phenols of sorghum. Cereal Chem. 63:4-8.

HAHN, D. H., FAUBION, J. M., and ROONEY, L. W. 1983. Sorghum phenolic acids, their high performance liquid chromatography separation and their relation to fungal resistance. Cereal Chem. 60:255-259.

HAHN, D. H., ROONEY, L. W., AND EARP, C. F. 1984. Tannins and phenols of sorghum. Cereal Foods World 29:776-779.

HAIKERWAL, M., and MATHIESON, A. R. 1971. Protein content and amino acid composition of sorghum grain. Cereal Chem. 48:690-699.

HAMAD, A. M., and FIELDS, M. L. 1979. Evaluation of the protein quality and available lysine of germinated and fermented cereals. J. Food Sci. 44:456-459.

HAMAKER, B. R., KIRLEIS, A. W., BUTLER, L. G., AXTELL, J. D., and MERTZ, E. T. 1987. Improving the in vitro protein digestibility of sorghum with reducing agents. Proc. Natl. Acad. Sci. USA 84:626-628.

HARISH KUMAR, P. M., VIRUPAKASHA, T. K., and VITHYATHIL, P. J. 1978. Sorghum proteinase inhibitors: Purification and some biochemical properties. Int. J. Pep. Protein Res. 12:185-196.

HEMANALINI, G., UMAPATHY, K. P., RAO, J. R., and SARASWATHI, G. 1980. Nutritional evaluation of sprouted ragi. Nutr. Rep. Int. 22:271-277.

HEUPEL, R. C., SAUVAIRE, Y., LE, P. H., PARTISH, E. J., and NES, W. D. 1986. Sterol composition and biosynthesis in sorghum: Importance to developmental regulation. Lipids 21:69-75.

HIBBERD, C. A., WAGNER, D. G., SCHEMM, R. L., MITCHELL, E. D., WEIBEL, D. E., and HINTS, R. L. 1982. Digestibility characteristics of isolated starch from sorghum and corn grain. J. Anim. Sci. 55:1490-1497.

HILU, K. W., and de WET, J. M. J. 1967. Domestication of *Eleusine coracana*. Econ. Bot. 30:199-208.

HODSON, M. J., and PARRY, D. W. 1982. The ultrastructure and analytical microscopy of silicon deposition in the aleurone layer of the caryopsis of *Setaria italica* (L.) Beauv. Ann. Bot. 50:221-228.

HORAN, F. E., and HEIDER, M. F. 1946. A study of sorghum and sorghum starches. Cereal Chem. 23:492.

HOSENEY, R. C., DAVIS, A. B., and HERBERS, L. H. 1974. Pericarp and endosperm structure of sorghum grain shown by scanning electron microscopy. Cereal Chem. 51:552-558.

HOSENEY, R. C., VARRIANO-MARSTON, E., and DENDY, D. A. V. 1981. Sorghum and millets. Ch. 3 in: Advances in Cereal Science and Technology, Vol. 4. Y. Pomeranz, ed. Am. Assoc. Cereal Chem., St. Paul, MN.

HOSENEY, R. C., ANDREWS, D. U., and CLARK, H. 1987. Sorghum and pearl millet. Ch. 9 in: Nutritional Quality of Cereal Grains: Genetic and Agronomic Improvement, R. A. Olson and K. Frey, eds. Am. Soc. Agron., Madison, WI.

HUBBARD, J. E., HALL, H. H., and EARLE, F. R. 1950. Composition of the component parts of the sorghum kernel. Cereal Chem. 51:825-829.

HULSE, J. H., LAING, E. M., and PEARSON, O. E., eds. 1980. Sorghum and the Millets: Their Composition and Nutritive Value. Academic Press, San Francisco, CA.

INDIRA, R., and NAIK, M. S. 1971. Nutrient composition and protein quality of some minor millets. Indian J. Agric. Sci. 41:795-797.

JAIN, A. K., and DATE, W. B. 1975. Relative amylase activity of some malted cereal grains. J. Food Sci. Technol. (India) 12:131.

JAMBUNATHAN, R., and MERTZ, E. T. 1973. Relationship between tannin levels, rat growth and distribution of proteins in sorghum. J. Agric. Food Chem. 21:692-696.

JAMBUNATHAN, R., MISRA, P. S., and MERTZ, E. T. 1972. Nutritive value and protein solubility characteristics of normal and pigmented sorghum. Fed. Proc. 31:695.

JAMBUNATHAN, R., SINGH, U., and SUBRAMANIAN, V. 1984. Grain quality of sorghum, pearl millet, pigeon pea and chick pea. Pages 47-60 in: Interfaces Between Agriculture, Nutrition and Food Science. Proc. Workshop, Hyberabad, India, 1981. Food Nutr. Bull. (Tokyo) Suppl. 9.

JAMBUNATHAN, E. M., MERTZ, E. T., and AXTELL, J. D. 1975. Fractionation of soluble proteins of high lysine and normal sorghum grain. Cereal Chem. 52:119-121.

JANSEN, G. R., DIMAIO, L. R., and HAUSE, N. L. 1962. Amino acid composition and lysine supplementation of teff. J. Agric. Food Chem. 10:62-64.

JAYATISSA, P. M., PATHIRANA, R. A., and SIVAYOGASUNDERAM, K. 1980. Malting quality of Sri Lankan varieties of sorghum. J. Inst. Brew. 86:18-20.

JELLUM, M. D., and POWELL, J. B. 1971. Fatty acid composition from pearl millet seed. Agron. J. 63:29.

JONES, R. W., BECKWITH, A. C., KHOO, U., and INGLETT, G. E. 1970. Protein composition of proso millet. J. Agric. Food Chem. 18:37-39.

KACED, I., HOSENEY, R. C., and VARRIANO-MARSTON, E. 1984. Factors affecting rancidity in ground pearl millet (*Pennisetum americanum* L. Leeke). Cereal

Chem. 61:187-192.

KAHADEVAPPA, V. G., and RAINA, P. L. 1978. Lipid profile and fatty acid composition of finger millet (*Eleusine coracana*). J. Food Sci. Technol. 15:100.

KAMBAL, A. E., and BATE-SMITH, E. C. 1976. A genetic and biochemical study on pericarp pigments in a cross between two cultivars of grain sorghum, *Sorghum bicolor*. Heredity 37:413-416.

KARIM A., and ROONEY, L. W. 1972. Characterization of pentosans in sorghum grain. J. Food Sci. 37:369-371.

KASSAM, A. H. 1976. Crops of the West African Semi-Arid Tropics. Int. Crops Res. Inst. Semi-Arid Tropics (ICRISAT), Patancheru, A. P., India.

KAZANAS, N., and FIELDS, M. L. 1981. Nutritional improvement of sorghum by fermentation. J. Food Sci. 46:819-821.

KHETARPAUL, N. 1988. Improvement of nutritional value of pearl millet by fermentation and utilisation of the fermented product. Ph.D. diss. Haryana Agricultral University, Hisar, India.

KHETARPAUL, N., and CHAUHAN, B. M. 1990. Effect of germination and fermentation on in vitro starch and protein digestibility of pearl millet. J. Food Sci. 55:883-884.

KIRLEIS, A. W. 1990. The prolamins of sorghum: Their role in protein digestibility. Pages 168-176 in: Proc. Int. Conf. on Sorghum Nutritional Quality. G. Ejeta, E. T. Mertz, L. Rooney, R. Schaffert, J. Yohe, eds. Purdue University, W. Lafayette, Indiana.

KLOPFENSTEIN, C. F., VARRIANO-MARSTON, E., and HOSENEY, R. C. 1981. Cholesterol lowering effect of sorghum diet in guinea pigs. Nutr. Rep. Int. 24:621-627.

KLOPFENSTEIN, C. F., HOSENEY, R. C., and LEIPOLD, H. W. 1983a. Goitrogenic effects of pearl millet diets. Nutr. Rep. Int. 27:1039-1047.

KLOPFENSTEIN, C. F., HOSENEY, R. C., and LEIPOLD, H. W. 1983b. Further studies on the goitrogenic effects of pearl millet diets. Nutr. Rep. Int. 28:1137-1143.

KLOPFENSTEIN, C. F., HOSENEY, R. C., and LEIPOLD, H. W. 1985. Nutritional quality of pearl millet and sorghum grain diets and a pearl millet weaning food. Nutr. Rep. Int. 31:287-297.

KLOPFENSTEIN, C. F., LEIPOLD, H. W., and CECIL, J. E. 1991. Semiwet milling of pearl millet for reduced goitrogenicity. Cereal Chem. 68:177-179.

KNABE, D. A. 1990. Sorghum as swine feed. Pages 95-107 in: Proc. Int. Conf. on Sorghum Nutritional Quality. G. Ejeta, E. T. Mertz, L. Rooney, R. Schaffert, J. Yohe, eds. Purdue University, W. Lafayette, Indiana.

KOCK, J. L. F., GROENEWALD, E. G., KRUGER, G. H. J. ELOFF, J. N., and LATEGAN, P. M. 1985. Extraction of polyphenols and hydrolysis of birdproof sorghum starch. J. Sci. Food Agric. 36:1140-1144.

KOETZ, R., and NEUKOM, H. 1977. Nature of bound nicotinic acid in cereals and its release by thermal and chemical treatment. Pages 305-350 in: Physical, Chemical and Biological Changes in Food Caused by Thermal Processing. T. Hoyden and O. Kuale, eds. Applied Science Publishers, London.

LAI, C. C., and VARRIANO-MARSTON, E. 1980a. Changes in pearl millet meal during storage. Cereal Chem. 57:275-277.

LAI, C. C., and VARRIANO-MARSTON, E. 1980b. Lipid content and fatty acid composition of free and bound lipids in pearl millets. Cereal Chem. 57:271-274.

LASZTITY, R., and LASZTITY, L. 1990. Phytic acid in cereal technology. Ch. 5 in: Advances in Cereal Science and Technology, Vol. 10. Y. Pomeranz, ed., Am. Assoc. Cereal Chem., St. Paul, MN.

LESTER, R. N., and BEKELE, E. 1981. Amino acid composition of the cereal teff and related species of *Eragrostis* (Graminae). Cereal Chem. 58:113-115.

LORENZ, K. 1977. Proso and foxtail millets—Scanning electron microscopy and starch characteristics. Lebebsm.-Wiss. Technol. 10:324-327.

LORENZ, K. 1983. Tannins and phytate content in proso millet (*Panicum miliaceum*). Cereal Chem. 60:424-426.

LORENZ, K., and DILSAVER, W. 1980a. Proso millets. Milling characteristics, proximate compositions, nutritive value of flours. Cereal Chem. 57:16-20.

LORENZ, K., and DILSAVER, W. 1980b. Rheological properties and food applicatons of proso millet flours. Cereal Chem. 57:21-24.

LORENZ, K., and HINZE, G. 1976. Functional characteristics of starches from proso and foxtail millets. J. Agric. Food Chem. 24:911-914.

LORENZ, K., and HWANG, Y. S. 1986. Lipids in proso millet (*Panicum miliaceum*) flours and brans. Cereal Chem. 63:387-390.

LORENZ, K., MacFARLAND, G., and HINZE, G. 1976. The mineral composition of proso and foxtail millets. Lebensm. Wiss. Technol. 9:357-359.

MALLESHI, N. G., and DESIKACHAR, H. S. R. 1986. Nutritive value of malted millet flours. Qual. Plant Foods Hum. Nutr. 36:191.

MANGAY, A. A., PEARSON, W., and DARBY, W. J. 1957. Millet (*Seteria italica*): Its amino acid and niacin content and supplementary nutritive value for corn (maize). J. Nutr. 62:377-393.

MAXSON, E. D., and ROONEY, L. W. 1972. Evaluation of methods for tannin analysis in sorghum grain. Cereal Chem. 49:719-729.

MAXSON, E. D., ROONEY, L. W. LEWIS, R. W., CLARK, L. E., and JOHNSON, J. W. 1973a. The relationship between tannin content, enzyme inhibition, rat performance, and characteristics of sorghum grain. Nutr. Rep. Int. 8:145-152.

MAXSON, W. E., SHIRLEY, R. L., BERTRAND, J. E., and PALMER, A. Z. 1973b. Energy values of corn, bird resistant and non-bird resistant sorghum grain in rations fed to steers. J. Anim. Sci. 37:1451-1457.

MCDONOUGH, C. M., and ROONEY, L. W. 1989. Structural characteristics of *Pennisetum americanum* using scanning electron and fluorescence microscopies. Food Microstr. 8:137-149.

MCDONOUGH, C. M., and ROONEY, L. W. 1990. Developmental study of six varieties of sorghum. (Abstr.) Cereal Foods World 35:836.

MCDONOUGH, C. M., ROONEY, L. W., and EARP, C. F. 1986. Structural characteristics of *Eleusine coracana* (finger millet) using scanning electron and fluorescence microscopy. Food Microstr. 5:247-256.

MEHANSHO, H., ANN, D. K., BUTLER, L. G., ROGLER, J. C., and CARLSON, D. M. 1987a. Induction of proline rich proteins in hamster salivary glands by isoproterenol treatment and an unusual growth inhibition by tannins. J. Biol. Chem. 262:12344-12350.

MEHANSHO, H., BUTLER, L. G., and CARLSON, D. M. 1987b. Dietary tannins and salivary proline-rich proteins: Interactions, induction, and defense mechanisms. Annu. Rev. Nutr. 7:423-440.

MERTZ, E. T. 1974. Breeding for improved nutritional value. Cereal Sci. Today, 19:385.

MOHAN, D. P., and AXTELL, J. D. 1975. Diethyl sulfate induced high lysine mutants in sorghum. Proc. Bienn. Grain Sorghum Res. Utilization Conf., 9th, Lubbock, TX.

MUINDI, P. J., and THOMKE, S. 1981. Metabolic studies with laying hens on Tanzanian sorghum grains of different tannin contents. Swed. J. Agric. Res. 11:17-21.

MUKURU, S., MONYO, E., EJETA, G., HASSEN, M., MERTZ, E. T., and AXTELL, J. D. 1988. Traditional food processing technologies in East Africa improve nutritive value of high tanin sorghums. Agron. Abstr. p. 163.

MURTY, D. S., SINGH, U., SURYAPRAKASH, S., and NICODEMUS, K. D. 1985. Soluble sugars in five endosperm types of sorghum. Cereal Chem. 62:150-152.

NAGARAJ., R. H., and PATTABIRAMAN, T. N. 1985. Isolation of an amylase inhibitor from *Setaria italica* grains by affinity chromatography on blue sepharose and its characterization. J. Agric. Food Chem. 33:646.

NAIK, M. S. 1968. Lysine and tryptophan content in protein fractions of sorghum. Indian J. Genet. Plant Breed. 28:142-146.

NARAYANASWAMI, S. 1953. The structure and development of the caryopsis in some Indian millets. I. *Pennisetum typhoideum* Rich. Phytomorphology 3:98-112.

NARAYANASWAMI, S. 1955. The structure and development of the caryopsis in some Indian millets. IV. *Echinochloa frumentacea* Link. Phytomorphology 5:161-171.

NARAYANASWAMI, S. 1956. The structure and development of the caryopsis in some Indian millets. VI. *Setaria italica.* Bot. Gaz. 112-122.

NEUCERE, N. J., and SUMRELL, G. 1980. Chemical composition of different varieties of grain sorghum. J. Agric. Food Chem. 28:19-21.

NEWTON, R. J., BALTUSKONIS, D. A., GOESCHL, J. D., MECKENSTOCK, D. H., and MILLER, F. R. 1980. Distribution and transformation of soluble carbohydrates during germination growth of sorghum. Crop Sci. 20:265-268.

NIP, W. K., and BURNS, E. E. 1969. Pigment characterization in grain sorghum. I. Red varieties. Cereal Chem. 46:490-495.

NIP, W. K., and BURNS, E. E. 1971. Pigment characterization in grain sorghum. II. White varieties. Cereal Chem. 48:74-80.

NURALIKRISHNA, G., PARAMAHANS, S. V., and THARANATHAN, R. N. 1982. Carbohydrate make up of minor millets. Starch/Staerke 12:397-401.

NWASARU, M. A., REICHERT, R. D., and MUKURU, S. Z. 1988. Factors affecting the abrasive dehulling efficiency of high-tannin sorghum. Cereal Chem. 65:171-174.

NWASIKE, C. C., MERTZ, E. T., PICKETT, R. C., GLOVER, D. V., CHIBBER, B. A. K., and VANSCOYOC, S. W. 1979. Lysine level in solvent fractions of pearl millet. J. Agric. Food Chem. 27:1329-1331.

NYMAN, M., SILJESTROM, M., PEDERSEN, B., BACH KNUDSEN, K. E., ASP, N. G., JOHANSSON, C. G., and EGGUM, B. O. 1984. Dietary fiber content and composition in six cereals at diffferent extraction rates. Cereal Chem. 61:14-19.

OKOH, P. N., KUBICZEK, R. P., NJOKU, P. C., and IYEGHE, G. T. 1989. Some compositional changes in malted sorghum (*Sorghum vulgare*) grain and its value in broiler chicken diet. J. Sci. Food Agric. 49:271-279.

OPOKU, A. R., OHENHEN, S. O., and EJIOFOR, N. 1981. Nutrient composition of millet (*Pennisetum thypoides*) grains and malt. J. Agric. Food Chem. 29:1247-1248.

OPOKU, A. R., OSAGIE, A. U., and EKPERIGIN,

E. R. 1983. Changes in the minor constituents of millet (*Pennisetum americanum*) during germination. J. Agric. Food Chem. 31:507-509.

OSAGIE, A. U. 1987. Total lipids of sorghum grain. J. Agric. Food Chem. 35:601-604.

OSAGIE, A. U., and KATES, M. 1984. The lipid composition of millet (*Pennisetum americanum*) seeds. Lipids, 19:958-965.

OSMAN, A. K. 1981. Bulrush millet (*Pennisetum typhoides*) a contributory factor to the endemicity of goitre in western Sudan. Ecol. Food Nutr. 11:121.

OSMAN, A. K., BASU, T. K., and DICKERSON, J. W. T. 1983. A goitrogenic agent from millet (*Pennisetum typhoides*) in Darfur province, western Sudan. Ann. Nutr. Metab. 27:14.

OSUNTOGUN, B. A., ADEWUSI, S. R. A., ADEWUSI, A., OGUNDIWIN, J. O., and NWASIKE, C. C. 1989. Effect of cultivar, steeping, and malting on tannin, total polyphenol, and cyanide content of Nigerian sorghum. Cereal Chem. 66:87-89.

PALMER, G. H. 1989. Cereals in malting and brewing. Pages 216-225 in: Cereal Science and Technology. G. H. Palmer, ed. Aberdeen University Press, Great Britian.

PALMER, G. H., ETOKAKPAN, O. U., and IGYOR, M. A. 1989. Review: Sorghum as brewing material. Mircen J. Appl. Microbiol. Biotechnol. 5:265-275.

PANASIUK, O., and BILLS, D. D. 1984. Cyanide content of sorghum sprouts. J. Food Sci. 49:791-793.

PARAMAHANS, S. V., and THARANATHAN, M. 1980. Carbohydrate composition of the millet varagu. Starch/Staerke 32:73-76.

PARAMAHANS, S. V., WANKHEDE, D. B., and THARANATHAN, R. N. 1980. Studies on varagu starch. Starch/Staerke 32:109-112.

PARVATHY, K., and SADASIVAM, S. 1982. Comparison of amylase activity and carbohydrate profile in germinating seeds of *Setaria italica, Echinochloa frumentacea* and *Panicum miliaceum*. Cereal Chem. 59:543-544.

PAULIS, J. W., and WALL, J. S. 1979. Distribution and electrophoretic properties of alcohol soluble proteins in normal and high lysine sorghums. Cereal Chem. 56:20-23.

PERTEN, H., BROTHERS, B., and SWITZERLAND, U. 1983. Practical experience in processing and use of millet and sorghum in Senegal and Sudan. Cereal Foods World 28:680-683.

PFLUGFELDER, R. L., ROONEY, L. W., and SCHAKE, L. M. 1986. The role of germination in sorghum reconstitution. Anim. Feed Sci. Technol. 14:243-254.

PORE, M. S., and MAGAR, N. G. 1979. Nutrient composition of hybrid varieties of finger millet. Indian J. Agric. Sci. 49:526-531.

PORTERES, R. 1976. African cereals: *Eleusine*, fonio, black fonio, teff, *Brachiaria, Paspalum, Pennisetum*, and African rice. Pages 409-452 in: Origins of African Plant Domestication. J. R. Harlan, J. M. J. de Wet, and A. B. L. Stemler, eds. Mouton Publ., The Hague.

PRICE, M. L., BUTLER, L. G., ROGLER, J. C., and FEATHERSTON, W. R. 1979. Overcoming the nutritionally harmful effects of tannin in sorghum grain by treatment with inexpensive chemicals. J. Agric. Food Chem. 27:441-445.

PRUTHI, T. D., and BHATIA, I. S. 1970. Lipids in cereals. 1. *Pennisetum typhoideum*. J. Sci. Food Agric. 21:419-422.

RACHIE, K. O., and MAJMUDAR. J. V. 1980. Pearl millet. Pennsylvania University Press, University Park, PA.

RAMACHANDRA, G., VIRUPUSKA, T. K., and SHADAKSHARASWAMY, M. 1978. Com- of the protein fractions of finger millet. Phytochemistry 17:1487-1490.

REDDY, V. P., FAUBION, J. M. and HOSENEY, R. C. 1986. Odor generation in ground, stored pearl millet. Cereal Chem. 63:403-406.

REICHERT, R. D. 1979. The pH sensitive pigments in pearl millet. Cereal Chem. 56:291-294.

REICHERT, R. D., and YOUNGS, C. G. 1979. Bleaching effect of acid on pearl millet. Cereal Chem. 56:287-290.

REICHERT, R. D., NWASARU, M. A., and MUKURU, S. Z. 1988. Characterization of colored grain sorghum lines and identification of high tannin lines with good dehulling characteristics. Cereal Chem. 65:165-170.

RING, S. H., AKINGBALA, J. O., and ROONEY, L. W. 1982. Variation in amylose content among sorghums. Pages 269-279 in: Proc. Int. Symp. on Sorghum Grain Quality, L. W. Rooney and D. S. Murty, eds. Int. Crops Res. Inst. Semi-Arid Tropics (ICRISAT), Patancheru, A. P., India.

ROONEY, L. W. 1978. Sorghum and pearl millet lipids. Cereal Chem. 55:584-590.

ROONEY, L. W., and CLARK, L. E. 1968. The chemistry and processing of sorghum grain. Cereal Sci. Today. 13:258-261.

ROONEY. L. W., and McDONOUGH, C. M. 1987. Food quality and consumer acceptance of pearl millet. Pages 43-61 in: Proc. Int. Pearl Millet Workshop, J. R. Witcombe and S. R. Beckerman, eds. Int. Crops Res. Inst. Semi-Arid Tropics (ICRISAT), Patancheru, A. P., India.

ROONEY, L. W., and MILLER, F. 1982. Variation in the structure and kernel characteristics of sorghum. Pages 143-162 in: Proc. Int. Symp. on Sorghum Grain Quality. L. W.

Rooney and D. S. Murty, eds. Int. Crops Res. Inst. Semi-Arid Tropics (ICRISAT), Patancheru, A. P., India.

ROONEY, L. W., and PFLUGFELDER, R. L. 1986. Factors affecting starch digestibility with special emphasis on sorghum and corn. J. Anim. Sci. 63:1607-1623.

ROONEY, L. W., and SERNA-SALDIVAR, S. O. 1991. Sorghum. Ch. 5 in: Handbook of Cereal Science and Technology. K. J. Lorenz and K. Kulp, eds. Marcel Dekker, New York.

ROONEY. L. W., KHAN, M. N., and EARP, C. F. 1980. The technology of sorghum products. in: Cereals for Food and Beverages: Recent Progress in Cereal Chemistry. E. Inglett and L. Munck, eds. Academic Press, New York.

ROONEY, L. W., EARP, C. F., and KHAN, M. N. 1982. Sorghums and millets. Page 123 in: CRC Handbook of Processing and Utilization in Agriculture, Vol. 2. I. A. Wolf, ed., CRC Press, Boca Raton, FL.

ROONEY, L. W., KIRLEIS, A. W., and MURTY, D. S. 1986. Traditional foods from sorghum: Their production, evaluation and nutritional value. Pages 317-353 in: Advances in Cereal Science and Technology, Vol. 8. Y. Pomeranz, ed. Am. Assoc. Cereal Chem., St. Paul, MN.

ROONEY, T. K., ROONEY, L. W., and LUPTON, J. R. 1992. Physiological characteristics of sorghum and millet brans in the rat model. Cereal Foods World 37:782-786.

ROST, T. L. 1971. Fine structure of endosperm protein bodies in *Setaria lutescens* (Gramineae). Protoplasma 73:475-479.

ROST, T. L., and LERSTEN, N. R. 1973. A synopsis and selected bibliography of grass caryopsis anatomy and fine structure. Iowa State J. Res. 48:47-87.

SALUNKHE, D. K., JADHAV, S. J., KADAM, S. S., and CHAVAN, J. K. 1982. Chemical, biochemical and biological significance of polyphenols in cereals and legumes. Crit. Rev. Food Sci. Nutr. 17:277-305.

SALUNKHE, D. K., CHAVAN, J. K., and KADAM, S. S. 1990. Dietary Tannins: Consequences and Remedies. CRC Press, Boca Raton, FL.

SAN BUENAVENTURA, M. L., DONG, F. M., and RASCO, B. A. 1987. The total dietary fiber contant of wheat, corn, barley, sorghum, and distillers' dried grains with solubles. Cereal Chem. 64:135-136.

SAUNDERS, E. H. 1955. Developmental morphology of the kernel in grain sorghum. Cereal Chem. 32:12-25.

SAUNDERS, J. A., and CONN, E. E. 1978. Presence of the cynaogenic glucoside dhurrin in isolated vacuoles from sorghum. Plant Physiol. 61:154-157.

SAWNEY, S. K., and NAIK, M. S. 1969. Amino acid composition of protein fractions of pearl millet and the effect of nitrogen fertilization on its protein. Indian J. Genet. Plant Breed. 29:395-406.

SECKINGER, H. L., and WOLF, M. J. 1973. Sorghum protein ultrastructure as it relates to composition. Cereal Chem. 50:455-465.

SEITZ, L. M., WRIGHT, R. L, WANISKA, R. D., and ROONEY, L. W. 1993. Contribution of 2-acetyl-1-pyrroline to odors from wetted ground pearl millet. J. Agric. Food Chem. 41:955-958.

SELL, D. R., ROGLER, J. C., and FEATHERSTON, W. R. 1983. The effects of sorghum tannin and protein level on the performance of laying hens maintained in two temperature environments. Poultry Sci. 62:2420-228.

SERNA-SALDIVAR, S. O., KNABE, D. A., ROONEY, L. W., and TANKSLEY, T. D. 1987a. Effects of lime cooking on energy and protein digestibilities of maize and sorghum. Cereal Chem. 64:247-252.

SERNA-SALDIVAR, S. O., TELLEZ-GIRON, A., and ROONEY, L. W. 1987b. Production of tortilla chips from sorghum and maize. J. Cereal Sci. 8:275-284.

SERNA-SALDIVAR, S. O., CANNETT, R., VARGAS, J., GONZÁLEZ, M., BEDOLLA, S., and MEDINA, C. 1988a. Effect of soybean and sesame addition on the nutritional value of maize and decorticated sorghum tortillas produced by extrusion cooking. Cereal Chem. 65:44-48.

SERNA-SALDIVAR, S. O., KNABE, D. A., ROONEY, L. W., TANKSLEY, T. D., and SPROULE, A. M. 1988b. Nutritional value of sorghum and maize tortillas. J. Cereal Sci. 7:83-94.

SERNA-SALDIVAR, S. O., MCDONOUGH, C. M., and ROONEY, L. W. 1991a. The millets. Ch. 6 in: Handbook of Cereal Science and Technology. K. Lorenz and K. J. Kulp, eds. Marcel Dekker, New York.

SERNA-SALDIVAR, S. O., ROONEY, L. W., and GREENE, L. W. 1991b. Effect of lime treatment on the bioavailability of calcium in diets of tortillas and beans: Rat growth and balance studies. Cereal Chem. 68:565-570.

SERNA-SALDIVAR, S. O., ROONEY, L. W., and GREENE, L. W. 1992. Effect of lime treatment on the bioavailability of calcium in diets of tortillas and beans: Bone and plasma composition in rats. Cereal Chem. 69:78-81.

SERNA-SALDIVAR, S. O., CLEGG, C., and ROONEY, L. W. 1994. Effects of parboiling and decortication on the nutritional value of sorghum (*Sorghum bicolor* L. Moench) and pearl millet (*Pennisetum glaucum* L.). J. Cereal Sci. 19:83-89.

SHEORAIN, B. S., and WAGLE, D. S. 1973. β-amylase activity in germinated bajra and barley varieties. J. Food Sci. Technol. (India) 10:184.

SHIVARAJ, B., and PATTABIRAMAN, T. N. 1981. Natural plant enzyme inhibitors. Biochem. J. 193:29-36.

SHULL, J. M., WATTERSON, J. J., and KIRLEIS, A. W. 1991. Proposed nomenclature for the alcohol soluble proteins (kafirins) of *Sorghum bicolor* (L. Moench) based on molecular weight, solubility and structure. J. Agric. Food Chem. 39:83-87.

SIMWEMBA, C. G., HOSENEY, R. C., VARRIANO-MARSTON, E., and ZELEZNAK, K. 1984. Certain B vitamin and phytic acid contents of pearl millet (*Pennisetum americanum* (L.) Leeke). J. Agric. Food Chem. 32:31-34.

SINGH, P., SINGH, U., EGGUM, B. O., KUMAR, K. A., and ANDREWS, D. J. 1987. Nutritional evaluation of high protein genotypes of pearl millet (*Pennisetum americanum* (L.) Leeke). J. Sci. Food Agric. 38:41-48.

SKOVRON, J., and LORENZ, K. 1979. Enzymatic actvities in proso millet. Cereal Chem. 56:559-562.

SUBRAMANIAN, V., and SURYAPRAKASH, S. 1981. Sugars of pearl millet (*Pennisetum americanum* (L.) Leeke). J. Food Sci. 46:1614-1615.

SUBRAMANIAN, B., JAMBUNATHAN, R., and SUURYAPRAKASH, S. 1980. Note on the soluble sugars of sorghum. Cereal Chem. 57:440-441.

SUBRAMANIAN, V., JAMBUNATHAN, R., and RAMAIAH, C. D. 1981. Physical and chemical characteristics of pearl millet grains and their relationship to roti quality. J. Food Sci. 51:1005-1008.

SULLINS, R. D., and ROONEY, L. W. 1974. Microscopic evaluation of the digestibility of sorghum lines that differ in endosperm characteristics. Cereal Chem. 51:134-142.

SULLINS, R. D., and ROONEY, L. W. 1975. Light and scanning electron microscopic studies of waxy and nonwaxy endosperm sorghum varieties. Cereal Chem. 52:361-366.

SULLINS, R. D., and ROONEY, L. W. 1977. Pericarp and endosperm structure of pearl millet (*Pennisetum typhoides*). Pages 91-109 in: Proc. Symp. on Sorghum and Millets for Human Food. D. A. V. Dendy, ed. Tropical Products Institute, London, England.

SWEAT, V. E., FAUBION, J. M., GONZALES-PALACIOS, L., BERRY, G., AKINGBALA, J. O., and ROONEY, L. W. 1984. Gelatinization energy and temperature of sorghum and corn starches. Trans. ASAE 27:1960-1963, 1969, 1984.

TAIRA, H. 1962. Studies on amino acid contents in plant seeds. II. Amino acid pattern of seed protein fractions of Graminae. Bot. Mag. Tokyo 75:273-277.

TAIRA, H. 1968. Amino acid composition of different varieties of foxtail millet (*Setaria italica*). J. Agric. Food Chem. 16:1025-1027.

TAIRA, H. 1984. Lipid content and fatty acid composition of nonglutinous and glutinous varieties of foxtail millet. J. Agric. Food Chem. 32:369-371.

TANNER, F. W., PFEIFFER, S. E., and CURTIS, J. J. 1947. B-complex vitamins in grain sorghums. Cereal Chem. 24:268-274.

TAYLOR, J. R. N. 1983. Effect of malting on protein composition of sorghum. J. Sci. Food Agric. 34:885-812.

TAYLOR, J. R. N., and SCHUSSLER, L. 1986. The protein composition of the different anatomical parts of sorghum grain. J. Cereal Sci. 4:361-369.

TAYLOR, J. R. N., NOVELLIE, L., and LIEBENBERG, N. W. 1984. Sorghum protein body composition and ultrastructure. Cereal Chem. 61:69-73.

TAYLOR, J. R. N., NOVELLIE, L., and LIEBENBERG, N. W. 1985. Protein body degradation in the starchy endosperm of germinating sorghum. J. Exp. Biol. 36:1287-1295.

THARANATHAN, R. N., PARAMAHANS, S. V., and WANKHEDE, D. B. 1980. Amylolytic susceptibility of native groundnut and ragi starch granules as viewed by scanning electron microscopy. Starch/Staerke 32:158-161.

TOMITA, Y., SUGIMOTO, Y., SAKAMOTO, S., and FUWA, H. 1981. Some properties of starches of grain amaranth and several millets. J. Nutr. Sci. Vitaminol. 27:471-484.

UDUPA, S. L., and PATTABIRAMAN, T. N. 1985. Isolation and characterization of a trypsin/chymotrypsin inhibitor from the millet *Echinochloa fruneutacea*. J. Agric. Food Chem. 33:642-646.

VARRIANO-MARSTON, E., and HOSENEY, R. C. 1980. Note on mineral content and location in pearl millet. Cereal Chem. 57:150-152.

VINCENT-MONTERIO, P., VIRUPAKSHA, T. K., and RAJAGOPOL RAO, D. 1982. Proteins of Italian millet: Amino acid composition, solubility fractionation and electrophoresis of protein fractions. J. Sci Food Agric. 33:1072-1079.

VIRUPAKSHA, T. K., RAMACHANDRA, G., and NAGARAJU, D. 1975. Seed proteins of finger millet and their amino acid composition. J. Sci. Food Agric. 26:1237-1246.

VIVAS, N. E., WANISKA, R. D., and ROONEY, L. W. 1987. Effect of tortilla production on proteins in sorghum and maize. Cereal Chem. 64:384-389.

VIVAS-RODRIGUEZ, N. E., SERNA-SALDIVAR,

S. O., WANISKA, R. D., and ROONEY, L. W. 1990. Effect of tortilla chip preparation on the protein fractions of quality protein maize, regular maize, and sorghum. J. Cereal Sci. 12:289-296.

WALL, J. A., and BLESSIN, C. W. 1970. Composition of sorghum plant and grain. Pages 118-166 in: Sorghum Production and Utilization, J. S. Wall and W. M. Ross, eds. AVI Publishing Co., Westport, CT.

WALL, J. S., and PAULIS, J. W. 1978. Corn and sorghum grain proteins. Ch. 3 in: Advances in Cereal Science and Technology, Vol 2. Y. Pomeranz, ed. Am. Assoc. Cereal Chem., St. Paul, MN.

WANG, C., MITCHELL, H. L., and BARHAM, H. N. 1959. The phytin content of sorghum grain. Trans. Kansas Acad. Sci. 62:208.

WANG, Y. D., and FIELDS, M. L. 1978. Germination of corn and sorghum in the home to improve nutritive value. J. Food Sci. 43:1113-1115.

WANISKA, R. D., POE, J. H., and BANDYOPADHHYAY, R. 1989. Effects of growth conditions on grain molding and phenols in sorghum caryopsis. J. Cereal Sci. 10:217-225.

WANISKA, R. D., KOTARSKI, S. F., and THURN, K. K. 1990. Starch digestibility in sorghum. Pages 177-190 in: Proc. Int. Conf. on Sorghum Nutritional Quality. G. Ejeta, E. T. Mertz, L. Rooney, R. Schaffert, J. Yohe, eds. Purdue University, W. Lafayette, Indiana.

WANKHEDE, D. B., SHEHNAZ, A., and RAGHAVENDRA RAO, M. R. 1979a. Carbohydrate composition of finger millet (*Eleusine coracana*) and foxtail millet (*Seteria italica*). Qual. Plant. Plant Foods Hum. Nutr. 28:293-303.

WANKHEDE, D. B., SHEHNAZ, A., and RAGHAVENDRA RAO, M. R. 1979b. Preparation and physiochemical properties of starches and their fractions from finger millet (*Eleusine coracana*) and foxtail millet (*Setaria italica*). Starch/Staerke 31:153-159.

WARSI, A. S., and WRIGHT, B. C. 1973. Effects of rates and methods of nitrogen application on the quality of sorghum grain. Indian J. Agric. Sci. 43:722-726.

WATSON, S. A. 1984. Corn and sorghum starches: Production. Ch. 12 in: Starch Chemistry and Technology, 2nd ed. R. L. Whistler, J. N. Bemiller, and E. F. Paschall eds. Academic Press, Orlando, FL.

WATSON, T. G., and NOVEILLE, L. 1975. The development of amylase and maltase during the malting of *Sorghum vulgare*. Agrochemophysica 7:61-64.

WATTERSON, J., SHULL, J. M., MOHAMED, A. A., REDDY, V., and KIRLEIS, A. W. 1990. Isolation of a high cysteine kafirin protein and its cross reactivity with gamma zein antiserum. J. Cereal Sci. 12:137-144.

WOOLARD, G. R., RATHBONE, E. B., and NOVELLIE, L. 1976. A hemicellulosic β-D-glucan from the endosperm of sorghum grain. Carbohydrate Res. 51:249-252.

WOOLARD, G. R., RATHBONE, E. B., NOVELLIE, L., and OHLSSON, J. T. 1977. Structural studies on the water soluble gums from the endosperm of sorghum grain. Carbohydrate Res. 53:95-100.

WU, Y. V., and WALL, J. S. 1980. Lysine content of protein increased by germination of normal and high lysine sorghum. J. Agric. Food Chem. 28:455-458.

YANEZ, G. A., and WALKER, C. E. 1986. Effect of tempering parameters on extraction and ash of proso millet flours, and partial characterization of proso starch. Cereal Chem. 63:164-167.

YOUNG, R., HAIDARA, M., ROONEY, L. W., and WANISKA, R. D. 1990. Parboiled sorghum: Development of a novel decorticated product. J. Cereal Sci. 11:277-289.

ZEE, S. Y., and O'BRIEN, T. P. 1971. Aleurone transfer cells and other structural features of the spikelet of millet. Aust. J. Biol. Sci. 24:391-395.

ZELEZNAK, K., and VARRIANO-MARSTON, E. 1982. Pearl millet (*Pennisetum americanum* (L.) Leeke) and grain sorghum (*Sorghum bicolor* (L.) Moench) ultrastructure. Am. J. Bot. 69:1306-1313.

CHAPTER 5

NUTRITIONAL PROPERTIES OF SORGHUM AND THE MILLETS

C. F. Klopfenstein and R. C. Hoseney
Department of Grain Science and Industry
Kansas State University
Manhattan, Kansas

INTRODUCTION

Sorghum and millet grains have been dietary staples for centuries in parts of India, Africa, and China. Today, these crops are significant contributors to the protein and energy requirements of millions of people. In the future, they will need to feed many more. World population is projected to increase by another billion people in the next 10 years, and much of the growth will be in the semiarid, developing countries where drought-tolerant sorghum and millets are major crops.

A comprehensive volume covering the nutritional aspects of sorghum and the millets was published in 1980 by Hulse et al. Since then, discussions of the nutritional properties of sorghum and millets have been included in reviews by Hoseney et al (1981, 1987), Varriano-Marston and Hoseney (1983), Rooney et al (1986), Rooney and Serna-Saldivar (1991), and Serna-Saldivar and Rooney (1991).

SORGHUM

Sorghum Carbohydrates

Carbohydrates constitute the largest percentage of sorghum grain, as they do all cereals (Table 1). The primary carbohydrate, starch, makes up about 60 to 80% of normal (nonwaxy) kernels (Horan and Heider, 1946; McNeill et al, 1975) and is contained in simple granules embedded in a protein matrix. Although sorghum and maize have similar chemical compositions, sorghum has often been shown to be the less efficient grain in feeding trials. The reason for sorghum's poorer performance may be inferior starch digestibility. The data presented by Sikabbubba (1989) confirm that this may be a factor. When she compared in vitro starch digestibilities (glucoamylase digestion, glucose determined enzymatically; AACC 1983) for isolated starch from 11 normal tannin-negative sorghum cultivars and eight normal

Contribution No. 91-400-B of the Kansas Agricultural Experiment Station.

TABLE 1
Comparison of Nutrients in 100-g Edible Portions of Various Cereals at 12% Moisture[a]

Cereal	Protein[b] (g)	Fat (g)	CHO[c] (g)	Crude Fiber (g)	Ash (g)	Energy (kcal)	Calcium (mg)	Iron (mg)	Thiamin (mg)	Niacin (mg)	Riboflavin (mg)
Wheat	11.6	2.0	71	2.0	1.6	348	30	3.5	0.405	5.05	0.101
Brown rice	7.9	2.7	76	1.0	1.3	362	33	1.8	0.413	4.31	0.043
Maize	9.2	4.6	73	2.8	1.2	358	26	2.7	0.378	3.57	0.197
Sorghum	10.9	3.2	73	2.3	1.6	329	27	4.3	0.300	2.83	0.138
Pearl millet	11.0	5.0	69	2.2	1.9	363	25	3.0	0.3	2.0	0.15
Foxtail millet	9.9	2.5	72	10.0	3.5	351	20	4.9	0.593	0.99	0.099
Finger millet	6.0	1.5	75	3.6	2.6	336	350	5.0	0.3	1.4	0.10
Kodo millet	11.5	1.3	74	10.4	2.6	353	35	1.7	0.15	...	...
Japanese barn-yard millet	10.8	4.5	49	14.7	4.0	...	22	18.6	...	...	...
Proso millet	10.6	4.0	70	12.0	3.2	364	8	2.9	0.405	4.54	0.279

[a] Data from Hulse et al (1980), NRC/NAS (1982), USDA/HNIS (1989), Serna-Saldivar and Rooney (1991).
[b] N × 6.25.
[c] Carbohydrates.

corn varieties, she found values ranging from 33 to 48% in sorghum compared with 53 to 58% in corn. In addition, in vitro starch digestibility was lower for whole ground sorghum than for ground corn. However, Hibberd et al (1982a,b), making in vitro comparisons (gas production after treatment with amyloglucosidase and yeast) of isolated sorghum and corn starches, observed higher enzymatic degradation of the sorghum starch. Starch digestibility is reportedly reduced in sorghums containing tannins (Dreher et al 1984). Isolated sorghum tannins from different bird-resistant varieties have been shown to inhibit α-amylase and to bind to wheat, corn, and sorghum starches to different degrees (Davis and Hoseney, 1979a). Sikabbubba (1989) reported a positive correlation between in vitro sorghum starch digestibility and endosperm texture ($r = 0.567$, $P = 0.068$), with floury sorghums being more digestible than corneous ones. Ground floury sorghums were shown to have smaller Fisher subsieve particle size than similarly ground corneous sorghums. Particle size was strongly correlated with endosperm texture ($r = 0.965$, $P = 0.007$) and percent starch digestibility of whole sorghum ($r = 0.848$, $P = 0.036$), implying that the greater surface area of particles from milled floury sorghums is related to their higher digestibility.

Porridges made from flours of corneous sorghums were less sticky than those made from floury sorghums (Cagampang et al, 1982). Stickiness of cooked sorghum flour was a function of starch gelatinization (Hoseney, 1986). Starch in the corneous endosperm portion of sorghum grain is embedded in protein, which has been shown to influence starch gelatinization in sorghum (Chandrashekar and Kirleis, 1988). Samples with lower capacities for starch gelatinization had more prolamin-containing protein bodies whose organization around the starch granule appeared to act as a barrier to starch gelatinization. Modification of the protein disulfide bonds with a reducing agent increased starch gelatinization.

Most sorghum starch contains 70 to 80% amylopectin and 20 to 30% amylose, but waxy varieties, in which the starch is nearly 100% amylopectin, also exist (Rooney and Serna-Saldivar, 1991). Amylose content of the grain decreased and in vitro digestibility increased with incremental doses of the recessive waxy gene (Lichtenwalner et al, 1978). Hibberd et al (1982b) also reported higher enzymatic digestibility of isolated waxy sorghum starch than of normal sorghum starch. Elmalik et al (1986a) found values for glucoamylase digestible starch to be significantly higher in ground whole grains of waxy sorghum than normal sorghums, normal corn, and normal pearl millet. Feeding studies carried out with cattle (Sherrod et al, 1969), sheep (Nishimuta et al, 1969), and rats (Elmalik et al, 1986a) have shown better nutritional performance for the waxy grains than the nonwaxy types. In a swine-feeding study, Cohen and Tanksley (1973) were unable to demonstrate significant differences in dry matter and gross energy digestibilities between waxy and normal sorghum with similar endosperm texture, but they reported a slight advantage in performance for animals fed grain with the waxy starch. In addition, studies with isolated sorghum starches have shown waxy starch to be more soluble in boiling water than normal starch (83% vs. 29–54%) (Craig and Stark, 1984). Initially, none of the starches dissolved in cold water, but after the starches were damaged by fine grinding, 91% of the waxy starch became cold-water soluble, whereas only 49 to 52% of the normal starches dissolved. In general, waxy starches heated in water have much greater swelling power than nonwaxy starches, indicating that amylose has a role in restricting starch swelling (Rooney and Pflugfelder, 1986). Starch-protein interactions and associations within plant tissues are another factor affecting starch digestibility (Sikabbubba, 1989; Lichtenwalner et al, 1978; Hibberd et al, 1982b). In normal sorghum and corn, the endosperm consists of a corneous peripheral area surrounding an interior floury portion. Sorghum generally has a higher proportion of peripheral endosperm than corn (Rooney and Sullins, 1973). The dense peripheral layer is resistant to water penetration, enzymatic digestion, and mechanical disruption, and appears to protect imbedded or underlying starch granules from degradation. The endosperm of waxy sorghum has a more uniform distribution of protein and starch, especially in the peripheral area, possibly contributing to the accessibility of the starch to amylases (Sullins and Rooney, 1974). Lichtenwalner et al (1978) suggested that those same traits might be responsible for the inferior agronomic characteristics of waxy sorghums, which have poorer germination and lower yields than normal types. When 12 sorghums with different grain characteristics were compared with regard to their suitability for use in tortillas, those prepared from waxy sorghum produced sticky masa and tortillas, whereas all those from nonwaxy sorghums had an acceptable texture (Almeida-Dominguez et al, 1991). Development of agronomically superior heterowaxy hybrids with higher than normal digestibility appears to be a promising way to improve the nutritional impact of sorghum grain.

Total dietary fiber (TDF; method of Asp et al, 1983) content of uncooked Dabar, a common, low-tannin, Sudanese, food-grade sorghum, was 7.6% and that of uncooked feterita, a higher-tannin sorghum, was 9.2% (Bach Knudsen and Munck, 1985). Acid detergent fiber (ADF) for the same samples was 2.9 and 3.6%, respectively. TDF values for both sorghums were lower than those for barley, rye, wheat, or maize. Most of the dietary fiber was water insoluble (6.5% in Dabar, 7.9% in feterita). The same study showed that when porridge was prepared from the whole-grain flours, energy availability decreased, and 60% of the loss could be explained by the formation of enzyme-resistant starch. Fermentation to pH 3.9 before cooking counteracted the formation of enzyme-resistant starch, but immediate acidification to pH 3.9 did not. This low pH also inhibited the formation of residues measured as lignin during cooking. The contribution of polyphenols to the lignin fraction of dietary fiber accounted for the higher TDF value for feterita than for Dabar sorghum. Much larger amounts of protein were associated with the TDF and ADF fractions of sorghum than of wheat, maize, rye, or barley. The amount of protein bound to TDF and ADF increased on cooking and was higher for the high-tannin sorghum. Fermentation or acidification to pH 3.9 inhibited this protein-binding effect. Bach Knudsen and Munck (1985) pointed out that these results are consistent with the observed beneficial nutritional effect of the traditional method of fermentation of sorghum in Sudan. Others have reported that the insoluble fiber in sorghum grain is composed mainly of cellulose (1.19–5.23% of grain; Kamath and Belavady, 1980) and pentosans (2.51–5.57% of grain; Karim and Rooney, 1972).

TDF values of 12.8 to 12.6% have been reported for whole sorghum flours prepared from white durra grown in Somalia (Ali and Harland, 1991). In that study, rats fed sorghum-containing diets, rich in fiber and phytate, had lower tibia iron and zinc concentrations than animals fed a synthetic diet containing cellulose fiber.

The effects of crude fiber on nitrogen and dry matter digestibility in 11 Cameroonian adults eating diets composed of 80% whole or dehulled cooked sorghum meal were measured by Cornu and Delpeuch (1981). The increased fiber intake during consumption of whole meal diets resulted in a significant rise in fecal matter excreted, including nitrogen and formic-insoluble substances. However, the urinary nitrogen losses were higher with the low-fiber diet. The apparent digestibility of nitrogen dropped from 65.4 for dehulled grain to 60.5% for whole grain, but apparent nitrogen balances remained positive for all subjects.

Sorghum is decorticated in the preparation of many African and some Indian foods (Rooney and Serna-Saldivar, 1991), which reduces not only the amount of fiber in the food but also minerals and higher quality proteins (Eggum et al, 1982; Pedersen and Eggum, 1983; Serna-Saldivar et al, 1987). Although nutrient digestibilities are higher for decorticated than for whole grains, nitrogen retention and protein quality are lower because of partial removal of the lysine-rich germ (Taylor and Schussler, 1986).

The presence of water-soluble (1-3),(1-4)-β-D-glucan in sorghum endosperm was reported by Woolard et al (1976, 1977). Using differential staining techniques, Earp et al (1983) identified the mixed-linked β-glucans in sorghum pericarp, aleurone, and endosperm. The physicochemical and nutritional properties of those complex glucose polymers were reviewed by Wood (1984), who described their tendency to form highly viscous, sticky, aqueous solutions. That property is very important in the malting of barley and brewing of beer. Barley can be quite high in β-glucans (Bamforth, 1982). During germination, access to starch by the amylolytic enzymes may be inhibited until the cell wall β-glucans are hydrolyzed. Filtration rates and yields of the wort are lowered by thick hydrated particles. Residual β -glucans in the beer may cause gel-like precipitates. Although β-glucan concentration is reportedly low in sorghum (0.015% in endosperm [Woolard et al, 1976]; approximately 1% of grain [Prentice et al, 1980]), data are scarce, and the range of concentrations is not known. Presence of sorghum β-glucans may be of particular interest in parts of Africa where much sorghum is used for making a type of beer, which contributes substantial quantities of nutrients to the diet (Hoseney et al, 1981; Novellie and de Schaepdrijver, 1986).

A compilation of traditional sorghum foods showed that the grain is used in a multitude of various types of porridges worldwide (Rooney and Serna-Saldivar, 1991). In some of those foods, stickiness and high viscosity are desirable traits, but in others, they are undesirable. The β-glucan content of a sorghum cultivar may determine the suitability of its use in such traditional products.

Numerous studies have shown extracted cereal β-glucans and glucan-rich grains (oats and barley) to have cholesterol-lowering effects in humans and animals (Wood, 1984; Klopfenstein, 1988). When β-glucan-rich fractions were extracted from oats, barley, wheat, and sorghum grains, then freeze-dried and added to a standard white bread formula (at 7%, based on flour weight), all rats fed the experimental breads had lower serum cholesterol levels than animals fed the control bread, which contained no added cereal β-glucans (Klopfenstein and Hoseney, 1987).

Sorghum Proteins

A 100-g portion of sorghum will provide about 11 g of protein (Table 1; USDA/HNIS, 1989). Values reported for sorghum protein are more variable than those for the more common grains, partly because of wide varietal differences, but largely because of the diversity of agronomic conditions under which the crop is grown. According to Martin (1984), sorghum protein content, usually reported as

N × 6.25, may range from 8 to 16%, although values as high as 21.5% have been reported (Stemler et al, 1976). The average protein content of sorghum was negatively correlated with grain yield ($r = -0.85$, Worker and Ruckman, 1968) and was affected by a number of factors, including planting date, cultivar, seed size, and air temperature (Burleson et al, 1956; Waggle et al, 1967). In general, protein content of sorghum is higher than that of rice, equal to or higher than that of maize, and roughly equivalent to that of wheat.

As for most other grains, nitrogen content of sorghum grain is positively correlated with nitrogen fertilizer application. Soil deficiencies of nutrients, such as potassium or phosphorus, as well as conditions of soil moisture stress, can also result in increased seed nitrogen because of decreased grain yields (Eppendorfer et al, 1985).

Protein Quality

A common method for expressing protein quality is based on the concept that the nutritive value of a protein depends primarily on its content of essential amino acids. The percent of an essential amino acid that is in greatest deficit as compared to the amount in a standard or reference protein is known as the amino acid or chemical score of a protein (Block and Mitchell, 1946). However, today it is more common for scores to be calculated as percent of adequacy rather than percent deficiency. Egg and human milk protein have often been used as reference proteins, but the World Health Organization recommends using a scoring pattern more realistic with regard to human needs (WHO, 1985).

As previously mentioned, nitrogen fertilizer application has been shown to increase total nitrogen content of sorghum grain (Eppendorfer et al, 1985). However, as nitrogen application rate increases, percentages of the essential amino acids lysine, methionine, cystine, threonine, and tryptophan decrease as a percent of the protein, whereas isoleucine, leucine, phenylalanine and glutamine increase. Although the gain in protein quantity with nitrogen fertilization was offset by loss of protein quality, Eppendorfer et al (1985) reported that lysine as a percentage of dry matter increases with increasing nitrogen levels.

Lysine is consistently the most limiting amino acid in cereal grains, and the deficiency appears to be more serious in sorghum (Table 2). Although several high-lysine Ethiopian sorghums have been identified (Singh and Axtell, 1973), transference of the trait to plump grains has not yet been accomplished. Average whole-kernel lysine concentrations of Ethiopian IS11158 and IS11167 mutants average 3.13 and 3.33 g/100 g of protein at 17.2 and 15.7% protein, respectively, in contrast to normal sorghums with lysine values of 2.1 g/100 g of protein at about 12% protein (Ejeta and Axtell, 1987). Protein efficiency ratios (PER) of the mutants were 1.78 and 2.06, compared with values ranging from 0.74 to 1.24 for common sorghum lines (Singh and Axtell, 1973).

A diethyl sulfate–induced mutant (P721 opaque), having 60% more lysine than normal sorghum, was developed at Purdue University in 1975 (Mohan 1975). The enhanced lysine content of that line results primarily from an increase in relatively lysine-rich glutelin proteins and a decrease in lysine-poor prolamins (Van Scoyoc et al, 1988). However, the trait is apparently unstable, and its transference to a corneous endosperm has had limited success (Rao et al, 1983). The corneous endosperm is agronomically desirable and most appropriate for traditional food applications.

When Ejeta and Axtell (1987) compared protein and lysine levels in developing kernels of normal sorghum with Ethiopian and P721 high-lysine sorghums, they reported that high-lysine varieties had higher endosperm lysine contents than normal sorghums, whereas germ lysine content was not much different. The highest endosperm lysine content of all genotypes was observed at the late dough stage (31 days after flowering), which coincides with the time when the locally-grown, high-lysine, Ethiopian sorghums are usually harvested, roasted, and consumed. Growers in areas where high-lysine varieties are grown recognize that total grain yield per hectare is lower for the mutants than for normal sorghums, so the two are seeded in mixtures. The season when the mutant types are consumed often corresponds to a period of food shortage, and consumption of the immature sorghum could be an important nutritional benefit (Ejeta and Axtell, 1987).

In general, the trends associated with lysine concentration during seed development were similar for P-721 and Ethiopian mutants (Ejeta and Axtell, 1987). However, P-721 opaque had a lower protein content than its normal parent in all stages of grain development, whereas the percent protein of the Ethiopian high-lysine varieties increased as lysine concentration increased. The combination of such factors as higher percentage of lysine-rich germ, higher absolute protein content, and production of relatively lysine-rich glutelin proteins and less lysine-deficient kafirins makes the Ethiopian high-lysine mutants nutritionally superior not only to normal varieties but also to a chemically induced mutant (Ejeta and Axtell, 1987). Until major improvements occur in yield and other agronomic characteristics, including development of vitreous endosperm types, the high-lysine sorghums probably will not make a major nutritional contribution.

Rapid methods for determining both lysine and total protein have been evaluated and recommended for routine screening of large numbers of samples (Jambunathan et al, 1983). Those authors reported that lysine values from a dye-binding procedure were highly and positively correlated with amino acid analyzer values (r = 0.993), and the Technicon Autoanalyzer yielded protein values that were strongly related to values obtained by the micro-Kjeldahl method (r = 0.933).

The amino acid threonine is also limiting in sorghum grain, as it is in wheat,

TABLE 2
Lysine Amino Acid Scores for Different Age Groups Based on the 1985 World Health Organization Index

Grain	Infant (<1 yr, milk)	Preschool (2–5 yr)	School (10–12 yr)	Adult
Wheat[a]	43	46	62	100+
Rice (husked)[a]	57	61	82	100+
Maize[a]	41	43	58	100+
Sorghum[b]	17–51	18–55	25–74	71–100+
Pearl millet[b]	26–69	28–74	38–100+	100+
Foxtail millet[b]	28–38	30–40	40–55	100+
Finger millet[b]	39–63	41–68	56–91	100+
Kodo millet[b]	46–52	48–55	65–74	100+
Japanese barnyard millet[b]	26	27	37	100+
Proso millet[b]	23–72	24–74	32–98	93–100+

[a] Scores from Pellet (1985).

[b] Lysine values used to calculate scores were the range of means from studies compiled by Hulse et al (1980).

maize, and some of the millets. In addition, concentrations of isoleucine, total sulfur-containing amino acids, and valine appear to be marginal.

Conversely, concentration of the branched-chain amino acid leucine is quite high in sorghum grain, ranging from 186 to 189% of the FAO/WHO (1973) reference value (Chung and Pomeranz, 1985). Protein of the Ethiopian high-lysine mutants contained a lower concentration of leucine than that of normal sorghum (Singh and Axtell, 1973; Nelson, 1976). The nutritional and health significance of sorghum's high leucine content has been a controversial subject for decades, and the issue is still not resolved. During the 1960s and 1970s, research in India indicated that this branched-chain amino acid imbalance might impair the niacin-sparing effect of the amino acid tryptophan in sorghum-eating populations (Belavady et al, 1963; Ghafoorunissa and Narasinga Rao, 1973; Krishnaswamy et al, 1976; Srikantia et al, 1968). Data from those studies and later investigations (Bender, 1983; Magboul and Bender, 1983) indicated that diets high in leucine could precipitate niacin deficiency in experimental animals. Epidemiological data are supportive of those results, because the niacin deficiency disease, pellagra, is endemic in areas of the world where sorghum grain is a dietary staple (Gopalan and Srikantia, 1960; Gopalan, 1968).

However, when Young and Fukagawa (1988) reviewed a number of amino acid interactions, they cited five studies carried out in the late 1970s and early 1980s that do not support an involvement of leucine in the etiology of pellagra (Manson and Carpenter, 1978a,b; Nakagawa and Sasaki, 1977; Nakagawa et al, 1975; Cook and Carpenter, 1987). The latter paper reported that niacin deficiency was not more severe in rats fed niacin-deficient diets plus additional leucine than in those fed the same diets without added leucine, and no major differences occurred in tryptophan metabolism between the two groups. Additional research apparently will be required to resolve this issue.

Protein Digestibility

In vitro studies and in vivo studies with livestock and laboratory animals indicate that sorghum proteins are generally less digestible than those of other cereals.

TABLE 3
Factors Affecting In Vitro Protein Digestibility of Tannin-Negative Whole Sorghum and Corn[a]

	Correlation Coefficient[b]	
Factor	**Sorghum**	**Corn**
Total protein content	−0.548*	−0.568**
Total prolamin content	−0.627**	0.179
Alpha prolamin content	−0.198	0.082
Beta prolamin content	−0.647**	−0.042
Total prolamin digestibility	0.202	−0.022
Alpha prolamin digestibility	0.107	−0.141
Beta prolamin digestibility	0.727***	−0.162
Adhering protein in extracted starch	−0.465	−0.136
Endosperm texture	0.570*	...

[a] Data from 11 sorghum and eight corn cultivars (Sikabubba, 1989).
[b] Significant at $^{*}P = 0.1$, $^{**}P = 0.05$, and $^{***}P = 0.01$.

Silano (1977) reported that different varieties, with essentially equal protein and lysine contents, varied from 30 to 70% in protein digestibility. It is well-documented that condensed polyphenols (tannins) present in some varieties have a negative impact on digestibility of sorghum protein (Butler et al, 1984; Hoseney et al, 1981; Hahn et al, 1984; Bach Knudsen et al, 1988a,b) and amino acids (Whitacre and Tanner, 1989). A number of other factors affecting the in vitro protein digestibility of sorghum and corn are shown in Table 3 (Sikabbubba, 1989). Protein digestibility in sorghum was negatively correlated with total protein content, total prolamin proteins (alcohol-soluble, method of Sastry and Virupaksha, 1969), cross-linked prolamins (beta-prolamins), and beta-prolamin digestibility (Sikabbubba, 1989). A similar relationship between protein content and digestibility existed in corn, but corn protein digestibility was not significantly affected by the content or digestibility of prolamin proteins, including α-prolamin (non-cross-linked). The amount of protein associated with extracted starch was higher for sorghum than for corn varieties, and the protein content of isolated starch appeared to be related to total protein content of the sorghum grain as well as starch digestibility. In rats, protein digestibility for sorghum with intermediate- or corneous-textured endosperm was shown to be 70.3 and 74.5%, compared with 78.5% for corn (Elmalik et al, 1986a).

EFFECT OF COOKING ON SORGHUM PROTEIN DIGESTIBILITY

Rat feeding studies also show that protein digestibilities of cooked sorghum gruels range from 76.5 to 83.0% (Axtell et al, 1981). In vitro (pepsin) digestibility of the whole raw sorghums tested in that study ranged from 88.6 to 93.0%, but cooking sharply decreased protein digestibilities from 45.3 to 56.7%. The adverse effect of cooking was also observed for the dehulled kernels. When gruels prepared from the same four sorghum flours (two of which were from high-lysine varieties) were fed to children, the average digestibility was only 46%, compared with values of 81%, 73%, and 66% for wheat, maize, and rice, respectively (MacLean et al, 1981). In that study, the high-lysine sorghums performed no better than normal sorghums. The in vitro digestibility measurement appeared to be a more representative assay than rat bioassays for determining the food value of cooked sorghum products in children (Axtell et al, 1981). A simple in vitro pepsin digestion method developed at Purdue University yielded protein digestibility values similar to those reported in child feeding studies (Mertz et al, 1984). Bookwalter et al (1987b) used the method to compare protein digestibilities of cooked and uncooked cereals and soy products used in the Food for Peace Program of the U.S. Agency for International Development. The adverse effect of cooking is apparent from those data (Table 4).

Earlier work had also given an indication of cooked sorghum's poor performance as a food for children and adults. When Kurien et al (1960) replaced rice with sorghum in the diets of seven normal boys (aged 10–11), protein digestibility of the diets fell from 75 to 55%. Other workers have reported similar results in 10–11-year-old girls (Daniel et al, 1966) and young men (Nicol and Phillips, 1978).

Hamaker et al (1986) also reported similar in vitro protein digestibilities for raw sorghum and maize, but after cooking, sorghum protein digestibility was significantly decreased, whereas corn protein digestibility was unchanged. Those authors found that cooked sorghum prolamin proteins (procedure of Landry and Moreaux, 1970) became much less soluble and less digestible by pepsin than cooked maize prolamins. Cooking sorghum in the presence of reducing agents increased protein digestibility to a level comparable to that of other cereals (Hamaker et al, 1987).

The effects of cooking, pH, and grain polyphenol content on nutritional quality of the thick West African sorghum porridge, *tô*, have been reported by Bach Knudsen et al (1988a). Digestible energy in rats was reduced to the same extent in porridges when the cooking conditions were neutral (pH 7–7.4, cooked in water), acidic (pH 4.7–5.0, cooked in tamarind extract), or basic (pH 7.9–9.4, cooked with added wood ash). True protein digestibility was affected in the same manner as energy digestibility, but net protein utilization (NPU) in rats was not reduced in any of porridges made from a decorticated white sorghum. With *tô* made from whole red sorghum flour, NPU was significantly lower than in the uncooked control. However, biological value was higher for neutral and alkali porridges than for acid *tô* and raw flours. The authors concluded that the net effect could be accounted for by a change in the excretory route of nitrogen from urine to feces.

Alkali processing. Cooking of sorghum and maize with alkali (e.g., wood ashes or lime) in the production of nixtamalized porridges and tortillas has been shown to cause small, but significant, decreases in protein quality and availability (Serna-Saldivar et al, 1987). Vivas et al (1987) showed that alkaline processing decreased the solubility of proteins in salt water and alcohol and increased the amount of unextractable proteins. Pepsin digestibility of protein in sorghum and maize decreased after processing, with sorghum having slightly lower digestibility than maize.

The loss in nutritive value during alkali processing of proteins has been recognized for some time (DeGroot and Slump, 1969). It is caused by the combined effects of reduced protein digestibility as a result of cross-linking of amino acids (e.g., lysinoalanine, lanthionine, and ornithoalanine), the destruction of some essential amino acids, and the racemization of biologically available L-amino acids to the less utilizable D-forms (Hayashi and Kameda, 1980; Liardon and Hurrell, 1983). In addition, D-serine has been reported to be nephrotoxic (Ganote et al, 1974).

Parboiling. Recently, methods have been developed for producing parboiled sorghum and millet grains (Young et al, 1990). Increased decortication yields were reportedly high enough to more than offset the additional fuel requirements of the process. The improvement in decortication yield was higher in sorghums with floury endosperm texture than in corneous varieties. The floury sorghums became

TABLE 4
Effect of Cooking on Pepsin Digestibilities of Cereals, Soy Flour, and Cereal–Soy Product Blends[a]

	Percent Change in Protein Digestibility	
Processed Cereals	**Without Soy Products**	**With Soy Products Added**[b]
Wheat flour	−4.1	−3.1
Bulgur	+0.6	−2.0
Rolled oats	−6.7	−2.7
Degermed corn meal	−0.3	−7.4
Sorghum grits	−16.0	−12.0
Soy flour	+1.0	. . .

[a] Adapted from Bookwalter et al (1987b).

[b] Wheat flour + 12% soy flour, bulgur + 15% soy grits, rolled oats + 15% soy flakes, cornmeal + 15% soy flour, sorghum grits + 15% soy grits.

more vitreous during the process, making them more acceptable for use in traditional food products. However, effects of the process on protein and energy digestibility have not been reported.

EFFECT OF FERMENTATION ON SORGHUM PROTEIN DIGESTIBILITY

Mertz et al (1984) reported in vitro protein digestibilities for the following cereals (%): wheat (85.5), maize (85.3), rice (83.8), sorghum P721-normal (59.0), sorghum P721-opaque (high-lysine, 63.2), pearl millet (74.8), and Dabar sorghum (59.8). In cooked gruel (*nasha*) prepared from fermented, drum-dried, whole-kernel Dabar sorghum and pearl millet flours, protein digestibility increased to 65.5 and 85.5%, respectively (Mertz et al, 1984).

Proteins of fermented Sudanese sorghum baked products (*kisra* and *abrey*) were shown to be more digestible by pepsin than proteins of unfermented cooked gruels (Axtell et al, 1981). Eggum et al (1983) demonstrated that lowering the pH of the gruels before cooking, either by adding mineral acid or by fermentation, prevented the negative effects of cooking on protein digestibility and energy values. Graham et al (1986) also found that the traditional Sudanese fermentation technique markedly enhanced the protein and energy digestibilities for children of two whole-grain sorghum flours, one of which had been very poorly digested by infants and children of the same age in an earlier study.

In Nigeria, the popular food *ogi* is traditionally prepared by steeping maize, millet, or sorghum grain in water for two to four days at room temperature, then wet-milling the fermented grain. Adeyemi (1988) has established that acceptable *ogi* flour quality can be obtained by controlled fermentation followed by dry-milling of sorghum grain, thereby allowing the possibly of producing instant *ogi* powder.

Similar improvement in sorghum nutritional value by fermentation has been measured using the protozoan, *Tetrahymena pyriformis* (Au and Fields, 1981; Kazanas and Fields, 1981). The effect of fermentation on thiamin, niacin, and riboflavin content of sorghum is less clear. Au (1979) reported decreased B-vitamin content (Au, 1979), and others reported no change or increased contents with fermentation (Kazanas and Fields, 1981; Aliya and Geervani, 1981). Chavan et al (1988) have shown that fermentation periods of one to five days significantly increased the amount of soluble protein, free amino acids, and in vitro protein digestibility in sorghum, green gram, and sorghum–green gram mixtures. The optimum fermentation period for improving protein availability with minimal undesirable souring of the product was 24 hr.

EFFECT OF GERMINATION ON SORGHUM NUTRITIONAL VALUE

Wu and Wall (1980) reported that germination of sorghum grain resulted in a substantial increase in the lysine-rich albumin protein fraction and a decrease in prolamin and cross-linked prolamin fractions. The prolamin fractions are low in lysine. Taylor (1983) reported a large decrease in the Osborne prolamin protein fraction (to 16% of original concentration in seven days) and a smaller decrease in glutelin protein (57% of original) during malting. Concentrations of some proteins of the albumin and globulin fraction decreased and others increased.

Scanning and transmission electron microscopy techniques have been used to study the patterns of modification that occur in sorghum grain during germination (Glennie et al, 1983). The modification began at the endosperm scutellum interface, then moved into the floury endosperm, with slight modification of the peripheral endosperm. The corneous endosperm was modified last. The matrix protein began to disappear first, and after it was disrupted, the starch granules and protein bodies were degraded simultaneously. The aleurone cells were extensively modified during germination, and their mineral content was greatly reduced. The cell walls were the only part of the endosperm that appeared visually unchanged after germination (Glennie et al, 1983). Low endo-β-glucanase activity might be partly responsible for the limited degradation of endosperm cell walls during malting of sorghum grain (Aisien, 1982).

Investigators have noted increased lysine, threonine, isoleucine, valine, phenylalanine, and methionine and reduced concentration of leucine as a result of five-day germination of waxy and normal varieties with a wide range of endosperm textures (Elmalik et al, 1986b). Taylor (1983) reported increases in all essential amino acids, and Wang and Fields (1978) noted increases in methionine, lysine, and tryptophan. In the study of Elmalik et al (1986b), the most favorable enhancement in essential amino acids occurred in a highly corneous cultivar. In vitro protein and starch digestibilities were higher in sorghum germinated for only two days than in ungerminated samples. However, it should be noted that these improvements were associated with a dry matter loss averaging about 15%, with the greatest loss being for intermediate-textured normal sorghum.

Obizoba (1988) concluded, on the basis of a rat feeding study, that malting and wet-milling of corn and sorghum for use in gruels consumed in West Africa seemed to improve nutritional value.

Scientists have recommended some caution in the use of germinated sorghum in foods, citing the potential risk of cyanide poisoning from hydrolysis of the cyanogenic glycoside, dhurrin, which apparently is formed during the germination process (Panasiuk and Bills, 1984; Osuntogun et al, 1989). Dada and Dendy (1987) confirmed that germinated feterita sorghum contains dangerously high levels of hydrocyanic acid, the concentration of which is not affected by drying sorghum at 50°C before it is ground into meal. However, HCN concentration in the germinated grain was reduced below the recommended maximum levels (200 ppm) either by dry heating to 180°C, by moist heating to 100°C, or by lactic acid or alcoholic fermentation. The authors caution that during processing of germinated sorghum by the methods described, dhurrin is decomposed or hydrolyzed, liberating HCN. Therefore, care should be taken to provide good ventilation of the processing area to allow removal of the HCN.

EFFECT OF EXTRUSION COOKING ON PROTEIN DIGESTIBILITY

Studies in young children (seven to 24 months old) fed extruded, decorticated sorghum showed nitrogen absorption and retention to be 81 and 21%, respectively (MacLean et al, 1983), compared with values (from a previous study) of only 46% and 14%. In vitro digestibility measurement on the same extruded grain by Mertz et al (1984) indicated that protein digestibility was greatly enhanced by extrusion processing. Fapojuwo et al (1987) reported that extrusion temperature affected sorghum protein digestibility (in vitro), with the greatest benefit achieved at a temperature of 200°C. Raising the pH with $Ca(OH)_2$ before extrusion further improved digestibility. Addition of boiling water to the extruded sorghum with mixing and no additional heat treatment produced a smooth, thin porridge, in which the improved protein digestibility attained after extrusion was retained. However, when the extruded product was mixed with boiling water and heated for 5 min to obtain a stiff porridge, protein digestibility was significantly reduced. The authors suggested that extruded samples containing calcium hydroxide were more digestible after boiling because of tannin deactivation.

Conversely, studies using extrusion-processed sorghum for feeding swine (Allee, 1976) or poultry (Deyoe et al, 1967; Sloan et al, 1971; Zhuge et al, 1990) have not reported enhanced nutritional quality compared with ground grain. Although starch and protein are generally found to be more digestible after extrusion at mild to moderate conditions, protein quality and biological value may suffer under more extreme processing conditions. Loss of bioavailable lysine during extrusion of wheat flour ranged from 0 to 40% and was affected by feed rate, screw speed, temperature, and moisture (Bjork et al, 1983; Bjork and Asp, 1984). The effects of extrusion cooking on the nutritional value of foods, particularly cereals, have been extensively reviewed (Bjork and Asp, 1983; Bjork, 1984; Asp and Bjork, 1984, 1989; Cheftel, 1986; Harper, 1988).

Sorghum Lipids

Crude fat content of sorghum averages about 3%, which is higher than that of wheat and rice but lower than that of maize and pearl millet (Table 1). The germ and aleurone layers are the lipid-rich fractions of the grain, with the germ contributing about 80% of the total oil (Rooney and Serna-Saldivar, 1991). Because the germ constitutes only 10% of normal sorghum but 14 to 16% of some naturally occurring high-lysine mutants, the lipid content of the mutants is 5.8–6.6% higher (Jambunathan, 1980). Fatty acid composition of sorghum oil is similar to that of corn oil, with high concentrations of linoleic (49%), oleic (31%), and palmitic acids (14%). In addition, the oil contains 2.7% linolenic acid, 2.1% stearic acid, and 0.2% arachidic acid (Rooney, 1978).

Minerals

Values reported for the various mineral components of sorghum grain vary widely, to a large extent reflecting the mineral composition of the soils and conditions where the plants are grown. Hulse et al (1980) compiled data from a multitude of sources, which emphasized the extreme variability in sorghum mineral content. Values for calcium and iron, which are often deficient in high-cereal diets, are similar in sorghum and other common grains (Table 1).

The bioavailability of iron in sorghum is negatively affected by the presence of polyphenols and phytates. Absorption of iron was much higher in Indian women consuming sorghum containing no condensed polyphenols than in those eating bird-resistant, high-tannin sorghum (Gillooly et al, 1984). When the phytate-rich outer layers of the polyphenol-negative sorghum were pearled away, iron absorption improved, as was also the case for the bird-resistant cultivar. Absorption of iron was not higher from porridge prepared from malted than from unmalted sorghums; however, addition of ascorbic acid to all the sorghum porridges improved iron absorption, whereas tea consumption reduced iron bioavailability (Giloolly et al, 1984).

Milling method and the addition of peanut paste have been shown to affect iron availability in sorghum and maize flour gruels (Mbofung and Ndjouenkeu, 1990).

Although total iron was not significantly different in gruels prepared from traditionally milled sorghum (moistened grain was dehulled using a mortar and pestle) than in mechanically dehulled sorghum, percentages of soluble and ionizable (in vitro bioavailable) iron were much higher in the mechanically milled sorghum. Both milling methods involved 12-hr soaking and 1-hr drying periods after dehulling, during which phytase presumably was active. The authors suggested that the increase in iron bioavailability with mechanical milling could have been the result of more efficient removal of the phytate-rich hulls and/or greater enzymatic destruction of phytate (300 vs. 630 mg/100 g).

Other minerals in sorghum include (db): chlorine (0.10%), magnesium (0.18%), phosphorus (0.33%), potassium (0.39%), sodium (0.03%), sulfur (0.15%), cobalt (0.18 mg/kg), copper (11 mg/kg), iodine (0.04 mg/kg), manganese (18 mg/kg), and zinc (19 mg/kg) (NRC/NAS, 1982). One report suggested that high dietary leucine, as is the case with sorghum-rich diets, enhances copper absorption (Krishnamacheri, 1974), but Klopfenstein et al (1983b) were unable to verify that finding. As in other cereals, much of the phosphorus present (48–70%) is in the form of phytates (Kurien et al, 1960), which probably reduces its bioavailability and may adversely affect the availability of other minerals as well. Sorghum germ contains 69% of the total minerals in the grain, the endosperm contains 21%, and the bran 11% (Hubbard et al, 1950). Pedersen and Eggum (1983) showed the effects of milling to different extraction rates on the mineral and phytate contents of sorghum flours (Table 5).

When Stuart et al (1987) compared absorption of radiolabeled iron and zinc by rats fed meals containing sorghum and maize gruels, they found no significant differences in iron absorption from acidic, basic, and neutral sorghum *tô*, maize gruel, or fermented sorghum *aceda*, with values ranging from 77.2 to 83.2%. However, a higher amount of zinc was absorbed by rats fed the fermented *aceda* (97%) than by those fed the maize gruel or other sorghum products (69–78%).

Bioavailability (chick assay) of minerals in sorghum and pearl millet cereals is compared with sesame seed meal in Table 6 (Nwokolo, 1987). Percent of available magnesium was higher in millet than in sorghum, but zinc, iron, and manganese were less available in millet. Only potassium was less available in sesame seed meal than in the cereals.

The significant mineral and vitamin contribution of South African beer derived from sorghum adjunct or maize grits is shown in Table 7 (Van Heerden, 1989). Concentrations of thiamin, nicotinic acid, ash, copper, manganese, iron, magnesium, potassium, and phosphorus were higher in beers brewed with sorghum ad-

TABLE 5
Effect of Milling on Mineral Content (dry basis) of Sorghum Flours[a]

	Extraction Rate (%)					
Mineral	**100**	**95**	**90**	**80**	**73**	**64**
Calcium (mg/g)	0.50	0.30	0.22	0.37	0.54	0.61
Phosphorus (mg/g)	4.0	3.9	3.4	2.4	1.9	1.5
Phytate phosphorus (mg/g)	3.1	3.0	2.7	1.6	1.2	0.4
Zinc (ppm)	36	33	30	21	10	10
Copper (ppm)	4.9	4.3	4.0	3.3	2.4	2.4
Iron (ppm)	179	104	65	83	76	54

[a] Adapted from Pedersen and Eggum (1983).

TABLE 6
Availability of Minerals in Sorghum, Pearl Millet, and Sesame Seed Meal[a]

	Mineral Availability (%)		
Mineral	**Sorghum**	**Pearl Millet**	**Sesame Meal**
Phosphorus	59.8 b[b]	61.1 b	65.7 a
Calcium	60.0 b	58.2 b	64.9 a
Magnesium	55.6 b	65.0 a	67.6 a
Potassium	62.1 a	61.2 a	36.2 b
Copper	58.2 b	61.4 b	79.9 a
Zinc	−2.6 c	−6.9 d	41.2 a
Iron	4.0 b	−12.4 c	58.6 a
Manganese	42.6 c	21.4 d	78.3 a

[a] Adapted from Nwokolo (1987).
[b] Means in the same row not followed by the same letter are significantly different at $P < 0.05$.

TABLE 7
Minerals and Vitamins in South African Beers Derived from Sorghum Adjunct or Maize Grits[a]

	Type of Beer			Contribution (%/L)	
Nutrient (mg/L)	**Maize Grits**	**Sorghum Adjunct**	**RDA[b] (mg)**	**Maize Grits**	**Sorghum Adjunct**
Zinc	1.63	1.94*[c]	12–15	11–14	13–16
Copper	0.17	0.27*	1.5–3.0	6–11	9–18
Manganese	0.88	1.83*	2.0–5.0	18–44	37–92
Iron	1.26	3.44*	10–15	8–13	23–34
Calcium	53	52	800	7	7
Magnesium	94	178*	280–350	27–34	51–64
Potassium	217	407*	2,000	11	20
Sodium	21	18	500	4	4
Phosphorus	150	305*	800	19	38
Thiamin	0.58	0.96*	1.1–1.5	39–53	64–87
Riboflavin	0.41	0.47	1.3–1.7	24–32	28–36
Nicotinic acid	3.03	5.58*	15–19	16–20	29–37

[a] Adapted from Van Heerden (1989).
[b] Values cover requirements for women and men aged 25–50 years (NRC/NAS 1989).
[c] $^{*}P = 0.1$.

junct than in those prepared from maize grits. In addition, Derman et al (1980) reported that iron absorption was more than 12 times greater from maize and sorghum beer than from a gruel made from the constituents used to prepare the beer. It is obvious that the beers can make substantial contributions to human dietary requirements.

The mineral composition of sorghum flakes and 10 or 50% wheat-flour-fortified sorghum flakes developed for use as breakfast food in Saudi Arabia was reported by Al-Kahtani (1989). Mineral composition of the flakes generally showed an abundance of sodium, potassium, magnesium, and iron but shortages of calcium and copper. The 50% wheat replacement resulted in higher iron but lower phosphorus and zinc contents.

Vitamins

Niacin, thiamin, and riboflavin concentrations in sorghum are similar to those in maize (Table 1). Niacin contents of both maize and sorghum are lower than those of wheat and rice, and the niacin deficiency disease, pellagra, is common in areas of the world where maize and sorghum are dietary staples. Just how much of the niacin present in the grain is biologically available has been the subject of controversy, with results being dependent to some extent on extraction solvents and procedures used in chemical assays and the organism employed in bioassays. Hulse et al (1980) reviewed the earlier (1940–1980) data. Carter and Carpenter (1981) showed clearly that niacin in sorghum is bound to high molecular weight entities. Niacin was biologically available to rats after alkali treatment of the grain but not after boiling. When Carter and Carpenter (1982) compared total niacin in several boiled grains, they reported the following values (in milligrams per kilogram): sorghum, 45.5; rice, 70.7; maize, 18.8; and wheat, 57.3. The percentage of total niacin available to the animals from the boiled grains was sorghum, 33; rice, 41; maize, 37; and wheat, 31. According to Wall and Carpenter (1988), little doubt remains that niacin bioavailability in cereals is limited.

Although sorghum is not deficient in the amino acid tryptophan, which is considered to be a contributing factor in the pellagragenicity of maize diets, its high concentration of leucine may interfere with the biochemical conversion of tryptophan to niacin in humans (Belavady and Gopalan, 1965).

Other vitamins present in sorghum grain include (dry basis[db]): vitamin D-2 (29 IU/g), vitamin E (12 mg/kg), vitamin K (0.2 mg/kg), carotene (1.0 mg/kg), biotin (0.42 mg/kg), folacin (0.2 mg/kg), pantothenic acid (12.5 mg/kg), and vitamin B-6 (5.0 mg/kg) (NRC/NAS, 1982). Biotin bioavailability as measured by turkey poult bioassay is reportedly higher in sorghum (29.5%) than in wheat (17.0%) and triticale (15.9%), but much lower than in maize (95.2%), soybean meal (76.8%), or canola meal (65.4%) (Misir and Blair, 1988).

Vitamin loss during the preparation of traditional sorghum breads in Saudi Arabia was small, except for pantothenic acid, which decreased about 40% (Khalil et al, 1984). Pantothenic acid and vitamin B-12 concentrations were significantly higher when bread was made from fermented (18 hr) dough. Taur et al (1984) showed that ascorbic acid in germinated sorghum increased from 5.00–6.87 to 8.12–10.00 mg/100 g after 72 hr of fermentation. With longer fermentation times, ascorbic acid concentration declined.

The ascorbic acid requirement was higher for guinea pigs fed sorghum diets than for those fed casein-supplemented diets (Klopfenstein et al, 1981a–c). Diets supplemented with 2 mg of ascorbic acid/day for nine weeks resulted in good growth rates in guinea pigs fed casein-based diets, and additional ascorbic acid (40 mg/day) did not improve guinea pig growth rates (Klopfenstein et al, 1981b). Guinea pigs were used as the test animal because, like humans, they require a dietary source of ascorbic acid. However, when the animals were fed isocaloric, isonitrogenous, nutritionally adequate, sorghum-based diets supplemented with 2 mg/day of ascorbic acid, the animals grew poorly. When the guinea pigs fed sorghum diets were supplemented with 40 mg/day of ascorbic acid, they grew as well as those fed casein-based diets supplemented with 2 or 40 mg/day. Higher levels of dietary ascorbic acid also appeared to have a niacin-sparing effect in sorghum-fed guinea pigs (Klopfenstein et al, 1981c). The authors suggested that increased need for ascorbic acid in sorghum diets might be related to sorghum's high leucine content. In addition, when guinea pigs fed sorghum diets were supplemented with adequate ascorbic acid for optimum growth, their blood and/or liver cholesterol levels were consistently lower than those receiving a lower, but supposedly adequate (Yew, 1975; Collins and Elvehjem, 1958) supplement (Klopfenstein et al, 1981a,b, 1983a).

Increased ascorbic acid requirements in grain-fed animals have also been reported by Majumder et al (1975), who noted that guinea pigs fed wheat, rice, maize, or pearl millet diets required about five times more ascorbic acid (about 10 mg/day) for optimum growth than did animals fed cereal diets enriched with 15% casein. Subramanian et al (1973) suggested that the increased requirement for ascorbic acid is the result of the need to detoxify elevated levels of histamine produced in response to stressful diets (Subramanian et al, 1973). A better understanding of the role of ascorbic acid in sorghum diets is particularly important, because in populations where sorghum is a dietary staple, dietary ascorbic acid intakes are often low.

Nutritional Benefits of Composite Foods

As previously discussed, sorghum protein is severely limiting in lysine. Addition of synthetic essential amino acids can improve sorghum diets (Daniel et al, 1966), but the practice is inefficient and impractical in populations where sorghum is a dietary staple. Therefore, combinations of sorghum with lysine-rich legumes have been used to enhance its nutritional contribution. The nutritional values of cereal-soy product combinations used in the Food for Peace Program were reported by Bookwalter et al (1987b). Soybean meal–fortified (8%) maize-sorghum (75:25) tortillas were organoleptically acceptable and were much improved nutritionally over sesame-fortified or unfortified products (Serna-Saldivar et al, 1988).

Essential amino acid content, PER, digestibility, and biological value in rats of sorghum-legume mixtures (2:1 based on protein contribution) were measured by Pushpamma and Devi (1979). Legumes tested with sorghum were red gram (*Cajanus cajan*), black gram (*Phaseolus mungo*), Bengal gram (*Cicer arietinum*), green gram (*Phaseolus radiatus*), cowpea (*Vigna sinensis*), horse gram (*Dolichos biflorus*), and soybean (*Glycine max*). PER ranking for the mixtures was soybean (2.47), black gram (2.44), Bengal gram (2.36), cowpea (2.31), horse gram (2.28), green gram (2.03), and red gram (1.92), compared with a value of 2.76 for skim milk powder. When similar combinations were fed to preschool children at a childcare facility at the International Crops Research Institute for the Semi-Arid Tropics (ICRISAT) near Hyderabad, India, balance studies showed nitrogen retention to be higher in children fed a rice–red gram diet than in those fed sorghum–red gram or sorghum–black gram diets (Pushpamma et al, 1979). However, comparison of the increase in anthropometric measurements showed that values for the experimental (sorghum-legume) group increased two to three times more than those of the control group (children living in the area not in childcare at ICRISAT).

Sorghum-wheat-legume formulations for baked products with high protein values were developed and tested by Okeiyi and Futrell (1983). Combinations with soybeans, cowpeas, and peanut butter were evaluated. PER was 2.37 for the sorghum-wheat-soy blend (not different from the casein control). The product met the FAO/WHO (1973) protein requirements for children and was well-accepted by African children.

Akinyele and Fasaye (1988) demonstrated significant nutritional improvement in a traditional fermented Nigerian food (*ogi*) by addition of cowpeas (70:30, cereal to cowpeas) to the sorghum or maize gruels. However, *ogi* from maize-cowpea mixtures was rated more highly acceptable (100%) as a food product than sorghum-cowpea *ogi* (70%). Mixtures of 30% sorghum with the under-utilized legume, bambara groundnut (*Voandzeia subterranea* (L.) Thouars), were used to produce inexpensive, nutritious, ready-to-eat dishes in Nigeria (Obizoba, 1989).

The nutritive value of some new and traditional Sudanese sorghum food products has been evaluated by Badi et al (1990). They found fermented sorghum (*kisra*) with added chickpeas or peanuts to be a well-balanced food for infants older than one year. Baby foods prepared from 58% sorghum flour, 25% sorghum malt, and 17% milk powder were also found adequate to provide the sole source of protein for children more than one year old.

The addition of 10 to 50% popped amaranth flour (which contains about twice as much lysine as cereals) to sorghum, maize, and wheat flours improved their protein quality without affecting energy utilization in rats (Pedersen et al, 1987).

Antinutrients in Sorghum Grain

TANNINS (CONDENSED POLYPHENOLS)

All sorghums contain phenolic compounds, including phenolic acids and flavonoids. Some contain condensed polyphenols called tannins. Those compounds can affect color, flavor, and nutritional quality of the grain and products prepared from it (Hahn et al, 1984). Little is known about the adverse nutritional effects of sorghum phenolic acids and flavonoids, but they are generally considered minor factors in the poor performance of brown, bird-resistant, high-tannin sorghums. Such sorghums are characterized by the presence of a pigmented inner integument (often called testa) located beneath the pericarp. Desirable agronomic characters of high-tannin sorghums are their resistance to premature sprouting and molding (Price and Butler, 1979). Reduced nutritional quality of those sorghums has been attributed mainly to their high content of condensed polymers of 5,7-hydroxyflavan-3-ol units. The antinutrient effects of tannins in sorghum grain have been thoroughly reviewed (Price and Butler, 1980; Salunkhe et al, 1982; Butler et al, 1984,1986; Salunkhe et al, 1990).

Tannins impart astringency to sorghum, which tends to reduce animal intake but is beneficial in providing protection of the crop from birds (Butler et al, 1984). Reduced weight gains in animals has been attributed to this reduced feed intake. In addition, tannins bind to and precipitate both endogenous and exogenous (including digestive enzyme) proteins, reducing the nutritional contribution of the grain (Asquith and Butler, 1986; Griffiths, 1985). Numerous trials have shown reduced digestibilities, growth rates, feed efficiencies, metabolizable energy, and amino acid bioavailability in chicks, rats, and various livestock fed high-tannin sorghums. High-tannin sorghum contains more tannin and is nutritionally inferior for experimental animals when it is immature (Rogler and Sell, 1984). Davis and Hoseney (1979b) showed that assayable tannins rose gradually in three sorghum varieties until six weeks before harvest, then remained unchanged in one variety and decreased in the others. Similar antinutritional effects of dietary tannins on human subjects is lacking, so it is difficult to assess their nutritional significance in human diets (Salunkhe et al, 1990). However, a study involving Indian women showed that iron absorption was significantly poorer from porridge prepared from bird-resistant sorghum than from similar food prepared with tannin-negative sorghum (Gillooly et al, 1984). Conversely, another Indian study indicated that tannins played only a minor role in determining iron availability, and that phytic acid in the grain was the major determinant of bioavailability (Radhakrishnan and

Sivaprasad, 1980). Fermentation of sorghum reportedly reduces the phytic acid content (Kazanas, 1979). Reddy et al (1989) addressed the nutritional consequences of phytates in cereals.

Much work has been reported on efforts to remove or inactivate tannins in sorghum (Salunkhe et al, 1990). Because tannins are concentrated in the testa or seed coat of the grain, removal of that layer should eliminate the compounds. However, using traditional milling processes, which often involve pounding in a mortar and pestle, complete removal of the pericarp and inner integument (testa) is difficult without considerable loss of endosperm as well. Several mechanical mills have been used to remove the outer layers of the grain, but flour yields were low, and protein quality suffered because of the soft endosperm of the high-tannin genotypes (Chibber et al, 1978; Reichert, 1982). Following a study to optimize conditions for abrasive dehulling of high-tannin sorghums, Mwasaru et al (1988) concluded that development of harder, rounder grain with the minimum tannin content required to confer bird resistance and other desirable agronomic characteristics will be required before high-throughput, commercial milling of high-tannin sorghum can become economically feasible. However, several high-tannin lines have been identified that could be abrasively dehulled and yield flour extraction levels of 70% or higher (Reichert et al, 1988).

When sorghum flour is mixed into a batter, the amount of assayable (i.e., extractable) tannins decreases and is further reduced on cooking (Price et al, 1980). However, rats fed the cooked or uncooked batter gain even less weight than those fed untreated high-tannin grain. Germination has also been shown to reduce assayable tannin levels in sorghum (Reichert et al, 1980a; Moharram, 1987; Osuntogun et al, 1989), but levels rise again when the seeds are dried (Butler, 1982). Those results emphasize the necessity of feeding trials in determining the value of various tannin-reduction procedures.

Nutritional improvement for rats, chicks, or pigs has been achieved with various degrees of success by treatment of high-tannin sorghum with ammonia (Price and Butler, 1979), Magadi soda solution ($Na_2CO_3 \cdot NaHCO_3 \cdot H_2O$; Muindi and Thomke, 1981), formaldehyde (Daiber and Taylor, 1982), gelatin (Butler et al, 1986), and polyethyleneglycol (Hewitt and Ford, 1982); by high-moisture reconstitution (Reichert et al, 1980a; Mitaru et al, 1983; Teeter et al, 1986); or by supplementing diets with orthophosphoric acid or dicalcium phosphate (Ibrahim et al, 1988) or sodium bicarbonate (Banda-Nyirenda and Vohra, 1990).

PROTEIN INHIBITORS OF DIGESTIVE ENZYMES

Inhibitors of trypsin, chymotrypsin, elastase, and other proteases have been identified in sorghum and many of the millets (Pattabiraman, 1985). Table 8 shows the relative inhibitory action on proteases by aqueous extracts of sorghum and some millets. Sorghum extract was a stronger inhibitor of elastase and chymotrypsin than of trypsin. Protease inhibitory activity for some millet extracts was much higher than in sorghum extract. The caseinolytic activity of human and bovine pancreatic extracts was found to be inhibited similarly by crude extracts of sorghum and several millets.

Protein inhibitors of α-amylase have also been measured in sorghum and some millets (Buonocore and Silano, 1985), but, as is the case with the protease inhibitors, little information is available regarding the nutritional implications of their presence.

TABLE 8
Relative Inhibition of Proteases by Sorghum and Millet Extracts[a]

Grain	Enzyme Inhibited (μg)/Protein (mg) of Seed Extract[b] Bovine Trypsin	Bovine Chymotrypsin	Porcine Elastase
Sorghum	2.59	6.48	7.41
Pearl millet	6.60	0.00	3.30
Foxtail millet	7.18	0.00	1.28
Kodo millet	0.83	1.30	3.47
Finger millet	33.30	12.50	8.33
Japanese barnyard millet	10.00	4.86	6.94

[a] Adapted from Pattabiraman (1985).
[b] Substrate used was casein.

LECTINS

The presence of lectins, which are carbohydrate-binding compounds, in some sorghum varieties has been reported by Neucere (1982). Again, their nutritional significance is not known with certainty, but probably is not great, because their inactivation by heat does not appear to have any effect on the nutritional quality of wheat, barley, and rice (Liener, 1981).

MOLDS AND MYCOTOXINS

Throughout the world, molds and their toxic metabolic products are responsible for the deterioration of much grain, including sorghum (Stoloff, 1976). The extent of human illness as a result of eating contaminated grains is not fully known. However, data show that mycotoxins are commonly found in food and may be involved in the etiology of certain diseases in some areas of the world (Mirocha et al, 1980). Storage fungi, mostly species of *Aspergillus* and *Penicillium*, grow when grain is stored with moisture contents above 13–14% (Sauer, 1988). Aflatoxins B_1, B_2, G_1, and G_2 were found in molded heads of sorghum in India (Tripathi, 1973), Uganda (Alpert et al, 1971), and the United States (Shotwell et al, 1969; Stoloff, 1976). If, at any point, cereals are stored above 23% moisture content, *Fusarium* species will grow, and the possibility for production of zearalenone and tricothecene toxins exists (Mirocha et al, 1980). The presence of high concentrations of zearalenone (5,600 ppb) in sorghum in the United States has been reported by Mirocha et al (1976). Although most problems with molds occur during storage of the grain, some preharvest damage takes place. McMillian et al (1983) reported contamination of sorghum with aflatoxin and zearalenone before the grain was harvested. During a wet growing season, Seitz et al (1983) identified *Alternaria* and *Fusarium* species in sorghum soon after the grain reached physiological maturity. Hulse et al (1980) cited several studies indicating that tannins in sorghum may provide some degree of protection against preharvest molding. Symptoms of human poisoning by tricothecenes include nausea, vomiting, diarrhea, dizziness, abdominal pain, and throat irritation (Scott, 1990). Alimentary toxic aleukia (ATA) is an often-fatal disease caused by tricothecene contamination of grain (Joffe, 1965).

THE MILLETS

The millets encompass a group of small-seeded annual grasses that are dietary staples in parts of Africa, Europe, and Asia (see also chapter 2). Major millet gen-

TABLE 9
Comparison of Sorghum and Millet Production in Countries with Developing Market Economies, 1988[a]

Item	Millets	Sorghum
Area harvested (10^3 ha)	37,764	40,601
Yield (kg/ha)	761	1,123
Production of (10^6 metric tons)	28,745	44,325

[a] Data from FAO (1990).

era are *Pennisetum, Eleusine, Setaria, Echinochloa, Panicum, Paspalum, Eragrostis,* and *Digitaria.* In general, millets have short growing seasons and are more likely than most other cereals to produce a crop when rainfall amounts are limited. In areas of northern China, southern Russia, and most of the semiarid tropics, the millets are consumed primarily as human food (Schery, 1963). Although the area of land devoted to millets is not much less than that for sorghum, the amount of millet harvested yearly is much lower because of the comparatively low yields (Table 9). In Africa, a constraint to increased production is the instability of the grain markets. Because of widely fluctuating market prices, rural families tend to produce only enough to meet their own needs, with little surplus to market (Hoseney et al, 1987).

Pearl Millet
(*Pennisetum glaucum*: Synonyms *P. typhoideum, P. americanum*)

The nutritional properties of pearl millet (bajra, bulrush, or cattail millet) have received more attention than those of the other common millets, because it is the largest-seeded, most widely grown type (Hoseney et al, 1987). Major areas of consumption are in Africa and India (Serna-Saldivar and Rooney, 1991). Representative contents of selected nutrients in pearl millet are compared with those in sorghum, other millets, wheat, rice, and maize in Table 1.

PEARL MILLET CARBOHYDRATES

Normal pearl millet examined by Singh et al (1987) contained about 70% starch, although lower starch values (56–65%) were found by Freeman and Bocan (1973) and Sheorain and Wagle (1973). Free sugar values reported range from 1.9 to 2.8% (Singh et al, 1987; Subramanian and Suryaprakash, 1981). In addition to sucrose (which constitutes about 66% of the free sugar), stachyose, raffinose, glucose, and fructose have also been identified in pearl millet (Subramanian et al, 1981). Reported amylose content of pearl millet starch ranges from 17% (Badi et al, 1976b) to almost 25% (Singh and Popli, 1973). Physical and chemical characteristics of pearl millet starch were described extensively by Beleia et al (1980). Alpha-amylase activity is eight to 15 times greater in pearl millet than in wheat (Sheorain and Wagle, 1973). Beleia and Varriano-Marston (1981a,b) have reported the properties of millet amylases and their activity toward intact and heated starch granules. Interestingly, millet amylases showed higher amylolytic action on wheat starch than on millet starch. High amylase activity in millet flour probably is responsible for its improving effect in wheat flour breads. When Badi et al (1976b) added 10% millet flour to a standard baking formula, loaf volume was increased, crumb grain was improved, and flavor was very good. The addition of 10% millet flour to a formula containing no malt or sugar gave a better loaf volume than the standard formula with the same wheat flour and containing 6% sugar. Adding 5 to 20% of sorghum flour to the standard formula decreased loaf volume. The findings are significant economically, especially in countries where millet is grown, where sugar is a cash crop for foreign exchange, and where malt and wheat are imported (Badi et al, 1976b).

PEARL MILLET FIBER

Total dietary fiber (as determined by the method of Southgate et al, 1978) in pearl millet is about 17% (Singh et al, 1987), compared with values (AOAC, 1990) of 14.1% for whole wheat, 15.0% for corn, 19.1% for barley, 11.2% for oat groats, and 3.9% for brown rice (USDA/HNIS, 1988, converted to dry basis). Neutral detergent fiber content was about 9% (Singh et al, 1987), indicating that the total dietary fiber in pearl millet contains about equal amounts of soluble and insoluble nonstarch polysaccharides.

PEARL MILLET PROTEIN

Protein contents ranging from less than 9% to more than 20% have been reported for the grain (Desai and Zende, 1979), but values close to 11% appear to be representative (Hulse et al, 1980). Results of a study of 22 pearl millet varieties showed that the variation in protein content of the same varieties among different locations was as wide (to 7%) as that among different varieties grown at one location (Sharada et al, 1986).

PROTEIN QUALITY

Although lysine is the most limiting amino acid in pearl millet protein, quite a range of concentrations has been reported, with values at the higher end of the range (3.6 g/100 g protein) as high as that for opaque-2 maize (Badi et al, 1976a). Table 2 shows that pearl millet has a higher potential to meet the lysine requirements for children than most other cereals. Except for its lysine deficit, pearl millet has well-balanced protein, with higher concentration of threonine and lower (but adequate) leucine than sorghum protein. Tryptophan levels are generally higher in pearl millet than in other cereals (Chung and Pomeranz, 1985). As is the case with other grains, data from chemical or biological assays of amino acids in millet were highly variable (Hoseney et al, 1987). Some of the variability in reported results comes from cultivar differences and some from differences in agronomic and/or storage conditions, but much of the variability arises from different analyzing and reporting methods of amino acids. What can be concluded about pearl millet protein quality is that it ranks quite high in comparison to other cereals. Pushpamma et al (1972) reported a PER of 1.84 for pearl millet, compared with PERs of 1.74, 1.46, and 1.36 for finger millet, sorghum, and maize, respectively. With the addition of lysine alone (0.3% lysine hydrochloride) to pearl millet diets for rats, the animals' growth response was 93% of that observed for animals fed casein diets (Howe and Gilfillan, 1970).

High protein quality in pearl millet is probably a reflection of the pattern of nitrogen distribution in the Landry-Moureaux protein fractions of the grain. When Nwasike et al (1979) compared pearl millet data with that for corn (Misra et al, 1976) and sorghum (Guiragossian et al, 1978), they reported that 22.3% of pearl millet's total nitrogen was in the albumin-globulin fraction, compared with 16.6%

for corn and only 10.0% for sorghum. Although all three cereals had about the same percentage (47.0–48.7%) of total prolamin nitrogen, only 6% of pearl millet's nitrogen was in cross-linked prolamins, similar to the value of 10.0% for corn, and much lower than the 31.3% reported for sorghum. Lysine contents of the albumin-globulin fractions were 5.9% for pearl millet, 4.3 to 6.3% for corn, and 3.2 to 4.8% for sorghum. Non-cross-linked prolamins in pearl millet contained 0.5% lysine compared with values of 0.1 to 0.5 for corn and 0.1 to 0.3 for sorghum. Lysine contents of cross-linked prolamins were 0.1% in pearl millet, 0.4 to 0.8% in corn, and 0.1 to 0.2% in sorghum (Nwasike et al, 1979).

Protein content and grain yield do not seem to be strongly associated in pearl millet (Kumar et al, 1983; Sharada et al, 1986). Studies on the relationship between nutritional quality and grain yield suggested that selection for increased protein content should be possible without sacrificing grain yield or protein lysine concentration (Kumar et al, 1983). In India, high-protein genotypes in pearl millet have resulted from the breeding programs at ICRISAT (Singh et al, 1984) and at Haryana Agricultural University at Hissar, India (Chauhan et al, 1986). The latter authors reported protein contents ranging from 10.5 to 14.8% for high-yielding varieties. Singh et al (1987) reported that the protein content of high-protein lines (14–19%) developed at ICRISAT was 60% higher than in parents; starch and soluble sugars were 40 and 20% lower, respectively, and total dietary fiber was increased by about 10%. A 20% decrease was noted for albumin proteins, with a corresponding increase in prolamins. Although percentage of lysine in the protein was decreased by 16%, total lysine in the sample increased by an average of 37% as a result of the higher total protein content.

DIGESTIBILITY

True protein digestibility in rats ranged from 94 to 97% and was not different for high- and normal-protein samples (Singh et al, 1987). Digestible energy content was high in both normal- and high-protein types (82.6 to 90.0), although the higher fiber content of the high-protein lines apparently reduced their digestibility somewhat (Singh et al, 1987). Because of the lower percent lysine in the protein, biological value was lower for the high-protein millet than for normal millet, but the amount of utilizable protein (protein × NPU/100) was 1.5 to two times higher in the high-protein millet (Singh et al, 1987).

Dassenko (1980) found protein and calorie digestibilities to be higher for pearl millet flour than for wheat flour. However, when chapatis were prepared from the flours, protein and calorie digestibility increased in the wheat chapati but not in the millet product. Millet protein quality decreased during processing into chapatis. PER for raw millet flour was higher than for the raw wheat flour. When chapatis were prepared from the wheat flour, PER increased. However, when chapatis were prepared from millet, PER decreased to a value lower than that of the wheat chapati. It was noted that longer cooking times were required to produce acceptable chapatis from pearl millet than from wheat.

Effect of Fermentation on Pearl Millet Digestibility. Fermentations by single and mixed cultures of yeasts (*Saccharomyces diastaticus, S. cerevisiae*) and lactobacilli (*Lactobacillus brevis, L. fermentum*) at 30°C for 72 hr improved both in vitro protein and starch digestibilities of pearl millet flour (Khetarpaul and Chauhan, 1990). Autoclaving the raw pearl millet (15 psi for 15 min) significantly increased protein digestibility. Although all yeast, lactobacillus, and mixed-culture fermentations increased protein digestibility in the autoclaved pearl millet, the greatest increase was seen with *S. cerevisiae*. Among all the single- and mixed-culture fermentations, *S. diastaticus* resulted in the highest starch digestibility. The authors suggested that improvements in protein and starch digestibilities resulted from 1) proteolytic and amylolytic action of the microorganisms and 2) reduction of phytate in the samples. Phytate is known to inhibit proteolytic (Knuckles et al, 1985) and amylolytic enzymes (Serraino et al, 1985; Yoon et al, 1983) and probably forms indigestible complexes with endogenous protein and starch. Khetarpaul and Chauhan (1990) noted significant negative correlations between phytic acid levels and protein and starch digestibilities in fermented millet. When various weaning mixtures were prepared from the fermented millet and evaluated by a panel of nine adults, the mixture containing fermented flour, roasted peanut, green gram dal, sesame, and sugar had the highest acceptability (>7 on a 9-point hedonic scale where 9 = like extremely and 1 = dislike extremely), but all products ranked above 6 (like slightly) on the scale (Khetarpaul and Chauhan, 1990).

In contrast to results reported above, when Aliya and Geervani (1981) compared total digestibilities (in rats) of legume and millet (*ambali*) products fermented by endogenous microorganisms for 24 hr, they noted increased digestibility for sorghum but not for pearl or finger millets or legumes. Although fermentation improved the biological value and net protein utilization for the legumes, sorghum, and finger millet, no improvement was noted for pearl millet (Aliya and Geervani, 1981). Dassenko (1980) reported that fermentation had no effect on protein and calorie digestibilities in wheat or pearl millet porridges, but PERs were lower for fermented porridges than for unfermented porridges.

Effects of Germination on Pearl Millet Protein and Energy Utilization. One of the major nutritional benefits of germinating pearl millet (and other common millets) lies in increased concentration of some essential amino acids in the malt (Malleshi and Desikachar, 1986). Data in Table 10 show higher concentrations of lysine and tryptophan in malted than in native millets, whereas concentrations of threonine and sulfur-containing amino acids are less affected by germination. PERs were higher for the native pearl millet diets than for finger or Italian millets. Roasting the cereal malts at 70°C for 45 min did not affect PER values (Malleshi and Desikachar, 1986).

TABLE 10
Effect of 48-hr Germination on Some Essential Amino Acids and Protein Efficiency Ratios (PER) in Millets[a]

	Amino Acids (g/100 g of protein)[b]					
Millet Type	**Lys**	**Trp**	**Thr**	**Met**	**Cys**	**PER**
Native pearl	3.7	1.5	3.7	2.6	1.6	1.02 b[c]
Malted pearl	4.3	1.9	3.7	2.5	1.5	1.11 b
Native finger	3.5	1.3	3.2	3.6	2.2	0.90 c
Malted finger	4.0	1.5	3.2	3.5	2.1	1.06 b
Native foxtail	3.0	1.1	3.1	2.1	1.5	0.85 c
Malted foxtail	3.3	1.2	2.9	2.1	1.4	1.02 b
Skim milk powder	8.4	1.6	4.7	2.7	0.8	3.30 a

[a] Adapted from Malleshi and Desikachar (1986).
[b] Lys = lysine, Trp = tryptophan, Thr = threonine, Met = methionine, and Cys = cystine.
[c] Means not followed by the same letter are significantly different at $P < 0.05$.

When Badi et al (1990) compared sorghum/sorghum malt/milk powder (58/25/17) and pearl millet/millet malt/milk powder (70/13/17) foods for babies, they noted that the millet-based food had a 20% higher protein level and better amino acid composition. The degree of extraction (100 vs. 80%) of the grains used in the formulas had little effect on protein and energy utilization for either grain. Pearl millet–based products provided a safe level of intake for children over nine months old, whereas sorghum-based products provided a safe level of intake only for children above the age of one year, based on FAO/WHO (1985) recommendations. In addition to providing increased levels of essential amino acids, the use of malted cereals resulted in baby foods with high solids content but reduced viscosity. These are desirable attributes of cereal-based baby formulas, which generally have low energy and nutrient densities (Badi et al, 1990).

PEARL MILLET LIPIDS

Lipid content is generally higher for pearl millet than for sorghum and most other common cereals (Table 1). As a result, the amount of energy per 100 g is greater than that for sorghum and nearly equal to that of brown rice. About 75% of the fatty acids in pearl millet are unsaturated, with a representative fatty acid composition of: palmitic (18.6%), palmitoleic (0.6%), stearic (4.7%), oleic (25.7%), linoleic (46.3), linolenic (3.3%), and arachidic (0.6%) acids (Osagie and Kates, 1984; Rooney, 1978; Jellum and Powell, 1971). Small amounts of myristic acid (0.2%) have also been reported (Agarwal and Sinha, 1964). The fatty acid profile is very similar to that of cottonseed oil (Holum, 1978). The amount and composition of total lipids and fatty acids in pearl millet varies with type of solvent used to extract the lipids. Hot-water-saturated butanol extracted 7.19% lipids, whereas *n*-hexane/ether (80:20) extracted 5.39% total lipids from the grain (Osagie and Kates, 1984). The butanol-extracted lipids consisted of 85% neutral lipids (mainly triacylglycerols with small amounts of mono- and diacylglycerols, sterols, and free fatty acids), 12% phospholipids (10 components), and 3% glycolipids (10 components) (Osagie and Kates, 1984).

Intact pearl millet kernels can be stored for long periods without major changes occurring in their lipids, but rancidity occurred in stored, ground pearl millet (Kaced et al, 1984; Lai and Varriano-Marston, 1980; Carnovale and Quaglia, 1973). Kaced et al (1984) reported that oxidative degradation of fatty acids was a minor factor contributing to rancidity in pearl millet. Commonly used mechanical or traditional milling practices resulted in extraction rates of 75 to 85%, with the flour retaining part of the lipid-rich germ (Lai and Varriano-Marston, 1980). Pearl millet's high lipid content has been implicated in causing reduced nutritive value of the flour after storage periods of only a few days (Dahiya and Kapoor, 1983; Boora and Kapoor, 1985; Kapoor and Kapoor, 1990). When whole-grain flour was stored for 10 or 20 days at 20 to 27°C and 60 to 70% relative humidity, protein quality decreased, as measured by chemical score and rat feeding (Dahiya and Kapoor, 1983). Samples stored in polyethylene bags were less severely affected than those stored in earthen pots or gunny bags. When whole-grain flours from two pearl millet varieties were stored in cans for eight days (25 ± 5°C), PER, true digestibility, biological value, net protein retention, and protein utilization in rats fell markedly (Boora and Kapoor, 1985). Decrease in protein quality was higher in the sample containing 8.05% fat than in that containing 6.03% fat. In a later experiment, defatted flour stored for 30 days under conditions similar to those above was found to be as nutritious as fresh flour (Kapoor and Kapoor, 1990). The addition of butylated hydroxyanisole (0.02%) or ascorbic acid (0.5%) as antioxidants reduced detrimental nutritional changes during storage of the meal, as did heating the meal in an oven at 100°C for 1 hr, but values for the chemically and heat-treated meals were still significantly lower than those for fresh or defatted flour (Kapoor and Kapoor, 1990).

In 1983, a simple roller-milling process was reported for the production of low-fat (1.2%) grits from pearl millet. Although tempering the grain to 22% moisture before milling greatly facilitated removal of the germ, yield of grits from whole grain was only 61% (Abdelrahman et al, 1983).

Pearl millet grits that have been wetted and then dried, developed a characteristic mousy, acidic odor. The odor was apparently the result of enzymatic degradation of a flavonoid component of the grain (Reddy et al, 1986).

MINERALS

Total mineral (ash) content of all the millets is often higher than that of sorghum and other cereals (Table 1). Values for individual minerals in pearl millet vary widely (Hulse et al, 1980) and are dependent to a large extent on the mineral composition of the soils in the growing areas. Hoseney et al (1987) have collected data from numerous sources showing the ranges of mineral values in pearl millet (expressed in milligrams per 100 g): sodium (2.2–4.9), potassium (366–543), calcium (7–117), magnesium (71–156), phosphorus (631–1353), iron (2.1–11.7), copper (0.42–0.58), manganese (0.66–0.92), zinc (0.10–3.8), and molybdenum (0.014–0.024). Mineral concentrations cited by Serna-Saldivar and Rooney (1991) fell within those ranges. Values for calcium, magnesium, and iron in pearl millet reported by Singh et al (1987) also were in the previously mentioned ranges, but zinc concentrations were higher and ranged from 3.88 to 5.38 mg/100 g.

Higher concentrations of minerals were reported in the grain's outer layers and germ than in the endosperm (Varriano-Marston and Hoseney, 1980), as is typical of cereals in general. Only about one-third of the calcium and iron remained in 75% extraction flour from pearl millet (De Wit and Schweigart, 1970). Dassenko (1980) reported significant losses of calcium, magnesium, and sodium, but not of iron or potassium when millet was milled to 67% extraction flour.

That calcium is more limiting than lysine or other nutrients in pearl millet was demonstrated when young rats fed millet supplemented with only calcium carbonate continued to grow well after seven weeks, whereas those fed unsupplemented pearl millet ceased growing after four weeks (Klopfenstein et al, 1985). Although absorption of magnesium was higher from pearl millet than from sorghum, absorptions of zinc, iron, and manganese were significantly lower from the millet (Table 6). Utilization of calcium, phosphorus, and other minerals in the grain may be adversely affected by phytates, the concentration of which is similar to that found in wheat (about 250 mg phytate P/100 g; Simwemba et al, 1984). Total phosphorus concentration in eight high-yielding pearl millet lines was 327 ± 32 mg/100 g dry matter (Chauhan et al, 1986). Over 60% of the total P was in the form of phytate P.

VITAMINS

Limited data are available concerning the vitamin content of pearl millet. Some representative values for thiamin, niacin, and riboflavin content compared with other grains are given in Table 1. In general, values are not much different from those for sorghum. Hulse et al (1980), citing multiple sources, listed a range of means for the following vitamins (in milligrams per 100 g, db): thiamin (0.34–0.50), niacin (1.54–3.88), riboflavin (0.19–0.24), pantothenic acid (0.87–1.31), total folic acid (0.063, one sample), and carotene (0.03–0.54). Chauhan et al (1986) reported thiamin amounts ranging from 0.29 to 0.40 mg/100 g, db, (mean ± SD = 0.34 ± 0.04) in eight high-yielding Indian varieties. Native millets do not contain measurable amounts of ascorbic acid, but after 48-hr germination, pearl, finger, and Italian millets developed 8, 5, and 6 mg/100 g ascorbic acid, respectively, and thiamin content also increased slightly (Malleshi and Desikachar, 1986). Opoku et al (1981) reported increases in riboflavin, thiamin, ascorbic acid, vitamin A, carotene, and tocopherol in pearl millet germinated for 84 hr, then kilned at 45°C. However, a significant decrease in niacin from 36.2 to 24.7 IU/100 g) was noted in that study.

Although some data indicate little loss of B-vitamins in the milling process (Dassenko, 1980), other results show losses of about 50% for thiamin, niacin, and riboflavin in 75% extraction flour (De Wit and Schweigart, 1970).

Increases in thiamin (90%) and riboflavin (85%) were observed in fermented pearl millet batters (Aliya and Geervani, 1981). However, steaming of the batters reduced thiamin (–64%) and riboflavin (–28%) to below levels in uncooked, unfermented batter. When the cooked product was again fermented for 24 hr, thiamin showed an increase of 152%, and riboflavin was 55% higher than in the original batter. Although the same pattern was noted for legumes, sorghum, and finger millet, the fermentation effect was greatest for pearl millet. The authors suggested that pearl millet may be a more favorable substrate for B-vitamin synthesis, or the organisms present in pearl millet may be different from those in the other grains and legumes tested. Dassenko (1980) also noted significant increases in niacin, thiamin, and riboflavin in fermented pearl millet flour (65% extraction) and whole meal compared with raw flour and meal. In that experiment, cooking the fermented grain did not adversely affect B-vitamin content.

ANTINUTRIENTS IN PEARL MILLET

Phytic Acid. The phytic acid content of pearl millet has been discussed earlier in relation to its adverse effects on nutrient bioavailability. It has been known for some time that phytic acid is destroyed during seed germination (Sebolev, 1962), when the enzyme phytase is synthesized and activated (Mandal and Biswas, 1970). Data from Malleshi and Desikachar (1986) show that after 48-hr germination with devegetation of the malt, the percentage of phosphorus as phytate phosphorus decreased from 38 to 20% in pearl millet, from 34 to 30% in finger millet, and from 31 to 29% in Italian millet. In that study, however, total phosphorus in the malted seeds decreased by 16, 12, and 9%, and calcium decreased by 23, 7, and 25% in pearl, finger, and Italian millets. When Nattress et al (1987) studied a weaning food mixture composed of germinated wheat, pearl millet, garbanzo bean, mung bean, and sesame in the proportions of 8:4:4:3:1, they found 4.39 mg/100 g of phytate, compared with 10.0 mg/100 g in the mixture prepared from ungerminated seeds. Vegetative matter was not removed from germinated seeds, and calcium, iron, and zinc concentrations were higher in the germinated than in the ungerminated mixture.

Fermentation of pearl millet also results in decreased phytic acid concentrations (Dhanker and Chauhan, 1987). When the traditional Indian food, *rabadi* (containing pearl millet flour and buttermilk), was fermented at different temperatures (23–50°C) for 3, 6, or 9 hr, phytic acid and polyphenol contents decreased as fermentation time increased, but temperature showed little effect on degradation (Dhanker and Chauhan, 1987). After 9 hr of fermentation, phytic acid and polyphenols were reduced by 27–30% and 10–12%, respectively.

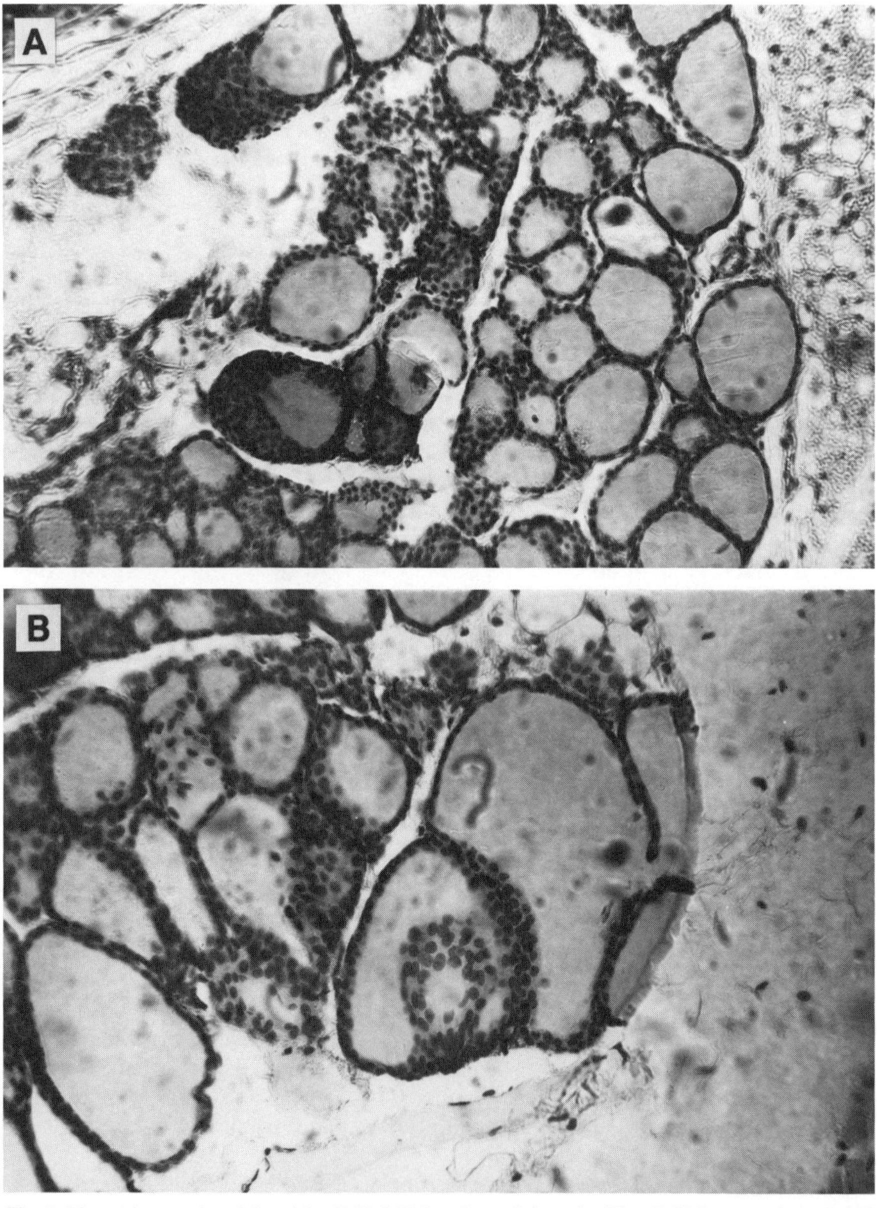

Fig. 1. Photomicrographs of thyroid colloid follicles of rats fed pearl millet. Follicles are enlarged, and epithelial cells are flattened. Hyperplasia (A) with infolding (B) is evident.

Oxalic Acid. Pearl millet, along with other grains, contains oxalic acid (Opoku et al, 1981), which forms an insoluble complex with calcium, thereby reducing biological availability of the mineral (Whitney et al, 1987). As noted previously, calcium concentration in pearl millet is generally quite low, and the presence of oxalate could exacerbate the deficiency. Soluble dietary oxalate reportedly is absorbed and must be excreted by the kidneys, where it may lead to the formation of kidney stones (Whitney et al, 1987). Opoku et al (1981) reported that total oxalate in pearl millet decreased from 0.619 to 0.433% when the grain was malted for 84 hr. Perhaps a more significant observation from a health standpoint was that soluble oxalate decreased from 0.502 to 0.068% with malting.

Goitrogens. Pearl millet was first implicated as contributing to the endemic goiter in some areas of Sudan when Osman and Fatah (1981) noted that, in villages where pearl millet made up a larger percentage of dietary calories, the incidence of goiter was higher than in urban areas where other grains were consumed. Goiter incidence in six African countries has been shown to be highly correlated with per capita pearl millet production ($r = 0.968$, $P = 0.002$; Klopfenstein et al, 1983c). Thyroids of rats fed pearl millet show goitrous histological changes, and serum thyroid hormone patterns are distorted (Osman, 1981; Klopfenstein et al, 1983b,c; Birzer et al, 1987; Birzer and Klopfenstein, 1988). Histological studies show hyperplasia (Fig. 1A) and infolding (Fig. 1B) of epithelial cells surrounding enlarged thyroid colloid follicles in rats. Autoclaving pearl millet can alleviate the goitrogenicity, but additional iodine supplementation and fermentation have no effect (Klopfenstein et al, 1983b,c). Antithyroid activity in extracts of pearl millet increases when they are boiled or stored (Gaitan et al, 1989). Strong evidence suggests that the phenolic flavonoid compounds (*C*-glycosylflavones) and their metabolites are the primary factors contributing to goitrogenicity of the grain (Birzer et al, 1987; Birzer and Klopfenstein, 1988; Gaitan et al, 1989). The compounds are probably responsible for color in pearl millet (Reichert, 1979) and have also been identified as causing off-odors in the grain (Reddy et al, 1986). Reichert et al (1980b) reported a steep gradient in *C*-glycosylflavone concentration from outer to inner portions of the seed, with much higher percentages in the outer layers. Because of the small kernel size and firmly embedded germ, efficient separation of the pearl millet bran and germ from the endosperm is difficult with a dry-milling process. However, when the grain was tempered to 26% moisture overnight (semiwet process), then milled (Buhler 202 laboratory mill), antithyroid properties were absent in the flour (Klopfenstein et al, 1991). Adverse effects on serum thyroid histopathology and hormone patterns were strongly correlated with *C*-glycosylflavone concentrations in the grains and milling fractions. Rats fed yellow-colored pearl millet had hormone levels closer to those of control animals than did animals fed brown or gray millets (Klopfenstein et al, 1991).

In addition to the *C*-glycosylflavones, small amounts of other possible goitrogens have been reported in pearl millet (Osman et al, 1983; Birzer et al, 1987).

Proso Millet
(*Panicum miliaceum*)

Proso millet (broomcorn, hershey, hog millet, common millet) is grown mainly in China, the former U.S.S.R., and parts of Africa (Serna-Saldivar and Rooney, 1991; Lorenz and Dilsaver, 1980; see also chapter 2). In the United States, proso millet has greater economic significance than other millets. The crop is grown in eastern Colorado and western Nebraska, where it is used in livestock rations and as a component of seed mixtures for wild birds (Lorenz and Dilsaver, 1980). The nutrient composition of proso millet is compared with that of other grains in Table 1, and its lysine amino acid score for various age groups is shown in Table 2. Protein content and quality for proso millet are similar to those of sorghum, but lipid, ash, and energy contents of proso millet are higher. The protein is limiting in threonine in addition to being deficient in lysine, and tryptophan levels are marginal (Chung and Pomeranz, 1985). Lorenz and Dilsaver (1980) successfully prepared flour from the whole grain (with no preconditioning) using a Quadrumat Junior mill. In the same study, they found that proso millet protein was less readily hydrolyzed (in vitro) than wheat protein, and that nitrogen solubilities were considerably lower in millet flours than in wheat flour. Crude fat contents of the 73 to 87% extraction flours ranged from 1.03 to 3.64%. Heating the grain for 1 hr at 80°C before milling reduced flour yields. Lorenz and Hwang (1986) reported that free lipids in flour ranged from 3.20 to 4.06%, whereas those in the bran were as high as 6.78%.

Bookwalter et al (1987a) also used a heat treatment before roller milling to inactivate lipid-degrading enzymes in proso millet. The 80% extraction flour contained 4.8% fat, high-quality germ protein, increased mineral content, and had extended storage life. No adverse changes occurred in functional properties or in the quality of couscous prepared from the flour.

Lipids in proso millet have been characterized by Lorenz and Hwang (1986), who reported that linoleic, oleic, and palmitic acids were the predominant fatty acids.

Phytate content of proso millet ranged from 0.17 to 0.47%, which is lower than that reported for corn and durum wheats (Lorenz, 1983). Dehulling of the millet resulted in a 27 to 53% reduction in phytate. Tannin content (as measured by the vanillin-HCl method) of some light- and dark-colored proso millets ranged from 0.055 to 0.178% catechin equivalents (Lorenz, 1983). It is not uncommon for sorghum to contain over 3% catechin equivalents (Hulse et al, 1980).

Foxtail Millet
(*Setaria italica*)

Foxtail millet (also known as Italian millet) is grown primarily in China and the Near East (Casey and Lorenz, 1977) and to some extent in southern and central India (Beghel et al, 1985; see also chapter 2). Representative contents of selected nutrients in foxtail millet are shown in Table 1. Protein content of the grain is generally slightly higher and lipid content is lower than those of maize. Beghel et al (1985) found the essential amino acid index to be similar for foxtail millet (10.79) and maize (10.71), and Table 2 shows that there is little difference in lysine amino acid scores for the two grains. Setarin, the alcohol-soluble protein of the grain, is relatively rich in methionine, but in plants raised under sulfur-deficient conditions, synthesis of the methionine-rich protein fraction was considerably reduced (Naren and Virupaksha, 1990). Hulse et al (1980) collected data from many sources showing the effects of various protein supplements in foxtail millet diets on rat weight gains and PERs. Combinations with horse gram, cowpeas, groundnut, chickpeas, and maize are included, as well as the effects of supplementation with lysine and tryptophan. Patwardhan (1961) reported that the PER for foxtail millet was low (0.8), and dehusking the grain reduced the value to 0.5. Supplementation of the dehusked grain with amaranth and chickpeas raised the PER to 2.2.

Crude fiber content of the grain is significantly higher than that of most other common grains, with the exception of Japanese barnyard millet (Table 1). Values as high as 10.8% have been reported (Beghel et al, 1985).

In parts of northern China, foxtail millet bran reportedly may contain up to 20% by weight of silica (O'Neill et al, 1982). That study suggested that silica fragments from dietary foxtail millet play a role in the etiology of esophageal cancer in the region, because silica has been found in high concentration in esophageal mucosa surrounding tumors.

Finger Millet (*Eleusine coracana*)

At present, 55 to 60% of the finger millet (ragi, bird's-foot millet, African millet) crop is grown in Central and South Africa; most of the rest is produced in India (Rao and Mushonga, 1985; see also chapter 2). Although the protein content of finger millet is usually lower than that of most other cereals (Table 1), protein quality is comparable to that of pearl millet (Table 2). Lysine is apparently the only limiting amino acid in finger millet (Chung and Pomeranz, 1985). Unusually high concentrations of calcium (350 mg/100 g) have been reported in the grain, as well as high levels of iron and phosphorus. For that reason, it is considered good food for lactating mothers (Rao and Mushonga, 1985). Some higher protein (8.0-12.1%), high-yielding hybrids have been developed (Babu et al, 1987), which are also very rich in calcium (294–390 mg/100 g). When Joseph et al (1959) replaced rice with finger millet in the diets of nine- to 10-year-old girls, a positive nitrogen balance was maintained. More calcium was retained from the diets containing finger millet, suggesting that its use could help overcome the calcium deficiency of rice diets.

The use of finger millet in child and infant feeding is limited, however, because of its low digestibility and poor availability of nutrients (Hemanalini et al, 1980). The growth-promoting effect (for rats) of flour from sprouted finger millet was found to be much higher than that of flour from ungerminated grain or flour prepared from germinated grain that was dried, roasted, then filtered through cloth (low-fiber flour). However, there was no significant difference in PER between flours from the sprouted and unsprouted grains (Hemanalini et al, 1980). The retention of calcium was the same for flours from both sprouted and unsprouted grain but was higher for the low-fiber flour.

In malted grains used in brewing, the saccharifying enzymes of finger millet were found to be more active than those of sorghum, pearl millet, or maize (Rao and Mushonga, 1985). A weaning food with low hot-paste viscosity and high caloric density was developed using devegetated, malted finger millet and dehulled green gram (70:30) (Malleshi and Desikachar, 1982). PER of the weaning food was 2.2, and NPU was 51.6. Values increased to 2.7 and 63, respectively, with the addition of 10% skim milk powder (Malleshi et al, 1986).

Puffing of cereals, sometimes followed by the addition of sweeteners or spices, is a common method for preparing ready-to-eat snack foods. Shukla et al (1986) reported that brown-seeded finger millet was more suitable for puffing than white-seeded varieties, but the puffed white-seeded millet was organoleptically superior.

Protein and fat content of the puffs (eight varieties tested) ranged from 5.61 to 7.20% and 1.08 to 1.47%. respectively, and phytic acid varied from 65 to 133 mg/100 g. Calcium content was high in all samples (316–442 mg/100 g; Shukla et al, 1986).

Ramachandra et al (1977) reported that condensed tannins in finger millet reduced protein digestibility (in vitro). Concentrations ranged from 0.04% (catechin equivalents) to levels as high as those found in some bird-resistant sorghums (3.47%). Protease inhibitory activity is quite high in finger millet compared with other common millets (Table 8) (Pattabiraman, 1985). Chandrasekher and Pattabiraman (1982) reported that finger millet inhibitors of caseinolytic activity by human and bovine pancreatic extracts were twice as active as those of sorghum and other millets. Finger millet contains two inhibitors. One is active against trypsin and chymotrypsin; the other appears unique in that it acts on trypsin and alpha-amylase (Shivaraj et al, 1982). Both are stable over a wide pH range and at a temperature of 100°C for 1 hr The trypsin-chymotrypsin inhibitor is also resistant to pepsin digestion.

When Dublish et al (1988) investigated the effects of extrusion cooking of cereal-legume blends, they reported that PERs of rice/finger millet/defatted soy (42.5:42.5:15) increased from 1.92 to 2.41. PER of the finger millet/defatted soy flour blend (85:15) increased from 1.81 to 2.23 after extrusion processing. Extrusion at 95°C inactivated 72.3 to 100% of the trypsin inhibitor present in the mixtures and decreased the amount of measurable tannins. The percentage of phosphorus as phytate phosphorus decreased by 11.1 to 29.1%. However, available lysine in the samples decreased by 3.6 to 12.8%, depending on extrusion conditions (Dublish et al, 1988). As has been noted for other grains, a significant increase in PER (from 0.90 to 1.06) was noted when finger millet was germinated for 48 hr (Table 10).

Japanese Barnyard Millet (*Echinochloa frumentacea*)

This millet is presently not a major food crop, and its nutritional properties have received relatively little attention. Protein and fat contents of Japanese barnyard millet are similar to those of proso millet, but crude fiber content is highest among the millets (Table 1). Although the grain is severely limited in lysine (Table 2), concentrations of other essential amino acids are not limiting (Monteiro et al, 1988).

Kodo Millet (*Paspalum scrobiculatum*)

In southern Asia, kodo millet (*varagu*) is used in the diets of livestock and humans. Protein content (Table 1) and quality (Table 2) are quite high in this grain compared with other common grains. In addition to lysine, kodo millet protein is also deficient in tryptophan (Chung and Pomeranz, 1985). As with other cereals, supplementation of the millet with legumes improves its nutritive value. Rajalakshmi and Majmudar (1966) have reported the improving effects of pigeon peas, chickpeas, moth bean, field bean, cowpeas, and peas in kodo millet diets for rats. PER of whole grain kodo millet increased from 0.7 to 1.9 when chickpeas and amaranth were added to rat diets (Patwardhan, 1961).

Teff
(*Eragrostis tef*)

Teff is an important food crop in Ethiopia, where it may account for half the total grain production (Hulse et al, 1980). The grain is used alone or in combination with sorghum to prepare the fermented flat bread, *injera*. Protein (9.5–12.6%; Platt, 1962; Jansen et al, 1962), crude fiber (2.5%; Platt, 1962), and ash (2.2%; Platt, 1962) contents of teff are similar to those of pearl millet, but lipid content (2.5%; Platt, 1962) of the grain is about half that of pearl millet. Jansen et al (1962) reported that the PER was slightly higher for teff than for pearl millet (1.95 vs. 1.83). When lysine hydrochloride was added to the diets, PERs increased to 3.27 for teff and 3.28 for pearl millet; control PER (casein) was 3.47. Threonine appears to be a second limiting amino acid in teff (FAO/WHO, 1973; Jansen, 1962). Although analyses of clean grain samples showed iron content of teff to be about 3.3 mg/100 g, soil contamination can raise concentrations to near 50 mg/100 g (Sufian and Pitwell, 1968), making the grain a significant dietary source of the mineral in some locations.

Fonio Millet
(*Digitaria exilis*)

This very small-seeded (2,000 seeds/g), yellow-colored millet is grown mainly in some parts of West Africa (Hulse et al, 1980). Fonio may be used alone or in combination with sorghum in making porridges and is also used in brewing beer. Another *Digitaria* species, *D. iburua*, is cultivated in Nigeria. A source cited by Hulse et al (1980) shows the proximate composition of *D. exilis* to be (db): protein, 8.7%; lipids, 3.5%; fiber, 8.5%; and ash, 3.8%. Limited data indicate that lysine is the only limiting essential amino acid in *D. exilis,* whereas both lysine and threonine are limiting in *D. iburua*.

CONCLUSION

Much of the world is just beginning to realize the enormous nutritional significance of sorghum and the millets in human diets. Support for research involving those grains still lags far behind that expended for the more common grains. As this review has shown, progress is being made in defining the nutritional qualities of sorghum and the millets, but much remains to be learned. Except for newly released, genetically improved varieties, ranges and means for chemical compositions of most cereal staples appear in the literature, although data are scarce for the minor millets. Effects of processing by both traditional and new methods are not well documented for many of the grains. More information about the presence and role of antinutrients in the seeds is needed, as well as research on methods to detoxify them. The health and lives of millions of people will depend on this research.

LITERATURE CITED

ABDELRAHMAN, A., HOSENEY, R. C., and VARRIANO-MARSTON, E. 1983. Milling process to produce low-fat grits from pearl millet. Cereal Chem. 60:189-191.

ADEYEMI, I. A. 1988. *Ogi* quality of sorghum flour dry-milled from fermented sorghum grains. J. Food Sci. 53:641-642.

AGARWAL, P. N., and SINHA, N. S. 1964.

Chemical composition of the oil from seeds of *Pennisetum typhoideum* (bajra). Indian J. Agron. 9:288-296.

AISIEN, A. O. 1982. Enzymic modification of sorghum endosperm during seedling growth and malting. J. Sci. Food Agric. 33:754-759.

AKINYELE, I. O., and FASAYE, O. A. 1988. Nutrient quality of corn and sorghum supplemented with cowpeas (*V. unguiculata*) in the traditional manufacture of *ogi*. J. Food Sci. 1750-1752,1755.

AL-KAHTANI, H. A. 1989. Processing of different wheat flour-supplemented sorghum flakes (sorghum and wheat of Saudi Arabia): Sensory, nutritional, and microbiological evaluation. Food Chem. 33:133-149.

ALI, H. I., and HARLAND, B. F. 1991. Effects of fiber and phytate in sorghum flour in weanling rats: A pilot study. Cereal Chem. 68:234-238.

ALIYA, S., and GEERVANI, P. 1981. An assessment of the protein quality and vitamin B content of commonly used fermented products of legumes and millets. J. Sci. Food Agric. 32:837-842.

ALLEE, G. L. 1976. Effect of processing methods on nutritional value of milo for weaned pigs. J. Anim. Sci. 43:248.

ALMEIDA-DOMINGUEZ, H. D., SERNA-SALDIVAR, S. O., and ROONEY, L. W. 1991. Properties of new and commercial sorghum hybrids for use in alkaline-cooked foods. Cereal Chem. 68:25-30.

ALPERT, M. E., HUTT, S. R., WOGAN, G. N, AND DAVIDSON, C. S. 1971. Association between aflatoxin content of food and hepatoma frequency in Uganda. Cancer 28:253-260.

AMERICAN ASSOCIATION OF CEREAL CHEMISTS. 1983. Approved Methods of the AACC, 8th ed. The Association, St. Paul, MN.

ASP, N.-G., and BJORK, I. 1984. The effect of extrusion-cooking on nutritional value. Pages 162-167 in: Thermal Processing and Quality of Foods. P. Zeuthen, J.-C. Cheftel, C. Eriksson, M. Jul, H. Leniger, P. Linko, G. Varela, and G. Vos, eds. Elsevier Applied Science Publ., London.

ASP, N.-G., and BJORK, I. 1989. Nutritional properties of extruded foods. Pages 399-434 in: Extrusion Cooking. C. Mercier, P. Linko, J. M. Harper, eds. American Association of Cereal Chemists, St. Paul, MN.

ASP, N.-G., JOHANSSON, C.-G., HALLMER, H., and SILJESTROM, M. 1983. Rapid enzymatic assay of insoluble and soluble dietary fiber. J. Agric. Food Chem. 31:476-482.

ASQUITH, T. N., and BUTLER, L. G. 1986. Interactions of condensed tannins with selected proteins. Phytochemistry 25:1591-1593.

ASSOCIATION OF OFFICIAL ANALYTICAL CHEMISTS. 1990. Official Methods of Analysis, 15th ed. The Association, Arlington, VA.

AU, P. M. 1979. Study of fermentation of grain sorghum to improve its amino acid and vitamin content. M.S. thesis, Univ. of Missouri, Columbia, MO.

AU, P. M., and FIELDS, M. L. 1981. Nutritive quality of fermented sorghum. J. Food Sci. 46:652-654.

AXTELL, J. D., KIRLEIS, A. W., HASSEN, M. M., D'CROZ MASON, N., MERTZ, E. T., and MUNCK, L. 1981. Digestibility of sorghum proteins. Proc. Natl. Acad. Sci. 78:1333-1335.

BABU, B. V., RAMANA, T., and RADHAKRISHNAN, T. M. 1987. Chemical composition and protein content in hybrid varieties of finger millet. Indian J. Agric. Sci. 57:520-522.

BACH KNUDSEN, K. E., and MUNCK, L. 1985. Dietary fiber contents and composition of sorghum-based foods. J. Cereal Sci. 3:153-164.

BACH KNUDSEN, K. E., KIRLEIS, A. W., EGGUM, B. O., and MUNCK, L. 1988a. Carbohydrate composition and nutritional quality for rats of sorghum *tô* prepared from decorticated white and whole grain red flour. J. Nutr. 118:588-597.

BACH KNUDSEN, K. E., MUNCK, L., and EGGUM, B. O. 1988b. Effect of cooking, pH and polyphenol level on carbohydrate composition and nutritional quality of a sorghum [*Sorghum bicolor* (L.) Moench] food, ugali. Br. J. Nutr. 59:31-47.

BADI, S. M., HOSENEY, R. C., and CASADY, A. J. 1976a. Pearl millet. I. Characterization by SEM, amino acid analysis, lipid composition, and prolamine solubility. Cereal Chem. 53:478-487.

BADI, S. M., HOSENEY, R. C., and FINNEY, P. L. 1976b. Pearl millet. II. Partial characterization of starch and use of millet flour in breadmaking. Cereal Chem. 53:718-724.

BADI, S., PEDERSEN, B., MONOWAR, L., and EGGUM, B. O. 1990. The nutritive value of new and traditional sorghum and millet foods from Sudan. Plant Foods for Hum. Nutr. 40:5-19.

BAMFORTH, C. W. 1982. Barley β-glucans—Their role in malting and brewing. Brewers Dig. 57:22-27.

BANDA-NYIRENDA, D. B. C., and VOHRA, P. 1990. Nutritional improvement of tannin-containing sorghums (*Sorghum bicolor*) by sodium bicarbonate. Cereal Chem. 67:533-537.

BEGHEL, R. P. S., NETKE, S. P., and BAJPAI, L. D. 1985. Nutritive value of kangni. Poult. Guide 22(5):28-29.

BELAVADY, B., and GOPALAN, C. 1965. Production of black tongue in dogs by feeding diets containing jowar (*Sorghum vulgare*). Lancet 2:1220-1221.

BELAVADY, B., SRIKANTIA, S. G., and GOPALAN, C. 1963. The effect of the oral adiministration of leucine on the metabolism of tryptophan. Biochem. J. 87:652-655.

BELEIA, A., and VARRIANO-MARSTON, E. 1981a. Pearl millet amylases. I. Properties of partially purified alpha-amylase. Cereal Chem. 433-437.

BELEIA, A., and VARRIANO-MARSTON, E. 1981b. Pearl millet amylases. II. Activity toward intact and heated starch granules. Cereal Chem. 58:437-440.

BELEIA, A., VARRIANO-MARSTON, E., and HOSENEY, R. C. 1980. Characterization of starch from pearl millets. Cereal Chem. 57:300-303.

BENDER, D. A. 1983. Effects of a dietary excess of leucine on the metabolism of tryptophan in the rat: A mechanism for pellagragenic action of leucine. Br. J. Nutr. 50:25-32.

BIRZER, D. M., and KLOPFENSTEIN, C. F. 1988. The pearl millet goitrogens. Cereal Foods World 33:229-231.

BIRZER, D. M., KLOPFENSTEIN, C. F., and LEIPOLD, H. W. 1987. Goiter-causing compounds found in pearl millet. Nutr. Rep. Int. 36:131-140.

BJORK, I. 1984. Nutritional properties of wheat products processed by HTST extrusion cooking. Thesis. University of Lund, Lund, Sweden.

BJORK, I., and ASP, N.-G. 1984. The effects of extrusion cooking on nutritional value—A literature review. J. Food Eng. 2:281-308.

BJORK, I., NOGUCHI, A., ASP, N.-G. CHEFTEL, J. C., and DAHLQVIST, A. 1983. Protein nutritional value of a biscuit processed by extrusion cooking: Effects on available lysine. J. Agric. Food Chem. 31:488-492.

BLOCK, R. J., and MITCHELL, H. H. 1946. The correlation of the amino acid composition of proteins with their nutritive value. Nutr. Abstr. Rev. 16:249-278.

BOOKWALTER, G. N., LYLE, S. A., and WARNER, K. 1987a. Millet processing for improved stability and nutritional quality without functionality changes. J. Food. Sci. 52:399-402.

BOOKWALTER, G. N., KIRLEIS, A. W., and MERTZ, E. T. 1987b. In vitro digestibility of protein in milled sorghum and other processed cereals with and without soy fortification. J. Food Sci. 52:1577-1579, 1619.

BOORA, P., and KAPOOR, A. C. 1985. Influence of storage on the protein quality of pearl millet flour. J. Sci. Food Agric. 36:59-62.

BUONOCORE, V., and SILANO, V. 1985. Biochemical, nutritional, and toxicological aspects of alpha-amylase inhibitors from plant foods. Pages 483-507 in: Nutritional and Toxicological Significance of Enzyme Inhibitors in Food. Advances in Exp. Med. Biol. v 199. M. Friedman, ed. Plenum Press, New York.

BURLESON, C. A., COWLEY, W. R., and OTEY, G. 1956. Effect of nitrogen fertilization on yield and protein content of grain sorghum in the Lower Rio Grande Valley of Texas. Agron. J. 48:524-525.

BUTLER, L. G. 1982. Polyphenols and their effects on sorhum quality. Pages 294-311 in: Proc. Int. Symp. Sorghum Grain Quality. L. W. Rooney and D. S. Murty, eds. Int. Crops Res. Inst. Semi-Arid Tropics (ICRISAT), Patancheru, A. P., India.

BUTLER, L. G., RIEDL, D. J., LEBRYK, D. G., and BLYTT, H. J. 1984. Interaction of protein with sorghum tannin: Mechanism, specificity and significance. J. Am. Oil Chem. Soc. 61:916-920.

BUTLER, L. G., ROGLER, J. C., MEHANSHO, H., and CARLSON, D. M. 1986. Dietary effects of tannins. Pages 141-157 in: Plant Flavonoids in Biology and Medicine: Biochemical, Pharmacological and Structural Activity Relationships. Alan R. Liss, New York.

CAGAMPANG, G. B., GRIFFITH, J. E., and KIRLEIS, A. W. 1982. Modified adhesion test for measuring stickiness of sorghum porridges. Cereal Chem. 59:234-235.

CARNOVALE, E., and QUAGLIA, G. B. 1973. Influence of temperature and humidity controlled preservation on the chemical composition of milling products from millet. Ann. Technol. Agric. 22:371-380.

CARTER, E. G. A., and CARPENTER, K. J. 1981. Bound niacin in sorghum and its availability. Nutr. Res. 1:571-579.

CARTER, E. G. A., and CARPENTER, K. J. 1982. The available niacin values of foods for rats and their relation to analytical values. J. Nutr. 112:2091-2103.

CASEY, P., and LORENZ, K. 1977. Millet—Functional and nutritional properties. Bakers Dig. 51(1):45-51.

CHANDRASHEKAR, A., and KIRLEIS, A. W. 1988. Influence of protein on starch gelatinization in sorghum. Cereal Chem. 65:457-462.

CHANDRASEKHER, G., and PATTABIRAMAN, T. N. 1982. Natural plant enzyme inhibitors. Protease inhibitors in millets. J. Sci. Food Agric. 33:447-450.

CHAUHAN, B. M., SUNEJA, N., and BHAT, C. M. 1986. Nutritional value and fatty acid composition of some high yielding varieties of bajra. Bull. Grain Technol. 24:44-49.

CHAVAN, U. D., CHAVAN, J. K., and KADAM, S. S. 1988. Effect of fermentation on soluble proteins and in vitro protein digestibility of sorghum, green gram, and sorghum-green gram blends. J. Food Sci. 53:1574-1575.

CHEFTEL, J. C. 1986. Nutritional effects of extrusion-cooking. Food Chem. 20:263-283.

CHIBBER, B. A. K., MERTZ, E. T., and AXTELL, J. D. 1978. Effects of dehulling on tannin content, protein distribution, and quality of high and low tannin sorghum. J. Agric. Food Chem. 26:679-683.

CHUNG, O. K., and POMERANZ, Y. 1985. Amino acids in cereal proteins and protein fractions. Pages 65-107 in: Digestibility and Amino Acid Availability in Cereals and Oilseeds. J. W. Finley and D. T. Hopkins, eds. American Association of Cereal Chemists, St. Paul, MN.

COHEN, R. S., and TANKSLEY, T. D., JR. 1973. Energy and protein digestibility of sorghum grains with different textures and starch types by growing swine. J. Anim. Sci. 37:931-935.

COLLINS, M., and ELVEHJEM. C. A. 1958. Ascorbic acid requirement of the guinea pig using growth and tissue ascorbic acid as criteria. J. Nutr. 64:503-511.

COOK, N. E., and CARPENTER, K. J. 1987. Leucine excess and niacin status in rats. J. Nutr. 117:519-526.

CORNU, A., and DELPEUCH, F. 1981. Effect of fiber in sorghum on nitrogen digestibility. Am. J. Clin. Nutr. 34:2454-2459.

CRAIG, S. A. S, AND STARK, J. R. 1984. A comparison of the molecular properties of sorghum starches of different origins. Starch/Staerke 36:127-131.

DADA, L. O., and DENDY, D. A. V. 1987. Preliminary study of the effect of various processing techniques on the cyanide content of germinated sorghum. Trop. Sci. 27:101-104.

DAHIYA, S., and KAPOOR, A. C. 1983. Effect of storage conditions on the protein quality of pearl millet flour. Nutr. Rep. Int. 28:1351-1359.

DAIBER, K. H., and TAYLOR, J. R. N. 1982. Effects of formaldehyde on protein extraction and quality of high- and low-tannin sorghums. J. Agric. Food Chem. 30:70-72.

DANIEL, V. A., LEELA, R., DORAISWAMY, T. R., JAJALAKSHMI, D., RAO, S. V., SWAMINATHAN, J, AND PARPIA, H. A. B. 1966. The effect of supplementing a poor kaffir-corn (*Sorghum vulgare*) diet with L-lysine and DL-threonine on the digestibility coefficient, biological value, and net utilization of proteins and retention of nitrogen in children. J. Nutr. Dietet. 3:10-14.

DASSENKO, S. 1980. Effect of milling, fermentation, and cooking on nutritive value of pearl millet [*Pennisetum americanum* (L.) Leeke]. Ph.D. diss. Kansas State University, Manhattan. (Diss. Abstr. 81-01736).

DAVIS, A. B., and HOSENEY, R. C. 1979a. Grain sorghum condensed tannins. I. Isolation, estimation, and selective adsorption by starch. Cereal Chem. 56:310-314.

DAVIS, A. B., and HOSENEY, R. C. 1979b. Grain sorghum condensed tannins. II. Preharvest changes. Cereal Chem. 56:314-316.

DE WIT, J. P., and SCHWEIGART, F. 1970. The potential role of pearl millet as a food in South Africa. S. Afr. Med. J. 44:364-366.

DEGROOT, A. P., and SLUMP, P. 1969. Effect of severe alkali treatment of proteins on amino acid composition and nutritive value. J. Nutr. 98:45-56.

DERMAN, D. P., BOTHWELL, T. H., TORRANCE, J. D., BEZWODA, W. R., MACPHAIL, A. P., KEW, M. C., SAYERS, M. H., DISLER, P. B., and CHARLTON, R. W. 1980. Iron absorption from maize (*Zea mays*) and sorghum (*Sorghum vulgare*) beer. Br. J. Nutr. 43:271-279.

DESAI, B. B., and ZENDE, G. K. 1979. Role of bajra (*Pennisetum typhoides*) in human and animal nutrition. Indian J. Nutr. Diet. 16:390-396.

DEYOE, C. W., SANFORD, P. F., and WAGGLE, D. H. 1967. Feeding expanded corn and grain sorghum in broiler diets. (Abstr.) Poult. Sci. 46:1252.

DHANKER, N., and CHAUHAN, B. M. 1987. Effect of temperature and fermentation time on phytic acid and polyphenol content of rabadi—A fermented pearl millet food. J. Food Sci. 52:828-829.

DREHER, M. L., DREHER, C. J., and BERRY, J. W. 1984. Starch digestibility of foods: A nutritional perspective. CRC Crit. Rev. Food Sci. Nutr. 20:47-71.

DUBLISH, R. K., CHAUHAN, G. S., and BAINS, G. S. 1988. Nutritional quality of extruded rice, ragi, and defatted soy flour blends. J. Food Sci. Technol. 25:35-38.

EARP, C. F., DOHERTY, C. A., FULCHER, R. G., and ROONEY, L. W. 1983. β-glucans in the caryopsis of *Sorghum bicolor* (L.) Moench. Food Microstruct. 2:183-188.

EGGUM, B. O., BACH KNUDSEN, K. E., and MUNCK, L., AXTELL, J. D., and MUKURU, S. S. 1982. Milling and nutritional value of sorghum in Tanzania. Pages 211-225 in: Proc. Int. Symp. Sorghum Grain Quality. L. W. Rooney and D. S. Murty, eds. Int. Crops Res. Inst. Semi-Arid Tropics (ICRISAT), Patancheru, A. P., India.

EGGUM, B. O., MONOWAR, L., BACH KNUDSEN, K. E., MUNCK, L., and AXTELL, J. 1983. Nutritional quality of sorghum and sorghum foods from Sudan. J.

Cereal Sci. 1:127-137.

EJETA, G., and AXTELL, J. 1987. Protein and lysine levels in developing kernels of normal and high-lysine sorghum. Cereal Chem. 64:137-139.

ELMALIK, M., KLOPFENSTEIN, C. F., HOSENEY, R. C., and BATES, L. S. 1986a. Digestibility and nutritional value of sorghum grain with contrasting kernel characteristics. Nutr. Rep. Int. 34:811-821.

ELMALIK, M., KLOPFENSTEIN, C. F., HOSENEY, R. C., and BATES, L. S. 1986b. Effects of germination on the nutritional quality of sorghum grain with contrasting kernel characteristics. Nutr. Rep. Int. 34:941-950.

EPPENDORFER, W. H., BILLE, S. W., and PATIPANAWATTANA, S. 1985. Protein quality and amino acid–protein relationships of maize, sorghum and rice grain as influenced by nitrogen, phosphorus, potassium, and soil moisture stress. J. Sci. Food Agric. 36:453-462.

FAO. 1990. Statistics Series No. 61. FAO Production Yearbook, Vol. 42. Food and Agriculture Organization of the United Nations, Rome.

FAO/WHO. 1973. Energy and Protein Requirements. Report of a Joint FAO/WHO Ad Hoc Expert Committee. Tech. Rep. Ser. 522. World Health Organization, Geneva. Also FAO Nutr. Rep. Ser. 52. Food and Agriculture Organization of the United Nations, Rome.

FAO/WHO. 1985. Energy and protein requirements. Tech. Rep. Ser. 724. World Health Organization, Geneva.

FAPOJUWO, O. O., MAGA, J. A., and JANSEN, G. R. 1987. Effect of extrusion cooking on in vitro protein digestibility of sorghum. J. Food Sci. 52:218-219.

FREEMAN, J. E., and BOCAN, B. J. 1973. Pearl millet: A potential crop for wet milling. Cereal Sci. Today 18:69-73.

GAITAN, E., LINDSAY, R. H., REICHERT, R. D., INGBAR, S. H., COOKSEY, R. C., LEGAN, J., MEYDRECH, E. F., HILL, J., and KUBOTA, K. 1989. Antithyroid and goitrogenic effects of millet: Role of *C*-glycosylflavones. J. Clin. Endocrinol. Metab. 68:707-714.

GANOTE, C. E., PETERSON, D. R., and CARONE, F. A. 1974. The nature of D-serine-induced nephrotoxicity. Am. J. Pathol. 77:269-282.

GHAFOORUNISSA, R., and NARASINGA RAO, B. S. 1973. Effect of leucine on enzymes of the tryptophan-niacin metabolic pathway in rat liver and kidney. Biochem. J. 134:425-430.

GILLOOLY, M., BOTHWELL, T. H., CHARLTON, R. W., TORRANCE, J. D., BEZWODA, W. R., MACPHAIL, A. P., DERMAN, D. P., NOVELLI, L., MORRALL, P., and MAYET, F. 1984. Factors affecting the absorption of iron from cereals. Br. J. Nutr. 51:37-46.

GLENNIE, C. W., HARRIS, J., and LIEBENBERG, N. V. D. W. 1983. Endosperm modification in germinating sorghum grain. Cereal Chem. 60:27-31.

GOPALAN, C. 1968. Leucine and pellagra. Nutr. Rev. 26:323-326.

GOPALAN, C., and SRIKANTIA, S. G. 1960. Leucine and pellagra. Lancet 1:954-957.

GRAHAM, G. G., MACLEAN, W. C., JR., MORALES, E., HAMAKER, B. R., KIRLEIS, A. W., MERTZ, E. T., and AXTELL, J. D. 1986. Digestibility and utilization of protein and energy from nasha, a traditional Sudanese fermented sorghum weaning food. J. Nutr. 116:978-984.

GRIFFITHS, D. W. 1985. The inhibition of digestive enzymes by polyphenolic compounds. Pages 509-516 in: Nutritional and Toxicological Significance of Enzyme Inhibitors in Foods. Adv. Exp. Biol. Med. 199.

GUIRAGOSSIAN, V., CHIBBER, B. A. K., VAN SCOYOC, G. E., JAMBUNATHAN, R., MERTZ, E. T., and AXTELL, J. D. 1978. Characteristics of proteins from normal, high-lysine, and high-tannin sorghums. J. Agric. Food Chem. 26:219-223.

HAHN, D. H., ROONEY, L. W., and EARP, C. F. 1984. Tannins and phenols of sorghum. Cereal Foods World 29:776-779.

HAMAKER, B. R., KIRLEIS, A. W., MERTZ, E. T., and AXTELL, J. D. 1986. Effect of cooking on the protein profiles and in vitro digestibility of sorghum and maize. J. Agric. Food Chem. 34:647-649.

HAMAKER, B. R., KIRLEIS, A. W., BUTLER, L. G., AXTELL, J. D., and MERTZ, E. T. 1987. Improving the *in vitro* protein digestibility of sorghum with reducing agents. Proc. Natl. Acad. Sci. 84:626-628.

HARPER, J. M. 1988. Effects of extrusion processing on nutrients. Pages 365-391 in: Nutritional Evaluation of Food Processing. E. Karmas and R. Harris, eds. Van Nostrand Reinhold Co., New York.

HAYASHI, R., and KAMEDA, I. 1980. Racemization of amino acid residues during alkali treatment of protein and its adverse effect on pepsin digestibility. Agric. Biol. Chem. 44:891-895.

HEMANALINI, G., UMAPATHY, K. P., RAO, J. R., and SARASWATHI, G. 1980. Nutritional evaluation of sprouted ragi. Nutr. Rep. Int. 22:271-277.

HEWITT, D., and FORD, J. E. 1982. Influence of tannins on the protein nutritional quality of food grains. Proc. Nutr. Soc. 41:7-17.

HIBBERD, C. A., WAGNER, D. G., SCHEMM, R. L., MITCHELL, E. D., JR., HINTZ, R. L., and WEIBEL, D. E. 1982a. Nutritive characteristics of different varieties of sorghum and corn grains. J. Anim. Sci. 55:655-672.

HIBBERD, C. A., WAGNER, D. G., SCHEMM, R. L., MITCHELL, E. D., JR., WEIBEL, D. E., and HINTZ, R. L. 1982b. Digestibility characteristics of isolated starch from sorghum and corn grain. J. Anim. Sci. 55:1490-1497.

HOLUM, J. R. 1978. Lipids. Pages 218-237 in Organic and Biological Chemistry. John Wiley & Sons, New York.

HORAN, F. E., and HEIDER, M. F. 1946. A study of sorghum and sorghum starches. Cereal Chem. 23:492-503.

HOSENEY, R. C. 1986. Pages 47-54 in: Principles of Cereal Science and Technology. Am. Assoc. of Cereal Chem., St. Paul, MN.

HOSENEY, R. C., VARRIANO-MARSTON, E., and DENDY, D. A. V. 1981. Sorghum and Millets. Pages 71-144 in: Advances in Cereal Science and Technology, Vol. 4. Y. Pomeranz, ed. Am. Assoc. Cereal Chem., St. Paul, MN.

HOSENEY, R. C., andREWS, D. J., and CLARK, H. 1987. Sorghum and pearl millet. Pages 397-456 in: Nutritional Quality of Cereal Grains: Genetic and Agronomic Improvement. R. A. Olsen and K. J. Frey, eds. American Society of Agronomy, Madison, WS.

HOWE, E. E., and GILFILLAN, E. W. 1970. Limiting nutrients in cereal grains for the growth of the laboratory rat. Indian J. Nutr. Dietet. 7:17-20.

HUBBARD, J. E., HALL, H. H., and EARLE, F. R. 1950. Composition of the component parts of the sorghum kernel. Cereal Chem. 27:415-420.

HULSE, J. H., LAING, E. M., and PEARSON, O. E. 1980. Sorghum and the Millets: Their Composition and Nutritive Value. Academic Press, New York.

IBRAHIM, S., FISHER, C. ALAILY, H. E. SOLIMAN, H., and ANWAR, A. 1988. Improvement of the nutritional quality of Egyptian and Sudanese sorghum grains by the addition of phosphates. Br. Poult. Sci. 29:721-728.

JAMBUNATHAN, R. 1980. Improvement of the nutritional quality of sorghum and pearl millet. Food Nutr. Bull. 2(1):11-16.

JAMBUNATHAN, R., RAO, N. S., and GURTU, S. 1983. Rapid methods for estimating protein and lysine in sorghum (*Sorghum bicolor* (L.) Moench). Cereal Chem. 60:192-194.

JANSEN, G. R., DIMAIO, L. R., and HAUSE, N. L. 1962. Cereal proteins. Amino acid composition and lysine supplementation of teff. J. Agric. Food Chem. 10:62-64.

JELLUM, M. D., and POWELL, J. B. 1971. Fatty acid composition from pearl millet seed. Agron. J. 63:29-33.

JOFFE, A. Z. 1965. Toxin production by cereal fungi causing toxic alimentary aleukia in man. Pages 77-85 in: Mycotoxins in Foodstuffs. G. N. Wogan, ed. MIT Press, Cambridge, MA.

JOSEPH, K., KURIEN, P. P., SWAMINATHAN, M., and SUBRAMANYAN, V. 1959. The metabolism of nitrogen, calcium and phosphorus in undernourished children. Br. J. Nutr. 13:213-218.

KACED, I., HOSENEY, R. C., and VARRIANO-MARSTON, E. 1984. Factors affecting rancidity in ground pearl millet [*Pennisetum americanum* (L.) Leeke]. Cereal Chem. 61:187-192.

KAMATH, M. V., and BELAVADY, B. 1980. Unavailable carbohydrates of commonly consumed Indian foods. J. Sci. Food Agric. 31:194-202.

KAPOOR, R., and KAPOOR, A. C. 1990. Biological evaluation of pearl millet protein: Effect of different treatments and storage. Plant Foods Hum. Nutr. 40:175-183.

KARIM, A., and ROONEY, L. W. 1972. Pentosans in sorghum grain. J. Food Sci. 37:369-371.

KAZANAS, H. 1979. Natural fermentation of sorghum. Ph.D. diss. University of Missouri-Columbia, Columbia.

KAZANAS, N., and FIELDS, M. L. 1981. Nutritional improvement of sorghum by fermentation. J. Food Sci. 46:819-821.

KHALIL, J. K., SAWAYA, W. N., SAFI, W. J., and AL-MOHAMMAD, H. M. 1984. Chemical composition and nutritional quality of sorghum flour and bread. Qual. Plant. Plant Foods Hum. Nutr. 34:141-150.

KHETARPAUL, N., and CHAUHAN, B. M. 1990. Fermentation of pearl millet flour with yeasts and lactobacilli: *In vitro* digestibility and utilisation of fermented flour for weaning mixtures. Plant Foods Hum. Nutr. 40:167-173.

KLOPFENSTEIN, C. F. 1988. The role of cereal beta-glucans in nutrition and health. Cereal Foods World 33:865-867.

KLOPFENSTEIN, C. F., and HOSENEY, R. C. 1987. Cholesterol-lowering effect of beta-glucan-enriched bread. Nutr. Rep. Int. 36:1091-1098.

KLOPFENSTEIN, C. F., VARRIANO-MARSTON, E., and HOSENEY, R. C. 1981a. Cholesterol-lowering effect of sorghum diet in guinea pigs. Nutr. Rep. Int. 24:621-627.

KLOPFENSTEIN, C. F., VARRIANO-MARSTON, E., and HOSENEY, R. C. 1981b. Effects of ascorbic acid in casein vs. sorghum grain diets in guinea pigs. Nutr. Rep. Int. 24:1017-1028.

KLOPFENSTEIN, C. F., HOSENEY, R. C., and VARRIANO-MARSTON, E. 1981c. Effects of ascorbic acid in guinea pigs fed millet and sorghum grain diets. Nutr. Rep. Int. 24:1099-1107.

KLOPFENSTEIN, C. F., HOSENEY, R. C., and

VARRIANO-MARSTON, E. 1983(a). Effects of ascorbic acid in sorghum-, high leucine-, and casein-fed guinea pigs. Nutr. Rep. Int. 27:121-129.

KLOPFENSTEIN, C. F., HOSENEY, R. C., and LEIPOLD, H. W. 1983b(formerly**a**). Goitrogenic effects of pearl millet diets. Nutr. Rep. Int. 27:1039-1047.

KLOPFENSTEIN, C. F., HOSENEY, R. C., and LEIPOLD, H. W. 1983c(formerly**b**). Further studies on the goitrogenic effects of pearl millet diets. Nutr. Rep. Int. 28:1137-1143.

KLOPFENSTEIN, C. F., HOSENEY, R. C., and LEIPOLD, H. W. 1985. Nutritional quality of pearl millet and sorghum grain diets and a pearl millet weanling food. Nutr. Rep. Int. 31:287-297.

KLOPFENSTEIN, C. F., LEIPOLD, H. W., and CECIL, J. E. 1991. Semiwet milling of pearl millet for reduced goitrogenicity. Cereal Chem. 68:177-179.

KNUCKLES, B. E., KUZMICKY, D. D., and BETSCHART, A. A. 1985. Effect of phytate and partially hydrolyzed phytate on in vitro protein digestibility. J. Food Sci. 50: 1080-1082.

KRISHNAMACHERI, K. A. V. R. 1974. Some aspects of copper metabolism in pellagra. Am. J. Clin. Nutr. 27:108-111.

KRISHNASWAMY, K., BAPURAO, S., RAGHURAM, T. C., and SRIKANTIA, S. G. 1976. Effect of vitamin B6 on leucine induced changes in human subjects. Am. J. Clin. Nutr. 29: 177-181.

KUMAR, K. A., GUPTA, S. C. ANDREWS, D. J. 1983. Relationship between nutritional quality characters and grain yield in pearl millet. Crop Sci. 23:232-235.

KURIEN, P. P., NARAYANARAO, M., SWAMINATHAN, M., and SUBRAHMANYAN, V. 1960. The metabolism of nitrogen, calcium and phosphorus in undernourished children. 6. The effect of partial or complete replacement of rice in poor vegetarian diets by kaffircorn (*Sorghum vulgare*) Br. J. Nutr. 14:339-345.

LAI, C. C., and VARRIANO-MARSTON, E. 1980. Changes in pearl millet meal during storage. Cereal Chem. 57:275-277.

LANDRY, J., and MOUREAUX, T. 1970. Heterogeneity of the glutellins of maize grain: Selective extraction and composition in amino acids of the three fractions isolated. (In French) Bull. Soc. Chim. Bio. 52:1021-1027.

LIARDON, R., and HURRELL, R. F. 1983. Amino acid racemization and alkali-treated proteins. J. Agric. Food Chem. 31:432-437.

LICHTENWALNER, R. E., ELLIS, E. B., and ROONEY, L. W. 1978. Effect of incremental doses of the waxy gene of sorghum on digestibility. J. Anim. Sci. 46:1113-1119.

LIENER, I. E. 1981. The nutritional significance of the plant lectins. Pages 143-157 in: Antinutrients and Natural Toxicants in Foods. R. L. Ory, ed. Food and Nutrition Press, Inc., Westport, CT.

LORENZ, K. 1983. Tannins and phytate content in proso millets (*Panicum miliaceum*). Cereal Chem. 60:424-426.

LORENZ, K., and DILSAVER, W. 1980. Proso millets. Milling characteristics, proximate compositions, nutritive value of flours. Cereal Chem. 57:16-20.

LORENZ, K., and HWANG, Y. S. 1986. Lipids in proso millet (*Panicum miliaceum*) flours and brans. Cereal Chem. 63:387-390.

MACLEAN, W. C., JR., LOPEZ DE ROMANA, G., PLACKO, R. P., and GRAHAM, G. G. 1981. Protein quality and digestibilility of sorghum in preschool children: Balance studies and plasma free amino acids. J. Nutr. 111:1928-1936.

MACLEAN, W. C., JR., LOPEZ DE ROMANA, G., GASTANADUY, A., and GRAHAM, G. G. 1983. The effect of decortication and extrusion on the digestibility of sorghum by preschool children. J. Nutr. 113:2171-2177.

MAGBOUL, B. I., and BENDER, D. A. 1983. The effect of dietary excess leucine on the synthesis of nicotinamide nucleotides in the rat. Br. J. Nutr. 49:321-329.

MAJUMDER, A. K., NANDI, B. K., SUBRAMANIAN, N., and CHATTERJEE, I. B. 1975. Growth and ascorbic acid metabolism in rats and guinea pigs fed cereal diets. J. Nutr. 105:233-239.

MALLESHI, N. G., and DESIKACHAR, H. S. R. 1982. Formulation of a weaning food with low hot-paste viscosity based on malted ragi (*Eleusine coracana*) and green gram (*Phaseolus radiatus*). J. Food Sci. Technol. 19:193-197.

MALLESHI, N. G., and DESIKACHAR, H. S. R. 1986. Nutritive value of malted millet flours. Qual. Plant. Plant Foods Hum. Nutr. 36:191-196.

MALLESHI, N. G., DESIKACHAR, H. S. R., and RAO, S. V. 1986. Protein quality evaluation of a weaning food based on malted ragi and green gram. Qual. Plant. Plant Foods Hum. Nutr. 36:223-230.

MANDAL, N. C., and BISWAS, B. B. 1970. Metabolism of inositol phosphates. Plant Physiol. 45:4-7.

MANSON, J. A., and CARPENTER, K. J. 1978a. The effect of a high level of dietary leucine on the niacin status of chicks and rats. J. Nutr. 108:1883-1888.

MANSON, J. A., and CARPENTER, K. J. 1978b. The effect of a high level of dietary leucine on niacin status in dogs. J. Nutr. 108:1889-1898.

MARTIN, F. W. 1984. Handbook of Tropical Food Crops. CRC Press, Boca Raton, FL.

MBOFUNG, C. M. F., and NDJOUENKEU, R. 1990. Influence of milling method and peanut extract on in vitro iron availability from maize and sorghum flour gruels. J. Food Sci. 55:1657-1659,1675.

MCMILLIAN, W. W., WILSON, D. M., MIROCHA, C. J., and WIDSTROM, N. W. 1983. Mycotoxin contamination in grain sorghum from fields in Georgia and Mississippi. Cereal Chem. 60:226-227.

MCNEILL, J. W., POTTER, G. D., RIGGS, J. K., and ROONEY, L. W. 1975. Chemical and physical properties of processed sorghum grain carbohydrates. J. Anim. Sci. 40:335-339.

MERTZ, E. T., HASSEN, M. M., CAIRNS-WHITTERN, C., KIRLEIS, A. W., TU, L., and AXTELL, J. D. 1984. Protein digestibility of proteins in sorghum and other major cereals. Proc. Natl. Acad. Sci. USA. 81:1-2.

MIROCHA, C. J., PATHRE, S. V., SCHAUERHAMER, B., and CHRISTENSEN, C. M. 1976. Natural occurrence of *Fusarium* toxins in feedstuff. Appl. Environ. Microbiol. 32:553-556.

MIROCHA, C. J., PATHRE, S. V., and CHRISTENSEN, C. M. 1980. Mycotoxins. Pages 159-225 in: Advances in Cereal Science and Technology, Vol. 4. Y. Pomeranz, ed., American Association of Cereal Chemists, St. Paul, MN.

MISIR, R., and BLAIR, R. 1988. Biotin bioavailability of protein supplements and cereal grains for starting turkey poults. Poult. Sci. 67:1274-1280.

MISRA, P. S., MERTZ, E. T., and GLOVER, D. V. 1976. Studies on corn proteins. IX. Comparison of the amino acid composition of Landry-Moureaux and Paulis-Wall endosperm fractions. Cereal Chem. 53:699-704.

MITARU, B. N. REICHERT, D., and BLAIR, R. 1983. Improvement of the nutritive value of high tannin sorghums for broiler chickens by high moisture storage (reconstitution) Poult. Sci. 62:2065-2072.

MOHAN, D. 1975. Chemically-induced high-lysine mutants in *Sorghum bicolor* (L.) Moench. Thesis. Purdue University, West Lafayette, Indiana, USA.

MOHARRAM, Y. G. 1987. Sorghum: Properties and tannin deactivation treatments. Alex. J. Agric. Res. 32:215-220.

MONTEIRO, P. V. SUDHARSHANA, L., and RAMACHANDRA, G. 1988. Japanese barnyard millet (*Echinochloa frumentacea*): Protein content, quality and SDS-PAGE of protein fractions. J. Sci. Food Agric. 43:17-25.

MUINDI, P. J., and THOMKE, S. 1981. The nutritive value for rats of high- and low-tannin sorghums treated with Magadi soda. J. Sci. Food Agric. 32:139-145.

MWASARU, M. A., REICHERT, R. D., and MUKURU, S. Z. 1988. Factors affecting the abrasive dehulling efficiency of high-tannin sorghum. Cereal Chem. 65:171-174.

NAKAGAWA, I., and SASAKI, A. 1977. Effect of an excess intake of leucine with and without additions of vitamin B_6 and/or niacin, on tryptophan and niacin metabolism in rats. J. Nutr. Sci. Vitaminol. 23:535.

NAKAGAWA, I., OHGURI, S., SASAKI, A. ET AL. 1975. Effects of excess intake of leucine and valine deficiency on tryptophan and niacin metabolites in humans. J. Nutr. 105:1241-1252.

NAREN, A. P., and VIRUPAKSHA, T. K. 1990. Effect of sulfur deficiency on the synthesis of α-setarin, a methionine-rich protein of Italian millet. Cereal Chem. 67:136-138.

NATTRESS, L. A., MEHTA, T., MITCHELL, M. E., and FINNEY, P. L. 1987. Formulation and nutritive value of weaning food from germinated food grains. Nutr. Res. 7:1309-1320.

NELSON, O. E. 1976. Interpretive summary and review. In: Genetic Improvement of Seed Proteins. Nat. Acad. Sci.: Washington, DC.

NEUCERE, J. N. 1982. Lectins in grain sorghum [*Sorghum bicolor* (L.) Moench]. J. Agric. Food Chem. 30:604-605.

NICOL, B. M., and PHILLIPS, P. G. 1978. The utilization of proteins and amino acids in diets based on cassava (*Manihot utilissimus*), rice or sorghum (*Sorghum vulgare*) by young Nigerian men of low income. Br. J. Nutr. 39:271-287.

NISHIMUTA, J. F., SHERROD, L. B., and FURR, R. D. 1969. Digestibility of regular, waxy and white sorghum grain rations by sheep. Proc. West. Sect., Am. Soc. Anim. Sci, p. 259.

NOVELLIE, L., and DE SCHAEPDRIJVER, P. 1986. Modern development in traditional African beers. Pages 73-157 in: Progress in Industrial Microbiology, Vol. 23. M. R. Adams, ed. Elsevier Science Publ., Netherlands.

NRC/NAS. 1982. United States–Canadian Tables of Feed Composition. Third Revision. National Academy Press, Washington, DC.

NRC/NAS. 1989. Recommended Dietary Allowances. 10th ed. National Academy Press, Washington, DC.

NWASIKE, C. C., MERTZ, E. T., PICKETT, R. C., GLOVER, D. V., CHIBBER, B. A. K., and VAN SCOYOC, S. W. 1979. Lysine level in solvent fraction of pearl millet. J. Agric. Food Chem. 27:1329-1331.

NWOKOLO, E. 1987. Composition and availability of nutrients in some tropical grains and oilseeds. Nutr. Rep. Int. 36:631-640.

O'NEILL, C. O., CLARKE, G., HODGES, G., JORDAN, P., NEWMAN, R., PAN, Q.-Q., LIU, F.-S., GE, M., CHANG, Y.-M., and TOULSON, E. 1982. Silica fragments surrounding oesophageal tumours in patients in northern China. Lancet 1:1202-1206.

OBIZOBA, I. C. 1988. Nutritive value of malted, dry- or wet-milled sorghum and corn. Cereal Chem. 65:447-449.

OBIZOBA, I. C. 1989. The nutritive value of mixtures of two varieties of bambarra groundnut and sorghum. Cereal Chem. 66:249-251.

OKEIYI, E. C., and FUTRELL, M. F. 1983. Evaluation of protein quality of formulations of sorghum grain flour and legume seeds. Nutr. Rep. Int. 28:451-461.

OPOKU, A. R., OHENHEN, S. O., and EJIOFOR, N. 1981. Nutrient composition of millet (*Pennisetum typhoides*) grains and malt. J. Agric. Food Chem. 29:1247-1248.

OSAGIE, A. U., and KATES, M. 1984. Lipid composition of millet (*Pennisetum americanum*) seeds. Lipids 19:958-965.

OSMAN, A. K. 1981. Bulrush millet (*Pennisetum typhoides*): A contributory factor to the endemicity of goitre in western Sudan. Ecol. of Food and Nutr. 11:121-128.

OSMAN, A. K., and FATAH, A. 1981. Factors other than iodine deficiency contributing to the endemicity of goiter in Darfur Province (Sudan). J. Hum. Nutr. 35:302-309.

OSMAN, A. K., BASU, T. K., and DICKERSON, J. W. T. 1983. A goitrogenic agent from millet (*Pennisetum typhoides*) in Darfur Province, Western Sudan. Ann. Nutr. Metab. 27:14-18.

OSUNTOGUN, B. A., ADEWUSI, S. R. A., OGUNDIWIN, J. O., and NWASIDE, C. C. 1989. Effect of cultivar, steeping, and malting on tannin, total polyphenol, and cyanide content of Nigerian sorghum. Cereal Chem. 66:87-89.

PANASIUK, O., and BILLS, D. D. 1984. Cyanide content of sorghum sprouts. J. Food Sci. 49:791-793.

PATTABIRAMAN, T. N. 1985. Trypsin/ chymotrypsin inhibitors from millets. Pages 439-448 in: Nutritional and Toxicological Significance of Enzyme Inhibitors in Foods. Advances in Experimental Medicine and Biology, Vol. 199.

PATWARDHAN, V. N. 1961. Nutritive value of cereal and pulse proteins. Pages 201-210 in: Progress in meeting protein needs of infants and preschool children. Publ. 843. National Academy of Sciences/National Research Council, Washington, DC.

PEDERSEN, B., and EGGUM, B. O. 1983. The influence of milling on the nutritive value of flour from cereal grains. 6. Sorghum. Qual. Plant. Plant Foods Hum. Nutr. 33:313-326.

PEDERSEN, B., HALLGREN, L., HANSEN, I., and EGGUM, B. O. 1987. The nutritive value of amaranth grain (*Amaranthus caudatus*). 2. As a supplement to cereals. Plant Foods for Hum. Nutr. 36:325-334.

PELLET, P. L. 1985. Amino acid scoring systems and their role in the estimation of the protein quality of cereals. In: Amino Acid Composition and Biological Value of Proteins, R. Lasztity and M. Hidvegi, eds. D. Reilel Publishing Co., Boston.

PLATT, ·B. S. 1962. Tables of representative values of foods commonly used in tropical countries. Med. Res. Counc. Spec. Rep. Ser. 302 (rev. ed. of SRS 253). H. M. Stationery Office, London.

PRENTICE, N., BABLER, S., and FALBER, S. 1980. Enzymic analysis of beta-D-glucans in cereal grains. Cereal Chem. 57:198-202.

PRICE, M. L., and BUTLER, L. G. 1979. Treatments of sorghum grain that reduce the assayable tannin content and their effect on the nutritional value of the grain. Pages 39-42 in: Polyphenols in Cereals and Legumes. J. H. Hulse, ed. International Development Research Center, Ottawa, Canada.

PRICE, M. L., and BUTLER, L. G. 1980. Tannins and Nutrition. Bull. 272, Purdue Univ. Agric. Exp. Stn., West Lafayette, IN.

PRICE, M. L., HAGERMAN, A. E., and BUTLER, L. G. 1980. Tannin in sorghum grain: Effect of cooking on chemical assays and on antinutritional properties in rats. Nutr. Rep. Int. 21:761-767.

PUSHPAMMA, P., and DEVI, C. A. 1979. Nutritional quality of sorghum and legume based food mixtures for infants and pre-school children–I. Nutr. Rep. Int. 19:635-641.

PUSHPAMMA, P., PARRISH, D. B., and DEYOE, C. W. 1972. Improving protein quality of millet, sorghum, and maize diets by supplementation. Nutr. Rep. Int. 5:93-100.

PUSHPAMMA, P., RATNAKUMARI, A., and GEERVANI, P. 1979. Nutritional quality of sorghum and legume based food mixture for infants and pre-school children–II. Nutr. Rep. Int. 19:643-648.

RADHAKRISHNAN, M. R., and SIVAPRASAD, J. 1980. Tannin content of sorghum varieties and their role in iron bioavailability. J. Agric. Food Chem. 28:55-57.

RAJALAKSHMI, R., and MAJMUDAR, N. 1966. Effect of different legume supplements to kodri (*Paspalum scrobiculatum*) on weight gains and body composition of albino rats. Baroda J. Nutr. 2:21-30.

RAMACHANDRA, G., VIRUPAKSHA, T. K., and SHADAKSHARASWAMY, M. 1977. Relationship between tannin levels and in vitro protein digestibility in finger millet (*Eleusine coracana* Gaertn.). J. Agric. Food Chem. 25:1101-1104.

RAO, S. A., and MUSHONGA, J. N. 1985. Traditional food crops in Zimbabwe. I. Finger millet. Zimb. Agric. J. 82:101-104.

RAO, V. J. M., DEOSTHALE, Y. G., RANA, B. S., RAO, K. V., and RAO, N. G. P. 1983. Nutritional quality in grain sorghum—Behaviour of crosses between high lysine sorghum and agronomically superior types. Indian J. Genet. 43:383-394.

REDDY, V. P., FAUBION, J. M., and HOSENEY, R. C. 1986. Odor generation in ground, stored pearl millet. Cereal Chem. 63:403-406.

REDDY, N. R., PIERSON, M. D., SATHE, S. K., and SALUNKHE, D. K. 1989. Nutritional consequences of phytate. Pages 81-110 in: Phytates in Cereals And Legumes. CRC Press, Inc., Boca Raton, FL.

REICHERT, R. D. 1979. The pH-sensitive pigments in pearl millet. Cereal Chem. 56:291-294.

REICHERT, R. D. 1982. Sorghum dry-milling. Pages 547-563 in: Proc. Int. Symp. Sorghum Grain Quality. L. W. Rooney and D. S. Murty, eds. Int. Crops Res. Inst. Semi-Arid Tropics (ICRISAT), Patancheru, A. P., India.

REICHERT, R. D., FLEMING, S. E., and SCHWAB, D. J. 1980a. Tannin deactivation and nutritional improvement of sorhgum by anaerobic storage of H_2O-, HCl-, or NaOH-treated grain. J. Agric. Food Chem. 28:824-829.

REICHERT, R. D., YOUNGS, C. G., and CHRISTENSON, D. A. 1980b. Polyphenols in *Pennisetum* millet. Pages 50-60 in: Polyphenols in Cereals and Legumes. J. H. Hulse, ed. IDRC-145e. International Development Research Center: Ottawa, ON.

REICHERT, R. D., MWASARU, AND MUKURU, S. Z. 1988. Characterization of colored-grain sorghum lines and identification of high-tannin lines with good dehulling characteristics. Cereal Chem. 65:165-170.

ROGLER, J. C., and SELL, D. R. 1984. Effects of stage of maturity on the tannin content of low and high tannin sorghum. Nutr. Rep. Int. 29:1281-1287.

ROONEY, L. W. 1978. Sorghum and pearl millet lipids. Cereal Chem. 55:584-590.

ROONEY, L. W., and PFLUGFELDER, R. L. 1986. Factors affecting starch digestibility with special emphasis on sorghum and corn. J. Anim. Sci. 63:1607-1623.

ROONEY, L. W., and SERNA-SALDIVAR, S. O. 1991. Sorghum. Pages 233-270 in: Handbook of Cereal Science and Technology. K. J. Lorenz and K. Kulp, eds. Marcel Dekker, New York.

ROONEY, L. W., and SULLINS, R. D. 1973. The feeding value of waxy and nonwaxy sorghum grains as related to endosperm structure. Page 15 in: Proc. Corn and Sorghum Res. Conf. 28th. Am. Seed Trade Assoc. Washington, DC.

ROONEY, L. W., KIRLEIS, A. W., and MURTY, D. S. 1986. Traditional foods from sorghum, their production, evaluation, and nutritional value. Pages 317-353 in: Advances in Cereal Science and Technology, Vol. 8. Y. Pomeranz, ed. Am. Assoc. Cereal Chem., St Paul, MN.

SALUNKHE, D. K., JADHAVE, S. J., KADAM, S. S., and CHAVAN, J. K. 1982. Chemical, biochemical and biological significance of polyphenols in cereals and legumes. CRC Crit. Rev. Food Sci. Nutr. 17:277-305.

SALUNKHE, D. K., CHAVAN, J. K., and KADAM, S. S. 1990. Dietary Tannins: Consequences and Remedies. CRC Press, Inc., Boca Raton, FL.

SASTRY, L. V. S., and VIRUPAKSHA, T. K. 1969. Alcohol-soluble proteins of grain sorghum. Cereal Chem. 46:284-293.

SAUER, D. B. 1988. Mold invasion in relation to grain damage. Cereal Foods World 33:489-490.

SCHERY, R. W. 1963. Plants for Man. Prentice-Hall, Englewood, NJ.

SCOTT, P. M. 1990. Tricothecenes in grain. Cereal Foods World 35:661-666.

SEBOLEV, A. M. 1962. Enzymatic hydrolysis of phytin *in vitro* and in germinating seeds. Soviet Plant Physiol. 9:263-269.

SEITZ, L. M., MOHR, H. E., BURROUGHS, R., and GLUECK, J. A. 1983. Preharvest fungal invasion of sorghum grain. Cereal Chem. 60:127-130.

SERNA-SALDIVAR, S. O., and ROONEY, L. W. 1991. The millets. Pages 271-300 in: Handbook of Cereal Science and Technology. K. J. Lorenz and K. Kulp, eds. Marcel Dekker, New York.

SERNA-SALDIVAR, S. O., KNABE, D. A., ROONEY, L. W., and TANKSLEY, T. D. 1987. Effect of lime cooking on energy and protein digestibilities of maize and sorghum. Cereal Chem. 64:247-252.

SERNA-SALDIVAR, S. O., CANETT, R., VARGAS, J., GONZALEZ, M., BEDOLLA, S., and MEDINA, C. 1988. Effect of soybean and sesame addition on the nutritional value of maize and decorticated sorghum tortillas produced by extrusion cooking. Cereal Chem. 65:44-48.

SERRAINO, M. E., THOMPSON, L. U., SOVOIE, L., and PARENT, G. 1985. Effect of phytic acid on the in vitro rate of digestibility of rapeseed protein and amino acids. J. Food Sci. 50:1689-1692.

SHARADA, D., DEVI, D. P. DEVI, AND SAILAKSHMI, N. 1986. Varietal differences in protein and calorie content of different varieties of bajra. Indian J. Nutr. Dietet. 23:66-68.

SHEORAIN, V. S., and WAGLE, D. S. 1973. Beta-amylase activity in germinated bajra and barley varieties. J. Food Sci. Technol. 10:184-186.

SHERROD, L. B., ALBIN, R. C., and FURR, R. D. 1969. Net energy of regular and waxy sorghum grains for finishing steers. J. Anim. Sci. 29:997-1000.

SHIVARAJ, B., RAO, H. N., and PATTABIRAMAN, T. N. 1982. Natural plant enzyme inhibitors. Isolation of a trypsin/alpha amylase inhibitor and a chymotrypsin/trypsin inhibitor from ragi (*Eleusine coracana*) grain by affinity chromatography and study of their properties. J. Sci. Food Agric. 33:1080-1091.

SHOTWELL, O. L., HESSELTINE, C. W., BURMEISTER, H. R., KWOLEK, W. F., SHANNON, G. M., and HALL, H. H. 1969. Survey of cereal grains and soybeans for the presence of aflatoxin. I. Wheat, grain sorghum, and oats. Cereal Chem. 46:446-454.

SHUKLA, S. S., GUPTA, O. P., SHARMA, Y. K., and SAWARKAR, N. S. 1986. Puffing quality characteristics of some ragi (*Eleusine coracana*) cultivars. J. Food Sci. Technol. 23:329-330.

SIKABBUBBA, R. M. 1989. The effect of alcohol-soluble proteins on the digestibility of sorghum. M.S. thesis. Kansas State University, Manhattan, KS.

SILANO, V. 1977. Factors affecting digestibility and availability of proteins in cereals. Pages 13-46 in: Nutritional Evaluation of Cereal Mutants. Proc. Advisory Group Meeting on a Nutritional Evaluation of Cereal Mutants. International Atomic Energy Agency, Vienna.

SIMWEMBA, G. C., HOSENEY, R. C., VARRIANO-MARSTON, E., and ZELEZNAK, K. 1984. Certain B-vitamins and phytic acid contents of pearl millet [*Pennisetum americanum* (L.) Leeke]. J. Agric. Food Chem. 32:31-34.

SINGH, P., SINGH, U., KUMAR, K. A., and ANDREWS, D. J. 1984. Nutritional quality improvement in pearl millet. Paper presented at the Symposium on Processing of Sorghum and Millets: Criteria for Quality of Grain and Products for Human Food, Vienna, Austria 4-8 June.

SINGH, P., SINGH, U., EGGUM, B. O., KUMAR, K. A., and ANDREWS, D. J. 1987. Nutritional evaluation of high protein genotypes of pearl millet [*Pennisetum americanum* (L.) Leeke]. J. Sci. Food Agric. 38:41-48.

SINGH, R., and AXTELL, J. D. 1973. High lysine mutant gene (hl) that improves protein quality and biological value of grain sorghum. Crop Sci. 13:535-539.

SINGH, R., and POPLI, S. 1973. Amylose content and amylolytic studies on high yielding varieties of bajra. J. Food Sci. Technol. 10:31-33.

SLOAN, D. R., BOWEN, T. E., and WALDROUP, P. W. 1971. Expansion-extrusion processing of corn, milo, and raw soybeans before and after incorporation in broiler diets. Poult. Sci. 50:257-261.

SOUTHGATE, D. A. T., GEOFFREY, J. H., ENGLYST, H. 1978. The analysis of dietary fiber—The choices for the analyst. J. Sci. Food Agric. 29:979-988.

SRIKANTIA, S. G., NARASINGA RAO, B. S., RAGHURAMULU, N., and GOPALAN, C. 1968. Pattern of nicotinamide nucleotides in erythrocytes of pellagrins. Am. J. Clin. Nutr. 21:1306-1310.

STEMLER, A. B. L., COLLINS, F. I., DE WET, J. M. J., and HARLAN, J. R. 1976. Variation in the levels of lipid components and protein in ecogeographic races of *Sorghum bicolor*. Biochem. Syst. Ecol. 4:43-45.

STOLOFF, L. 1976. Occurrence of mycotoxins in foods and feeds. In: Mycotoxins and Other Fungal Related Food Problems. J. V. Rodricks, ed. Adv. Chem. Ser. 149. Amer. Chem. Soc., Washington, DC.

STUART, M. A., JOHNSON, P. E., HAMAKER, B. E., and KIRLEIS, A. 1987. Absorption of zinc and iron by rats fed meals containing sorghum food products. J. Cereal Sci. 6:81-90.

SUBRAMANIAN, N., NANDI, B. K., MAJUMDER, A. K., and CHATTERJEE, I. B. 1973. Role of L-ascorbic acid on detoxification of histamine. Biochem. Pharmacol. 22:1671-1673.

SUBRAMANIAN, V., and SURYAPRAKASH, S. 1981. Sugars of pearl millet [*Pennisetum americanum* (L.) *Leeke*]. J. Food Sci. 46:1614-1615.

SUFIAN, S., and PITTWELL, L. R. 1968. Iron content of teff (*Eragrostis abyssinica*). J. Sci. Food Agric. 19:439.

SULLINS, R. D., and ROONEY, L. W. 1974. Microscopic evaluation of the digestibility of sorghum lines that differ in endosperm characteristics. Cereal Chem. 51:134-142.

TAUR, A. T., PAWAR, V. D., and INGLE, U. M. 1984. Effect of fermentation on nutritional improvement of grain sorghum. Indian J. Nutr. Dietet. 21:89-96.

TAYLOR, J. R. N. 1983. Effect of malting on the protein and free amino nitrogen composition of sorghum. J. Sci. Food Agric. 34:885-892.

TAYLOR, J. R. N., and SCHUSSLER, L. 1986.

The protein composition of the different anatomical parts of sorghum grain. J. Cereal Sci. 4:361-369.

TEETER, R. G., SARANI, S., SMITH, M. O., and HIBBERD, C. A. 1986. Detoxification of high-tannin sorghum grains. Poult. Sci. 65:67-71.

TRIPATHI, R. K. 1973. Aflatoxins in sorghum grains infected with head moulds. Indian J. Exp. Biol. 11:361-362.

USDA/HNIS. 1988. Provisional Table on the Dietary Fiber Content of Selected Foods. HNIS/PT-106. U.S. Government Printing Office, Washington, DC.

USDA/HNIS. 1989. Composition of Foods: Cereal Grains and Pasta. U.S. Dep. Agric. Agric. Handb. 8-20. U.S. Govt. Printing Office, Washington, DC.

VAN HEERDEN, I. V. 1989. The nutritive content of African beers brewed with maize grits or sorghum adjunct. J. Inst. Brew. 95:17-20.

VAN SCOYOC, S. W., EFETA, G., and AXTELL, J. D. 1988. Kernel characteristics and protein fraction changes during seed development of high-lysine and normal sorghums. Cereal Chem. 65:75-80.

VARRIANO-MARSTON, E., and HOSENEY, R. C. 1980. Note on the mineral content of location in pearl millet. Cereal Chem. 57:150-152.

VARRIANO-MARSTON, E., and HOSENEY, R. C. 1983. Barriers to increased utilization of pearl millet in developing countries. Cereal Foods World 28:392-395.

VIVAS, N. E., WANISKA, AND ROONEY, L. W. 1987. Effect of tortilla production on proteins in sorghum and maize. Cereal Chem. 64:384-389.

WAGGLE, D. H., DEYOE, C. W., and SMITH, F. W. 1967. Effect of nitrogen fertilization on the amino acid composition and distribution in sorghum grain. Crop Sci. 7:367-378.

WALL, J. S., and CARPENTER, K. J. 1988. Variation in availability of niacin in grain products. Food Technol. 42(10):198-204.

WANG, Y. D., and FIELDS, M. L. 1978. Germination of corn and sorghum in the home to improve nutritive value. J. Food Sci. 43:1113-1115.

WHITACRE, M. E., and TANNER, H. 1989. Methods of determining the bioavailability of amino acids for poultry. Pages 129-141 in: Absorption and Utilization of Amino Acids, Vol. 3. M. Friedman, ed. CRC Press, Boca Raton, FL.

WHITNEY, E. N., CATALDO, C. B., and ROLFES, S. R. 1987. Understanding Normal and Clinical Nutrition, 2nd ed. West Publ. Co., New York.

WHO. 1985. Energy and Protein Requirements. Report of a Joint FAO/WHO/UNU Expert Consultation. Tech. Rep, Ser. 724. World Health Organization, Geneva.

WOOD, P. J. 1984. Physicochemical properties and technological and nutritional significance of cereal β-glucans. Pages 35-78 in: Cereal Polysaccharides in Technology and Nutrition, V. F. Rasper, ed. American Association of Cereal Chemists, St. Paul, MN.

WOOLARD, G. R., RATHBONE, E. B., and NOVELLIE, L. 1976. A hemicellulosic β-D-glucan from the endosperm of sorghum grain. Carbohydrate Res. 51:249-252.

WOOLARD, G. R., RATHBONE, E. B., NOVELLIE, L., OHLSSON, J. T. 1977. Structural studies on the water-soluble gums from the endosperm of sorghum grain. Carbohydrate Res. 53:101-108.

WORKER, G. F., JR., and RUCKMAN, J. 1968. Variations in protein levels in grain sorghum grown in the southwest desert. Agron. J. 60:485-488.

WU, Y. V., and WALL, J. S. 1980. Content of protein increased by germination of normal and high-lysine sorghums. J. Agric. Food Chem. 28:455-458.

YEW, M. S. 1975. Biological variation in ascorbic acid needs. Ann. NY Acad. Sci. 258:451-972.

YOON, J. H., THOMPSON, L. U., JENKINS, D. J. A. 1983. The effect of phytic acid on in vitro rate of starch digestibility and blood glucose response. Am. J. Clin. Nutr. 38:835-842.

YOUNG, R., HAIDARA, M., ROONEY, L. W., and WANISKA, R. D. 1990. Parboiled sorghum: Development of a novel decorticated product. J. Cereal Sci. 11:277-289.

YOUNG, V. R., and FUKAGAWA, N. K. 1988. Amino acid interactions: A selective review. Pages 27-71 in: Nutrient Interactions. IFT Basic Symp. Ser. C. E. Bodwell and J. W. Erdman, eds. Marcel Dekker, New York.

ZHUGE, Q., KLOPFENSTEIN, C. F., and BEHNKE, K. C. 1990. Nutritional value of sorghum grain depends on processing. Feedstuffs 62(9):18,55.51

CHAPTER 6

STORAGE OF SORGHUM AND MILLETS: INCLUDING DRYING FOR STORAGE, WITH PARTICULAR REFERENCE TO TROPICAL AREAS AND THE MYCOTOXIN PROBLEM

J. A. McFarlane, A. E. John, and R. C. Marder
Natural Resources Institute
Chatham Maritime
Chatham, United Kingdom

INTRODUCTION

This chapter focuses on storage in tropical regions and, especially, on developing countries in those regions including, notably, Africa and India, where the production and utilization of these food grains is vastly greater than elsewhere. The Food and Agriculture Organization of the United Nations (FAO) statistics for the years 1981–1985 (Anonymous 1988b, 1990; see also Chapter 2) show that for sorghum, approximately 75% of world production is in developing countries, mostly in Africa and Asia; the annual yield growth rate in these countries is nearly double that in developed countries, although areas planted are increasing more rapidly in the latter case. For millets, 96% of world production for food uses is in developing countries, predominantly in Africa and Asia (especially India), while 60% of the much lower production for animal feed and 89% of the production for other uses is also in these countries. The per capita food use in developing countries is also much greater (by a factor of nearly nine) than in developed countries, with the highest levels (above 10 kg/year) in Africa (seven countries: notably those in francophone West Africa, especially Niger, but also Ethiopia and Uganda) and India.

Furthermore, it is in these same regions that the infestation problems of food grain storage are greatest—for climatic reasons and, also, because of the relative dependence upon unsophisticated storage structures and techniques that may serve very well in stable traditional systems but which often serve less well under the stresses and strains of modern agricultural development and other socioeconomic forces.

The treatment of the information presented in this chapter is, on the whole, general, because it would be inappropriate to deal otherwise with regions in which situations and circumstances are extremely diverse. The general approach is intended to provide guidelines and to indicate various options, so that those who are in the position to assess situation-specific problems may be guided towards suitable choices and reasoned decisions or, at least, be assisted in the definition of those problems that may require further on-the-ground research or external intervention.

GENERAL CONSIDERATIONS

Scientific attention to the storage of sorghum and, especially, millets has been considerably less than that devoted to the storage and preparation for storage of other cereal grains. The main reason for this is that sorghums and millets, in the tropics, have been generally regarded as minor grain crops despite their relative importance as food staples in many subsistence-farming areas. A further and rather more valid reason, as noted by Richter (1988), is that farmers in arid and semiarid countries (i.e., the regions where sorghum and millets are commonly important) achieve quite impressive performances in grain storage by relatively simple traditional methods (Anonymous, 1984). However, the introduction of new varieties and the more recent focusing of scientific attention upon the role of small grains in food security, together with the problems associated with the processing and utilization of sorghums and millets (Anonymous 1987), has led to a considerable increase of interest in the storage characteristics of these grains. Available information on recorded storage losses for sorghum and millets is given in *Postharvest Food Losses in Developing Countries* (Anonymous, 1978), which includes a summary of observations by Guggenheim (1977) and Spencer et al (1975). Information on storage and marketing of grain sorghum in the United States is contained in Bennett et al (1990).

Sorghum and millets may be stored, after drying and threshing, as loose grain in bags or bulk containers. They are quite commonly left on the field, prior to threshing, in stacks or piles of harvested plants. In some countries (the United

TABLE 1
Basic Principles of Storage Management

Storage site and handling facilities	Must be fit for the purpose, i.e., adequate for the maintenance of grain quality, quantity, and availability in-line with realistic objectives and at an economic cost. Note: Fitness includes sanitation.
Stored commodities	Must be fit for storage, i.e., sufficiently dry and free from damage and impurities to permit satisfactory storage for the required period under local climatic conditions.
Insect pests	Should be controlled (especially in tropical climates) by measures that are suited to the situation and circumstances and are cost-effective in practice.
Rodent and bird pests	Should be excluded from storage structures by sanitation and by suitable barriers that are locally cost-effective.
Storage period	Should be limited to the shortest period possible in the circumstances. Note: Long-term storage calls for higher levels of fitness in the commodity and storage facilities.
Storage losses (including losses due to spillage and theft)	Must be identified (and accounted for), with regard to both cause and value, and restricted to economic (and sociologically acceptable) levels.

States, for example) this practice has now been almost entirely superseded by combine harvesting (Bennet et al, 1990). The detached heads may also be stored away from the field, in exposed stacks (sometimes very carefully constructed) or in traditional storage containers. The essential prerequisites for storage (Table 1) are the same as for all cereal grains and other stored produce. A comprehensive general account of the storage of cereal grains and their products is given in Sauer (1992).

Sorghums and millets, however, differ from most other cereal grains in some of their postharvest characteristics. Similarities and differences are set out in Table 2. Several of the differences, in particular small grain size and the consequent reduced susceptibility to infestation by storage insects (Table 3), are most extreme in some of the millets: notably *Eragrostis abyssinica* (teff) and *Digitaria* spp. (fonio). Teff, for example, has been shown to be almost completely immune to damage by several primary and secondary insect pests of stored grain (McFarlane and Dobie, 1972). The exception, in that study, was *Tribolium castaneum,* which was able to

TABLE 2
Characteristics for Sorghums and Millets as Stored Cereal Grains

Similarities to other stored grains
Suitability for storage and eventual product quality are influenced by prestorage handling and processing, e.g., by cleaning and drying.
Grain moisture content is isothermally related to ambient humidity. Note: Water activity (and equilibrium relative humidity) increase with temperature.
Mold growth and intrinsic deterioration are negligible when the commodity is sufficiently dry. Note: 70% relative humidity (a_w 0.70) is marginally safe for storage; 60–65% (a_w 0.60–0.65) is preferable.
Storage life is inversely related to temperature and humidity in storage; quality can be maintained by reducing storage temperature and humidity or moisture content (or all three factors).
Varietal improvements to increase nutritive value and/or grain yield generally result in an increased susceptibility to storage insects.
Highly susceptible to damage by commensal rodents.
Differences from other stored grains
Grain size is relatively small and in some millets very small; this (especially with millets) tends to reduce susceptibility to infestation by many storage insects but to increase the likelihood of loss by spillage and seepage (from sacks) and the difficulties of assessing percentage grain damage and consequent weight loss. Grain cleaning (by sieving and winnowing) may also be less efficient and more wasteful.
Grain value is sometimes relatively low, especially in comparison with rice; this reduces the apparent benefits of pest control measures and other storage improvements.
Production areas are generally in semiarid zones; this increases cultivation problems (although millets are relatively drought resistant) and the risk of preharvest attack by some pests but reduces (or even eliminates) the problems of drying for storage.
Natural barriers against insect infestation (sheathing leaves, husks, etc.) are very varied among the diverse species: some may be relatively resistant, others less so.
Varietal improvements to increase grain size, as well as quality and yield, are frequent objectives and are particularly likely to increase susceptibility to infestation by storage insects.
The storage consequences of preharvest grain damage by birds may be a particular problem, especially with the compact-panicle sorghums and bulrush or pearl millet. Here again, varietal improvements (especially for grain palatability) are likely to increase this pest problem.

multiply to some extent on undamaged whole teff by feeding upon the grain embryos, much as it does when attacking other whole cereal grains.

In contrast, the grain size of most sorghums allows infestation by all the common primary pests (weevils and grain moths), and many of the larger-grained millets (e.g., pearl millet) are sufficiently large for the very destructive pests *Rhyzopertha dominica* and *Trogoderma granarium* to thrive on them. For this reason, the popular concept of the millets, as grains that are hardly susceptible to damage by storage insects, is erroneous; only the very small-grained millets, like teff and fonio, have very low susceptibility. Others may have reduced susceptibility to infestation by *Sitotroga cerealella,* the larvae of which are typically internal grain-feeders and prefer good-sized grains (although pearl millet, stored on the head, may be heavily attacked), and some may be virtually immune to damage by *Sitophilus* spp., the larvae of which are obligate internal grainfeeders. However, they remain susceptible to damage by *R. dominica* and *T. granarium* and, consequently, to further damage by secondary beetle pests, notably *Tribolium castaneum* and *Cryptolestes* spp. Moreover, whole-grain pearl millet, and possibly some others, are also susceptible to damage, at the site of the grain embryo, by the larvae of warehouse moths, notably (in Mali, for example) *Corcyra cephalonica,* but probably *Ephestia cautella* also. The damage done by these larvae is similar to moth damage on stored wheat grains, where it is more conspicuous because of the larger grain size.

A further factor contributes to the idea that millets, in general, are less susceptible to insect infestation than sorghum and other cereals. The millets are most commonly grown in semiarid climates, where the stored grain is typically very dry, often at moisture contents in equilibrium with humidities well-below 40% rh. In such conditions, the warehouse moths and most secondary beetle pests do not thrive. However, the major pests *R. dominica* and *T. granarium* are relatively well adapted to extremely dry environments and can cause serious damage in these cir-

TABLE 3
Important Insect Pests of Cereal Grains Stored in the Tropics[a]

Order Family	Specific Name	Common Name
Coleoptera		
Bostrichidae	*Rhyzopertha dominica* (Fabricius)	Lesser grain borer
	Prostephanus truncatus (Horn)[b]	Larger grain borer
Cucujidae	*Cryptolestes* spp.	Flat grain beetles
Curculionidae	*Sitophilus oryzae* (L.)	Rice weevil
	S. zeamais Motschulsky	Maize weevil
Dermestidae	*Trogoderma granarium* Everts[c]	Khapra beetle
Silvanidae	*Oryzaephilus surinamensis* (L.)	Sawtoothed beetle
Tenebrionidae	*Tribolium castaneum* (Herbst)	Bran beetle (or Rust-red flour beetle)
Lepidoptera		
Gelechiidae	*Sitotroga cerealella* (Olivier)	Angoumois grain moth
Pyralidae	*Ephestia cautella* (Walker)	Tropical warehouse moth
	Plodia interpunctella (Hubner)	Indian-meal moth
	Corcyra cephalonica (Stainton)	Rice moth

[a] Adapted from McFarlane (1989).
[b] Now established in Africa as well as in the Americas.
[c] Common only on very dry grain, especially in Sahelian North Africa and the Indian subcontinent.

cumstances. Recent observations in Togo indicate that some varieties, at least, are also susceptible to postharvest infestation by *Prostephanus truncatus*, which is more commonly associated with dry maize both pre- and postharvest.

Current options for insect pest control on cereals stored in tropical climates are listed in Tables 4 and 5 and discussed in McFarlane (1989), which contains references to more detailed information. For sorghum and millets, particular attention should be given to the maintenance or further development of pest-resistant varieties (the penultimate option in Table 5) and to the resolution of conflicts between this and other plant breeding objectives (see, for example, Rogers and Mills, 1974).

SPECIFIC NOTES ON STORAGE

Sorghum

The general pattern and spectrum of storage infestation of sorghum are similar to those of maize. Varietal differences in susceptibility occur (Doggett, 1957; Giles, 1965), with traditional varieties commonly more resistant to damage by storage insects and to preharvest damage, by birds, which accelerates storage deterioration.

TABLE 4
Current Options for Grain Storage[a]

Storage Container	Associated Pest Control Methods
Small-scale bulk storage	
Cribs	
In open air	Smoking over an open fire Admixing a grain protectant
Inside a dwelling place	As above
Underground pits	Self-disinfesting if sufficiently airtight
Small storage bins	As above Admixing a grain protectant Intermittant "sunning" and winnowing
Large-scale bulk storage	
Sealed bunkers	Admixing a grain protectant Disinfesting by fumigation
Unwelded metal bins	As above Cooling by aeration
Flexible (collapsible) bins	As above, or self-disinfesting if sufficiently airtight and protected against insolation and rodent damage
Concrete bins or pits	Self-disinfesting if sufficiently airtight Admixing a grain protectant Disinfesting by fumigation or controlled atmospheres
Welded steel bins	As above
Storage in bags or small containers	
Sealed pots, gourds, etc.	Self-disinfesting if sufficiently airtight
Conventional bags (stacked or unstacked)	Admixing a grain protectant Fumigation under sheets, with or without protection against reinfestation Partially self-disinfesting, by insect heat, if tightly stacked
Wrapped bags in sealed pits or bunkers	Self-disinfesting if sufficiently airtight
Insect-proof packages	Fumigation or irradiation

[a]Adapted from McFarlane (1989).

TABLE 5
Pest Control Techniques: Current Options

Control Method and Major Uses	Advantages	Disadvantages
Fumigation with penetrating gases, for disinfestation in storage or prior to export	Pest mortality is almost immediate and can be total.	No residual protection No control of pests outside the enclosure May induce pest resistance
Space treatments with contact insecticide, to control flying insects especially moths	Spectacular effects Can be useful as an adjunct to fumigation	Affect only a small part of the total population
Residual treatments with contact insecticides, for long-term protection of food stocks	Effective protection for some time after treatment, especially with insecticide admixture; less effective with surface treatments	Effectiveness varies with circumstances. Highly dependent upon the availability of suitable high-quality formulations May induce pest resistance
Heat treatment for disinfestation before storage or export	Pest mortality is almost immediate and can be total.	Relatively advanced technology needed to ensure success and to avoid adverse effects. Solar heaters may offer a cheaper alternative.
Cooling for sustained pest control during storage	Highly effective and economic in some situations	Not usually feasible in the low-altitude tropics without expensive equipment
Sealed ("hermetic") storage in traditional containers, underground pits (semi-airtight), or completely airtight containers	Highly effective for dry grain (including seed) in well-sealed containers and can give satisfactory control in semiairtight conditions	Inadequate sealing, partial loading, or the use of relatively small semiairtight containers or pits permits substantial damage by insects and molds. Completely hermetic structures may be too expensive.
Entoleters to disinfest flour before storage or despatch	Convenient in modern processing systems	Not always practical
Physical barriers to protect uninfested or disinfested goods in storage and transport	Can be cheap and very effective if well designed and managed	Enclosed goods may need to be drier than usual or specially aerated if the barriers impede moisture vapor movements.
Controlled atmospheres for long-term storage	Potentially highly effective	Suitable storage structures are usually expensive.
Trapping to control rodents or monitor pest populations	Can be inexpensive, simple, and useful	Ineffective for control unless combined with other methods
Pheromones to monitor pest populations or reduce them by behavioral disruption	Very effective for monitoring low-density populations	Currently available for relatively few pests Rather expensive
Food attractants to monitor or reduce pest populations	Simple and effective for population monitoring	Ineffective for control unless combined with other methods
Insect growth regulators, for long-term control	Highly effective and persistent control has been demonstrated.	Currently experimental and few readily available for use
Pest pathogens to eliminate or reduce populations	Relatively pest specific	Pest may show high natural tolerance.
Predators and parasites to reduce populations	Natural occurrence	Ineffective until pests are well established
Resistant varieties to reduce the rate of population growth	Effective throughout the storage period	May conflict with other plant-breeding objectives
Sterile insects to reduce population growth rates	No proven advantages	Technique requires sophisticated equipment and management.

[a]Adapted from McFarlane (1989).

Resistance to storage insects appears to be determined largely by endosperm hardness, in particular the thickness of the vitreous ("corneous") shell, and by grain size (Doggett, 1957; Davey, 1965), with tannin content having little or no effect (Fadelmula and Horber, 1983). The presence of adherent glumes reduces damage by grain weevils (Rogers and Mills, 1974) but facilitates damage by the grain moth *Sitotroga cerealella* (Wongo, 1990). Sorghum is often stored unthreshed at the small farm level, and the advantages and disadvantages of this practice in particular situations and circumstances may warrant more investigation. As with maize and rice, infestation by *S. cerealella* is reduced by storing the threshed grain, in this case not only by the reduction of intergranular spaces but also by the removal of glumes, but infestation by other insects, especially *Sitophilus* spp., may be favored (Giles, 1965; Wongo and Pedersen, 1990).

Storage in underground pits, often reasonably successful when the pits are of large capacity, is common in Sudan, eastern Ethiopia, and Somalia (Gilman and Boxall, 1974). On-farm storage losses elsewhere, also are often low (Richter, 1988) especially where traditional varieties are stored. Golob (1981) found not more than 3% food loss in up to nine months' storage in Malawi, and P. S. Hindmarsh (*personal communication*) observed similarly low losses in Zambia. Ratnadass (1990) reports lower losses (generally less than 1% weight loss) in up to 10 months' storage in the Central African Republic. In arid regions, *Trogoderma granarium* is the major pest and, although it is not usually a problem in on-farm storage, it sometimes causes very great losses to bagged sorghum in commercial warehouses. There is a clear need, in some cases, for changes in the conventional storage and marketing system to promote higher standards of pest and quality control. Adaptive research to devise and demonstrate appropriate systems for the cost-effective use of available control techniques is also needed. Fumigation is likely to remain as an essential component, but physical barriers for protection against reinfestation could displace the use of residual spray treatments, which are commonly ineffective. Hermetic storage systems, cooling by aeration, or controlled atmosphere storage might also be considered. The natural dryness of the stored grain in these arid regions would favor the adoption of these various techniques (Calderon et al, 1989).

Pearl Millet

Pearl millet is probably the most widely grown millet in Africa. It generally keeps well in storage (Spencer et al 1975; Guggenheim, 1977), at least partly because of the dry climate in the production areas and consequent low grain moisture content. Losses by spillage or seepage from jute sacks and contamination by sand and other foreign matter are common problems (Rowley, 1984). However, insect and rodent infestation also occur; in the typical dry storage conditions, the insects *T. granarium, R. dominica,* and *T. castaneum* are the major pests. Traditional storage methods include sealed storage in clay pots, and, less commonly (in Botswana, for example; P. S. Tyler, *personal communication*), admixture with beans, which reduces the intergranular spaces between the beans, thereby impeding infestation of the beans by bruchid beetles and optimizing the use of storage space.

Proso, *Setaria*, and *Eleusine* (Finger) Millets

Specific information on the storage of these grains is lacking.

Teff

Teff is peculiar to Ethiopia, where it is the most popular and highly priced cereal used for the preparation of the traditional unleavened bread (*injera*). The tiny seeds are almost immune to insect damage, but rodent damage and spillage are problems. In traditional domestic storage, these problems are largely overcome by the use of well-made clay jars.

Fonio

Small quantities of fonio are produced in Sahelian North Africa and, in Mali for example, are often stored in closely woven jute bags. As with teff, the very small grain size precludes most insect infestation, but spillage and rodent damage are likely to be considerable problems.

DRYING FOR STORAGE

Thorough drying of cereal grains, to prevent mold damage, is an essential basic requirement for storage in almost all circumstances. The only possible exceptions to this are those circumstances in which grain can be stored in hermetic containers, with controlled atmospheres of sufficiently antibiotic gases, or in conventional structures after suitable treatment with fungicides such as propionic acid. However, such methods for the storage of high-moisture grain are still generally regarded as appropriate only for grain to be used for animal-feeding purposes. Grain so stored may be technically fit to eat, but its quality for cooking purposes is usually impaired.

The problem of drying for storage, which can be a major constraint upon the production of cereals grown in climatic zones with regular high rainfall or unseasonable rains in the postharvest period, is much reduced in the regions where sorghum and millets are most commonly produced. Millet grains, especially, are commonly dry enough for storage at the time of harvest and, in most situations, both sorghum and millets can be further dried by traditional field-drying or sun-drying procedures if necessary. For all stored products, the maximum "safe" moisture content for storage is that which is in equilibrium with 70% rh, but lower levels (in equilibrium with 65% rh) are advisable if quality loss is to be minimized (Table 6). The value varies between different commodities, and even between varieties, and is temperature-dependent (Pixton, 1982). In climates where the harvested grain is typically in equilibrium with ambient humidities below 50% rh, as for example in the millet-producing regions of Sahelian Africa, the stored grain is initially quite safe from damage by molds and other microorganisms. However (see General Considerations), it is not entirely safe from insect attack, which can cause significant moisture increase and may induce mold growth.

In situations where there may be a need for supplementary drying procedures to permit the safe storage of sorghum and millets, the techniques developed for drying other cereal grains can be applied, with modification where necessary. Traditional methods and improved technologies for both sorghum and millet drying with

TABLE 6
Equilibrium Moisture Contents (% wb) for Some Tropical Cereal Grains for Adsorption (Ads.) and Desorption (Des.)[a,b]

Relative Humidity (%)	Species and Variety									
	Oryza sativa Millyang (dehusked) (27° C)		*Pennisetum typhoides* Nigerian Red (27° C)		*Sorghum bicolor* Thengalamanga (26° C)		*Zea mays* UCA (26° C)		Triticale (unnamed) (27.5° C)	
	Ads.	Des.	Ads.	Des.	Ads.	Des.	Ads.	Des.	Ads.	Des.
20	7.5	7.5	6.2	6.2	...	...	6.8	6.8	6.0	6.0
30	8.8	8.8	7.8	7.8	8.0	...	8.0	8.0	7.5	8.0
40	10.0	10.0	9.4	9.4	9.5	...	9.0	9.3	9.3	10.0
50	11.5	11.5	11.0	11.0	11.0	11.5	10.3	10.8	11.0	11.8
60	12.5	12.5	12.8	12.8	12.0	12.5	11.6	12.3	12.0	12.8
70[c]	14.8	14.8	15.0	15.0	14.0	14.5	13.3	14.0	14.8	15.8
80	16.0	16.0	18.0	18.0	15.0	15.5	15.7	16.1	17.5	18.0
90	18.5	18.5	24.6	24.5	18.0	18.3	19.0	19.0	24.5	24.5

[a] Adapted from Gough and King (1980).

[b] Values are for temperatures in the range 26–27.5° C. Corresponding values for temperatures in the range 36–37.5° C would be approximately 0.6% lower. Corresponding values at other temperatures would differ similarly pro rata, being lower at higher temperatures and higher at lower temperatures due to the temperature-dependence of the equilibrium between moisture content and relative humidity.

[c] The values given for 70% rh represent the maximum for "safe" storage (i.e., safe from significant mold damage) at 26–27.5° C. In practice, however, lower moisture contents should be preferred. For example, Bennet et al (1990) recommend storage of grain sorghum at 11–13% moisture unless proper aeration is provided, when 15–16% may be safe.

reference to developing countries, particularly in Africa, have been described (Anonymous, 1988a).

Sorghum Drying

FIELD DRYING

In many developing countries, where traditional varieties (open panicle types) are grown, the crops tend to ripen after the rainy season. Thus, the drying operation is relatively free of problems.

By far the most common method of drying in these countries is to stack bundles of panicles in the field, leaving them to dry in the sun before threshing. However, in a few areas, particularly where nontraditional varieties are grown, the climate during the harvesting season does not permit stack drying. Bundles of panicles are therefore dried over and around the fire in the farmhouse. Where the harvested crop is stacked in the field for some time to allow drying to take place, important considerations are that the grain be adequately exposed to the prevailing wind and also that the stack be built off the ground. Stacks 0.3-m off the ground, 2–3 m high and 0.6–2.0 m wide are in most cases suitable to dry the grain quickly with minimum infestation. Bundles laid in the first row are placed with their heads facing out of the stack. The second row is laid perpendicular to the first, and the third the same way as the first (Vogel and Graham, 1978).

MECHANICAL DRYING

In those countries with temperate climates and where labor costs are high, such as in North America, sun-drying is not always a suitable option. Mechanical dryers are therefore employed. Such dryers consist of a fan for moving the air, a bin for holding the threshed grain, an air-distribution system, and, in many but not all cases, air heaters. There are, in principle, three systems (Sorensen and Person, 1970):

1) unheated air-drying, in which ambient air is forced through the drying bin;

2) air-drying with supplementary heat, in which heated air is used but only during periods when the ambient air is of high relative humidity (usually greater than 75%);

3) heated air-drying, in which the air is constantly heated by fuel burners or electricity.

Unheated and supplementary heated air-drying are used for drying deep beds (2–3 m) of grain and are particularly suited for drying in storage for on-farm situations. Heated air-drying is primarily used for shallower depths (up to 0.5 m) and operations that demand a high drying capacity. With all systems, the air flow is usually directed up through the grain bed. This is for two main reasons: 1) the wettest grain, predominantly first harvest, is generally first to be loaded into the bin and first to be dried; and 2) sampling during drying is more easily done from the top layer of grain, which is the last to dry.

UNHEATED AIR-DRYING

Comparatively, these systems demand the lowest investment in equipment, are simple to operate, and require minimum attention to provide good uniform drying. However, their operation is weather-dependent, and during persistent wet weather and conditions of high relative humidity (greater than 75%) they are relatively ineffective, so drying times are often long. It is therefore important to select airflow rates carefully to optimize drying and prevent mold growth due to prolonged storage at high moisture contents. Recommended rates are from 0.02 m^3/sec per cubic meter of sorghum at 14% moisture content (wet basis) up to 0.04 m^3/sec per cubic meter of sorghum at 20% moisture content (wet basis) to provide the most economic drying.

AIR-DRYING WITH SUPPLEMENTARY HEAT

Supplementary heaters are usually used to improve the performance of unheated air-drying systems, so drying can be completed regardless of weather. Heat is supplied during prolonged periods of high relative humidity, cold and wet weather, using either oil or gas burners or electric heaters. The selection of the heat source depends on cost, availability and level of control required. Such systems necessitate greater attention to prevent overheating, which can lead to grain damage and fires. Drying temperatures should not exceed 40°C, since higher temperatures can lead to overdrying at the air inlet. Compared with unheated air systems, the drying times are significantly reduced and the risk of mold infestation is lessened. Heat for low-temperature drying can also be supplied from solar energy. This has been reported in use both in the United States (Foster et al, 1980) and India (Bhinde et al, 1987) using covered flat plate and inflatable tunnel solar collectors. Covered flat plate collectors consist of a shallow and wide duct with a blackened base and a translucent (either glass or plastic) cover through which air is blown. Inflatable tunnel collectors comprise a semicircular clear plastic duct with a black plastic base. Drying temperatures up to a maximum of 15°C above ambient can be obtained using such designs, but they require a relatively large area of land if constructed separately. However, in many cases solar collectors can be incorporated into the roofs and walls of existing dryer or storage buildings, whereas the tunnel types are collapsible. Drying times using solar collectors can be significantly reduced compared with unheated systems but are of course dependent upon the level of insolation.

HEATED AIR-DRYING

Drying systems employing constantly heated air are of high capacity and can be operated in a batch or continuous mode. Their designs are relatively sophisticated with much higher investment costs. These systems require that grain be handled twice, once into the dryer and again into storage. They can be operated at temperatures in the range 60–100°C employing much greater airflow rates. Typical airflows are in the range of 0.5–1.5 m^3/sec per cubic meter of sorghum for batch dryers and 1.5–2.5 m^3/sec per cubic meter of sorghum for continuous dryers. Drying times therefore are very short.

Millet Drying

Information on the drying of millet is scanty, but field drying methods, similar to those described for sorghum, are known to be used in developing countries. Millet grains are smaller than sorghum, so dry much quicker. They are quite often harvested during rainy periods and left to dry in the field for up to two weeks. Further drying if required is completed after threshing on mats laid in the sun. Many African farmers consider that foods prepared from rain-beaten grains have improved quality and palatability (Vogel and Graham, 1978). Where mechanical drying might be required, similar principles would apply as with sorghum. It may be necessary, however, to use higher airflow rates or reduced grain bed depths to prevent overdrying of grains at the air inlet.

ACQUIRED TOXINS

Extraneous contamination may be acquired by most commodities in various ways and from diverse sources. Pesticide residues, resulting from deliberate applications for pest control purposes or from accidental contacts, are well known, but many other materials, of varying degrees of noxiousness, can be involved. Most risks of contamination can usually be avoided by careful management of storage and transport. Contamination by mycotoxins, which are the metabolic by-products of various fungi, is a more difficult problem. Infection by toxigenic fungi can occur both preharvest and postharvest, and while the risks can be reduced by good commodity management, including all harvesting and postharvest treatments, they cannot be entirely precluded. Moreover, following infection, the development of the toxins may proceed, in some cases, through inconspicuous mycelial growth on grains which are, nominally, dry enough for storage.

As part of the general crop improvement plan, traditional varieties of sorghum are being replaced by photo-insensitive varieties, which give higher yields of grain. However, these cultivars may set seed during periods of rainy or wet weather, when they are more prone to invasion by toxigenic fungi. Efficient commodity management, including correct harvesting, drying and storage procedures, becomes increasingly important if mycotoxin problems are to be minimized.

Mycotoxins in Sorghum

A survey of fungi colonizing Indian and African sorghum samples (M. E. Kalisz, R. E. Strange, V. F. P. Medlock, and R. Bandyopadhyah, *unpublished*) showed that principal species isolated were *Alternaria tenuissima, Fusarium moniliforme, Curvularia lunata,* and *Phoma sorghina.* These samples represented preharvest contamination, and the isolates are known toxigenic fungi. Different cultivars of sorghum showed different susceptibilities to contamination by the various organisms.

Several mycotoxins have been isolated from sorghum samples, and these include alternariol, alternariol monomethylether (Seitz et al, 1975), altenuene and altertoxin I (Sauer et al, 1978), patulin, trichothecenes, zearalenone, and aflatoxins B_1, B_2, G_1, and G_2. In Indian samples of sorghum (Ansari et al, 1990) tenuazonic acid was found as an additional contaminant. This mycotoxin was also reported by Kalisz et al (*unpublished*) where strains of *A. tenuissima*, isolated on rice medium, were shown to be toxigenic to brine shrimps. A major proportion (64%) of this toxicity was due to the tenuazonic acid, which has been implicated in a human hematological disorder, known as *onyalai*, found in southern Africa. The source of the toxin in this latter instance was suspected to be sorghum grain infected by *Phoma sorghina* (Rabie et al, 1975). A recent review (Jewers and John, 1990) highlights the toxic metabolites of *Alternaria* and quantifies the levels at which they occur in sorghum and other food commodities.

During a survey of sorghum samples collected from marketplaces from three states of India, results showed that 50% of samples analyzed were contaminated

TABLE 7
Occurrence of Aflatoxin B_1 in Sorghum Grain

Location	Level (μg/kg)	Reference
Australia	8,000	Connole and Hill (1970)
India	3–60	Mall et al (1986)
	0–60	John (1990)
Nigeria	100	Salifu (1981)
United States	3–19	Shotwell et al (1969)
	1–90	Schroeder and Boller (1973)
	7–148	McMillan et al (1983a)
	2–6	Hagler et al (1987)
Uganda	>1,000	McMillan et al (1983b)

with aflatoxin at levels up to 60 μg/kg (John, 1990). Other samples of sorghum for which natural contamination with aflatoxins has been recorded are shown in Table 7. Aflatoxins may be present in processed sorghum such as the sorghum malt used in the brewing industry (Ruffel and Trinder, 1990).

Ergot alkaloids have also been found as contaminants of sorghum. The alkaloids are contained in the sclerotia of *Claviceps sorghi* Kulkani, Seshadri & Hedge, and their presence has been demonstrated in Africa and Asia, particularly in Botswana, India, Nigeria, and South Africa (Bandyopadhyah et al, 1990).

Mycotoxins in Millets

Relatively few studies have been made on the mycotoxin content of millets. *Aspergillus terreus*, a producer of patulin, has been isolated from pearl millet (Girisham et al, 1986), as have a number of other common, toxigenic fungi. *Claviceps fusiformis* Loveless, which produces ergot alkaloids, is widely distributed on pearl millet and causes major reductions in crop yield and quality (Chahal et al, 1985). The fungal infection is particularly serious in commercial F_1 hybrids in India, since characteristics such as highly synchronized tillering and flowering and genetic homogeneity make the hybrids more vulnerable than open pollinated varieties (Thakur et al, 1985). The information that is available on fonio suggests that it is susceptible to contamination with zearalenone and aflatoxins (Gbodi et al, 1986). Kodo millet (*Paspalum scrobiculatum* L.) has been found to be contaminated with *Aspergillus tamari* Kita. This fungus also has been shown to produce ergot alkaloids, and symptoms observed after consumption of kodo (unconsciousness, delirium, violent tremors of the voluntary muscles, emesis, excessive perspiration, and difficulty in swallowing) closely match the toxic effects of these alkaloids (Tanarii, 1982). No specific information is available on the fungal flora and subsequent mycotoxin content of proso millet, *Setaria*, *Eleusine* millet, and teff. However, it is likely that these cereals are prone to the same pre- and postharvest fungi as others and will therefore be subject to similar mycotoxin problems.

LITERATURE CITED

ANONYMOUS. 1978. Postharvest Food Losses in Developing Countries. Natl. Acad. Sci. Washington, DC.

ANONYMOUS. 1984. Enquête sur la conservation traditionelle du maïs et du sorgho au Togo (Survey on traditional methods of storage of maize and sorghum in Togo). Bull. Protect. Végétaux 8/84. Republique Togolaise, Service

Nationale de la Protection des Végétaux, Cacaveli-Lomé.

ANONYMOUS. 1987. GASGA Exec. Sem. Ser. No. 1, Research and Development issues in Grain Post-harvest Problems in Africa. Group for Assistance on Systems relating to Grain After Harvest. GTZ, Eschborn, Germany.

ANONYMOUS. 1988a. Cereal Processing, U.N. Dev. Fund for Women, Food Cycle Technol. Source Book 3.

ANONYMOUS. 1988b. Structure and characteristics of the world sorghum economy. Report of the Committee on Commodity Problems, Intergovernmental Group on Grain, 23rd Session. FAO, Rome.

ANONYMOUS. 1990. Structure and characteristics of the world sorghum economy. Report of the Committee on Commodity Problems, Intergovernmental Group on Grain, 24th Session. FAO, Rome.

ANSARI, A. A., and SHRIVASTAVA, A. K. 1990. Natural occurrence of *Alternaria* mycotoxins in sorghum and ragi from Bihar, India. Food Addit. Contam. 7(6):815-820.

BANDYOPADHYAH, R., MUGHOGHO, L. K., MANOHAR, S. K., and SATYANARANGA, M. V. 1990. Stroma development, honeydew formation and conidial production in *Claviceps sorghi*. Phytopathology 80:812-818.

BENNETT, W. F., TUCKER, B. B., and MAUNDER, A. B. 1990. Modern Grain Sorghum Production. Iowa State Univ. Press, Ames.

BHINDE, V. G., TAKWALE, M. G., PAWAR, R. S., HENRIKSSON, R., and SVENSSON, L. 1987. Solar Drying of Agricultural Products. Swedish Univ. Agric. Sci., Lund.

CALDERON. M., DONAHAYE. E., NAVARRO, S., and DAVIS, R. 1989. Wheat storage in a semi-desert region. Trop. Sci. 29:91-110.

CHAHAL, S. S., RAO, V. P., and THAKUR. R. P. 1985. Variation in morphology and pathogenicity in *Claviceps fusiformis*, the causal agent of pearl millet ergot. Trans. Br. Mycol. Soc. 84:325-332.

CONNOLE, M. D., and HILL, M. W. H. 1970. *Aspergillus flavus* contaminated sorghum grain as a possible cause of aflatoxicosis in pigs. Aust. Vet. J. 46, 503-505.

DAVEY, P. M. 1965. The susceptibility of sorghum to attack by the weevil *Sitophilus oryzae* (L). Bull. Entomol. Res. 56, 287-297.

DOGGETT, H. 1957.. The breeding of sorghum in East Africa. I. Weevil resistance in sorghum grains. Emp. J. Exp. Agric. 25:1-9.

FADELMULA, A., and HORBER, E. 1983. Resistance of sorghum varieties to the rice weevil *Sitophilus oryzae* (L.) and the Angoumois grain moth *Sitotroga ceralella* (Oliv.). Proc. Int. Conf. Stored Prod. Entomol. 3:418-434.

FOSTER, G. H., BAKER, K. D., and PEART, R. M. 1980. Methods of Drying Grain with Solar Energy. Pages 22-33 in: Proc. Annu. Mtg. Int. Solar Energy Soc. Phoenix, USA, 2-6 June 1980.

GBODI, T. A., NWUDE, N., ALIU, Y. O., AND IKEDIOBI, C. O. 1986. The mycoflora and some mycotoxins found in acha (*Digitaria exilis* Stapf) in plateau state Nigeria. Food Chem. Toxicol. 24(4):339-342.

GILES, P. H. 1965. Control of insects infesting stored sorghum in Northern Nigeria. J. Stored Prod. Res. 1:145-58.

GILMAN, G. A., and BOXALL, R. A. 1974. The storage of food grains in traditional underground pits. Trop. Stored Prod. Inf. 28:19-38.

GIRISHAM, S., AND REDDY, S. M. 1986. Interaction of different seed-borne fungi of pearl millet (*Pennesitum americanum*) and its effect on patulin production by *Aspergillus terreus*. Curr. Sci. 55(15):730-732.

GOLOB, P. 1981. A practical appraisal of on-farm storage losses and loss assessment methods in Malawi. 1: The Shire Valley agricultural development area. Trop. Stored Prod. Inf. 40:5-13.

GOUGH, M. C., and KING, P E 1980. Moisture content/relative humidity equilibria of some tropical cereal grains. Trop. Stored Prod. Inf. 39:13-17.

GUGGENHEIM, H. 1977. Grain storage in the Fifth Region of Mali: Problems and solutions. Rep. to USAID. Wunderman Foundation, New York.

HAGLER, W. M.. JR., BOWMAN, D. T., BABADOOST, M., HANEY, C. A., and SWANSON, S. P. 1987. Aflatoxin, zearalenone, and deoxynivalenol in North Carolina grain sorghum. Crop Sci. 27:1273-1278.

JEWERS, K., and JOHN, A. E. 1990. *Alternaria* mycotoxins—Possible contaminants of new sorghum varieties? Trop. Sci. 30:397-409.

JOHN, A. E. 1990. Mycotoxins in sorghum grain, with particular reference to the toxic metabolites of *Alternaria* spp. Ph.D. thesis. CNAA, Portsmouth Polytechnic, Portsmouth, U.K.

MALL, O. P, PATERIA, H. M., and CHAUHAN, S. K. 1986. Mycoflora and aflatoxin in wet harvest sorghum. Indian Phytopathol. 39:409-413.

McFARLANE, J. A. 1989. Guidelines for pest management research to reduce stored food losses caused by insects and mites. Overseas Dev. Nat. Resour. Inst. Bull. 22.

McFARLANE, J. A., and DOBIE, P. 1972. The susceptibility of t'ef (*Eragrostis abyssinica* Schrad.) to infestation by some insect pests of stored grain. J. Stored Prod. Res. 8:177-182.

McMILLAN, W. W, WILSON, D. M., MIROCHA, C. J., and WIDSTROM, N. W. 1983a. Mycotoxin contamination in grain sorghum from fields in Georgia and Mississippi. Cereal Chem. 60:226-227.

McMILLAN, W. W, WILSON, D. M., and WIDSTROM, N. W. 1983b. Aflatoxin and zearalenone contamination in grain sorghum samples collected in Georgia fields in 1982. Sorghum Newsl. 29:122.

PIXTON, S. W. 1982. The importance of moisture and equilibrium relative humidity in stored products. Trop. Stored Prod. Inf. 43:16-28.

RABIE, C. J., VAN RENSBERG, S. J., VAN DER WATT, J. J., and LUBEN, A. 1975. Onyalai—The possible involvement of a mycotoxin produced by *Phoma sorghina* in the aetiology. S. Afr. Med. J. (Nutr. Suppl.) 49:1647-1650.

RATNADASS, A. 1990. Les pertes occasionnées par les insectes au sorgho stocké dans les villages de l'Ouham-Pendé et de l'Ouham (Republique Centrafricaine). Agron. Trop. 45(1):31-41.

RICHTER, A. 1988. Storage methods in relation to post-harvest losses in cereals at farm and village level. Insect Sci. Appl. 9(6):755-762.

ROGERS, R. R., and MILLS, R. B. 1974. Evaluation of a world sorghum collection for resistance to the maize weevil *Sitophilus zeamais* Motsch. (Coleoptera:Curculionidae). J. Kans. Entomol. Soc. 47:36-41.

ROWLEY, J. Q. 1984. An assessment of losses during handling and storage of millet in Mali. Trop. Stored Prod. Inf. 47:21-33.

RUFFEL, P. L., and TRINDER, D. W. 1990. A mini-column screening test for aflatoxin in sorghum malt. J. Inst. Brew. 96:7-10.

SALIFU, A. 1981. Mycotoxins in short season sorghums in northern Nigeria. Samaru J. Agric. Res. 1:83-88.

SAUER, D. B., ed. 1992. Storage of Cereal Grains and Their Products, 4th ed. Am. Assoc. Cereal Chem., St. Paul, MN.

SAUER, D. B., SEITZ, BURROUGHS, R., MOHR, H. E., WEST, J. L., MILLERET, R. J., and ANTHONY, H. D. 1978. Toxicity of *Alternaria* metabolites found in weathered sorghum grain at harvest. J. Agric. Food Chem. 26:1380-1383

SCHROEDER, H. E., and BOLLER, R. A. 1973. Aflatoxin production of species and strains of the *Aspergillus flavus* group isolated from field crops. Appl. Microbiol. 25:835-839.

SEITZ, L. M., SAUER, D. B., MOHR, H. E., BURROUGHS, R., and PANKSTELIS, J. V. 1975. Metabolites of *Alternaria* in grain sorghum. Compounds which may be mistaken for zearalenone and aflatoxin. J. Agric. Food Chem. 23:1-4.

SHOTWELL, O. L., HESSELTINE, C. W., BURMEISTER, H. R., KWOLEK, W. F., SHANNON, G. M., and HALL, H. H 1969. Survey of cereal grains and soy beans for the presence of aflatoxins: 1 Wheat, grain sorghum and oats. Cereal Chem. 46:446-454

SORENSON, J. W., and PERSON, N. K. 1970. Drying, storing and handling sorghum grain. Pages 469-506 in: Sorghum Production and Utilization. J. S. Wall and W. M. Ross, eds. AVI Publ. Co., Westport, CT.

SPENCER, W. P., PFOST, D. L., and PEDERSEN, J. R. 1975. Grain storage and preservation in Senegal. Rep. 5. Food and Feed Grain Institute, Kansas State Univ. Manhattan.

TANARII, A. 1982. Isolation of ergot alkaloids from *Aspergillus* species—Ergot poisoning connected with consumption of kodo millet (*Paspalum scrobiculatum*). Central Institute of Medicinal and Aromatic Plants (CIMAP) Newsl. 9(3):1.

THAKUR, R. P., RAO, V. P., WILLIAMS, R. J., CHAHAL, S. S., MATHUR, S. B., PAWAR, N. B., NAFADE, S. D., SHETTY, H. S., SINGH, G., and BANGAR, S. G. 1985. Identification of stable resistance to ergot in pearl millet. Plant Dis. 69:982-985.

VOGEL, S., and GRAHAM, M. 1978. Sorghum and Millet: Food Production and Use: Rep. Workshop, Nairobi, Kenya. International Development Research Centre (IDRC), Ottawa, Canada.

WONGO, L. E. 1990. Factors of resistance in sorghum against *Sitotroga cerealella* (Oliv.) and *Sitophilus oryzae* (L.) Insect Sci. Appl. 11(2):179-188.

WONGO, L. E., and PEDERSEN, J. R. 1990. Effect of threshing different sorghum cultivars on *Sitotroga cerealella* (Oliv.) and *Sitophilus oryzae* (L.) (Lepidoptera:Gelechiidae and Coleoptera:Curculionidae). J. Stored Prod. Res. 26(2):89-96.

CHAPTER 7

TRADITIONAL USES OF SORGHUM AND MILLETS

D. S. Murty
International Crops Research Institute for the Semi-Arid Tropics (ICRISAT)
West African Sorghum Improvement Program
Kano, Nigeria

K. A. Kumar
ICRISAT Sahelian Center
Niamey, Niger

INTRODUCTION

Sorghum (*Sorghum bicolor* (L.) Moench.), pearl millet (*Pennisetum glaucum* (L.) R. Br.), and the various small or minor millets are important food crops in Asia and Africa; in the other continents they are grown mostly for forage and feed. Sorghum is the fifth major cereal crop in the world in terms of production and utilization and is widely cultivated for food in India and Africa. Pearl millet is the principal food crop in low-rainfall zones of sub-Saharan Africa and northwestern India. Proso millet (*Panicum miliaceum* L.) is an ancient crop of considerable importance in India, central and southeast Asia, the eastern and central parts of the former U.S.S.R., and the Middle East (Purseglove, 1972). It was the most important cereal in China before the introduction of wheat and barley (Li, 1970). Foxtail millet (*Setaria italica* (L.) P. Beauv.) is an important food crop in China, Japan, Bangladesh, Sri Lanka, and India and is also grown in southeastern Europe (Rachie, 1975). Finger millet (*Eleusine coracana* (L.) Gaertn.) is popularly grown for food in southern India, Sri Lanka, and east and central Africa, especially in those countries surrounding Lake Victoria (Rachie and Peters, 1977). Teff (*Eragrostis tef* (Zucc.) Trotter) is the foremost staple food crop in Ethiopia while fonio (*Digitaria exilis* (Kippist) Stapf.) is grown as a complementary cereal exclusively in West Africa. Fonio is especially an important crop in the Fouta Djallon of Guinea and the Plateau State of Nigeria (Porteres, 1976). The Japanese barnyard millet (*Echinochloa frumentacea* Link) and kodo millet (*Paspalum scrobiculatum* L.) are of minor importance in southeast Asia and India.

Approved as ICRISAT Journal Article No. 1131.

Sorghum and millets were the most important cereal crops in Africa until a few centuries ago when maize was introduced. Maize displaced sorghum and millets, particularly in the wetter and cooler parts of Africa. However, sorghum and millets, which are drought resistant and can survive under adverse climatic and fertility conditions, continue to be important food crops in Africa. The semiarid tropics of the world (regions in which evapotranspiration exceeds rainfall for more than half the year) is the principal home of sorghum and millets, and many subsistence farmers in this region cultivate them as staple food crops for consumption at home. Arnould and Miche (1971) reviewed the economy and utilization of sorghum and millets in the world and concluded that in developing countries 82.5% of the sorghum and millet production is utilized for human food. In contrast 95.8% of the sorghum and millets production in the developed nations is utilized as animal feed. In the developing countries only a small proportion of the sorghum and millet grains produced enter the urban grain markets. It has been estimated that less than 20% of sorghum and millet grains produced in West Africa and India reach the grain markets (Morris, 1982; Von Oppen and Rao, 1982).

Sorghum and millet food products are not commonly found in urban and semi-urban markets, probably because of the drudgery involved in their domestic processing and the low prestige attached to them. Sorghum and millets are considered coarse grains, since they are hard and difficult to process. The flour or meal from sorghum and millets is coarse in feel and produces gritty products with a characteristic aroma. The grains are processed at home, mostly using traditional household methods, and the techniques employed to prepare the food are intended to mitigate the gritty character of the product. Appropriate milling technologies have been developed to mechanically process sorghum and pearl millet, but, for various socioeconomic reasons, these technologies have not yet become established in many developing countries, where these grains are used as staple food (Hulse et al, 1980). Thus, the food uses of sorghum and millets are still mostly traditional.

Documentation of the traditional processing methods and food uses of sorghum and millets was poor until the 1960s. In the last two decades, the International Crops Research Institute for the Semi-Arid Tropics (ICRISAT), International Development Research Centre (IDRC), the USAID Collaborative Research Support Program on Sorghum and Pearl Millet (INTSORMIL), and Study Group 32 of International Association for Cereal Science and Technology (ICC) have made conscientious efforts to collate available information on sorghum and millets (Vogel and Graham, 1979; Hulse et al, 1980; Hoseney et al, 1981; Rooney and Murty, 1982a,b; Rooney et al, 1986; Rooney and McDonough, 1987; Dendy et al, 1988). However, very little information is available on the traditional food uses of small millets such as proso millet, finger millet, foxtail millet, browntop millet, etc. (Rachie, 1975). Much of the literature is published in journals and reports with limited circulation or other inaccessible sources. Numerous names, mostly synonyms, of sorghum and millets and their food products, in several languages and regional dialects add to the difficulties in compilation. Many of the available reports lack details of processing techniques and accurate descriptions of the food products.

The objective of this chapter is to review the traditional food uses of sorghum and millets. An attempt will be made to describe the milling methods used to process the grains and domestic techniques used to produce some of the major foods of importance in the daily life of the people in sorghum and millet growing areas in different parts of the world. Information on grain and food quality attributes preferred or desired by the consumers is reviewed, and implications in plant breeding programs will be discussed.

DIVERSITY OF TRADITIONAL SORGHUM AND MILLET FOODS

Generally, sorghum and millet grains are decorticated partially or completely by traditional methods before further processing and consumption. Whole grains might as well be directly dry-milled to give a range of products: broken or cracked grains, grits, coarse meal, and fine flour. The flour thus obtained is used in the preparation of a whole range of simple to complex food products. Sorghum and millet flour can be mixed with flours of other cereals such as wheat and rice or legumes to improve palatability, nutritional value, and acceptance of the product.

Sorghum and millet flour is commonly used to prepare a range of porridges of varying consistencies; the flour may also be fermented or soured before making the porridge. The flour may be processed into a dough or a batter that can be used to prepare a variety of pancakes. Fermented or malted grains are also used to prepare a range of native alcoholic and nonalcoholic beverages.

Sorghum and millet grains are cooked to give ricelike products. Steam-cooking, deep-frying, and several other processing techniques are used to make many kinds of snacks and other popular foods. These diverse processes affect the flavor, nutritional value, and utilization properties of the food product. Several of these foods and beverages have social significance and are part of the traditional life, festivals, and religious rites of people in Africa and India.

Surveys conducted in a few states of India revealed literally scores of sorghum and millet products that are commonly consumed in villages and semi-urban areas (Subramanian and Jambunathan, 1980). Vogel and Graham (1979) summarized the proceedings of a workshop on sorghum and millet uses that was attended by food scientists and socio-economists and recognized seven broad classes of food products widely consumed across countries. There are minor differences in details of recipes for each class of food between households, communities, and regions due to individual preferences and needs. The traditional art of food preparation is not standardized, and the routine procedures have been passed on to the women through generations. Based on Vogel and Graham's (1979) preliminary classification, Rooney and Murty (1982b), Rooney et al, (1986) and Rooney and McDonough (1987) described nine major food categories of sorghum and millets: thick porridges, thin porridges, steam-cooked products, fermented breads, unfermented breads, boiled ricelike products, alcoholic beverages, nonalcoholic beverages, and snacks.

MILLING

Decortication

Sorghum and millet grains are traditionally decorticated by women. They pound moistened grain in a mortar with a pestle to remove the pericarp. The decorticated grain is later crushed to grits, flour, etc., as desired (Fig. 1). The traditional grain processing technique is a slow and strenuous job for the women every day.

In south India, sorghum and millet grains are normally decorticated before further processing and food preparation. Mortars made of stone are commonly used;

pestles are made of wood, and some have a metal ring at the pounding end. The stone mortar is sometimes fixed in the ground and appears as a pit or hole in the ground. Sorghum grain, usually less than a kilogram, is washed and placed in a mortar, and the moist grain is vigorously pounded with the pestle for approximately 5 to 6 min. The pounded grain is collected in a large vessel and washed in excess water. Depending on the extent of pericarp removal desired, the grain may be returned to the mortar for further pounding for a few more minutes. Normally millet grains require much more time to dehull than sorghum grains. The bran is washed again, and the clean endosperm recovered is left for sun-drying. The washes contain bran and a major portion of the germ, which gets broken and removed from the kernel during the process. These are collected and given to animals.

Carr (1961) and Muller (1970) described the traditional process of sorghum milling in Africa. The pestle and mortar are carved from wood. The mortar is about 60- to 70-cm high with a diameter of about 30 cm. The pestle weighs about 3 kg, is about 1.2-m long and 6-cm in diameter, and has bulbous ends. Approximately 2–3 kg of clean grain is placed in the mortar, and at least 250 ml of water is added. The water is mixed with the grain and then the moist grain is pounded vigorously. The grain is pounded by one, two, or three women, each working at about 60 strokes a minute. There are three types of pounding strokes: 1) up and down; 2) down, across, and up; 3) down, rotary scrape, and up. The latter two types of strokes knock the side of the mortar and prevent bridging of damp grain. The abrasive ac-

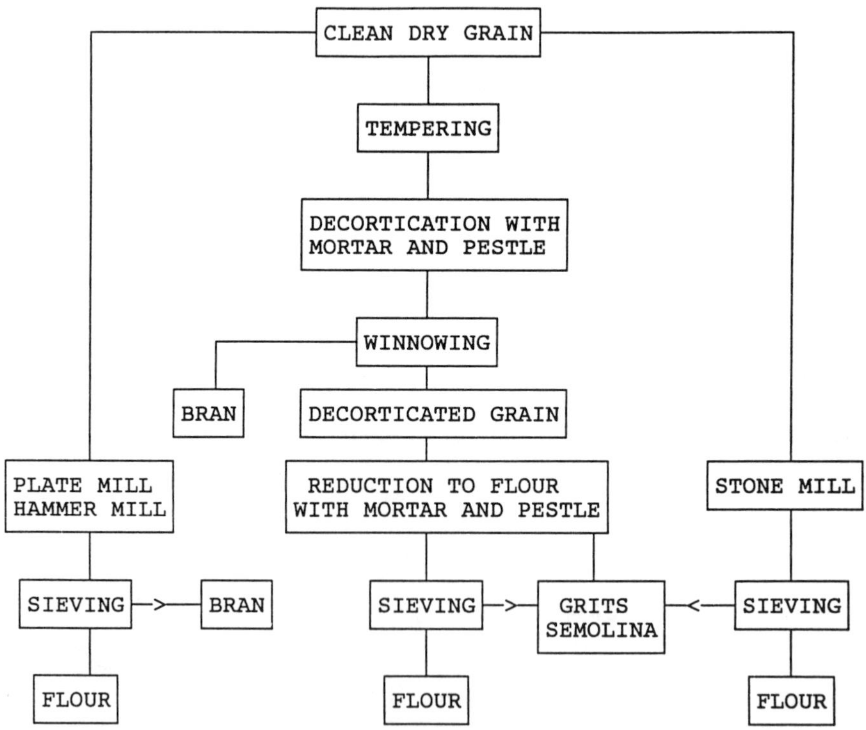

Fig. 1. Traditional milling methods of sorghum and millets.

tion generated due to strong interaction between the grains, and between the grains and the equipment, frees the pericarp from the endosperm, normally at the interface of the mesocarp. The moisture doesn't normally penetrate deep into the endosperm, since the grains are pounded immediately a few minutes after moistening the grain, which facilitates the peeling of the pericarp from the endosperm. The decorticated grain is removed from the mortar, sun-dried for a few minutes, and winnowed to separate the bran. During the process of pounding, additional water might be added if necessary. Sometimes the pounded and winnowed grain is returned to the mortar for further decortication to obtain an acceptable product. The decorticated grain is washed in excess water and sun-dried. Carr (1961) noted that when red sorghum was traditionally dehulled, the original grain produced 66% meal, 29% offal, and 5% waste. Favier et al (1972) recorded endosperm recoveries of 76.8 and 73.9% in two villages of Cameroon after traditional dehulling of sorghum. Faure and Muchnik (1989) opined that in sorghum, endosperm recoveries after traditional dehulling could vary between 70 and 75%. It was also estimated that 11 to 21 min are required to decorticate 3 kg of grain, which could amount to an energy equivalent of 20–62 kJ/kg of grain.

Consumers prefer high recovery of clean endosperm with minimal losses due to breakage. The physical and structural properties of the grain (i.e., shape, pericarp thickness, and hardness of endosperm) affect the traditional dehulling process. Scheuring et al (1982a) found that less time was required to hand decorticate sorghum with a thick pericarp than grain with a thin pericarp and that decorticated grain yields decreased as milling time increased. Endosperm recoveries ranged from 66.3 to 71.7%, and the best results were obtained with local Malian sorghums possessing a vitreous endosperm and a thick pericarp. Murty et al (1984b) studied the traditional dehulling properties of 25 sorghum cultivars of diverse origin and physical grain attributes, grown over two years. The manual effort required to de-

TABLE 1
Traditional Dehulling Quality of 25 Sorghum Cultivars Evaluated over Two Years in India[a]

	1980			1981		
Property	**Range**	**Mean**	**SEM**	**Range**	**Mean**	**SEM**
Grain volume, cm^3/100	1.93–5.40	2.68	0.134	1.85–5.15	2.46	0.131
Grain mass, g/100	2.46–6.90	3.31	0.173	1.96–5.98	2.90	0.162
Breaking strength, kg[b]	6.2–14.0	8.4	0.35	4.6–14.0	7.3	0.37
Shape[c]	1.0–5.0	2.8	0.24	1.5–5.0	2.6	0.18
Pericarp thickness[d]	1.0–5.0	3.3	0.26	1.0–5.0	3.2	0.27
Endosperm texture[e]	1.0–5.0	2.6	0.22	1.0–5.0	2.6	0.21
No. of pounding strokes[f]	315–537	448	11.4	264–594	456	16.8
Pericarp removal[g]	1.0–5.0	2.8	0.24	1.0–5.0	2.6	0.24
Endosperm recovery, %	67.2–89.6	80.2	1.15	52.0–82.6	73.4	1.48
Brokens (<U.S. no. 10 mesh), %	1.2–40.5	11.76	1.703	2.8–50.7	19.66	2.910

[a] Source: Murty et al (1984b), with modifications.
[b] Breaking strength was measured with a Kiya hardness tester.
[c] Grain shape was scored on a scale of 1 to 5, where 1 = round and 5 = flat.
[d] Pericarp thickness was scored on a scale of 1 to 5, where 1 = thin and 5 = very thick.
[e] Endosperm texture was scored on a scale of 1 to 5, where 1 = 81 to 100% corneous and 5 = 0 to 20% corneous.
[f] A sample of 400 g of grain was pounded using a traditional mortar and pestle, and the number of pounding strokes required for satisfactory pericarp removal was recorded.
[g] Pericarp removal was scored on a scale of 1 to 5 where 1 = complete removal and 5 = very poor removal.

hull a standard quantity of grain varied between cultivars as reflected by the number of pounding strokes (Table 1). The extent of pericarp removal among the cultivars ranged from complete to very poor. Further pounding of grains that had poor initial pericarp removal resulted in higher levels of broken endosperm. In general, grains with highly corneous endosperm yielded the least brokens and the highest endosperm recovery, whereas grains with floury endosperm yielded the highest percent brokens. Grains with a thin pericarp and highly corneous endosperm exhibited a relatively poor pericarp removal compared with grains having a thick pericarp and highly corneous endosperm. Pericarp removal of turtle-shaped caudatum type grains was slower and incomplete, whereas round or oval grains exhibited a relatively cleaner product. Cultivars originating from West Africa with a thick pericarp and highly vitreous endosperm yielded the cleanest endosperm and produced least brokens. Grains with a subcoat or brown pericarp exhibited very poor dehulling qualities. This is one of the major reasons why farmers prefer to grow hard endosperm grain types with a thick pericarp, particularly in Africa, where traditional dehulling is a common practice.

Endosperm recoveries from three improved high-yielding sorghum cultivars were compared with that of the local guinea cultivar grown in research farm trials in Burkina Faso (ICRISAT, 1987). It was noted that relatively poor endosperm recoveries from the high-yielding cultivars could offset their yield advantage over the local cultivar, which gave much higher and cleaner endosperm recovery. Thus selection for efficient dehulling quality should be an important criterion in breeding programs.

In pearl millet, Kante et al (1984) compared bran losses of five Malian cultivars during traditional decortication in four villages. They observed that grains from Sanio type cultivars gave maximum endosperm recovery in relatively less time and with less effort than Souna type cultivars (Table 2). The Sanio millet types possessed a thick pericarp and globular grains, whereas Souna grains possessed a thin pericarp and elongated to lanceolate shape. It was concluded that millet grains with a thick pericarp and globular shape offer relative ease in decortication, lose less endosperm, and yield a cleaner endosperm.

Adrian et al (1967) obtained an edible portion of 39% coarse semolina, 16% fine semolina, and 30% flour after traditional mortar-pestle grinding of pearl millet. Data from traditional millet milling studies in Senegal showed that an average family consumed 2.5 kg of grain per day (Chinsman, 1984). It required two women

TABLE 2
Endosperm Recoveries (%) After Traditional Decortication of Five Millet Cultivars at Four Test Locations in Mali[a,b]

	Cultivar				
Location	**Sanio 1**	**Sanio 2**	**Sanio 3**	**Souna 1**	**Souna 2**
Banamba	66.7	64.7	73.3	65.3	70.0
Sikasso	60.3	63.7	61.7	61.2	55.8
Koporo	45.5	75.3	73.0	53.8	50.6
Sinebougou	51.1	61.3	65.3	60.7	53.0
Mean[c]	55.9 b	66.3 a	68.1 a	60.3 a	57.6 b

[a] Source: Kante et al (1984).
[b] Data averaged over three replications. Standard error of mean for cultivars was 7.26.
[c] Values followed by the same letter are not significantly different at $P = 0.05$.

to decorticate the grain for 1.5 hr to obtain 1.9 kg of decorticated endosperm and 0.3 kg of bran, indicating a loss of 0.3 kg (12%).

The minor millets, i.e., foxtail millet, proso millet, and kodo millet, must be dehulled before consumption. Pushpamma (1989) reported that traditional processing of these millets takes 20 to 22 min per kilogram of grain. The percent extraction of the edible portion was 71% in foxtail millet and 54% in kodo millet. Thirumala Rao (1969) compared the traditional and mechanical milling qualities of foxtail millet and obtained similar results by both methods: 68% kernel, 28% bran, and 4% losses. In India, finger millet is dehulled by the traditional mortar-pestle technique before grinding (Malleshi, 1989). Roasting or malting of finger millet before grinding is common.

Hulse et al (1980) reviewed the information on the nutritional status of milling products obtained from processing sorghum and millets by traditional versus mechanical methods. In general, the traditionally milled products retained relatively more nutrients. However, the flour products from mechanical methods were more acceptable and had an improved shelf life.

Grinding

The traditional equipment used in Africa for reducing the decorticated grain or whole grain into flour involves either the mortar and pestle or a stone mill (Fig. 2). These tools are very ancient and were used in Egypt from at least 2500 B.C. (Muller, 1970). The flour reduction is a much more laborious and time-consuming process than decortication. In the mortar and pestle system, the decorticated grain is placed in the mortar and continuously pounded with the pestle. The flour or semolina thus obtained is sieved, and the overs are returned to the mortar for further pounding. Locally fabricated sieves of various mesh sizes are used to obtain the flour particle size appropriate to the food product desired. The flour product is usually sun-dried and cooked preferably within 48 hr. Its high moisture content may allow partial fermentation during storage.

The stone mill consists of a base plate and a roller. The base plate is placed on three leveling stones so that it slopes away from the operator, who sits, squats, or kneels behind the mill. A dish or a cloth is placed in front of the base plate to receive the ground material. The roller stone is oval or round in cross section but becomes flattened after some use. In Burkina Faso, some villages have a common flour-grinding facility for the community. About a dozen or more grinding base stones are arranged or fixed on a circular platform of about 1-m height. Individual operators can stand before each base stone and grind their raw material independently or collectively. Murty (1987) studied the flour particle size of decorticated grain from eight sorghum cultivars ground on traditional stone mills in Burkina Faso and recorded an average of 62% flour that passed through a 200-μm mesh sieve.

In India, the traditional stone mill usually consists of a stationary wheel-shaped basal plate with a wooden or stone axis at the center (Fig. 3). The top stone (which is the moving plate) has a hole in the center through which the axis of the basal plate protrudes and also provides space to feed the grain into the mill. The top stone is pivoted by the axis and is rotated on the basal stone with a handle. The mill could be used for dry milling of either whole or dehulled grain. If the flour product is required immediately after traditional dehulling of the grain, the mortar

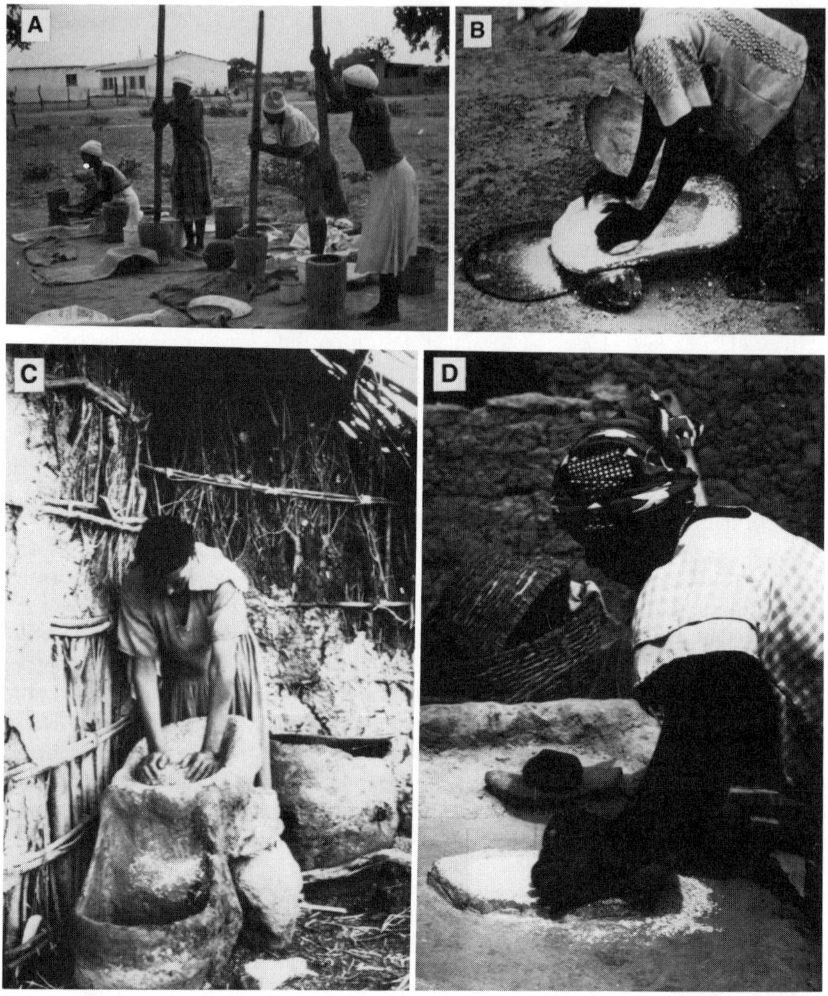

Fig. 2. (A) Dehulling of sorghum by women in Botswana with mortars and pestles. (B) Reduction of dehulled grains to flour on stone in Nigeria. (C) Flour grinding on an ancient saddle stone in Ethiopia. (D) Traditional grinding of dehulled grains on stone in Burkina Faso.

and pestle are used. The flour thus obtained is sieved and further processed to the required particle size. In recent years, however, power driven plate mills have become commonly used for reducing decorticated grains to flour. The plates are sometimes made of stone. More commonly, whole grain is dry-ground in these mills to grits, semolina, or flour, and the product is sieved to discard the bran. These power driven mills normally yield flour containing about 50% particles sized <75 μm (Murty et al, 1984c).

In Africa, the traditional process of decortication and grinding still persists in the rural areas. However, in the urban areas there is now an increasing preference for the use of diesel or power driven hammer mills or plate mills for grinding the decorticated grain into flour. In countries such as Nigeria, the bulk of sorghum

Fig. 3. (A) Traditional stone mill of India operated by hand. (B) Power driven *chakki* in India. The grinding plates of the mill are made of stone.

grain is custom-milled to whole flour, although several customers still bring traditionally decorticated grain to the mill for reduction into flour (Ngoddy, 1990).

THICK PORRIDGES

The most common and simplest food prepared from sorghum and millets is porridge. Stiff (thick) porridges are consumed in almost all countries where sorghum and millets are cultivated for food (Table 3). The basic difference between thick and thin porridges is the concentration of flour. Generally, thick porridges are "solid" and can be eaten with the hand, while thin porridges are "fluid" and are amenable to drinking or are eaten with a utensil. Stiff porridges are an important part of the daily meal and are consumed at least one or two times a day. The basic procedure for the preparation of stiff porridge involves adding flour to boiling water in increments accompanied by vigorous stirring. The flour is cooked until it forms a thick, homogeneous and well-gelatinized mass devoid of lumps (Fig. 4). However, there are several variations in the details of procedures adapted in various regions.

Flour is normally obtained from decorticated grains. Flour from whole grains, fermented grains, or roasted grains is also used in some regions. Flour particle size can vary from very fine to coarse meal depending upon the milling procedures used and the consistency of the product desired. The pH of the cooking medium varies depending upon the ingredients added to the water used. For example, in Burkina Faso, a thick porridge called *tô* is prepared by cooking sorghum flour in

TABLE 3
Traditional Thick Porridges Prepared from Sorghum and Millets

Common Name	Country or Region	Description and Variations
sankati, kali	Southern India	Coarse flour from dehulled/whole sorghum, pearl millet, finger millet, and foxtail millet
mudde	Southern India	Coarse flour made into a stiff porridge and later molded or shaped
nsima	Zambia, Malawi	Neutral pH stiff porridge made from dehulled sorghum or millet flour
mafo	Somalia	Neutral pH, stiff porridge
sadza	Zimbabwe	Neutral pH, stiff porridge
ugali	Kenya, Uganda, and Tanzania	Neutral pH, stiff porridge
mosokwane (*bogobe*)	Botswana	Neutral pH, stiff porridge
tuwo	Nigeria	Neutral pH, stiff porridge
tuo	Ghana	Neutral pH, stiff poridge
boule	Mauritania, Chad	Neutral pH, stiff porridge
bita	Niger	Neutral pH, stiff porridge
tô	Burkina Faso	Acid pH, stiff porridge made by adding tamarind or lemon extract to boiling water
tô	Mali	Alkaline pH, stiff porridge made by adding wood ash leachates or potash to boiling water
dalaki	Nigeria	Wet-milled fermented grains are used
aceda	Sudan	Fermented batter cooked into stiff porridge
ting (*bogobe*)	Botswana	Fermented batter cooked into stiff porridge
umqo	South Africa	Sour, fermented stiff porridge
umphokoqo, phuthu	South Africa	Sour, fermented stiff porridge

water to which tamarind extract or lemon juice is added, which results in an acid medium (Da et al, 1982). In Mali, leachates from wood ashes or potash are added to the water before *tô* preparation, thus producing an alkaline porridge (Scheuring et al, 1982b).

In countries such as Nigeria, Ghana, Zimbabwe, Tanzania, and some parts of south India, normal water without addition of any other ingredients is used in the preparation of thick porridges, and these porridges possess a neutral pH. In Botswana and Sudan the flour is soured and fermented for at least 18 hr before cooking, resulting in a fermented porridge. In some parts of Nigeria, wet-milled sorghum grains are used to make a fermented, thick porridge called *dalaki* (Obilana, 1982). In the northern and eastern parts of Uganda, western Tanzania, and Kenya, brown sorghum grains are mixed with small quantities of dry cassava, finger millet, or maize, and the mixture is ground to flour to make *ugali* (Mukuru et al, 1982). In India, flour from pearl millet, finger millet, and foxtail millet is also used to make thick porridges called various names, such as *sankati, mudde,* and *kali* (Murty et al, 1982a). Sometimes the flour is cooked in sour milk or in a fermented broth from sorghum or washes from rice (Subramanian and Jambunathan, 1980; Vogel and Graham, 1979). Thick porridges are sometimes molded or shaped into balls. Chen (1989) reported that foxtail millet and proso are used in China to

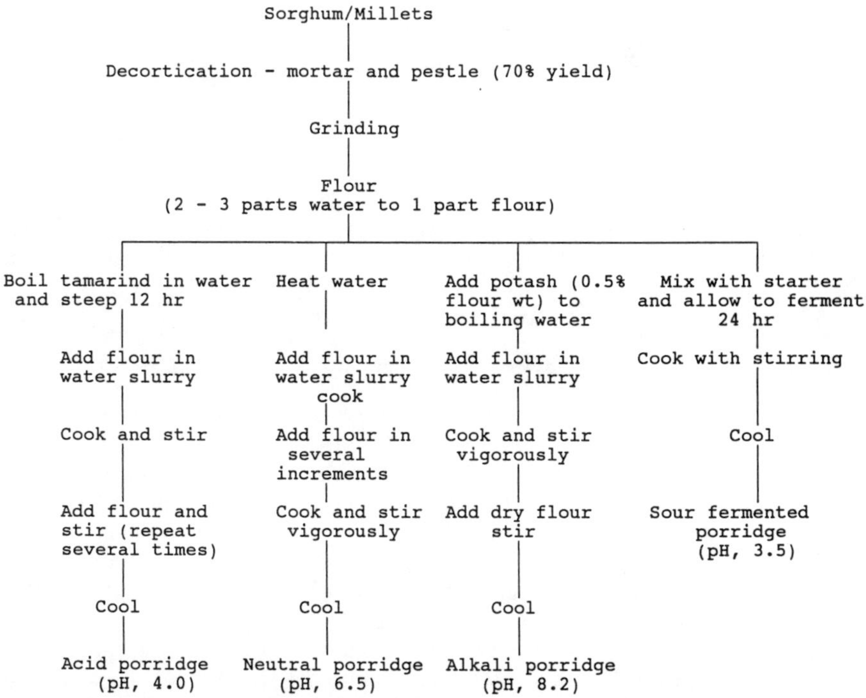

Fig. 4. Traditional process for producing thick porridges. Cooking time varies, but after water is boiling and the process starts, up to 30 min may be required to cook 2 kg of flour. Part of the thin porridge formed in the initial cooking phase is retained and mixed into the porridge along with dry flour to produce a smooth texture. (Source: Rooney et al, 1986).

make a gruel of pleasant flavor. In the Kaduna and Plateau States of Nigeria, whole meal from fonio grain (also known as *acha* and *findi*) is used in the preparation of a thick porridge called *tuwo*.

Tô

Tô is the principal cereal recipe consumed by millions of people in the 400–900 mm rainfall zone of West Africa. Sorghum *tô* is more commonly consumed in the wetter areas, and millet is used in the drier areas. The traditional procedure to prepare *tô* begins with decortication followed by grinding of the decorticated grains to flour. Scheuring et al (1982b) described the procedures of *tô* preparation in Mali. Approximately 4 L of water is brought to boil in a pot. Wood ash extract (10 g) is mixed with about 650 ml of cold water in a calabash or bowl, and about 750 g of sorghum flour is added. The mixture is stirred until homogeneous and then swirled into the boiling water in the pot. The boiling gravy is stirred for about 8 min, and at this point the thin solution is called *bouillie*. The heat is reduced by removing part of the burning wood. About 1,300 ml of *bouillie* is removed from the pot and set aside in a calabash.

About 1,125 g of sorghum flour is added, a handful at a time, to the boiling *bouillie* in the pot. After each addition of flour, the paste is vigorously stirred with a flat wooden spoon. When the paste is too thick for stirring, a small amount of

Fig. 5. (A) *Tô* prepared from millet in Mali. (B) Millet *furah* and sour milk on sale in Nigeria.

bouillie from the calabash is added. The process of addition of flour followed by stirring and further addition of *bouillie* continues for about 9 min until all the *bouillie* from the calabash is used and the paste is thick and homogeneous. The fire under the pot is reduced, and the thick paste is allowed to cook for about 12 min. The *tô* is now ready and is scooped into a large bowl where it is allowed to cool for at least an hour before eating (Fig. 5A).

Tô is usually prepared in the afternoon and served as an evening meal. The left-over *tô* is stored overnight in a bowl covered with a piece of cloth. The following morning it is eaten cold or reheated. The leftover *tô* should still be firm enough to withstand reheating without falling apart. *Tô* is consumed by tearing off a handful and dipping it into sauce made with vegetables such as tomatoes, onions, chilies, okra, garlic, cowpeas, peanuts, and baobab leaves. Okra contains mucilaginous gums that facilitate the swallowing of the porridge.

Ugali

The stiff porridge prepared from any cereal or cereal mixture in Kenya, Uganda, and Tanzania is commonly called *ugali* (Kiswahili language; Mukuru et al, 1982). There are several other synonyms for *ugali* in various dialects in eastern Africa. *Ugali* prepared from maize is currently popular in the region, but a similar product made from sorghum, pearl millet, and finger millet still enjoys considerable importance, particularly among the poor. In the case of sorghum, a wide range of grain types is used in the preparation of *ugali*, but the white corneous endosperm types are certainly preferred. In Uganda, Rwanda, and parts of western Tanzania, brown grain types are commonly cultivated and are also used for *ugali*. In finger millet also, white grain types are preferred for *ugali* preparation (Esele, 1989). Sorghum or millet flour is generally obtained by the traditional decortication and grinding methods described earlier; however, soft and brown grain types are usually pounded dry and winnowed before grinding to flour. In urban areas, hammer mills introduced for milling maize are also being used for grinding sorghum and millets. The cooking procedures for *ugali* are similar to those described for *tô,* except that normal water is commonly used for cooking the flour. Thus, the *ugali* porridge generally exhibits a neutral pH. However, in northern Uganda, tamarind water or mango or lemon juice are added to the boiling water to improve the taste and flavor. In Kenya, lemon juice or milk may be added to the cooking water. In Tanzania, the dehulled grain is sometimes soaked in water for one to two days, after which it is dried and ground into flour. This procedure is believed to improve the whiteness of the flour.

A good *ugali* should not contain any lumps. While eating *ugali*, one should be able to take portions of it with fingers without the product sticking to the fingers or teeth. After the *ugali* is prepared and cooled, a thin crust forms on the surface, but the porridge under the crust should be soft but not sticky, and should keep a similar texture for up to 24 hr.

Quality Criteria

The most important qualities of thick porridge desired by consumers are a thick and firm texture, nonstickiness, and keeping quality. In sorghum, the color of the pericarp, presence or absence of subcoat, and the texture of the endosperm affect the quality of thick porridges (Rooney and Murty, 1982a). The pH of the solution used to prepare the porridge can also affect the porridge quality (Scheuring et al, 1982b). Certain cultivars produced an acceptable porridge when cooked in an acid medium but not in an alkaline medium. Porridge made with tamarind extract or acid medium is generally firmer in texture and lighter in color than that made with alkaline extract (Da et al, 1982; Table 4). In general a white or yellow porridge is preferred. Laboratory evaluation of a range of thick porridges using a similar set of grain samples indicated that sorghum grains with a highly corneous endosperm and good traditional dehulling qualities are suitable for preparation of thick porridges (Murty et al, 1982d). Normally, grains with floury or soft endosperm texture produce porridges with a sticky and mushy texture and are also not suitable for decortication. Good gelling properties are preferred in the porridges. The keeping quality is of major importance, because the porridge is stored overnight, reheated, and consumed. Some cultivars of sorghum produce fresh porridge with acceptable eating quality but exhibit poor storage quality. Porridges with poor keeping quality become mushy, sticky, and lose water. In general, the actual taste of the porridge is masked by the sauces and stews consumed along with the porridge, and thus might not be as important as other qualities.

TABLE 4
***Tô* Softness and Stickiness of 24 Sorghum Cultivars from International Sorghum Food Quality Trials, 1979 and 1980**[a,b]

	Softness				Stickiness			
	Fresh		Stale		Fresh		Stale	
	Mean ± SD	Range	Mean ± SD	Range	Mean ± SD	Range	Mean ± SD	Range
Acid *tô*								
1979	76.5 ± 9.3	65.8–95.5	70.5 ± 10.2	56.5–93.3	2.3 ± 0.6	1.5–4.1	2.2 ± 0.8	1.0–4.4
1980	63.2 ± 9.4	49.0–80.3	58.5 ± 6.9	46.8–79.0	2.5 ± 0.9	1.5–5.4	2.3 ± 0.6	1.3–3.8
Alkali *tô*								
1979	85.5 ± 8.8	67.5–96.8	83.5 ± 9.4	64.8–103.0	4.1 ± 0.6	2.9–5.7	2.8 ± 0.7	1.6–4.2
1980	76.6 ± 8.1	61.3–94.3	67.8 ± 8.5	56.3–83.8	3.2 ± 1.0	2.0–6.9	2.7 ± 0.7	2.1–5.3

[a] Source: Da et al (1982), with modifications.

[b] *Tô* softness was measured by a precision penetrometer with a scale calibrated in 0.1-mm units. Stickiness was measured in millimeters by the double-pan balance method.

THIN PORRIDGES

Various unfermented and fermented thin porridges are traditionally prepared from sorghum and millets (Table 5). They may be prepared from dry-milled flour from whole, dehulled grains, soaked grains, or germinated or malted grains (Vogel and Graham, 1979). Composite flours of sorghum, millets, rice, and cassava are commonly used. A simple unfermented porridge like *ugi* is prepared by making a thick flour suspension and then adding it to boiling water. The thin boiling solution is stirred with a ladle for about 5 min, when the color changes indicating the completion of the gelatinization process. The initial flour suspension can be left aside overnight for fermentation and used next day to make a fermented *ugi*. Thin porridges have a mild sorghum-millet flavor and are normally consumed as breakfast food. Sugar and milk are also frequently added for improving the taste and acceptability. The gelling properties of cooled *ugi* vary among cultivars, and a good *ugi* produces a thick gel after overnight cooling (Murty and House, 1980).

In Africa, thin fermented porridges are more popular than unfermented porridges. The objective of fermentation is to produce acceptable flavors and mask undesirable ones (Odunfa, 1985). The process of fermentation also improves storability. In India, flours from pearl millet, finger millet, and foxtail millet are usually fermented overnight in water and cooked into a porridge the next day (Malleshi, 1989).

Ogi

The most important fermented thin porridge that is consumed in Nigeria and parts of Ghana is *ogi*. In the northern parts of these countries, *ogi* is prepared using sorghum and millets. Muller (1970), Banigo and Muller (1972), Akingbala et al (1981), and Obilana (1982) described the *ogi* product. *Ogi* is called *kamu* and *akamu* in parts of Nigeria and is the most extensively consumed traditional breakfast food. It is prepared by soaking whole sorghum grain in water at room temperature for two or three days. The steeped grain is washed and crushed in a slurry of water and sieved through a thin cloth to remove the bran. The throughs contain en-

TABLE 5
Thin Porridges Prepared from Sorghum and Millets

Common Name	Country or Region	Description and Variations
ugi	East Africa	Fermented or unfermented thin porridge
nasha,	Sudan	Thin porridge from fermented batter
obushera	Uganda	Thin porridge from germinated grains
edi	Uganda	Unfermented thin porridge
ogi, kamu, akamu	Nigeria	Fermented thin porridge
eko, kafa, agidi	Nigeria	Fermented thin porridge in gel form
kunu	Nigeria	Thin porridge from sorghum or fonio
kunu tzamia	Nigeria	Thin porridge in tamarind water
koko	Ghana	Fermented thin porridge
ambali	India	Thin porridge
puttu	India	Sweet pudding from finger millet flour
rabdi	India	Porridge from millet fermented in sour milk
isidudu	South Africa	Unfermented thin porridge that gels when cool
imbila	South Africa	Fermented thin porridge

dosperm chunks and are allowed to ferment for several hours. Odunfa and Adeyele (1985) reported on the microbial populations responsible for sorghum *ogi* fermentation. Most of the water is decanted, and the sediment that settles at the bottom is called *ogi*. Later the *ogi* is cooked in fresh water, or sometimes the same decanted water, to produce *ogi* porridge. It is free-flowing and has a creamy consistency and smooth texture. A light colored *ogi* with a slightly sour taste is preferred. *Ogi* is the preferred food for the elderly, the sick, and young children and is often cooked with milk. Traditionally, *ogi* is consumed on the same day it is made. However, when it is stored overnight or cooled it solidifies into a gel. The product is wrapped in leaves and stored up to two days in baskets or large calabashes. The gel products are called *kafa* and *agidi*. A good *kafa* retains its texture, consistency, and flavor during storage. These products are consumed along with sauce or stew.

Banigo and Muller (1972) compared the *ogi*-making properties of maize, sorghum, and millet and concluded that both maize and sorghum gave better yields than millet. In sorghum, Akingbala et al (1981) obtained mean yields of 72.9, 15.9, and 7.5% of *ogi*, bran, and solubles, respectively. They also found that brown grain cultivars produced unacceptable *ogi*. According to Muller (1970), the Ghanaian *koko* is prepared in much the same way as *ogi* except that it is fermented after wet grinding as a dough (called *mbor*) and not as a suspension.

Other Thin Porridges

In Sudan, sorghum flour is fermented as a batter and then cooked in boiling water to give a fermented porridge called *nasha* or *madida*. It is seasoned with tamarind, cardamom, or apricot and is served to children and lactating women, particularly during Ramadan (the Islamic fasting month) (Badi et al, 1988). When pearl millet is used, the flour may or may not be fermented, and the porridge is served with milk or honey.

In Nigeria, a fermented thin porridge called *kunu tzamia* made from sorghum and pearl millet is popular (Obilana, 1982).

Obushera is a thin porridge made from germinated sorghum grains processed in a unique traditional method. In Uganda, where brown sorghum types are commonly cultivated, the grain is mixed with wood ashes, germinated, cleaned, dried, and ground into flour. The flour from such malted grain is made into the thin sweet porridge called *obushera*. Mukuru (1988) observed that steeping the grain in wood ashes detoxifies the brown high-tannin grains, and the process results in a nutritionally acceptable porridge.

In general, sorghum cultivars that produce good thick porridges will make acceptable thin porridges. The thin fermented porridges have the same properties as acid porridges (Rooney et al, 1986).

STEAM-COOKED PRODUCTS

Flour from sorghum and millet grains is used in a variety of steam-cooked food products (Vogel and Graham, 1979; Subramanian and Jambunathan, 1980).

In India, fermented sorghum-millet flour is cooked over steam to make a product called *kudumu*. Wet-milled sorghum-millet batter can be mixed with wet-milled black gram batter, fermented overnight, and poured into small cakes or molds which are steam-cooked to give a product called *idli*. *Idli* is soft, moist, and spongy with a slight sour taste. Rice is the preferred cereal for *idli* preparation, but sorghum and millets may be used in drier areas where rice is scarce (Desikachar, 1977).

Muller (1970) and Vogel and Graham (1979) described dumplings made from sorghum, millet, and maize that are also steamed and are known by several names in various dialects of Ghana and Nigeria. In northern Nigeria and Niger, a traditional pearl millet product called *furah* is popular and is commonly sold in the market. Slightly fermented and humid flour from millet is shaped into balls of approximately 9-cm diameter. These balls are placed in a small quantity of water and steam-cooked for about an hour, after which they are pounded in a mortar to give a thick and highly viscous paste. The paste is reshaped into small balls of about 4- to 5-cm diameter. These balls are rolled in fine millet flour on board and sold in the market along with sour milk. They are crushed in a calabash with sour or fresh milk into a thin paste and eaten. *Furah* can be stored for about two to seven days. A similar product made in Mali is called *degué*.

Steam-cooked products made from sorghum are popular in China. Sorghum dough is steam-cooked to make products called *wowo, yuyu,* and *fa gao*. Sorghum noodles are an important food product in China (Niu Tian Tang and Zhang Fu Yao, 1990, *personal communication*).

Couscous

In much of northern Africa a steamed, granulated product called *couscous*, made from cereal flours (mostly wheat), is highly popular. In West Africa, sorghum, pearl millet, maize, and fonio are used to prepare couscous, although pearl millet is preferred.

Galiba et al (1987) reviewed the literature on sorghum and millet couscous. Sidibe et al (1982) described the traditional methods of couscous preparation in Mali. Flour is obtained from dehulled sorghum or millet grains and sifted with a traditional sieve of about 1-mm size, and the throughs are recycled to ensure uniform flour particle size (Fig. 6). The flour is wetted in a calabash with cool water and agglomerated into small particles gently with fingers. The flour aggregates are passed through a 1.5-mm opening sieve. The agglomerated flour particles thus obtained are steamed in a perforated and covered pot placed over another pot containing boiling water. The joint between the pots is carefully sealed with a wet cloth wiped with okra powder to ensure an efficient steaming process. After about 15 min of steaming, the agglomerated flour particles form a loose chunk, which is retrieved, broken into pieces, and further steam-cooked for another 15 min. These agglomerated and partially steamed particles are taken out of the pot and again broken into small particles and sieved through 2.5-mm screen. At this stage of the process, the product can be sun-dried and stored for future use.

For immediate use, the product is sprinkled with water and mixed thoroughly after adding baobab (*Adansonia digitata* L.) leaf powder or okra powder. The mucilages and gums from these powders prevent desiccation of the product and improve palatability. This flour aggregate mixture is again steamed for 25 min to give the product couscous. It is allowed to cool and served with a sauce containing vegetables, beans, and other foods. The sun-dried product can be stored even for six months. Dried couscous is reconstituted in milk or resteamed and served with sauce.

Couscous is an excellent and convenient food while traveling and is thus popular in West Africa. In countries such as Senegal, couscous is a commercial product. The flour for couscous production is sometimes fermented for less than 12 hr before granulation and steaming. In Burkina Faso, Mali, and Niger, couscous is reserved for the evening meal, while in Senegal it is consumed in the morning as well as evening. Couscous is called by several other names in West African dialects. Desirable consumer qualities of couscous include a high yield for a given

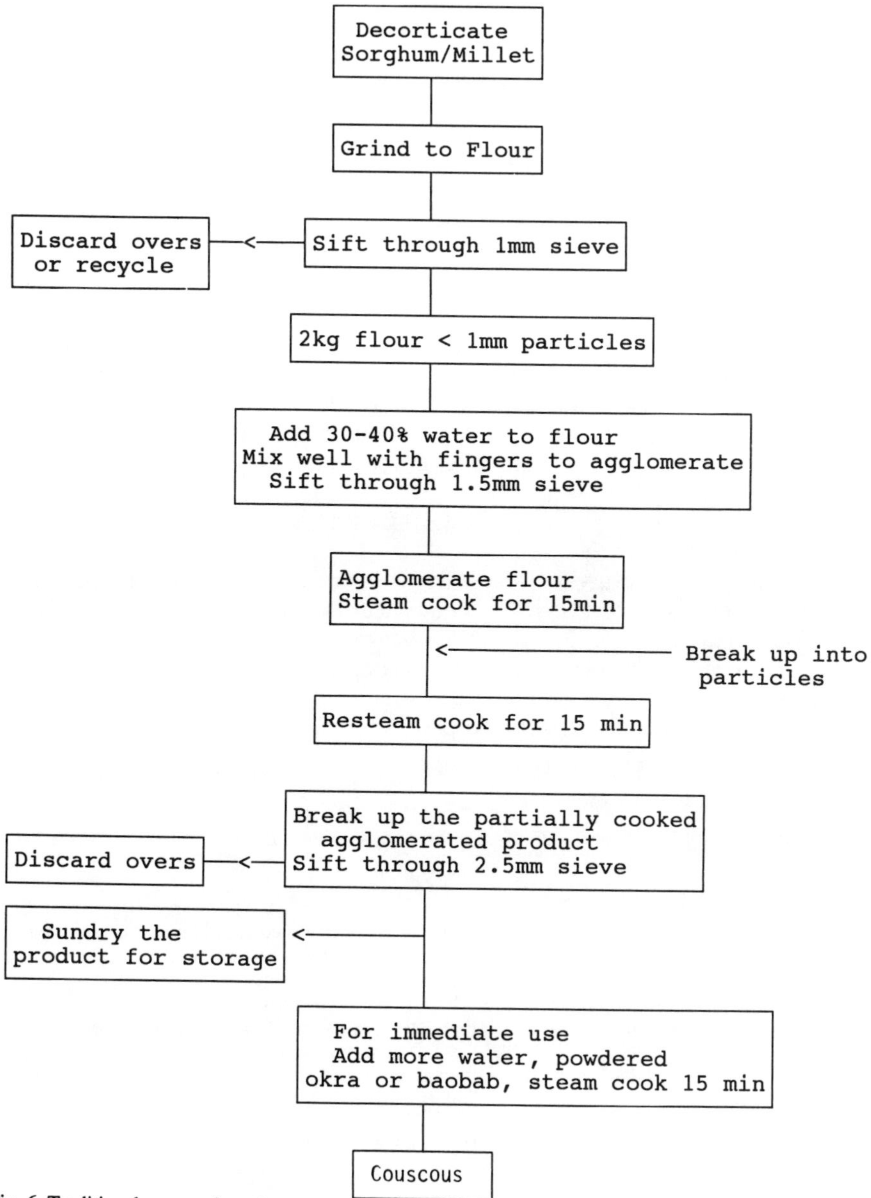

Fig. 6. Traditional preparation of couscous.

TABLE 6
Properties of Couscous Prepared from Sorghum and Millet from Burkina Faso[a,b]

	Weight (g)			Moisture (%)		Color Values[c]		
Cultivar	Agglomerated Flour	Couscous	Overs of Sieve	Couscous	Overs	*L*	*a*	*b*
Sorghum								
Local 1	137.0 a	135 a	4.3 a	36.4 a	45.3 a	73.0 a	−0.9 c	14.4 b
Local 2	139.3 a	139 a	3.6 a	36.6 a	48.4 a	63.9 c	1.6 a	12.2 c
Millet								
Souna	138.2 a	133 a	5.8 a	36.4 a	51.8 a	50.0 d	−0.3 b	10.9 d
Sanio	137.5 a	136 a	3.5 a	36.0 a	47.4 a	64.8 b	−0.8 c	19.0 a

[a] Source: Galiba et al (1987).
[b] Means in the same column with the same letter are not different at the 5% level of significance.
[c] Color values: L = lightness, $-a$ = green, $+a$ = red, $-b$ = blue, $+b$ = yellow.

quantity of flour and nonsticky particles. The flour quality should be such that, in spite of higher water uptake and swelling, the particles are firm and granular. In Mali, *dagafara* sorghum cultivars of *keninke* type are preferred for couscous production because they yield a relatively higher quantity of the product per unit quantity of flour (Sidibe et al, 1982).

Galiba et al (1987) compared the milling and couscous properties of sorghum and millet and concluded that yields of couscous from a unit quantity of flour were the same for both these cereals (see Table 6). However, the milling yields of pearl millet were lower than those from sorghum. Couscous made from pearl millet had a darker color and stronger flavor but was better gelatinized than that from sorghum.

FERMENTED BREADS

In India and Africa, sorghum and millet flours are fermented and used to make a wide range of pancakes (Table 7). Fermented breads are extremely popular in Ethiopia, Sudan, southern India, and Sri Lanka. The art of preparing traditional fermented breads is exemplified by products such as *injera, kisra,* and *dosa.* In West Africa, pancakes such as *massa* and *galettes* are prepared from sorghum and millets, but these are only consumed occasionally (Faure and Muchnik, 1989).

TABLE 7
Fermented and Unfermented Breads Traditionally Prepared from Sorghum and Millets

Common Name	Country or Region	Description and Variations
injera	Ethiopia	Fermented batter is baked into large thin pancakes
kisra	Sudan	Fermented batter is baked into large thin pancakes
kisar	Chad	Fermented batter is baked into large thin pancakes
dosa, thosai	Southern India, Sri Lanka	Fermented cereal batter is mixed with black gram batter and baked into thin pancakes
massa, maasa, waina	Nigeria, Ghana, Niger	Fermented batter from soaked grains is pressed into small cakes and baked on a dimpled pan with oil
galettes	Burkina Faso	Fermented batter is baked into large thin pancakes
roti, *rotte bhakri, rotla*	India	Unfermented dry pancake
tortilla	Mexico	Unfermented bread, lime-cooked grains are ground to *masa* and pressed into cakes, which are baked dry on a pan

Injera

The traditional cereal bread made in Ethiopia is called *injera* (Amharic language). Teff grain is reputed to make the best quality *injera* and is the staple diet in Ethiopia. However, in sorghum growing areas, sorghum is the preferred cereal for *injera* preparation, next after teff. Finger millet can also be used. Gebrekidan and

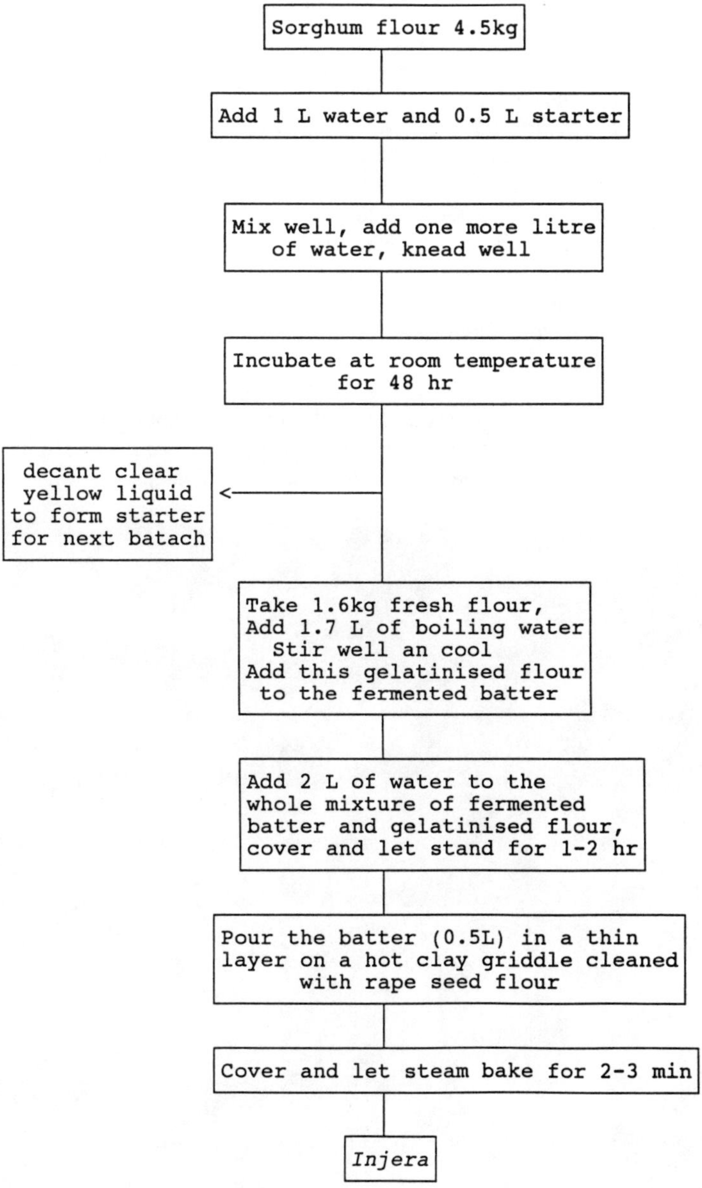

Fig. 7. Traditional method of *injera* preparation (Adapted from Gebrekidan and Gebrehiwot, 1982).

Gebrehiwot (1982) described traditional methods of sorghum *injera* production and stated that 80% of the sorghum produced in Ethiopia is consumed in the form of *injera*. Sorghum flour is sometimes mixed with that of teff or barley to improve acceptance and storability. Traditional preparation of *injera* involves fermentation of the flour with a starter from a previous batch (Fig. 7). The flour is fermented for about 48 hr. A fresh batch of flour is gelatinized by adding boiling water, and the gelatinized flour is added to the fermented batter to hasten a secondary fermentation. Alternatively, a small part of the fermenting flour is prepared into a gruel (*absit*) and added to the fermenting batter a day before *injera* preparation. The fermented batter is baked on a hot clay griddle called *metad*. Three large stones are placed near the fireplace to form a triangle, and the *metad* rests on the stones. The *metad* is heated underneath with fire from sorghum stalks or other firewood. Each time *injera* is baked, the *metad* is sprinkled with rapeseed flour and cleaned with a cloth. To bake *injera*, 0.5 L of batter is poured on the hot *metad* in a centrifugal motion from the edge of the griddle to the center. When small holes (eyes) start appearing on the top of *injera,* it is covered with a griddle lid and left to bake for 2–3 min. A typical *injera* is a circular pancake of about 60-cm diameter and 6-mm thick. The front side has uniformly spaced honeycomb like "eyes" each measuring about 4 mm in diameter and about four per square centimeter of *injera* surface. The underside is smooth, without any holes, and is nonsticky. The color is white to brown depending on the grain color of the cultivar used for flour.

A good quality *injera* has a soft and pliable texture, enabling the consumer to pick up *wat* (stew) by hand in a piece of *injera*. A good *injera* tastes slightly sour, is spongy, can be folded or rolled without cracking, and keeps its pliable texture

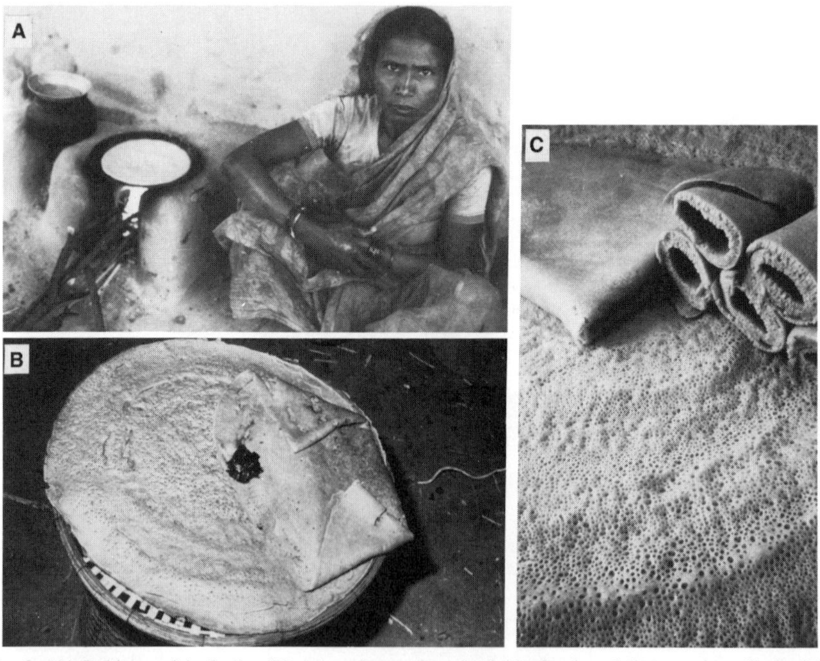

Fig. 8. (A) Baking roti in India. (B) *Injera* is a soft and pliable flat bread that can be rolled without cracking. (C) Close-up view of *injera* showing the numerous "eyes."

for three days (Fig. 8). On the contrary, poor *injera* is dry and brittle, does not exhibit uniform "eyes," and cannot be stored overnight. Insufficient fermentation produces a sweetish taste, and such *injera* is called *aflegna injera*. The quality of *injera* is influenced to a large extent by the fermentation process and the length of fermentation time. In teff, the primary agent of fermentation has been identified as the yeast, *Candida guilliermondii* (Stewart and Getachew, 1962).

Striking cultivar differences for sorghum *injera* quality were reported (Gebrekidan and Gebrehiwot, 1982). In general, decorticated sorghum grain produces *injera* of better quality and a lighter color. Soft endosperm sorghum types produce better quality *injera,* but such grains cannot be decorticated well. More information is required on the type of sorghums suited for *injera* preparation.

Kisra

Traditionally, *kisra* is the staple diet of Sudan and is prepared from either sorghum or millet, depending on regional preference. *Kisra* is eaten by one and all at least once a day. Clean grain is reduced to a fine flour on a traditional grinding stone or in attrition mills, which have recently become popular. A thick paste, *ajin*, of approximately two parts flour and three parts of water is prepared and allowed to stand for fermentation overnight after adding a starter (Ejeta, 1982). The following day, the batter is diluted with water to the required consistency and taste. Normally, a pH of 3.85 is preferred (Badi et al, 1988). The batter is baked on a fired clay griddle or metal plate. The griddle or metal plate is heated and then rubbed with a damp oily cloth dipped in sesame oil or bone marrow. The batter is poured with a small container along the length of the pan and then it is swiftly spread into a thin layer by using a *gargariba* (dry strip of palm leaf). The *kisra* is ready in 30–40 sec, after which it is lifted by hand and stacked in a straw tray.

A good quality *kisra* should peel off the baking pan as a smooth wafer in one piece, without holes. It should be supple and flexible to fold but not brittle. *Kisra* is normally consumed with stew or sauce. Good *kisra* should be soft and moist but not spongy in texture and should keep these qualities until the next day. *Kisra* quality depends mainly on cultivar, flour granulation, fermentation process, and baking conditions.

Cultivars with soft endosperm are more suitable for *kisra* preparation. Sorghum cultivars with white or yellow grain color are preferred, but local Sudanese sorghums with soft endosperm and a brown color also make acceptable *kisra*. Ejeta (1982) observed that Sudanese cultivars such as Dabar, Tetron, and Wad Fehal, which possess a relatively less vitreous endosperm and no subcoat, produce the best quality *kisra*.

Dosa

In southern India and Sri Lanka, fermented cereal batter is mixed with legume batter and is baked into a thin pancake called *dosa*. Rice is the preferred cereal but in dry areas where rice is expensive, sorghum and millets are used to make *dosa*.

Dehulled sorghum or millet grains are soaked, wet-milled, and then fermented overnight. The next day a small proportion of wet-milled black gram batter is mixed with the fermented cereal batter to give the right consistency. Alternatively millet grains and black gram are mixed in a ratio of 3:1 by volume and are wet-ground. The mixed batter is fermented overnight. The batter is poured in small quantities onto a hot metal plate greased with oil and baked for about a minute. *Dosa* is a soft, or sometimes slightly crisp, but yet flexible pancake. It can be thin or thick (2 to 4 mm) and is normally 20 to 25 cm in diameter. It is consumed with curd, vegetables, chilies, and other sauces.

UNFERMENTED BREADS

Roti

Roti is an unfermented dry pancake made in India from wheat, sorghum, pearl millet, small millets, and maize flour. In central and northwestern parts of India where sorghum and pearl millet are staple foods, these grains are used for roti preparation and are preferred. Murty and Subramanian (1982) described the traditional methods of roti production and estimated that 70% of sorghum grain produced in India is consumed in the form of roti. Roti is also known as chapati, *bhakri, rotla,* and *rotte* in various regional languages of India. It is eaten at breakfast, lunch, and dinner and is frequently stored overnight. It is consumed with vegetables, dal, meat, milk, curd, sour milk, pickles, and other sauces.

Whole sorghum or millet grains are milled to flour in a traditional stone mill. However, during the last few decades, electrical or diesel-operated disk or plate mills (*chakki*) have become widely popular in the villages, and sorghum and millet grains are custom dry-milled. The whole flour is sifted through a screen (850 μm), and the bran is rejected. Normally 95% or more of the flour is extracted from the grain. The flour particle size varies depending upon the mill and cultivar used. On average, about 43% of the flour will have particles of <75 μm. Murty et al (1984a) noted that milling in the *chakki* caused significantly higher starch damage and that it could improve dough quality. Flour obtained from these mills is not stored for more than a week.

Housewives normally take about 50 g of sorghum flour and mix it with approximately 45 ml of warm water, added in increments. The flour-water mixture is kneaded by hand on a wooden board into a cohesive dough. The dough is made into a ball and pressed by hand into a circular disk. The dough disk is placed on the wooden board and flattened by fast and deft hand strokes into a thin and circular sheet called *roti.* Small quantities of dry flour are used while flattening the disk to prevent the dough sticking to the hands. The size of the roti can vary from 12 to 25 cm in diameter and 1.3 to 3.0 mm in thickness. A creamy white and sticky dough with a characteristic aroma is preferred. Sorghum flour is sometimes mixed with wheat flour or legume flour in variable proportions to suit taste and palatability.

In rural areas, baking is done on an earthen or iron pan placed on three stones set in a triangle. The fire underneath is manipulated to provide suitable heat and baking temperatures of around 300–325°C. Permanent semicylindrical hearths made out of mud are also used. The roti is placed on the hot pan. A small quantity (less than 5 ml) of water is sprinkled on the roti, or the roti is moistened with a wet cloth. After the underside is cooked (30–40 sec), the roti is turned over with the fingers. The roti is removed from the pan after a minute and placed on hot coals near the fire, with the unmoistened side of the roti exposed to the heat, which leads to "puffing" of the roti.

There are several minor variations to the general procedure for roti preparation outlined here. Some housewives use hot water to make dough, whereas others use cold water. Water for the dough is occasionally replaced by sugarcane juice, milk, etc., in some regions. Sometimes finely sliced vegetables, spices, and salt are mixed into the roti dough. Sometimes a small quantity of flour cooked in water is added to the dry flour during dough preparation. The gelatinization of part of the flour improves the elasticity and cohesiveness of the dough and produces roti with improved quality (Desikachar and Chandrasekhar, 1982). In some restaurants, a dough ball is kept overnight in water, and small portions of it are used the next day to add to dry flour while making fresh dough. A good dough can be rolled into a sheet with a wooden rolling pin, although during domestic preparation the dough is flattened with the hands. Olewnik et al (1984) proposed that millet roti dough is a powder–water system in which the dough is held together by a continuous water phase between the flour particles. Water absorption of flour and rolling quality of dough are relatively poor in pearl millet compared with sorghum (Subramanian et al, 1986).

A good sorghum roti should possess a creamy white or light yellow color, soft and flexible texture, and minimal burning due to baking. Good rotis puff well during baking. Rotis are generally stacked in a pile wrapped in cloth and stored in perforated baskets. Occasionally they are sun-dried and stored for more than a week. In such instances, they are softened with milk or buttermilk before eating. Keeping quality is an important factor of rotis. Dough from certain cultivars cannot be flattened or rolled. Poor dough quality leads to tough rotis. Good quality sorghum rotis have excellent texture, taste, and color properties comparable to wheat rotis. Consumers prefer white to pale yellow, dense and round grains free from colored spots for roti preparation.

Acceptance of improved high-yielding cultivars by farmers and consumers is influenced by the roti-making quality of the grain (Von Oppen and Rao, 1982).

TABLE 8
Grain, Dough, and Roti Quality Characters of 15 Sorghum Cultivars with White Grains[a,b]

	Grain		Dough		Roti[c]			
Cultivar	Breaking Strength (kg)	Water Absorption (%)	Water Required (ml)	Rolling Quality (cm)	Taste	Texture	Aroma	Keeping Quality
Maldandi 35-1	7.7	18.1	26.1	23.7	1.3	1.5	1.0	1.5
CSH-8	6.5	26.4	25.6	23.0	2.0	2.0	1.8	2.0
CSH-6	9.7	20.7	26.5	22.0	1.7	2.0	1.2	2.7
CSV-5	7.6	23.8	25.1	20.5	2.0	2.6	1.6	3.5
BG-30	7.0	32.4	25.2	21.3	2.9	2.8	2.5	3.5
IS 7943	9.6	37.0	26.1	22.8	3.0	1.8	1.5	3.2
T-SS-47	8.6	23.8	28.3	20.3	2.3	2.2	1.4	4.1
IS12611	11.0	22.7	30.9	22.2	1.6	1.5	1.2	3.0
P721	6.8	32.7	25.7	18.1	3.7	3.3	2.7	4.0
M36082	10.0	22.7	30.3	22.8	1.4	2.0	1.0	3.0
CK 60 B	7.0	20.9	28.0	20.5	2.4	2.1	1.8	3.0
IS 1098	6.0	37.5	24.3	20.7	2.5	3.2	1.5	2.5
IS 2328	12.4	29.0	26.1	22.6	2.2	2.7	1.7	3.7
PJ 33 K	8.8	25.1	27.1	22.2	2.2	1.9	1.4	2.7
IS 11024	11.0	28.9	24.4	22.6	2.4	2.4	1.5	3.5

[a] Source: Murty and Subramanian (1982).
[b] Evaluated by standard methods described by Murty and Subramanian (1982).
[c] Roti characters were scored by a trained taste panel of five members on a scale of 1 to 5, where 1 = good and 5 = poor.

Indian sorghum cultivars such as Maldandi and Dagadi are popular for roti. Murty et al (1982b) evaluated several hundred sorghum cultivars for dough quality and roti characteristics and concluded that good roti-producing grains exhibit a thin and colorless pericarp and 60% corneous endosperm. Factors such as pericarp color, endosperm type, endosperm texture, grain weathering, and head bug damage to the grain affect roti quality significantly. Significant variation for dough and roti quality characters was observed even among visually similar white grain cultivars (Table 8).

Tortillas

An unfermented bread, called *tortilla*, is generally prepared from alkali-cooked and steeped maize, and is consumed as a staple food in Mexico and other Central American countries. Sorghum is used for tortilla preparation either alone or in combination with maize in Honduras, Nicaragua, Guatemala, El Salvador, and Mexico (Bressani et al, 1977; Futrell et al, 1982; De Walt, 1982). Since maize is drought susceptible, sorghum is a relatively more dependable cereal crop in marginal areas, and its importance as a staple food has increased in recent decades.

The procedure for sorghum tortilla preparation is very similar to that for maize except that sorghum requires lower levels of lime and a shorter cooking time. Sorghum grains are cleaned and washed in water. The grain is cooked in lime water in a ratio of 1:3, with about 0.8% concentration of lime in the solution. The cooking process normally takes about 40 min. The mixture called *nejayote* is poured out on the floor and allowed to cool. The lime-cooked grain is called *nixtamal.* The nixtamal is washed and rubbed in water to remove the bran. It is washed again in water two or three times and then ground to a paste in a metal or stone grinder. The paste is kneaded to a dough, called *masa.* The *masa* is pressed into a thin circular tortilla measuring about 10 to 12 cm in diameter. The tortilla is baked on each side on a hot griddle for about a minute and half and is then ready to eat.

Light yellow or white maize tortillas are preferred to sorghum tortillas because the phenolic compounds and tannins present in sorghum pericarp lead to light greenish off-colors. However, improved tortillas with acceptable color and texture can be made with grain from recently identified sorghum cultivars that have a tan plant color, white pericarp without a subcoat, intermediate endosperm texture, and low levels of color precursors. Decorticated sorghum can be used to produce excellent tortillas because decortication reduces leaching of colors from the pericarp and the cooking time. (Iruegas et al, 1982; Bedolla et al, 1983; Choto et al, 1985).

BOILED RICELIKE PRODUCTS

Sorghum and millets are often cooked to make boiled ricelike products. In regions where rice is used but scarce, sorghum and millets are used as a substitute. Whole grains, dehulled grains, cracked grains, and grits are boiled to give ricelike products.

In southern India, boiled sorghum is called *jonna annam* or *soru* (Subramanian et al, 1982). Traditionally, sorghum grains are dehulled before cooking. In recent decades, however, in order to circumvent the time-consuming and cumbersome dehulling process, whole grains are taken to plate or disk mills for obtaining coarse grits. The product is winnowed and sieved to separate the coarse bran pieces, and the remainder is used for cooking. Alternatively, whole or dehulled grains may be soaked overnight and then cooked the next morning. Soaking the grain reduces cooking time and color intensity of the product and improves keeping quality, probably due to partial fermentation. The dehulled grain is cooked in water in a proportion of 1:3. Sometimes, spices and legumes such as mung bean, chickpea, and groundnut are added. The cooked product should not become a sticky paste. It has to be soft and fluffy, white or light yellow in color, and should resemble boiled rice. The freshly prepared product is consumed either as lunch or dinner with dal, onions, chilies, vegetables, and sauces. Sometimes, the product is stored overnight in water and the next day is consumed with buttermilk.

Boiled ricelike products are also prepared from sorghum in China (*kaoliang mi fan*), Botswana (*lehata wagen*), Sudan (*balila*), Ethiopia (*nufro*), Burkina Faso (*wesla*), Niger (*gumba, dao*), and Tanzania (*kande*). Sorghum grits cooked like rice is a major food in Haiti (Rooney et al, 1986). In Mali and Sierra Leone, grains from specific sorghum cultivars belonging to the race margaretiferum are used to prepare ricelike products. These grains are small and exhibit a very hard endosperm.

Sorghum requires a much longer cooking time than rice, but less than that of maize. The cooked grain texture is usually firmer than rice. Murty et al (1984b) observed that cooking time of dehulled grains was reduced by half, whereas the increase in volume and weight of the cooked product was more than 100% of that of whole grains (Table 9). Kernels with intermediate texture exhibited better cooking quality characters. All the minor millets are generally cooked after dehulling in a

TABLE 9

Cooking Quality Attributes of Whole Grain and Dehulled Grain from 25 Sorghum Cultivars[a]

Attribute	Mean ± SE	Range
Water absorption, %[b]		
Whole grain	37.95 ± 0.93	30.62–48.52
Dehulled	24.70 ± 0.99	17.40–40.23
Cooking time		
Whole grain	45.24 ± 2.17	33.00–79.00
Dehulled	86.32 ± 1.56	71.00–111.00
Cooked weight		
Whole grain	74.80 ± 1.24	62.50–88.15
Dehulled	50.26 ± 0.60	44.04–52.50
Cooked grain volume		
Whole grain	70.18 ± 1.28	57.50–82.00
Dehulled	45.92 ± 0.60	40.00–52.50
Increase in weight, %		
Whole grain	273.60 ± 6.17	212.00–340.00
Dehulled	151.20 ± 2.82	130.50–181.50
Increase in volume, %		
Whole grain	362.40 ± 8.79	283.00–465.50
Dehulled	197.00 ± 4.94	142.50–273.50
Cooked grain texture[c]		
Whole grain	2.33 ± 0.11	1.00–3.57
Dehulled	2.55 ± 0.12	1.12–3.75

[a] Source: Murty et al (1984b).

[b] Percent water absorption was determined by soaking a sample of 100 grains in water for 5 hr at room temperature, and the percent increase in weight was recorded.

[c] A 20-g sample of grain was cooked in excess water under controlled conditions. Cooked grain texture was rated on a scale of 1 to 5, where 1 = very soft and 5 = very hard.

manner similar to rice (Pushpamma, 1989). In Nigeria and Mali, fonio grain is used to prepare boiled ricelike products. In the Plateau State of Nigeria, a crude meal of fonio is used to prepare *acha jollof.*

ALCOHOLIC BEVERAGES

A detailed discussion on alcoholic beverages made from sorghum and millets is outside the scope of this chapter. Only a brief survey of the traditional beverages popular in Africa will be attempted. Among many African communities, sorghum beer and wine play an important sociological role, being served on festive occasions, at social gatherings, and during community work projects. Malted sorghum and millet grains have been widely used in Africa for centuries to make alcoholic and nonalcoholic fermented beverages. Unmalted grains are used as a starch substrate or as adjuncts. It is difficult to hazard a guess on what proportion of sorghum and millet production is used for traditional alcoholic beverages. According to Winter (1964), among certain communities in Cameroon, fermented sorghum beverages represent 20% of total cereal consumption. Apparently, a fairly high proportion of sorghum and millet production in Africa is utilized in the making of traditional beers. In West Africa, use of sorghum and millets in beer production is

TABLE 10
Traditional Alcoholic Beverages Made by Using Sorghum and Millets

Common Name	Country or Region	Description and Variations
amgba	Cameroon	Sorghum beer, relatively clear
affouk	Cameroon	Sorghum wine, sour taste
bili bili	Tchad, Central African Republic	Sorghum beer, relatively clear, Sweet-sour
cochate	Tchad	Millet beer
chapalo	Niger	Sorghum beer
dolo	Burkina Faso, Mali	Sweet-sour relatively clear beer
burukutu, pito	Nigeria, Ghana	Sorghum beer
merissa	Sudan	Sorghum beer
busaa	Kenya	Beer from maize and sorghum malt or finger millet malt
sibamu	Zambia	Sorghum-millet beer
katata	Zambia	Finger millet malt-maize beer
kachabu	Zambia	Strong beer
mbege	Tanzania	Beer from millet malt and banana
pombe	Tanania	Sorghum beer
kwete	Uganda	Millet and maize beer
omulamba	Uganda	Beer from sorghum malt and banana-maize substrate
warangi (*waragi*)	Uganda	Strong beer, often contains banana
talla	Ethiopia	Sorghum beer
Bantu beer, *chibuku*	South Africa, Zimbabwe	Opaque beer from sorghum or millet malt and maize substrate
kibuku	Zaire	Sorghum beer
kasi kasi	Zaire	Sour beer from sorghum malt and banana juice
biere de mil	Senegal	Millet beer
dam	Togo	Millet beer
bojalwa	Botswana	Sorghum beer

more prevalent in the higher rainfall zones and is not common in areas where Islam is widely practiced. A wide variety of cereals and millets is available in the relatively high rainfall zones of Africa. The red and brown types of sorghum grains produced in these areas are mostly used for malting. Sorghum is the major raw material for alcoholic beverage production in China (Niu Tian-Tang and Zhang Fa Yao, 1990, *personal communication*). Names of some popular alcoholic beverages traditionally prepared from sorghum and millets are listed in Table 10.

Rooney et al (1986) distinguished two types of traditional beers: the sweet, relatively clear and nonsour type with low solids, common in West Africa; and the sour pinkish brown, opaque, effervescent type resembling a thin porridge, common in southern Africa. *Dolo* is an example of the former type of beer while Bantu beer is an example of the latter type. The main distinction between the two types is the absence (or relatively short period) of an extended souring process or bacterial (lactic acid) fermentation in the sweet and less sour beers. Bantu beer has relatively much more solids, a thick body, and tastes very sour. The traditional methods of Bantu beer production have been modernized and commercialized in South Africa (Schwartz, 1956; Daiber, 1988). The traditional procedure begins with malting of sorghum, followed by boiling a separate starch substrate, usually flour or grits of maize, sorghum, or millet, into a slurry. Cold water is added to ground malt, and the malt solution is poured into the substrate. The mixture is allowed to sour (one to two days), then boiled before adding some more malt, and allowed to ferment (with starter from a previous batch) for at least two days. Finally, the liquid is filtered through a coarse mesh to obtain the sour opaque beer. The beer consists of 2–4% alcohol (w/v), 0.3–0.6% lactic acid, 4–10% total solids and has a pH of 3.3–3.5. It is effervescent and is drunk in an actively fermenting state. Opaque sorghum beer is an important source of nutrients, because it contains vitamins, minerals, proteins, and carbohydrates that have been solubilized during malting and brewing (Novellie, 1977, 1982a). Another opaque beer, *busaa,* made from maize and finger millet malt, is popular in Kenya (Nout, 1980). In Sudan, an opaque beer called *merissa* is made from sorghum (Dirar, 1978).

Chevassus-Agnes et al (1976) described the methods of producing *amgba*, a sorghum beer popular in northern Cameroon. Ekundayo (1969) and Faparusi (1970) described the production of *pito* and *burukutu* beers traditionally made from sorghum in Nigeria and Ghana. Griffon (1985) summarized considerable information on the production of traditional tropical beers, wherein sorghum and millets are frequently used either as a source of malt or as adjuncts. The sources of starch range from banana to cassava, maize, sorghum, and the millets. Malts from pearl millet and finger millet are frequently used in eastern and southern Africa. In Uganda, finger millet malt is used alone or in a mixture with sorghum malt to prepare traditional local beer (Esele, 1989). Traditional production of beer is an art usually practiced by women, and they are known by the quality of the beer they produce. Malting and brewing are processes uniquely suited for overcoming the lower nutritional value of brown sorghum grain. In general, as white grain sorghums are preferred for porridges, breads, or other foods, the most available, least expensive grain for beer production is brown sorghum. Ideally, the best sorghums for brewing have a red pericarp without a pigmented subcoat, intermediate to soft texture, and produce a high level of amylases. The brown bird-resistant sorghums are undesirable for use in modern industrial sorghum malting and brewing (Novellie, 1984).

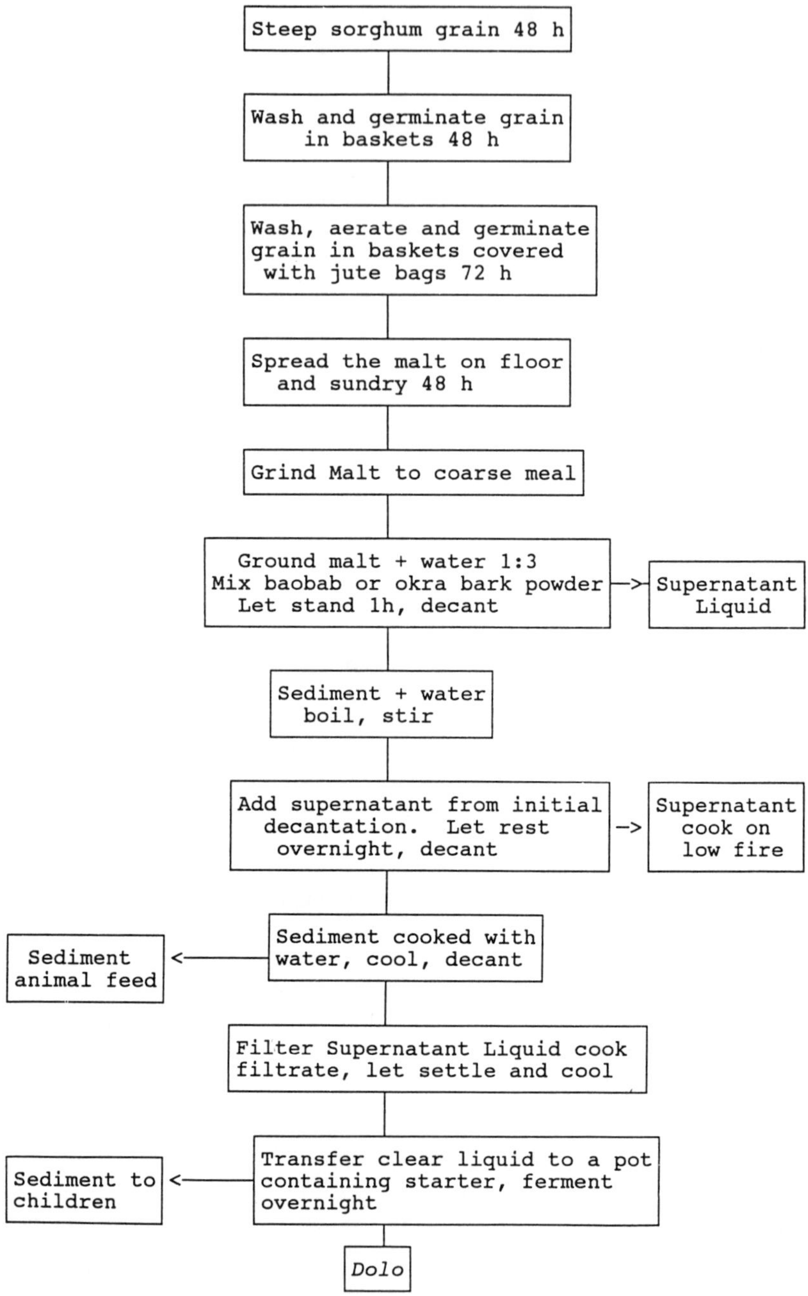

Fig. 9. Traditional production of *dolo,* a sorghum beer, in Burkina Faso.

Dolo

In Burkina Faso and Mali, a light red, relatively clear, alcoholic, sweet-sour beer with a fruity smell called *dolo* is made from sorghum malt (ICRISAT, 1987). The traditional preparation of *dolo* involves malting red or brown sorghum grain followed by cooking, filtration, and fermentation of the wort (Fig. 9). Sorghum grains are steeped in excess water for 48 hr and then washed with fresh water and germinated in large baskets covered with jute bags. The germination continues for five to seven days, during which the germinating grains are aerated and washed two to three times. Later, the germinating grains are spread on the floor as a thick layer and covered with gunny bags, followed by open sun-drying for two more days. The dried malt is ground using traditional methods or taken to a hammer mill for custom grinding.

The ground meal is taken in a large earthen pot and water is poured into it. A small quantity of powder from baobab or okra bark is thoroughly mixed into the solution. These mucilaginous substances aid the clarification process. The mixture is allowed to stand for an hour, and the supernatant liquid is decanted. The sediment is cooked with additional water for about 5 hr with vigorous stirring. The cold extracted supernatant liquid is added again to the hot paste, and the mixture is left to settle overnight. The next morning the supernatant liquid is decanted and heated on a low fire. The substrate or sediment is again cooked for 5 or 6 hr with additional water, cooled, and decanted. Finally, the sediment is removed and used as animal feed. The supernatant is filtered with the aid of straw mats. The filtrate is cooked again, the supernatant from the previous stage is added, and the whole mixture is allowed to cool and settle. This process results in the production of a sterilized wort. The dark brown and viscous liquid at the bottom of the wort is removed and is normally fed to children. The clear supernatant liquid is poured into a large pot in which the yeast from the previous batch is present. The liquid is allowed to ferment overnight. Next morning a bubbling liquid with froth, *dolo,* is ready for serving. *Dolo* is normally served within 24 hr. It is commonly sold in the villages and semiurban areas of Burkina Faso and Mali (Fig. 10).

NONALCOHOLIC BEVERAGES

In Africa, several nonalcoholic beverages are prepared from sorghum and millets. A great majority of these nonalcoholic fermentations are sourings (Novellie, 1982a,b). Lack of precise microbiological control under local conditions makes it difficult to conduct pure alcoholic or nonalcoholic fermentation. Nevertheless some degree of control is practiced in certain processing techniques, and many of the soured beverages produced have negligible or very low amounts of alcohol in them. Ground sorghum, sometimes mixed with malt of sorghum and millets, is added with boiling water, and the dough is held at hot and elevated temperatures for several hours. Thermophilic and mesophilic bacteria produce lactic acid. The process of malting encourages the development of a variety of microorganisms that hasten the souring process. The difference between a thin, sour porridge and a sour beverage is in the concentration of cereal present. Beverages normally contain much less than 10% total solids. A typical example of a nonalcoholic beverage is *mahewu* (also called *amahewu, marewu*, and *magou*), traditionally prepared from sorghum in South Africa. In recent years, maize has been more frequently used,

Fig. 10. (A) A typical dolotarium in a village in Burkina Faso showing malt from brown sorghum grain in the foreground. (B) A calabash containing *dolo,* a sorghum beer. (C) Popping sorghum in Ethiopia.

and the process has been modernized and commercialized (Novellie, 1982a,b). *Mahewu* tastes sour, and lactic acid is the chief souring agent.

In Sudan, a popular nonalcoholic beverage called *hulu mur* is prepared from sorghum (Vogel and Graham, 1979). One kilogram of malted sorghum flour is mixed with 2 kg of whole normal sorghum flour and 6 L of water, and the mixture is allowed to ferment. The fermented batter is baked into a thin pancake similar to *kisra* and then crumbled into small flakes. The flakes can be stored for a few weeks. These flakes are soaked in water, strained, and served as a beverage during the fasting period in Ramadan. The residue from the straining process is used as a feed. Another similar nonalcoholic beverage made in Sudan from sorghum is *abrey*. A normal flour and water mixture is allowed to ferment for three to five days and baked into *kisra*. The thin flakes from such *kisra* are soaked in water to give a white and slightly sour beverage. *Huswa* is another nonalcoholic Sudanese beverage made from malted sorghum flour.

In Uganda, Zambia, and parts of Tanzania, dried or roasted finger millet malt is used to make nonalcoholic drinks.

SNACK FOODS

There is an endless variation in traditional snack foods prepared from sorghum and millets. Vogel and Graham (1979) and Subramanian and Jambunathan (1980) describe a variety of snacks traditionally made from sorghum and millets in India, Sudan, Nigeria, and other countries. Several of these are prepared by deep-frying the dough in fat. Thick porridges or fermented doughs can be extruded or molded and then deep-fried or sun-dried and then deep-fried. Sugar, salt, spices, and numerous other ingredients may also be mixed in the dough to suit taste and flavor. Since flour from sorghum and millets does not give an ideal dough, legume flours, which also add to the nutritional value and palatability of the snacks, are frequently mixed in the dough. Popped or malted grain or flour are also used in various snacks. However, these snack foods make up a less significant portion of total sorghum and millet production.

SPECIAL USES

Popping

Certain cultivars of sorghum and millets have been chosen by the traditional farmers, possibly over centuries, to suit their special tastes and purposes. For example, sorghum grains are popped in India and Africa to give pop sorghum, which is considered to be a delicacy. In central parts of India, popped sorghum grains are consumed by one and all and are used in various snack foods.

Traditionally, popping is done by putting small quantities of grain in a hot pan or bowl over a steady fire. Popping of grain occurs within a minute with a hissing and splitting noise. The popped grains are removed immediately, and a fresh batch of grain is popped. Popped sorghum is considered to be superior to popped corn as it is tender, has less hull, does not stick between the teeth, and makes less noise when eaten. Popped grains are considered a crunchy, porous, and precooked product with good flavor.

Murty et al (1982c) identified 36 accessions of sorghum that exhibited superior popping quality among 3,682 Indian sorghum collections. Most pop sorghum grains have a small grain size, white color, medium thick pericarp, and a hard endosperm. They have a very low germ-endosperm ratio, and the embryo is located at a corner of the hylar region. Damon (1962) reported that in Ethiopia, Fendisha sorghums were popped like popcorn. Popped grain is used in various snacks. Flour from popped grain is made into a porridge or other delicacies and is fed to children and old people. Popping quality of sorghum grain is highly heritable (Murty et al, 1988).

In India, popped finger millet flour mixed with brown sugar and milk is called *hurihittu*. Malleshi and Desikachar (1985) reported varietal differences in popping quality of millets.

Parching

In several parts of India and Africa, sorghum heads are parched at the dough stage and the threshed grain is consumed. Sorghum cultivars mostly used for this purpose in India are called *vani jowar* (*vani* sorghums) (Prasada Rao and Murty, 1982). In the state of Maharashtra, a snack called *hurda* is prepared by parching *vani jowar*. The whole earhead is harvested at the dough stage and brought to a roasting pit. The heads are buried in hot coals and ashes for a few minutes to be parched, after which the grain is beaten off the panicle and winnowed.

The practice of eating sorghum at the dough stage either raw or in a roasted form also exists in Ethiopia and Sudan (Damon, 1962). The sorghum cultivars Bsenga and Muyera of Ethiopia are used for this purpose. The high-lysine sorghums Wotet Beguncha and Merchuke, cultivated in the Wollo province of Ethiopia, are popularly consumed during the dough stage after parching (Rameshwar and Axtell, 1973). A sorghum cultivar Maleek is cultivated in the backyards in the Kassala province of Sudan and is used for parching. The *vani* and Merohuke sorghums exhibit a shrunken endosperm at maturity and contain a relatively higher level of soluble sugars than normal grains. These characters are inherited through independent recessive genes (Murty et al, 1985; Murty and Nicodemus, 1987).

Scented Sorghums

In Maharashtra and Madhya Pradesh states of India, scented sorghum cultivars are grown by the farmers in small areas (Prasada Rao and Murty, 1982). The grains, stalk, and leaves of these cultivars emit an aroma typical of basmati rice. Basmati sorghum grains are consumed in the form of a porridge and are considered a delicacy.

Sweet Stalk or Chewing Sorghums

Sorghum land races possessing sweet stalks are cultivated in India and Africa and are used for chewing in a manner similar to sugarcane, particularly in dryland areas. Prasada Rao and Murty (1982) reported on the variation among 253 sweet stalk accessions available in the sorghum world collection. Some of the accessions (for example IS 19674) exhibited Brix values higher than 18 (Seetharama et al, 1987).

Chewing sorghums are popular in Botswana, Cameroon, Ethiopia, Nigeria, India, South Africa, Sudan, Uganda, Zambia, and Zimbabwe. In Ethiopia, sweet sorghums are called Tinkish and besides chewing are used for confections (Damon, 1962). In Sudan, sorghum cultivars called Ankolib are cultivated for their sweet stalks. Similarly, in Malawi and Nigeria, sorghum cultivars known as Misale and Takanda, respectively, are used for chewing. Sweet sorghum stalks are also sold in southern Africa. In the United States, sweet sorghum syrup has been produced since colonial days and several improved cultivars were identified (Freeman and Broadhead, 1986).

FUTURE RESEARCH NEEDS

A lot of progress has been made during the last 15 years in identifying, describing, and categorizing the traditional food products made from sorghum and pearl millet. However, very little information about the food uses of minor millets has been documented. There is an urgent need for collection and dissemination of detailed information about the traditional uses and processing methods of minor millets. Millions of poor farmers, particularly in the dry regions of several developing countries, depend on these minor millets for their daily diet, and a good knowledge of their current use is essential for their improved utilization. Home economists and nutritionists should conduct more specific surveys on the traditional processing methods and consumer requirements of these millets so that crop improvement programs can apply such information in their respective programs.

In several developing countries, despite the development of adequate mechanical processing technologies for sorghum and millets, the traditional methods and uses continue to be practiced. Plant breeders should take note of this fact while selecting cultivars for release and production. For example, in West Africa, where sorghum and millets are grown mainly for domestic consumption, experience has shown that high-yielding cultivars of recent origin have not become successful, partly because of their poor processing qualities when compared with traditional local controls. In India, the first commercial sorghum hybrid did not have acceptable roti quality and received consumer "resistance". However, the latest releases of commercial hybrids with improved consumer quality quickly became popular and successful. Processing and food quality traits pertaining to traditional uses of sorghum and millets should receive adequate attention early in breeding programs. Generally, plant breeding programs exploit considerable exotic germ plasm, which may not possess locally desired traditional consumer traits. Information on the consumer quality attributes of parents used for crossing is highly valuable.

Rooney et al (1986) reviewed the information available on the desired grain quality characters for various sorghum foods. It was found that factors like grain size, shape, pericarp color, type and texture of endosperm, presence or absence of testa, processing method, flour quality, production environment, and damage to grain caused by insects and molds significantly affected processing and food quality. Among all, endosperm texture was found to be the most important factor. Information of similar nature is lacking in millets. It is possible that quality differences among cultivars of some millets are only of minor importance. Preliminary evaluation of several cultivars of diverse origin and grain type for their processing and food quality should reveal any perceptible differences from the consumer point of view. Further detailed studies by food technologists on a chosen set of samples will be highly informative.

A general description of the traditional food uses of sorghum and millets and their classification is available. However, the desired consumer quality attributes of these processes and foods are known only in empirical terms. A precise definition of their quality in terms of physical and chemical parameters is still incomplete. Biochemical, nutritional, and microbiological changes that take place during the processing and storage of these products need critical investigations. Such information will be useful in the development of laboratory techniques essential for screening breeders' material. In general, there is a dearth of simple and rapid laboratory screening techniques to predict the food quality of small grain samples.

Traditional processing methods of sorghum and millets are intended to make the food products from these coarse grains more palatable, acceptable, and digestible. Although these techniques are simple, crude, and uncontrolled, they may have great significance from a nutritional point of view. Some of these processes and foods have good scope for commercialization, like Bantu beer in southern Africa. A commercialized traditional product has a greater chance of being popular and culturally acceptable than an exotic or novel product. It is encouraging to note that some progress in this direction is being made in Nigeria, where import of cereal grains such as wheat and barley has been banned (Murty, 1990). Collaborative research between home economists, food technologists, biochemists, nutritionists, and breeders is indeed necessary.

ACKNOWLEDGMENT

Thanks are due to J. O. Akingbala, Department of Food Technology, University of Ibadan, Nigeria, for going through a preliminary draft of the manuscript.

LITERATURE CITED

ADRIAN, J., FRANGNE, R., DAVIN, A., GALLANT, D., and GAST, M. 1967. Le probléme de la moutre du mil intérét nutritionnel des procédés artisanaux africains et essai de mécanisation. Agron. Trop. 22:687-698.

AKINGBALA, J. O., ROONEY, L. W., and FAUBION, J. M. 1981. A laboratory procedure for the preparation of *ogi*, a Nigerian fermented food. J. Food Sci. 46:1523-1526.

ARNOULD, J. P., and MICHE, J. C. 1971. Apercu sur l'économie et l'utilisation des mils et sorghos dans le monde. Agron. Trop. 26:865-887.

BADI, S. M., BURENG, P. L., and MONAWAR, L. Y. 1988. Commercial production: A break through in *kisra* technology. Pages 31-45 in: Proc. Quad. Symp. Sorghum Millets 4th, Int. Cereal Bread Congr. 8th. ICC, Lausanne, Switzerland.

BANIGO, E. O. I., and MULLER, H. G. 1972. Manufacture of *ogi* (a Nigerian cereal porridge): Comparative evaluation of corn, sorghum and millet. Can. Inst. Food Sci. Technol. 5:217-221.

BEDOLLA, S., GONZALEZ DE PALACIOS, M., ROONEY, L. W., DIEHL, K. C., and KHAN, M. N. 1983. Cooking characteristics of sorghum and corn for tortilla preparation by several cooking methods. Cereal Chem. 60:263-268.

BRESSANI, R., ELIAS, L. G., PAREDAS, A. E. A., and HUEZO, M. T. 1977. Processing of sorghum by lime-cooking for the preparation of tortillas Pages 57-58 in: Proc. ICC Symp. Sorghum and Millets for Human Food, Vienna, 1976. Tropical Products Institute, London.

CARR, W. R. 1961. Observations on the nutritive value of traditionally ground cereals in Southern Rhodesia. Br. J. Nutr. 15 :339-343.

CHEN, J. 1989. Utilization of millets in China. Pages 347-350 in: Small Millets in Global Agriculture. A. Seetharam, K. W. Riley, and G. Harinarayana, eds. Oxford and IBH Publishing Co., New Delhi.

CHEVASSUS-AGNES, S., FAVIER, J. C., and JOSEPH, A. 1976. Technologie traditionnelle et valeur nutritive des biéres de sorho au Cameroun. Cah. Nutr. Diet. 11:89-104.

CHINSMAN, B. 1984. Choice of technique in sorghum and millet milling in Africa. Pages 83-92 in: Proc. Symp. Process. Sorghum Millets: Criteria for Quality of Grains and Products for Human Food. ICC Congr. 11th, Vienna.

CHOTO, C. E. MORAD, M. M., and ROONEY,

L. W. 1985. The quality of tortillas containing whole sorghum and pearled sorghum alone and blended with yellow maize. Cereal Chem. 62:51-54.

DA, S., AKINGBALA, J. O., ROONEY, L. W., and MILLER, F. R. 1982. Evaluation of *tô* quality in a sorghum breeding program. Pages 11-23 in: Proc. Int. Symp. on Sorghum Grain Quality. L. W. Rooney and D. S. Murty, eds. Int. Crops Res. Inst. Semi-Arid Tropics (ICRISAT), Patancheru, A. P., India.

DAIBER, K. H. 1988. Malted and treated sorghum for the industrial production of beer, feed and food. Pages 15-30 in: Proc. Quad. Symp. Sorghum Millets 4th, Int. Cereal Bread Congr. 8th. ICC, Lausanne, Switzerland.

DAMON, E. G. 1962. The cultivated sorghums of Ethiopia. Exp. Stn. Bull. 6. Imperial Ethiopian College of Agriculture and Mechanical Arts, Addis Ababa.

DENDY, D. A. V., GOMEZ, M. I., HOUSE, L. R., and ROONEY, L. W. 1988. Uses of sorghum and millets. Summary Proc. Int. Workshop on Policy, Practice, and Potential Relating to Uses of Sorghum and Millets, Bulawayo, Zimbabwe. Int. Crops Res. Inst. Semi-Arid Tropics (ICRISAT), Patancheru, A. P., India.

DESIKACHAR, H. S. R. 1977. Processing of sorghum and millets for versatile food uses in India. Pages 41-45 in: Proc. ICC Symp. Sorghum and Millets for Human Food, Vienna, 1976. D.A.V. Dendy, ed. Tropical Products Institute, London.

DESIKACHAR, H. S. R., and CHANDRASEKHAR, A. 1982. Quality of sorghum for use in Indian foods. Pages 262-268 in: Proc. Int. Symp. on Sorghum Grain Quality. L. W. Rooney and D. S. Murty, eds. Int. Crops Res. Inst. Semi-Arid Tropics (ICRISAT), Patancheru, A. P., India.

DeWALT, K. 1982. Sorghum use in Honduras—A case stury. Proc. Sorghum Quality Workshop for Latin America. INTSORMIL/ICRISAT/INIA/CIMMYT, El Batan, Mexico.

EJETA, G. 1982. *Kisra* quality: Testing new sorghum varieties and hybrids. Pages 67-98 in: Proc. Int. Symp. on Sorghum Grain Quality. L. W. Rooney and D. S. Murty, eds. Int. Crops Res. Inst. Semi-Arid Tropics (ICRISAT), Patancheru, A. P., India.

EKUNDAYO, J. A. 1969. The production of *pito*—A Nigerian fermented beverage. J. Food technol. 4:217-218.

ESELE, J. P. E. 1989. Processing and utilization of finger millet in Uganda. Pages 351-353 in: Small Millets in Global Agriculture. A. Seetharam, K. W. Riley, and G. Harinarayana, eds. Oxford and IBH Publishing Co., New Delhi.

FAPARUSI, S. I. 1970. Sugar changes during the preparation of *burukutu* beer J. Sci. Food Agric. 121:70-74.

FAURE, J., and MUCHNIK, J. 1989. Mil, Mais, Sorgho—Techniques et Alimentation au Sahel. L'Harmattan, Paris.

FAVIER, J. C., CHEVASSUS-AGNES, S., JOSEPH, A., and GALLON, G. 1972. Traditional technology of sorghum in Cameroun-influence of pestling on nutritional value. (In French.) Ann. Nutr. Aliment. 26:221-250.

FREEMAN, K. C., and BROADHEAD, D. M. 1986. Sweet sorghum culture and syrup production. U.S. Dep. Agric. Agric. Handb. 611.

FUTRELL, M., McCULLOCH, E., and JONES, R. 1982. Use of sorghum as food in southern Honduras. Pages 100-102 in: Proc. Int. Symp. on Sorghum Grain Quality. L. W. Rooney and D. S. Murty, eds. Int. Crops Res. Inst. Semi-Arid Tropics (ICRISAT), Patancheru, A. P., India.

GALIBA, M., ROONEY, L. W., WANISKA, R. D., and MILLER, F. R. 1987. The preparation of sorghum and millet couscous in West Africa. Cereal Foods World 32:878-884.

GEBREKIDAN, B., and GEBREHIWOT, B. 1982. Sorghum *injera* preparations and quality parameters. Pages 55-66 in: Proc. Int. Symp. on Sorghum Grain Quality. L. W. Rooney and D. S. Murty, eds. Int. Crops Res. Inst. Semi-Arid Tropics (ICRISAT), Patancheru, A. P., India.

GRIFFON, D. 1985. Contribution a l'etude des alternatives technologiques de saccharification enzymatique des substrats amylaces tropicaux. Thesis. L. Institut National Polytechnique de Lorraine, France.

HOSENEY, R. C., VARRIANO-MARSTON, E., and DENDY, D. A. V. 1981. Sorghum and millets. Pages 71-144 in: Advances in Cereal Science and Technology, Vol. 4. Y. Pomeranz, ed. Am. Assoc. Cereal Chem., St. Paul, MN.

HULSE, J. H., LAING, E. M., and PEARSON, O. E. 1980. Sorghum and the Millets: Their Composition and Nutritive Value. Academic Press, New York.

INTERNATIONAL CROPS RESEARCH INSTITUTE FOR THE SEMI-ARID TROPICS (ICRISAT). 1987. ICRISAT West African Programs Annual Report 1986. ISC, Niamey, Niger.

IRUEGAS, A., CEJUDO, H., and GUIRAGOSSIAN, V. 1982. Screening and evaluation of tortilla from sorghum and sorghum-maize mixtures. Pages 92-99 in: Proc. Int. Symp. on Sorghum Grain Quality. L. W. Rooney and D. S. Murty, eds. Int. Crops Res. Inst. Semi-Arid Tropics (ICRISAT), Patancheru, A. P., India.

KANTE, A., COULIBALY, S., SCHEURING, J. F., and NIANGADO, O. 1984. The ease of decortication in relation to the grain characteristics of Malian pearl millets. Pages 44-47 in: Proc. ICC Symp. Processing of Sorghum and Millets: Criteria for Quality of Grains and Products for Human Food. ICC, Vienna.

LI, H. 1970. The origin of cultivated plants in south east Asia. Econ. Bot. 24:3-19.

MALLESHI, N. G. 1989. Processing of small millets for food and industrial uses. Pages 325-339 in: Small Millets in Global Agriculture. A. Seetharam, K. W. Riley and G. Harinarayana, eds. Oxford and IBH Publishing Co., New Delhi.

MALLESHI, N. G., and DESIKACHAR, H. S. R. 1985. Milling, popping and malting characteristics of some minor millets. J. Food Sci. Technol. 22 :400.

MORRIS, W. M. M. 1982. Consumer preferences and the adoption of new cultivars in Sahelian West Africa. Page 747 in: Sorghum in the Eighties, Vol. 2. L. R. House, L. K. Mughogho, and J. M. Peacock, eds. Int. Crops Res. Inst. Semi-Arid Tropics (ICRISAT), Patancheru, A. P., India.

MUKURU, S. Z. 1988. Traditional technologies in small grain processing. Page 17 in: Uses Of Sorghum and Millets. Summary Proc. Int. Workshop. D. A. V. Dendy, M. I. Gomez, L. R. House, and L. W. Rooney, eds. SADCC/ICRRISAT, Bulawayo, Zimbabwe and ICRISAT, Patancheru, A. P., India.

MUKURU, S. Z., MUSHONGA, J. N., and MURTY, D. S. 1982. Sorghum *ugali*. Pages 39-44 in: Proc. Int. Symp. on Sorghum Grain Quality. L. W. Rooney and D. S. Murty, eds. Int. Crops Res. Inst. Semi-Arid Tropics (ICRISAT), Patancheru, A. P., India.

MULLER, H. G. 1970. Traditional cereal processing in Nigeria and Ghana. Ghana J. Agric. Sci. 3:187-195.

MURTY, D. S. 1987. Sorghum breeding for the 900–1,200 mm rainfall zone of West Africa, Annu. Rep. ICRISAT/USAID/SAFGRAD Project. Int.Crops Res. Inst. Semi-Arid Tropics (ICRISAT), Ouagadougou, Burkina Faso.

MURTY, D. S., ed. 1990. Industrial utilization of sorghum. Summary Proc. Symp. Current Status and Potential of Industrial Uses of Sorghum in Nigeria. Int.Crops Res. Inst. Semi-Arid Tropics (ICRISAT), Patancheru, A. P., India.

MURTY, D. S., and HOUSE, L. R. 1980. Sorghum food quality: Its assessment and improvement. Joint Mtg. UNDP-CIMMYT-ICRISAT Policy Advisory Committee, 5th. Int. Crops Res. Inst. Semi-Arid Tropics (ICRISAT), Patancheru, A. P., India.

MURTY, D. S., and SUBRAMANIAN, V. 1982. Sorghum Roti: I. Traditional methods of consumption and standard procedures for evaluation. Pages 73-78 in: Proc. Int. Symp. Sorghum Grain Quality. L. W. Rooney and D. S. Murty, eds. Int. Crops Res. Inst. Semi-Arid Tropics (ICRISAT), Patancheru, A. P., India.

MURTY, D. S., and NICODEMUS, K. D. 1987. Interallelic relations among endosperm variants in sorghum. J. Hered. 78:391-394.

MURTY, D. S., PATIL, H. D., and HOUSE, L. R. 1982a. *Sankati* quality evaluation of sorghum cultivars. Pages 36-38 in: Proc. Int. Symp. Sorghum Grain Quality. L. W. Rooney and D. S. Murty, eds. Int. Crops Res. Inst. Semi-Arid Tropics (ICRISAT), Patancheru, A. P., India.

MURTY, D. S., PATIL, H. D., and HOUSE, L. R. 1982b. Sorghum Roti: II. Genotypic and environmental variation for roti quality parameters. In: in: Proc. Int. Symp. Sorghum Grain Quality. L. W. Rooney and D. S. Murty, eds. Int. Crops Res. Inst. Semi-Arid Tropics (ICRISAT), Patancheru, A. P., India.

MURTY, D. S., PATIL, H. D., PRASADA RAO, K. E., and HOUSE, L. R. 1982c. A note on screening the Indian sorghum collection for popping quality. J. Food Sci. Technol. 19:79-80.

MURTY, D. S., ROONEY, L. W., PATIL, H. D., and HOUSE, L. R. 1982d. Report on the International Sorghum Food Quality Trials (ISFQT). Int. Crops Res. Inst. Semi-Arid Tropics (ICRISAT), Patancheru, A. P., India.

MURTY, D. S., RAO, N. S., and SINGH, U. 1984a. Evaluation of sorghum grain quality as influenced by milling methods. Pages 24-28 in: Proc. ICC Symp. Processing of Sorghum and Millets: Criteria for Quality of Grains and Products for Human Food. ICC, Vienna.

MURTY, D. S., PATIL, H. D., and HOUSE, L. R. 1984b. Processing and cooking quality characters in Sorghum. Pages 156-170 in: Nutritional and Processing Quality of Sorghum. D. S. Salunkhe, J. K. Chavan, and S. J. Jadhav, eds. Oxford and IBH Publishing Co., New Delhi.

MURTY, D. S., PATIL, H. D., and HOUSE, L. R. 1984c. Studies on flour particle size and endosperm texture in sorghum. J. Food Sci. Technol. (India) 21:359-363.

MURTY, D. S., SINGH, M., and NICODEMUS, K. D. 1988. A genetic study of popping quality in sorghum (*Sorghum bicolor* L. Moench). Euphytica 37:5-8.

MURTY, D. S., SINGH, U., SURYAPRAKASH, S., and NICODEMUS, K. D. 1985. Soluble sugars in five endosperm types of sorghum. Cereal Chem. 62:150-152.

NGODDY, P. 1990. Sorghum milling in Nigeria: A review of industrial practice and research innovations. Page 12 in: Industrial Utilization of Sorghum: Summary Proc. Symp. Current Status and Potential of Industrial Uses of Sorghum in Nigeria. Int.Crops Res. Inst. Semi-Arid Tropics (ICRISAT), Patancheru, A. P., India.

NOUT, M. J. R. 1980. Microbiological aspects of the traditional manufacture of *busaa*, a Kenyan opaque maize beer. Chem. Mikrobiol. Technol. Lebensm. 6:137-142.

NOVELLIE, L. 1977. Beverages from sorghum and milets. Pages 73-77 in: Proc. Symp. Sorghum and Millets for Human Food. D. A. V. Dendy, ed. Tropical Products Institute, London.

NOVELLIE, L. 1982a. Fermented beverages. Pages 113-120 in: Proc. Int. Symp. Sorghum Grain Quality. L. W. Rooney and D. S. Murty, eds. Int. Crops Res. Inst. Semi-Arid Tropics (ICRISAT), Patancheru, A. P., India.

NOVELLIE, L. 1982b. Fermented porridges. Pages 121-128 in: Proc. Int. Symp. Sorghum Grain Quality. L. W. Rooney and D. S. Murty, eds. Int. Crops Res. Inst. Semi-Arid Tropics (ICRISAT), Patancheru, A. P., India.

NOVELLIE, L. 1984. Sorghum quality for malting and brewing. Pages 59-63 in: Proc. ICC Symp. Processing of Sorghum and Millets: Criteria for Quality of Grains and Products for Human Food. ICC, Vienna.

OBILANA, A. T. 1982. Traditional sorghum foods in Nigeria: Their preparation and quality parameters. Pages 45-54 in: Proc. Int. Symp. Sorghum Grain Quality. L. W. Rooney and D. S. Murty, eds, Int. Crops Res. Inst. Semi-Arid Tropics (ICRISAT), Patancheru, A. P., India.

ODUNFA, S. A. 1985. African fermented foods. Pages 155-191 in: Microbiology of fermented foods, Vol. 2. B. J. B. Wood, ed. Elsevier Applied Science Publishers, London.

ODUNFA, S. A., and ADEYELE, S. 1985. Microbiological changes during the traditional production of *ogi-baba*, a West African fermented sorghum gruel. J. Cereal Sci. 3:173-180.

OLEWNIK, M. C., HOSENEY, R. C., and VARRIANO-MARSTON, E. 1984. A procedure to produce pearl millet rotis. Cereal Chem. 61:28-33.

PORTERES, R. 1976. African cereals: *Eleusine*, fonio, black fonio, teff, *Bracharia, Paspalum, Pennisetum,* and African rice. Pages 409-452 in: Origins of African Plant Domestication, J. R. Harlan and J. M. J. de Wet, eds. Mouton Publishing, The Hague.

PRASADA RAO, K. E., and MURTY, D. S. 1982. Sorghum for special uses, Pages 129-134 in: Proc. Int. Symp. Sorghum Grain Quality. L. W. Rooney and D. S. Murty, eds. Int. Crops Res. Inst. Semi-Arid Tropics (ICRISAT), Patancheru, A. P, India.

PURSEGLOVE, J. W. 1972. Tropical Crops—Monocotyledons. Longman Group Limited, London.

PUSHPAMMA, P. 1989. Utilization of small millets in Andhra Pradesh (India). Pages 321-324 in: Small Millets in Global Agriculture. A. Seetharam, K. W. Riley, and G. Harinarayana, Oxford and IBH Publishing Co., New Delhi.

RACHIE, K. O. 1975. The Millets—Importance, Utilization and Outlook. Int. Crops Res. Inst. Semi-Arid Tropics (ICRISAT), Patancheru, A. P., India.

RACHIE, K. O., and PETERS, L. V. 1977. The Eleusines—A review of the world literature. Int. Crops Res. Inst. Semi-Arid Tropics, Patancheru, A. P., India.

RAMESHWAR, S., and AXTELL, J. D. 1973. High lysine mutant gene (*hl*) that improves protein quality and biological value of grain sorghum. Crop Sci. 13:535-539.

ROONEY, L. W., and MURTY, D. S., eds. 1982a. Proc. Int. Symp. on Sorghum Grain Quality. Int. Crops Res. Inst. Semi-Arid Tropics (ICRISAT), Patancheru, A. P., India.

ROONEY, L. W., and MURTY, D. S. 1982b. Evaluation of sorghum food quality. Pages 571-588 in: Sorghum in the Eighties. L. R. House, L. K. Mughogho, and J. M. Peacock, eds. Int. Crops Res. Inst. Semi-Arid Tropics (ICRISAT), Patancheru, A. P., India.

ROONEY, L. W., and MCDONOUGH, C. M. 1987. Food quality and consumer acceptance of pearl millet. Pages 43-61 in: Proc. Int. Pearl Millet Workshop.Int. Crops Res. Inst. Semi-Arid Tropics (ICRISAT), Patancheru, A. P., India.

ROONEY, L. W., KIRLEIS, A. W., and MURTY, D. S. 1986. Traditional foods from sorghum: Their production, evaluation and nutritional value. Pages 317-353 in: Advances in Cereal Science and Technology, Vol. 8. Y. Pomeranz, ed. Am. Assoc. Cereal Chem., St. Paul, MN.

SCHEURING, J. F., SIDIBE, S., ROONEY, L. W., and EARP, C. F. 1982a. Sorghum pericarp thickness and its relation to decortication in a wooden mortar and pestle. Cereal Chem. 60:86-89.

SCHEURING, J. F., SIDIBE, S., and KANTE, A. 1982b. Sorghum alkali *tô*: Quality considerations. Pages 24-31 in: Proc. Int. Symp. Sorghum Grain Quality. L. W. Rooney, and D. S. Murty, eds. Int. Crops Res. Inst. Semi-Arid Tropics (ICRISAT), Patancheru, A. P., India.

SCHWARTZ, H. M. 1956. Kaffircorn malting and brewing studies. I. The Kaffir beer brewing industry in South Africa. J. Sci. Food Agric. 7:101-105.

SEETHARAMA, N., PRASADA RAO, K. E., SUBRAMANIAN, V., and MURTY, D. S. 1987. Screening for sweet stalk sorghums and environmental effect on stalk sugar concentra-

CHAPTER 8

NEW MILLING TECHNOLOGIES AND PRODUCTS: WHOLE PLANT UTILIZATION BY MILLING AND SEPARATION OF THE BOTANICAL AND CHEMICAL COMPONENTS

Lars Munck
Department of Dairy and Food Science
The Royal Veterinary and Agricultural University
Frederiksberg C
Copenhagen, Denmark

SCOPE

In subsistence plant husbandry, all parts of crops like sorghum and millets were originally needed for survival and were, therefore, carefully utilized—the grain for food and beverage, and the stalk, leaves, roots, and offal for cattle feed, thatch, brooms, and crafts or for fuel. The life of the subsistence farmer was hard and implied heavy drudgery. Introduction of a monetary economy ended the isolation of the village, and now, for better or worse, it was integrated in a national and later into an international economy, leading to increased competition and specialization, breaking the closed circuits of the production chains of the village.

New technology based on exploitation of the nonrenewable fossil fuels supported this change, which still goes on in an ever accelerating pace, giving a range of options and drawbacks (Munck, 1993).

Thus, in the United States, for example, sorghum hybrids can now give yields of up to 8–10 metric tons (t) per hectare and are competitive with maize in dry areas because of their higher yield at a lower input of resources, primarily those based on fossil fuels such as fertilizers, pesticides, and machinery.

The cheap fossil fuels, however, downgraded the stalk, which in today's industrialized agriculture is uneconomic to utilize and burned as garbage in the field. On the other hand, in the arid parts of Africa, where sorghum originates, societies were unable to organize themselves to absorb modern technology to the same extent as in industrialized countries and were thus not able to burn off fossil fuels for their benefit at the same rate. On the contrary, these countries became importers of wheat and maize from the surpluses of western agriculture. Artificial pricing of grain by subsidy of currency was a disincentive to the local African farmers (see chapter 2), whose yields still lag behind those of the industrialized world at levels of about 0.2–1.5 t/ha.

Technological-physical or chemical separation of an inhomogeneous entity like seed and straw provides an important value-added input because the fractions in each case of application have an improved quality compared to the combined commodity. In Figure 1 (Hallgren et al, 1992) the fractionation of the whole sorghum plant into its botanical and chemical components for different uses is outlined setting the scope of this chapter.

Compared to other cereal crops like wheat, rice, and maize, where specialized milling technologies have been developed to produce finished products of high acceptance, the potential to industrially utilize sorghum and millets for food products in urban areas has been little developed. However, the local knowledge about suitable varieties and processes is still surviving in the households of the remote villages. Like barley, sorghum and millets carry among sophisticated people the stigma of being considered a coarse grain intended for feed (van Vuuren, 1993). This is very unfortunate because even now in Africa and India, for example, a range of fine foods and beverages is made by hand in rural areas (Rooney and Murty, 1982) (see chapter 7). With regard to industrial milling, sorghum has been much more investigated than the diverse group of small seeded millets where pearl millet has been studied more than finger millet, foxtail millet, proso millet and several other minor millet species.

Because of the pressure in the United States for self-sufficiency and resource optimization induced by the Second World War, a diversified sorghum dry and wet milling industry was built up (Hahn, 1970; Watson, 1970). When wartime short-

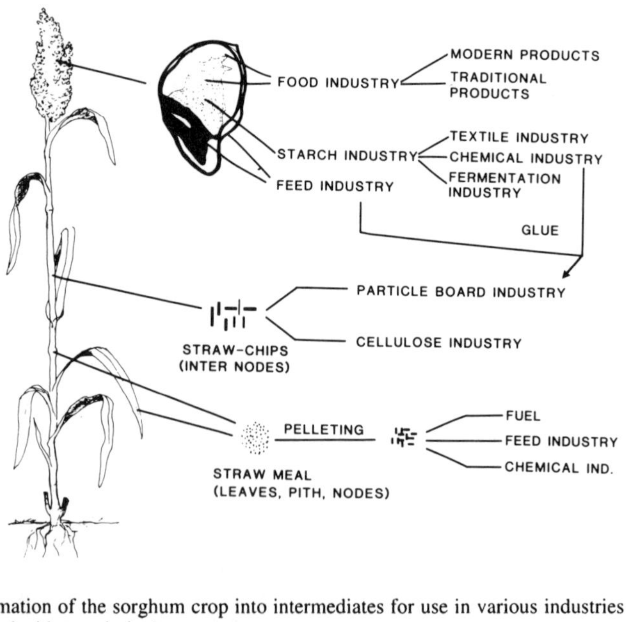

Fig. 1. Transformation of the sorghum crop into intermediates for use in various industries (Hallgren et al, 1992; reprinted with permission).

ages of maize necessitated other starch sources, waxy sorghum replaced imported cassava, and the brewing industry consumed large quantities of sorghum grits. Hahn (1970) reported six dry milling plants and Watson (1970) one wet milling industry in the United States based on sorghum. In 1970, Hahn finalized his inspiring review on sorghum dry milling in the following words about the future:

> The greatest significance of grain sorghum lies in its major components and their properties. In the future, the combined efforts of all scientific disciplines will be applied to exploit every aspect of grain sorghum. Engineers will develop new dry milling techniques for making improved separations; chemists will provide new chemical derivatives; and geneticists will produce new types of sorghum with improved milling properties and greater utility. Grain sorghum production is increasing and the supply is strong and dependable. Compared to other cereals, grain sorghum is a low cost raw material, especially for markets in the sorghum growing area. Thus, all factors seem to point to a bright and developing future for sorghum dry milling.

Today, however, sorghum milling seems to have lost the battle against maize milling in the United States. The sorghum milling alternative was obviously not so competitive as Hahn suggested. We will revert to find a possible explanation later in this chapter. As of now, the simple hammer mill processing alternative for feed manufacturing seems to be almost 100% prevalent, dominating other more sophisticated milling technologies. However, as Hahn rightly forecast, since 1970 much effort has been devoted to develop milling and separation techniques for the utilization of sorghum grains. In order to make sorghum and millets competitive with wheat and maize in urban areas in developing countries, a wide range of universities and a few companies became engaged in research aimed at producing refined food products either based on local traditional or on introduced technologies. The farmer should thus ideally be able to retain his traditional foods without drudgery and demand sorghum alternatives of new foods when he moves to the city (Dendy, 1993; Gomez and Munck, 1993). Alternatively, partially or totally sorghum-based westernized foods such as bread and biscuits should be made available. In this way a market for sorghum could be developed for the farmers who remain in the countryside, where the environment in those countries often is ideal for sorghum production but hostile to wheat and maize. It would thus for many developing countries be possible to increase a sustainable production of food grains as well as to alleviate the great burden of the drain in the national hard currency cash flow caused by importing wheat and maize. On the other hand, the present discussion as outlined in the United Nations "Brundtland" Report (1987) focuses on the necessity in the long-term perspective for a complete utilization of the renewable sorghum plant resources as indicated in Figure 1, including the stalk, both for food and nonfood purposes.

In the following, the technological options for all these possibilities will be briefly outlined, finally focusing on a discussion of their feasibility in alternative scenarios.

SORGHUM AND MILLET CROPS AS RAW MATERIALS FOR MILLING AND SEPARATION PROCESSES FOR SPECIFIC PRODUCTS

Sorghum is one of the most variable and challenging cereal species in the world, suitable for utilization in food, feed, and nonfood products. The less-studied millets are as a group even more variable. Research on sorghum and millets raw materials and products has advanced considerably throughout the last 20 years, mainly supported in the food area by the ICRISAT/INTSORMIL research program (Rooney and Murty, 1982; House et al, 1982) as presented at the historical sorghum meetings in Hyderabad, India, in 1981, and at the quadrennial sorghum and millets symposia arranged by the International Association for Cereal Science and Technology (ICC) and Natural Resources Institute (NRI) (Dendy, 1976, 1984, 1988, 1992a). The work before 1980 on sorghum and millets is documented in the extensive review of Hulse et al (1980) and by Faure and M'Benque (1991). A review from a workshop meeting in Bulawayo, Zimbabwe, in 1988 (Gomez et al, 1992) and the ICC meeting in South Africa (Taylor et al, 1993) give the most recent summaries. Today, this research is fundamental to the advancement of sorghum food processing through milling and separation technology on a semi-industrial and industrial scale, aiming at supplying traditional and new products of high quality as indicated by consumer acceptance. In marketing sorghum and millets food products in competition with products of wheat and maize, there is no room for compromises in obtaining the highest qualities. This often implies white products without specks for the consumer and an equally high product yield for economy. Sorghum is ideally a nutritious starch-rich cereal, high in available energy content comparable to wheat. However, the lysine content of the protein, which is the nutritionally limiting amino acid, is the lowest of all cereals, especially in milled flour, where the lysine-rich germ (about 3% of the grain in weight, but containing about 30% of the seed lysine) has been eliminated. The sorghum food, therefore, has to be eaten together with legumes, vegetables, and animal protein, so that the protein can be utilized nutritionally. The storage proteins in grain sorghum are highly cross-linked and hydrophobic, which affects water absorption, processing conditions, and digestibility of the milled sorghum flour. These problems are further exaggerated in high-tannin varieties containing polyphenols, which in large quantities are toxic (e.g., bird-resistant varieties) and impair digestibility.

The key raw material characters for grain milling quality are the size, form, and structure of the seed, including the development of its outer (bran) layers and the endosperm hardness. Also of great importance for quality is the distribution of the above-mentioned antinutritional factors as well as of other pigments among the botanical tissue components that are removed in the milling and separation processes. Millet species in general have a much more nutritionally favorable amino acid composition than sorghum, including a relatively high level of lysine. Due to their small grain size, milling is often more complex than with sorghum. This point will be touched on in the dehulling section of this paper.

In chapter 4 the aspects of sorghum and millet structure and chemistry crucial to milling were described. The pericarp consists of three sublayers (pericarp, mesocarp, and endocarp). The latter layer is important in decortication because it can take up moisture, and it is here that breakage points are induced.

Generally, the most simple sorghum grains to mill by both traditional techniques or by machinery, are grains with a pearly or white appearance having a thin pericarp or with a starchy mesocarp, the latter giving a white chalky appearance. These grains should lack a colored testa layer with all its polyphenols. The ideal seed for milling should also have a reasonably hard endosperm in order to get the cleanest separation of the botanical tissue components, including a high yield of endosperm in the final product. There are sorghums that display a testa with moderate levels of polyphenols like the Feteritas from Sudan as well as those from central Africa, for example, which are so high in polyphenols that they are toxic to birds. Some of these local sorghums have the spreader gene for polyphenols, which causes them to diffuse from the testa throughout the entire endosperm during grain formation and makes separation of polyphenols by dry milling impossible. The high-tannin, bird-resistant sorghum varieties are used mainly for brewing, a process which apparently has a detoxifying effect. They are also used as grains of "last resort" in droughts. As thoroughly described by Rooney and Miller (1982), the color of the pericarp layers and the condition of the testa (the latter implying pholyphenols) are strictly genetically based. Thus, the color of the pericarp is inherited separately from the testa trait. The color of the sorghum grain is thus not at all indicative for tannins, the dark color of which could be effectively masked by a starchy mesocarp and a white pericarp, for example.

In the hard, dense, vitreous endosperm, protein is tightly bound to starch, whereas in the soft, floury endosperm parts there is more of a loose binding, leaving air spaces. As judged by cutting the seed by a knife, the hardest part of the endosperm is most often situated peripherally in the endosperm, enclosing the soft

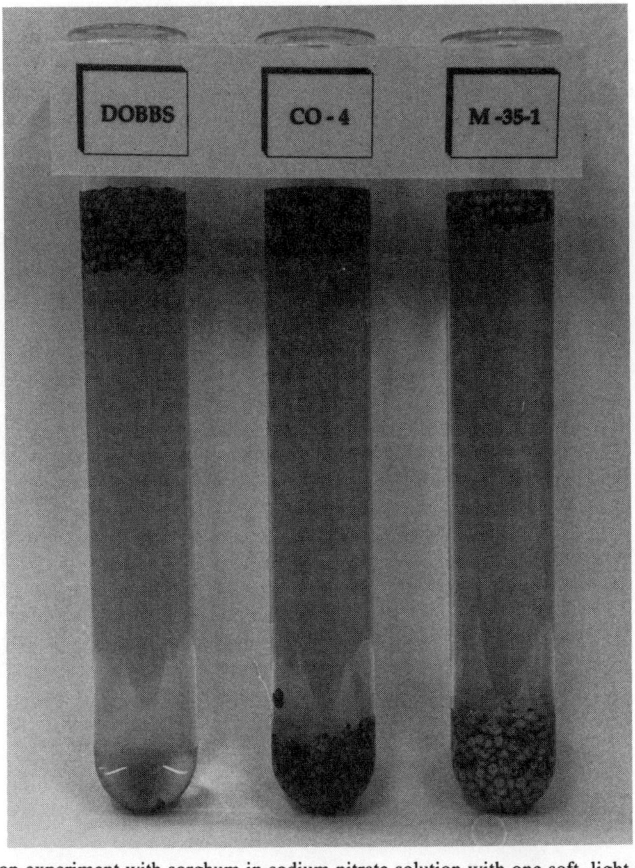

Fig. 2. Flotation experiment with sorghum in sodium nitrate solution with one soft, light type (Dobbs), one hard, heavy type (M-35-1), and one intermediate (CO-4). Method according to Hallgren and Murty (1983).

area. On a light table, hard seeds are translucent and soft seeds are opaque (dark). A convenient way to grade soft and hard seeds according to density, which is well correlated with hardness, is to float them in a solution of sodium nitrate and water solution (Fig. 2; Hallgren and Murty, 1983).

As pointed out by Doggett (1982), there is a clear relation between resistance of sorghum seeds to molds and insects and the degree of hardness of the grain where local varieties grown under tropical moist conditions need to be harder than the varieties grown in the cooler and drier highlands.

Similarly, the high-tannin trait improves resistance to birds, insects, and molds. Such varieties are most often very soft, even when found in the lowlands, because they have not been required to develop the hard grain trait when equipped with the protection given by the polyphenols.

Figure 3 (Hallgren et al, 1992) shows a range of traditional sorghum grain products relating to their dependence on hard and soft endosperm, leading in milling to high and low starch damage, respectively, influencing water absorption and rheology. Subramanian and Jambunathan (1988) observed a varietal variation (by milling in a Udy laboratory mill) in starch damage from flour finer than 150 μm as high as from 12% (soft) to 46.1% (hard). *Ugali* (Fig. 3) is a typical stiff porridge from eastern and southern Africa. *Kisra*, roti, *injera,* and tortilla are flat breads from Sudan, India, Ethiopia, and Mexico. Composite sorghum flour is white sorghum flour mixed with wheat flour to obtain a leavened raised bread. Pearled sorghum is dehulled grains, which are cooked as rice in Africa and India, while brewers grits and starch are products mainly aimed for industry. The traditional uses are further discussed in chapters 7 and 10.

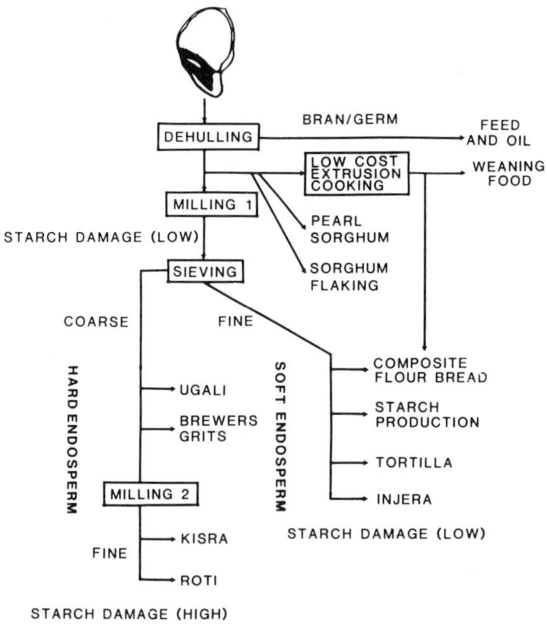

Fig. 3. Separation of sorghum hard and soft endosperm material for different purposes by dehulling, milling, and sieving (Hallgren et al, 1992; reprinted with permission).

GRAIN PREPARATION FOR MILLING

A sound, dry uniform grain, uninfected by molds and insects, of a defined variety and quality is the prerequisite for obtaining high-quality products by milling. While it is possible to obtain large sorghum deliveries in countries such as the United States, this is often impossible in developing countries where a wide range of farmers delivers small amounts of sorghum grain in sacks to the market, often showing a great variation between landraces, officially approved bred sorghums, and imported varieties. The grain contains impurities, such as broken kernels, other seeds, glumes, sticks, chaff, dirt, and stones, which must be removed by aspiration, sieving, and destoning. In some regions, stones and lumps of earth the same size and gravity as the sorghum seeds create difficulties in grain cleaning.

As indicated earlier, a reasonably hard seed is important in milling sorghum and millets; it is possible to upgrade a mixed sorghum seed lot of an otherwise acceptable quality on a density gradient table to concentrate the prevalent hard grains, provided that the resulting surplus of soft grains can be sold, e.g., for feed. As with wheat and maize, it is often advantageous to condition sorghum and millet seeds by adding water, administered in a mixing screw, followed by a resting period in a bin before milling. Conditioning toughens the outer bran parts of the seed and makes them less prone to fine disintegration in milling, facilitating separation of the coarse bran particles from the endosperm, which due to its crystalinity, is more finely milled. While in traditional industrial wheat milling about 2–5% of water is added in the conditioning step, local hand pounding of sorghum and millets often uses up to 20%. As will be discussed next, these "semi-wet" conditions may also be applied in machine milling for sorghum and millets.

TRADITIONAL MILLING TECHNIQUES COMPARED WITH MECHANICAL VILLAGE MILLS AND INDUSTRIAL MILLING PLANTS

Grain sorghum suitable for flour has a hard texture, and a considerable amount of energy and drudgery is involved in producing products that are suitable for cooking and for human consumption by hand decortication and milling. The traditional equipment is the mortar and a pestle and the stone hand mill (quern) consisting of a flat stationary stone and a moveable stone. Almost without exception, milling is done by women. Care is taken to prepare the milling products and cook them freshly every day, because keeping quality of flour is low in a tropical climate (Vogel and Graham, 1978).

The most simple type of processing is to grind the whole grain in a stone hand mill, a village stone mill, or a hammer mill driven by a diesel engine, after which the most coarse bran particles are sieved away, giving about 95% extraction of product for food use from the raw material. White-seeded sorghum lacking the testa layer and polyphenols is the preferred raw material. This technique is practiced in India for flour for baking roti and in Sudan for *kisra,* unleavened, and leavened bread, respectively (Murty and Subramanian, 1982; Ejeta, 1982). However, in rural areas of both India and Africa, home decortication in a stone or wooden mortar with a pestle is often preferred because it gives a cooked product of higher quality. Thus, an investigation in India in 1981 indicated that 78% of the households consuming sorghum in Andra Pradesh dehull daily (Pushpamma and Rao,

1981). There are no small-scale machine dehullers available in India. From a visit to Sudan in 1981, the writer noted that in the city of Khartoum most sorghum for food (as *kisra* bread) was milled whole on machine-driven stone mills followed by removal of the coarse bran particles by sieving. Principal varieties were white, low-tannin sorghums such as Dabar, but the variety Feterita displaying a testa but with a moderate polyphenol content was also used. Flour from the latter gave rise to a lightly red colored *kisra*. However, for special occasions like weddings, a hand-decorticated flour from white varieties like Dabar is still preferred to obtain a very thin and white *kisra*.

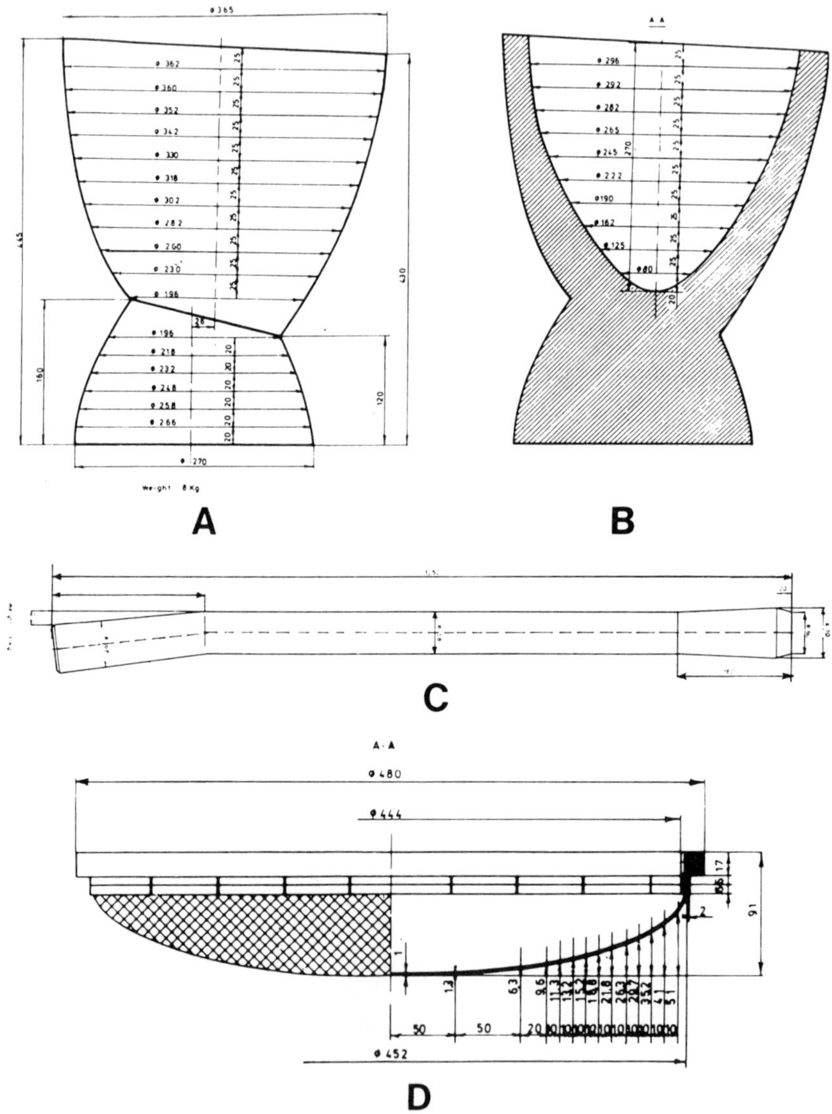

Fig. 4. Technical documentation of mortar (A, B), pestle (C), and winnowing basket (D) from the Dodoma district in Tanzania (Eggum et al, 1982; reprinted with permission).

In other countries in Africa such as Tanzania and Nigeria, sorghum is regularly dehulled in a wooden mortar with a pestle in a "semi-wet" process, followed by winnowing in a basket, and finally by hand-milling in a stone mill (further described with photographs in chapter 7). In Figure 4, typical dehulling (made of hard wood) and winnowing equipment from the Dodoma district of Tanzania is depicted with actual dimensions (Eggum et al, 1982). In Tanzania, most of the local sorghum landraces used for making grits for porridge (*ugali*) are extremely hard, with a thin pericarp and without testa, giving a pearly white impression. A range of decortication trials was performed with the cooperation of a Tanzanian housewife (Eggum et al, 1982). Grain was poured into the mortar to form a 10–15-cm deep layer. Water was then added, and the grains were pounded for 5 min. Every blow directed the pestle towards the center of the mortar, causing an attrition effect that loosened the bran in large flakes. The grains were then skillfully winnowed to remove the bran by throwing them up in the air on a basket, creating airstreams that moved the light bran to the ground while the purified, decorticated grains were falling back into the basket. Following the addition of more water, the decortication was repeated. The total water addition was estimated to be about 20% of the original grain weight. Normally two to three decortications are made to obtain a satisfactory product from the local hard (soft endosperm 3.9–13.6%) sorghum cultivars with a recovery from 72–86 % of the raw material (Table 1). The winnowed bran on the ground was picked up by hens and chickens. The decorticated grains were then milled by hand in a stone mill. The total processing time to flour amounted to about 1 hr for 2 kg of raw material, the three-step dehulling procedure taking about 30 min, including winnowing. This is much longer than in other investigations (Vogel and Graham, 1978, Pushpamma and Rao, 1981) but is explained by the extreme hardness of the local Tanzanian varieties, which are difficult to process by hand, especially in the milling step.

In plant breeding programs for developing countries in Tanzania, for example, the importance of a suitable milling quality of the new lines was not thought of at the onset. Thus, the white dwarf line Lulu D, displaying a compact panicle low in

TABLE 1
Milling Yield, Starch and Protein Contents, Agtron Reflectance of Raw Materials, and Final Yield of Nutrients from Locally Decorticated Sorghum Varieties in Tanzania and After Milling in the United Milling Systems (UMS) Sorghum Milling Process[a]

	Percent Soft Endo-sperm	No. of Decorti-cations	Milling Yield (%)	Starch (% dm)	Protein (% dm)	Yield: Starch (%)	Yield: Protein (%)	Blue Agron Reflec-tance
Five local varieties	3.9–13.6	...	...	...	...	...	...	...
Raw material	...	...	...	72.4	11.7	...	...	29–39
Hand decorticated	...	3	78	85.5	11.2	92	75	50–55
Lulu D								
Raw material	30.7	...	...	69.7–71.8	13.8–12.5	...	...	26–32
Hand decorticated	...	4	50[b]	85.4	12.2	61	44	44
Village machine mill decorticated	...	2	50[b]	80	12.3	58	45	40
UMS sorghum milling system	...	1	80	83.0	12.0	93	77	51

[a]Source: Munck et al (1982b); used by permission.
[b]Approximate figures calculated on the basis of starch in hull flour.

polyphenols, is too soft (soft endosperm 30.7%) for hand-decortication and milling (Table 1; Munck et al, 1982b). If a fine flour of comparable whiteness (measured by reflectance in an Agtron apparatus) is desired, the "modern" variety Lulu D yields in hand-pounding only 50%, while yields from local types range from 73 to 86%, being strongly negatively correlated to percentage of soft endosperm. Also, recovery of nutrients is very low for Lulu D (starch 61% and protein 44%) compared with the local varieties (on average 92% for starch and 75% for protein). It is, therefore, clear that at present, most introduced sorghums are unsuited for hand-pounding. When decorticated to a yield of 80%, which is a reasonable figure given the food situation in rural Tanzania, the flour product from Lulu D is much darker than the products made from local hard varieties, thereby paving the way for competitive maize products. In Tanzania there is no efficient machine equipment which can dehull sorghum. Even after conditioning, a local rice huller of the attrition type gave results just as bad (Table 1) as hand-pounding for the soft Lulu D sorghum, while an industrial process also involving conditioning (2 t/hr) with recovery of endosperm fragments from the hull fraction, gave a high yield and whiteness for Lulu D comparable to the products made from the local varieties (see Table 1). The latter process will be discussed later in this chapter.

In Table 2 (Eggum et al, 1982), the effect of decortication on the protein, lysine, and threonine content of the decorticated products is elucidated for five locally processed Tanzanian varieties and for a hand-dissected endosperm from a sorghum hybrid and from a winter wheat. By removing the germ by hand-pounding, the lysine level of 2.1 g/16 g of N of the whole sorghum seed drops to 1.4 g/16 g of N, almost on the level of the hand-dissected sorghum endosperm (1.2 g/16 g of N). Removing the embryo produces a much more pronounced decrease in lysine (59% remaining of the whole seed) for sorghum than for wheat (74% remaining), emphasizing the extremely low levels of lysine in sorghum endosperm. The decreases in the other essential amino acids, such as threonine, by decortication and removing the embryo are not more drastic than in wheat, and the loss of protein is much lower in sorghum (96% yield after hand-dissection) than in wheat (67%). Hand-decortication of a low-tannin, white sorghum by a skilled Tanzanian housewife (Table 3) drastically reduced fat, fiber, and ash and lowered lysine by 40% (also see chapter 5).

Table 4 presents results from rat trials with the same diets as in Table 3; the biological value of whole low-tannin white sorghum was only 55.9% due to its low

TABLE 2
Amino Acid Composition (g/16 g N) of Whole Grain, Decorticated Grain, and Hand-Dissected Endosperm of Sorghum and Wheat[a]

	Tanzanian Sorghum[b]			U.S. Hybrid Sorghum			Winter Wheat		
	Whole Grain	Decorticated Grain		Whole Grain	Hand Dissected Endosperm		Whole Grain	Hand Dissected Endosperm	
	(g)	(g)	%	(g)	(g)	%	(g)	(g)	%
Protein	10.3	9.9	96	13.5	13.1	97	13.0	8.7	67
Lysine	2.1*[c]	1.4**	67	2.1	1.2	59	3.4	2.5	74
Threonine	2.9	2.8	97	2.7	2.6	96	3.1	2.5	81

[a]Source: Eggum et al (1982); used by permission.
[b]Means of 5 local varieties (Tables 1, 2 and 3).
[c]* = Variation 2.2–1.9; ** = variation 1.5–1.4.

lysine content. In maize, wheat, and rice, the corresponding biological values are 59.8, 62.6, and 70.8%. The digestibility for protein as well as of energy is, however, in the upper range of values for cereals. As an effect of decortication of sorghum, true protein digestibility increased from 89.6 to 96.4%. For barley, wheat, and rice endosperm, the corresponding figures for protein digestibility are 99.6, 97.0, and 99.7% (Eggum et al, 1982). The biological value was reduced by decortication from 55.9% in the raw material to 47.4% in the decorticated grain, and the daily gain in weight of the rats from 1.25 g to 0.55 g (Table 4). This reduction is due to the pronounced decrease of lysine from 2.1 to 1.2% through decortication. In Figure 5, lysine consumption (mg/animal per day) is plotted against weight gain (g/animal per day) for the sorghum products presented in Table 3 together with

TABLE 3
Chemical Composition (% dry matter base) of 2K × 17/B/1 Whole Grain Sorghum, Decorticated Grain, and *Ugali* Prepared in Tanzania and in Denmark in the Rat Feeding Experiments[a]

Sample	Crude Fiber	Fat	Ash	Protein (N × 6.25)	Lysine (g/16 g N)	Starch + Sugar	Tannin
Whole grain	2.27	4.52	1.52	12.4	2.0	74.1	0.41
Decorticated grain	0.34	1.15	0.20	10.8	1.2	86.1	0.23
Tanzanian *ugali* (decorticated grain)	0.57	1.08	0.47	10.1	1.2	83.7	0.12
Laboratory *ugali* (whole grain)	2.45	4.22	1.66	12.1	1.9	70.2	0.34

[a]Source: Eggum et al (1982); used by permission.

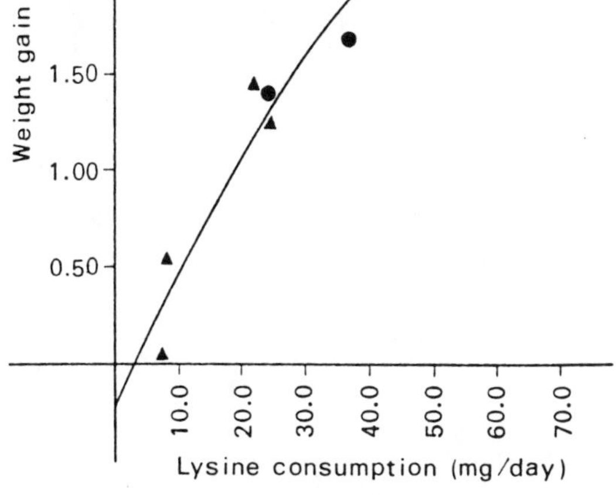

Fig. 5. Relationship between lysine consumption and weight gain of rats fed on sorghum and barley fractions (Eggum et al, 1982; reprinted with permission).

four barley milling fractions (Eggum et al, 1982). It is seen that the values for sorghum and barley fall on the same regression line. This strongly indicates that the poor weight gain for rats fed on low-tannin sorghum (Table 4) can be fully explained by the low lysine content alone.

Cooking *ugali* porridge reduced protein digestibility from 100 to 91.8% in decorticated grain and in whole grain from 94.7 to 90.0%, which is still an acceptable figure although it is in the lower end of the range in similar products from other cereals. In in vitro experiments, Axtell et al (1981) found a negative effect of cooking sorghum flour on digestibility, while in experiments with small children, MacLean et al (1982) found very low apparent retention of nitrogen as well as energy. However, when sorghum was fed in a balanced ration with legumes to preschool children (Pushpamma et al, 1979), protein digestibility as well as nitrogen retention were almost as high as for a rice-legume mixture. In conclusion (Axtell et al, 1982), low-tannin sorghum is poorly utilized as a food source because it lacks lysine, and this is especially true if the sorghum seeds are decorticated and the embryos are removed. In combination with other foods that are rich in lysine, the protein and energy from a low-tannin sorghum source will be effectively utilized to an extent comparable with other cereal products when properly processed. In milling sorghum, a high extraction rate of about 85% should be obtained, which avoids loss of the nutritious embryo while removing the pericarp and testa, which are poor in nutrients and rich in antinutritional factors. In high-tannin sorghum, the polyphenols interact with digestion and absorption of both proteins (amino acids) and carbohydrates (Axtell et al, 1982), which challenges the miller and the processor to remove the testa layer and extract or transform the polyphenols.

Traditional methods of dehulling sorghum caused losses from 32 to 38% in thiamin, 7 to 23% in riboflavin, and 8 in 16% of niacin in an Indian study (Pushpamma and Rao, 1981). However, when fermenting a sorghum dough for making *kisra* bread in Sudan (Vogel and Graham, 1979), thiamin, riboflavin, and niacin contents increased in comparison with the whole grain starting values.

There is a wide range of traditional pretreatments before traditional milling that is typical for the local area as discussed in chapter 7 and by Vogel and Graham (1979). The grains may be steamed and dried, soaked in water, alkali, or acid, and roasted or germinated. All these treatments combined with milling give rise to a wide range of fascinating food products from sorghum, which are further described in chapter 7.

TABLE 4
Feed Consumption in Rat Trials and Nutritional Quality of Whole Sorghum Grain, Decorticated Grain, and *Ugali* Prepared in Tanzania and in Denmark[a]

Diet	*n*	Feed Consumption (g/day)	Weight Gain (g/day)	TD[b] (%)	BV[b] (%)	NPU[b] (%)	DE[b] (%)
Whole grain	4	9.7	1.26	94.7	55.9	53.1	89.6
Decorticated grain	3	6.3	0.55	100.3	47.4	47.5	96.4
Tanzanian cooked *ugali*	4	6.1	0.04	91.8	49.4	45.3	94.0
Laboratory whole-grain *ugali*	4	9.8	1.40	90.0	55.7	50.2	88.0

[a]Source: Eggum et al (1982); used by permission.
[b]TD = True digestibility of protein, NPU = net protein utilization. BV = biological value of protein, DE = digestible energy.

MILLING MACHINERY

Historical Development in Sorghum-Eating Countries

The labor of milling cereal grains for food gave a great motivation for inventions that could alleviate this problem. The traditional hand-driven stone mill (quern) was first combined with the wheel in animal-, water-, and wind-driven horizontal stone mills consisting of two circular stones, one stationary and one rotating, where the grains were introduced in the center through the hollow axis. Spiral patterns engraved in the stones transported the milled material to the periphery, where it was collected for subsequent sieving and grading. These traditional mills moved quite slowly compared with the more recent diesel engine–driven versions where hand-cut stones have been replaced by artificial stones, often made from carborundum grits embedded in magnesium oxychloride or other cement. Such mills are today available in Sudan for milling sorghum for *kisra* and in India for roti.

From this basic design, vertical plate mills were developed where the milling surfaces consist of cast iron or hard metal pieces inserted to form rugged surfaces of different shapes which can at the same time mill the seeds and transport the milled material through the machine. Traditional plate mills yield a coarse product and are not much used for dry milling. They are, however, the standard mill for milling wet and fermented grains, e.g., for *ogi* in Nigeria (see chapter 7).

The machine-driven stone and plate mills function either as custom (toll) mills, where the owner of the grain brings her seed material, which she relies on, to the mill for milling against payment, either in the form of money or as a percentage of the final product. In the towns, such a business often coexists with, for example, large diesel-driven stone mills that are run as factories. They buy the raw material and sell the milled product from the factory door directly to the consumer. Also, hammer mills are used for crude milling of whole sorghum for food. They are cheap to purchase, easy to operate and run, and are extensively used in markets which are accepting hammer milled products.

In developing countries today, the milling techniques described above, together with traditional hand-milling, constitute the by far most used means of milling sorghum for food, employing mostly low-tannin white varieties. However, in attempts to improve the separation of bran, technical ideas have been transferred from milling techniques which are today well established on a large, industrial scale in other areas such as roller milling of wheat and abrasive dehulling of rice.

Equipment for Dry and Semi-Wet Milling of Sorghum and Millets

The purpose of the milling step is first to reduce the grain in a dry or semi-wet state in such a way that the different botanical components can be separated as cleanly as possible by sifting. The separated fractions such as the endosperm can then, if too coarse, be finely milled in a simple processing step without the need for a subsequent sophisticated separation process. The mechanical forces are applied either by accelerating the grain against a metal surface (impact) or by utilizing the squeezing and shearing forces of rollers or rotating disks or pressurized cylinders (attrition), where not only are there grinding effects due to metal-to-seed action but also the seed surfaces rub against each other. By introducing millstone elements or sieves (abrasion) as well as brushes, the friction forces are considerably increased, leading to special effects. Even when milling machinery is referred to as exploiting just one of these mechanical principles, in practice there are several principles working together in the milling process.

Roller Milling

TRADITIONAL ROLLER MILLING

The industrialization of wheat milling has been achieved through the roller mill principle where, after toughening the bran by water conditioning, the seed is squeezed and abraded through the shearing forces of two corrugated metal rollers of slightly different speeds (first break). Next, the flattened, large seed particles are carefully scraped off to remove the exposed endosperm by passing through a great number of corrugated and smooth rollers, each 0.75–2 m in width. The flour is sifted away in each milling step, each contributing a small percent to the total flour yield. It is natural that when sorghum milling technology on an industrial scale was introduced in the United States (reviewed by Hahn, 1970) the experience gained from roller milling wheat was applied to sorghum. Hahn (1970) concludes that "colour and speckiness of roller-milled (sorghum) flours due to incomplete bran separation limit their usefulness. Some improvement in colour is made by lowering extraction and proper tempering".

Perten (1976) considered plain roller milling unfit for sorghum processing because of the relatively high processing costs due to the elaborate milling process, low extraction rate, and quality, which led to problems with consumer acceptance as tested in Senegal, Chad, and Sudan. It was argued that even after conditioning the sorghum bran is much more brittle than wheat bran, which results in a less clean fractionation result with roller mills. On the other hand, the round sorghum grains lack the deep crease of the wheat seed, which should favor an approach opposite to that of roller milling, dehulling, where the bran and germ are removed first, and thereafter the dehulled more or less whole endosperm could be separated from the bran particles and milled separately.

SEMI-WET ROLLER MILLING

Recent studies at the Natural Resources Institute (NRI) in England (Cecil, 1986) revealed that roller milling results may be improved by prolonged steeping at high water contents under semi-wet conditions.

Table 5 (Cecil, 1986) shows the results from semi-wet milling with a Bühler laboratory mill of one low-tannin (JS 76) and one high-tannin (Serena) sorghum variety. To avoid development of toxic cyanides due to sprouting (which is peculiar to sorghum) steeping at 60–70°C for a short time (maximum 6 hr) was provided. The optimal milling results were obtained at 20 and 27% water addition. The yields of flour were in both cases 56%. JS 76 flour was soft and dense with no specks and grittiness. 99.8% passed a 125-μm screen. It had a pleasant off-white color and produced a bread with no grittiness felt in the mouth when mixed 30 to 70 with wheat. The flour from Serena displayed similar characteristics except it had a pastel pink color. The Serena flour was greatly reduced in tannins compared with the raw material (from 8.2 to 0.2 units). The roller-milled flour was compared with flour first prepared by milling pearled sorghum (65% extraction) and then processed in the roller mill without tempering. The flour obtained (50.7% yield)

TABLE 5
Semi-Wet Roller Milling at Optimal Water Addition with Low-Tannin, White Sorghum JS76 and High-Tannin Variety Serena, and Broomcorn (variety Comun)[a,b]

	JS 76[c]				Serena[d]				Comun	
	Raw Material	Flour	Middlings	Bran	Raw Material	Flour	Middlings	Bran	Raw Material[e]	Flour
Dry substance yield, % grains	...	56	30	14	...	56	17	27.0	...	48 (63)[f]
Composition, % dm										
Fat (Soxhlet)	3.2	2.6	2.9	6.2	2.4	1.2	2.6	2.4	3.6	1.0
Fiber	1.9	0.14	1.2	9.5	2.3	0.3	1.5	2.3	5.6	0.5
Tannin	Trace	Trace	Trace	0.3	1.7	0.8	1.9	1.7	8.4	0.2
Protein	14.2	9.7	23.6	12.2	13.0	10.3	24.7	13.0	14.2	10.2
Starch	74.8	81.4	70.0	59.0	69.2	80.2	62.1	69.2	...	...
Ash	1.6	1.5	1.5	2.3	2.0	1.0	2.0	2.0	2.6	1.4
Water added, %	27	...	...	...	20	...	...	...	10.4[g]	...
Total moisture in grain, %	ND[h]	...	...	...	26	...	...	...	ND	...

[a] Data from Cecil (1986) and Dada and Dendy (1987).
[b] Milling on a Bühler MLU 202 laboratory mill for wheat with six passes combined with a centrifugal sifter.
[c] Conditioning at 20 min in 0.3% sodium metabisulfite at 70°C.
[d] Conditioning 6 hr at 60°C.
[e] With glumes.
[f] Based on deglumed seeds.
[g] Moisture in grain before conditioning to 14.3%.
[h] ND = no data.

was thus darker in the Pekar test than the semi-wet milled flour. Breads and biscuits with pink to chocolate color without any specks or grittiness and with appealing taste were obtained from the semi-wet milled Serena flour. The color of these products may be objectionable in some cultures. However, in Sudan the variety Feterita which has a higher content of tannins than the above semi-wet milled Serena flour is used for *kisra*-baking when milled whole.

Obviously, the sorghum seed takes up water much more slowly than wheat, partly because of differences in bran structure and composition but also because of the hydrophobic kafirin proteins of the endosperm, which are much more slowly hydrated than the wheat gliadins and glutelins. The hydrophobic quality of the endosperm thus makes prolonged conditioning or even soaking in water of sorghum seeds possible, because the difference in water content between bran and endosperm can be maintained even under extreme conditions, securing a maximal difference in physical properties, e.g., an elastic and tough bran and a hard and crystalline endosperm. Commercial full-scale roller milling trials with semi-wet milling of brown sorghum were carried out in Zimbabwe (Cecil, 1989) and were proved to give products with improved quality compared with those from roller milling with a low amount of water conditioning. The experiments were made on a maize roller mill at a rate of 2.0 t/hr to produce grits with 70% extraction.

The NRI team also carried out trials of co-milling of four species of millet (Crabtree and Dendy, 1979) with wheat. The products were used for composite flour experiments. The bread was of satisfactory quality up to 10% millet, with 20% as the limit.

Dada and Dendy (1987) demonstrated the superiority of the semi-wet roller milling laboratory method over that of conventional rice and carborundum dehullers in a preliminary trial to obtain relatively white flours of low tannin from broomcorn sorghum suitable for making biscuits (Table 5). The great variability of the sorghum genus, discussed in chapter 3, is further demonstrated by the occurrence of broomcorn, which is mainly grown to make brooms for domestic use. The black, high-tannin seeds to which glumes adhere are sometimes used for livestock feed but frequently considered valueless and are, therefore, left in the fields. The semi-wet roller mill trials demonstrate that it is possible to obtain a flour product of reasonable quality from this previously valueless asset, which should be useful for the local producers. Further investigations regarding implementation of this technology, with the broomcorn-growing farmers in South America, for example, should be worthwhile. Gomez (1993) investigating different sorghum lines for plant breeding, found roller milling under semi-wet conditions superior to dry milling.

Perten (1984) found that when using the Bühler laboratory roller mill with a small amount of water conditioning, the yield of the flour from pearl millet was comparatively low, and the structure was coarse. The level of fat in the fine fractions was higher than that in the coarser fractions. He concluded that the outer layer of the seed must be abraded away prior to reducing the endosperm to flour. However, Cecil (1986) (Table 6) using the same semi-wet milling roller process with pearl millet as described earlier for sorghum now at 10% water addition, achieved a soft, light-colored flour (at about 50% yield) which was superior to a product obtained by an abrasion milling technique followed by hammer milling, where the flour product was both gritty and gray. Proso millet milled exceptionally well (Cecil, 1986) and gave a flour yield of 66.9% by the semi-wet roller mill process.

TABLE 6
Semi-Wet Roller Milling of Pearl Millet with 10% Water Addition to a Raw Material of 10% H_2O/ds[a,b]

	Raw Material	Middlings	Flour	Bran
Dry substance yield[c]	...	...	48.5	16
Composition, % dm				
Fat	5.8	4.7	3.0	16.4
Fiber	1.2	1.4	0.15	3.9
Protein	10.6	12.1	6.3	18.9
Starch	76.5	75.6	88.0	43.7
Ash	1.8	1.5	0.8	5.1

[a] Data from Cecil (1986).
[b] The pearl millet was bleached overnight in 7% (w/w) of 18% concentrated H_2SO_4 in water.
[c] On a 10% grain moisture basis.

The pearl millet was bleached overnight in 7%, w/w, of 18% concentrated H_2SO_4 in water.

Dendy (1992, *personal communication*) commented on the proper use of semi-wet milling as follows:

> It was observed that the high levels of water addition permit cleaning of the grain by wet separation methods, which are more efficient and much less expensive than dry separation when adequate supplies of clean water are available." However, the use of water for cleaning grain can transfer viable bacteria, yeasts, and molds from the surface of the grain through cracks and other passages into the interior. Although such transfers can occur in dry conditions, the presence of free water may vastly accelerate the movement. The absence of objectionable microorganisms (e.g., "rope"-producing *Bacillus subtilis*) on the surface cannot be assumed. Other milling systems remove a greater or lesser proportion of the external surface on which most of these microorganisms reside. Abrasive milling removes most of this surface; hammer milling and traditional milling remove very little. Avoiding recirculation of flume water would minimize the transfer of these microorganisms into the grain but would increase losses of soluble material, and would also increase the quantity of water needed and the cost of heating and/or treating it. Treating the flume water is possible (e.g., with acid or metabisulfite or both). Treatment is not necessarily very expensive, but the efficiency of the treatment, its cost, and the possibility of flavoring the products must all be considered.
>
> At high moisture levels, the respiration rate of viable seeds will increase rapidly, producing heat. Holding a large volume of damp grain overnight might lead to germination in the center of the mass and rapid growth of microorganisms unless adequate steps are taken to control such activity. As overnight tempering of millet is not essential (5 hr tempering is adequate), problems associated with it can be avoided very easily.

Klopfenstein et al (1991) showed that semi-wet milling of pearl millet at 75% extraction rate removed the goitrogenic factor, as demonstrated by patterns of serum thyroid hormones and thyroid histopathology. The goitrogenic *C*-glycosylflavones, which also cause a musty flavor (Hoseney et al, 1992), are concentrated in the bran, and it was suggested that conventional dry debranning would not remove all of this. It must be remembered that the goitrogenic factor is only important in areas where the diet is iodine-deficient.

It may be concluded that semi-wet processing with roller mills has great advantages with regard to product quality, especially in milling of high-tannin sorghum, but extraction rates of flour are lower than usually commercially acceptable in milling low-tannin sorghum and millets.

Decortication

In applying the decortication principle to sorghum and millets, rice and barley dehullers, decorticators, and polishers have been the major sources of inspiration to be used as such or as the basis for further optimized development to serve for grain sorghum (see reviews by Hulse et al, 1980; Reichert, 1982).

ATTRITION-TYPE DECORTICATOR

The operating elements in this type of dehuller are several, such as cylinders, often in the form of sieves, which can be pressurized with tightly packed seeds fed by a screw. The pressure is regulated with a valve in the outlet featuring a counter weight. Another element is rotating metal disks with grinding elements, most often one stationary and one moving, tuned for coarse milling (distance 2–4 mm). The seeds are introduced in the center and brought by centrifugal forces to the periphery where the treated seeds are collected.

The Palyi compact mill described by Reichert (1982) combines both of these principles (Fig. 6). The hulling action is provided by the two attrition plates, one

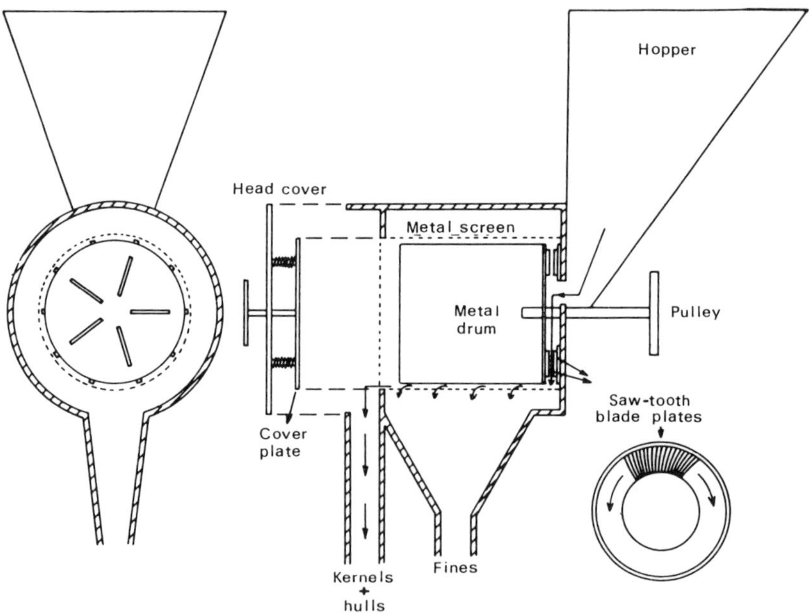

Fig. 6. Schematic diagram of the front view (left) and side view (right) of the Palyi compact mill (Reichert, 1982; reprinted with permission).

stationary and one fixed, fitted with sawtooth blades. The distance between the plates is adjustable, providing variable ratios between coarse (partly decorticated seeds) and finer particles (bran plus broken endosperm). After the partly decorticated grains leave the plates, further attrition rubbing occurs when they pass the narrow opening between a rotating drum, which is a prolongation of the outer rotating attrition plate, and the surrounding stationary cylindrical sieve, which is fixed in the outer casting. Fine particles pass through the cylindrical screen peripherally, and decorticated seeds emerge through the cross section. A cover plate with springs produces a counter pressure on the exiting decorticated seeds. The cover plate pressure can be regulated, which determines the degree of decortication together with the speed of the motor and the quality of attrition surfaces. Two machine types, DHA 400 (11 kW) and 600 (22 kW), with throughputs of about 400 and 1,400 kg/hr are produced and marketed by United Milling Systems A/S, Denmark.

In rice milling, the attrition principle is widely exploited in a range of milling machines produced by different milling equipment manufactures in Europe, Japan, and the United States. Cylindrical sieves are pressurized by screws and internal rotors such as in the Squires rice dehuller, which was used by Anderson (1969) for decortication of tempered sorghum. Another example is the United Milling System type DVA dehuller for sorghum, which is built on similar principles (Munck et al, 1982a,b). The sorghum or millet kernels (Fig. 7) are conveyed by a screw into the

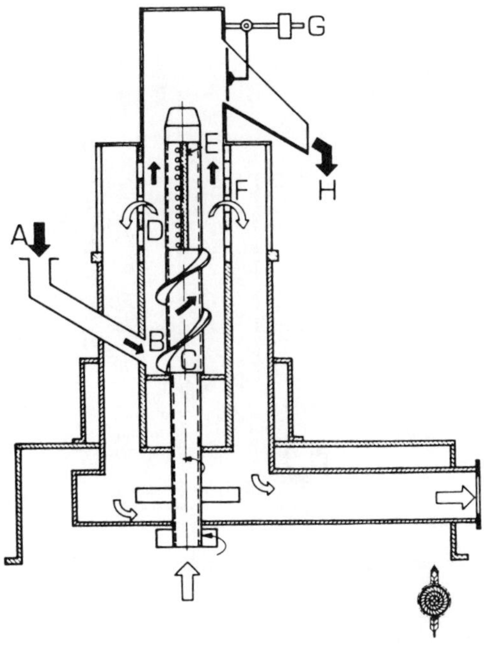

Fig. 7. Principle of the DVA sorghum decorticator (United Milling Systems A/S, Copenhagen, Denmark). Sorghum or millet kernels are conveyed by a screw (C) into the decortication chamber (D), where a steel rotor rotates the grain mass towards the cylindrical screen (F). The pressure can be controlled by the counterweight (G) in the outlet (H). (Munck et al, 1982a,b; reprinted with permission.)

decortication chamber, where a steel rotor rotates the grain mass towards the cylindrical screen. The pressure between the seeds during decortication can be controlled by the counter weight in the outlet. The hulls and the endosperm fragments from the cracked kernels are discharged through the screen by an air current at high pressure. This fraction—the screen flour—is sucked out from the bottom of the machine into a cyclone. The capacity of the machine (22 kW) is up to 2,000 kg/hr depending on decortication effect.

ABRASION-TYPE DECORTICATOR

Abrasion milling utilizes carborundum or other abrasive surfaces mounted on a vertical or horizontal rotor in order to abrade, by friction grinding, the outer layers of the grain. Also here, there are several rice whiteners that utilize this principle either in the form of vertical core decortication or as in horizontal emery pearlers, where the newest types often give the least breakage in sorghum milling (Desikachar, 1982). In barley milling, the horizontal (Bühler Miag, Uzvil, Switzerland) and the vertical (Decomatic, Bernhard Keller AG, Zurich, Switzerland and F.H. Schüle, GmbH, Hamburg, Germany) machines are marketed for industrial use. The Decomatic machinery has been thoroughly studied by the United Nations Food and Agriculture Organization (FAO) program for sorghum milling by Perten (1976, 1984). For semi-industrial and village use, the International Development Research Centre (IDRC), Canada, has developed smaller horizontal carborundum dehullers such as the PRL dehuller (Reichert, 1982). Another type of abrasive dehuller that has been tested for sorghum is the FAO Eurafric dehuller distributed by

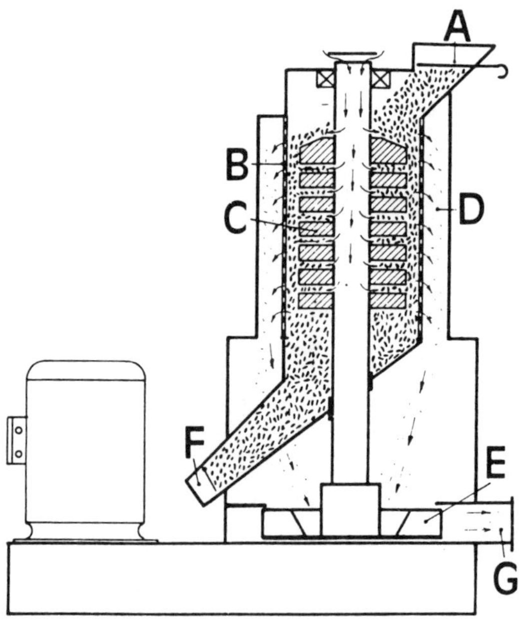

Fig. 8. Design of a vertical abrasive polishing machine. The sorghum flows by gravity from (A) into the vertical rotor surrounded by a cylindrical screen (B) and equipped with abrasive grinding stones (C). Bran and endosperm particles are removed with a pneumatic system (D). The degree of polishing is regulated by controlling the flow of the outlet (F). (Munck et al, 1982a,b; reprinted with permission.)

Société Comia-FAO SA, Vitre, France. In Figure 8, a typical vertical barley dehuller similar to the Decomatic is shown.

The sorghum flows by gravity from above into the vertical rotor surrounded by a cylindrical screen and equipped with segments of abrasive grinding stones. While the rotor runs with a peripheral speed of 18–25 m/sec, the kernels are abraded against the rough surface of the carborundum stones, against the peripheral screen, and against each other. Bran and endosperm particles are removed with a pneumatic system and collected in a cyclone. The degree of polishing can only be regulated by controlling the flow of the outlet at the bottom of the machine. Thus, the time during which the kernels stay in the polisher can be regulated. Cracking of seeds should be avoided in order not to lose endosperm to the bran fraction. The Decomatic vertical dehuller (15 kW) has a capacity of about 500–1,000 kg/hr of grain depending on the grade of decortication.

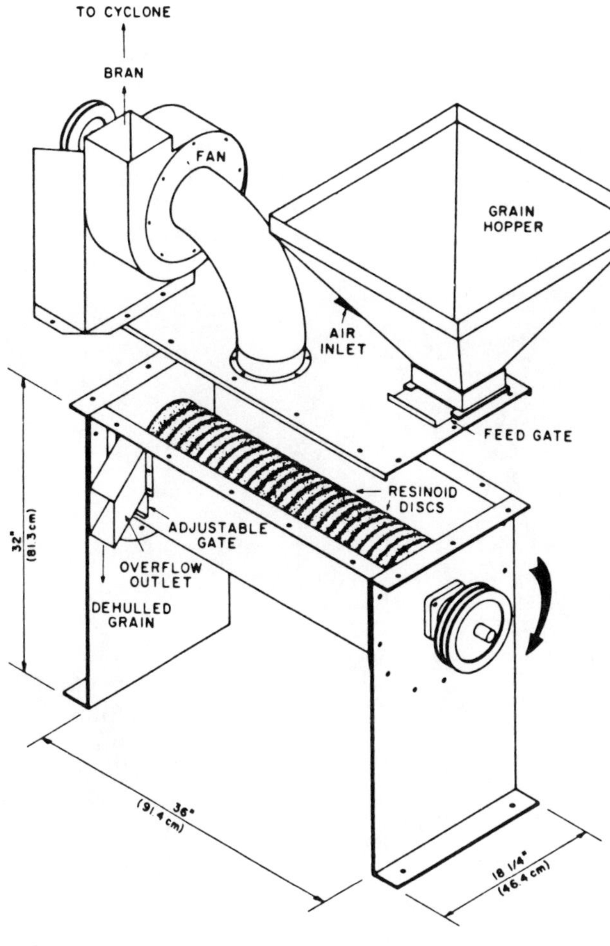

Fig. 9. Diagram of the PRL dehuller (Reichert, 1982; reprinted with permission).

The PRL dehuller used in the IDRC village development project in Africa (Bassey and Schmidt, 1989) is described by Reichert (1982; and Fig. 9). The abrasive action is provided by 13 carborundum stones (30.5 cm in diameter × 3.2 cm thick) or 27 resinoid discs (30.5 in diameter × 0.64 cm thick) mounted on a horizontal axis. When the feed gate is opened, grain flows from the hopper into the body of the machine. Fines generated by the action of the abrasive surface are taken off by the fan and bagged at a cyclone. Dehulled grains leave the machine via an overflow outlet and the throughput is adjusted by regulating the feed gate and the adjustable gate on the overflow outlet. Sorghum throughputs were about 300 and 550 kg/hr at extraction rates of 70 and 85%, respectively, when a stone speed of 1,050 rpm was used. Using a softer sorghum, throughputs in the order of 900 kg/hr were obtained with a stone speed of 1,300 rpm and an extraction rate of 77% (6 kW). The PRL dehuller was modified into either a continuous or batch operation design. A 7-kg batch-driven mini-dehuller was designed on the same principle, working in dehulling periods of up to 5 min (2.2 kW). The machine has proved very successful in Botswana (Eastman, 1980).

The main difference between the attrition and the abrasion decorticators is that bran and embryos in the first type of machine mainly come out as more or less coarse bran flour, whereas the abrasion hullers produce, by carborundum friction, a much more homogenous fine bran flour, also containing the embryos milled down. While in attrition decortication it is highly recommended to condition the seeds by water treatment prior to milling, this is often not feasible in abrasion milling (Perten, 1976), first because water tends to clog the fine flour and block the outlet in the cylindrical sieve arrangements, depressing capacity, and second because the friction of the grinding stones diminishes with water addition also conflicting with

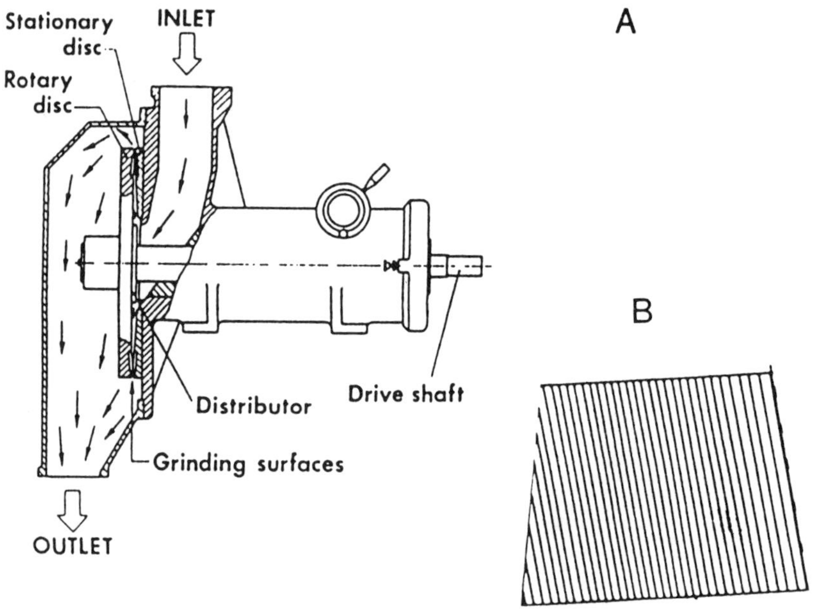

Fig. 10. Cross section of the UMS disk mill MHA 600 (A) with a hard metal disk section (B). (Andersen, 1987; Gram et al, 1989).

decortication rate in relation to capacity. Finally, many grinding stones are made by pressing stone material together with adhesive material that is dissolved by water.

PLATE (DISK) MILLS

The above-mentioned plate mill techniques can give reasonable results with carefully conditioned sorghum grain with the plates set rather close (e.g., 0.5 mm). After sieving in one step, a coarse bran with a fine color and of fair quality can be obtained as demonstrated by Desikachar (1982). At Carlsberg (1978–1988) the milling surface of the UMS MHA 600 disk mill was further developed from saw blades to radial corrugated milling surfaces made of hard metal (Fig. 10; Andersen, 1987; Gram et al, 1988). In light of the experiences with semi-wet conditioning of Desikachar and Cecil, the modified UMS MHA 600 mill should be further tested with regard to semi-wet disk milling of sorghum.

LABORATORY MILLING

To screen for milling quality in sorghum, a wide range of laboratory analyses (Subramanian and Jambunathan, 1992) and equipment has been developed, which in small scale (samples from 0.5 g to 2 kg) aims at simulating the decortication and milling process for testing, e.g., in plant breeding. Rooney and Sullins (1969) used a Strong Scott barley pearler and adapted it with a wire brush instead of the disks to dehull 100–200 g samples batchwise. The bran flour was separated from the decorticated seeds by sieving, followed by a registration of the weights of the fractions. A Satake grain-testing mill based on the same principle as the barley pearler was used by Rooney et al (1972) for sorghum. Malleshi and Desikachar (1985) used a combination of a Satake sheller for dehusking and a McGill rice polisher for debranning of foxtail, proso, barnyard, little, and kodo millets, reaching yields of polished grains of 69–77, 67–69, 64–68, 75, and 64%, respectively. Most millets except finger millet can be decorticated with equipment designed for rice or processed in roller mills (Malleshi, 1988b). These methods were rather slow due to the batch process and to the relatively large amount of samples to be tested. Shepherd (1979) modified the Udy cyclone mill, which is based on abrasion, to provide a gentle action and combined it with a staining procedure specific for pericarp, testa, and aleurone to study decortication on small sorghum and pearl millet samples (de Francisco et al, 1982). The differences in decortication rate were largely related to endosperm softness. Oomah et al (1981) and Reichert et al (1982) (Fig. 11) developed an abrasion technique to remove successive layers from sorghum and other grains by a multisample tangential abrasive dehulling device (TADD).

A circular carborundum stone is mounted horizontally on the shaft of an electric motor. A dehulling head plate holds in the periphery six stainless steel cups with open bottoms for 5-g seed samples. While running, the cups are covered with a cover plate of rubber. The motor is run at 1,725 rpm for 1–4 min as indicated by a timer. Fines are collected in a bag by means of a built-in fan. The dehulled samples are removed with a vacuum sample collector. Results are presented as percent kernels removed and related to retention time, flour color after milling, and other attributes. The reproducibility is high, and about 40 test samples can be milled per hour. Subramanian et al (1988) compared dehulled grain recovered in sorghum material milled by the traditional method (pounding) with grains treated with the Scott barley pearler and the TADD abrasion equipment. Correlation coefficients were +0.78 ($P < 0.05$) and +0.89 ($P < 0.01$), respectively. Seed density measured as percent floaters was significantly negatively correlated to recovered grain with all three milling methods.

Large laboratory-pilot mills are available, like the Schüle LUSM carborundum dehuller, that are copies of the full-scale machine Decomatic mill (Munck et al, 1982a). In both abrasion and attrition dehullers, the seeds are highly pressurized, and the friction between the grains is an important feature in the dehulling work. Thus seed size is highly positively correlated to milling capacity in the Schüle LUSM carborundum dehuller ($r = +0.75$ [$P < 0.001$]; Hallgren et al, 1988), reflecting the increasing seed-to-seed contact surface when seed size is diminished. In the previously described laboratory dehullers—the Strong Scott pearler, the modified Udy cyclone mill by Shepherd, and the TADD abrasion mill—the friction forces between the seeds are minimized, and the seeds are dehulled in a gentle way, especially for the two latter mills, a feature that is untypical for commercial dehullers. Hallgren et al (1988) tested the Kett micro scale dehulling unit, which resembles a coffee grinder (Kett Electronics Laboratories, Tokyo, Japan) (Fig. 12).

The dehulling chamber is equipped with a stationary metal file and a metal wall as well as a rugged, circular agitating file, which rotates at high speed. Sample size may vary between 3 and 5 g and dehulling time between 0.5 and 5 min. The ex-

Fig. 11. Component parts of the TADD II laboratory mill illustrating the Carborundum (a), dehulling headplate (b), sample cups (c), cover plate (d), timer (e), fines collection bag (f), and fan (g). (Reichert, 1982; reprinted with permission.)

traction rate was determined from the weight of the starting material and the polished material collected in the outgoing sample holder.

The Kett micro scale dehuller, which was first used for sorghum decortication by Wills and Ali (1983), gives the seeds a much rougher treatment than the modified Udy cyclone mill and the TADD abrasion dehuller and is probably in this respect more comparable with large-scale industrial decorticators such as the Decomatic abrasion and the UMS DVA attrition machine. Hallgren et al (1988) compared the laboratory Kett and the industrial UMS DVA machine with the Schüle LUSM pilot decorticator using sorghum material of 15 varieties from India,

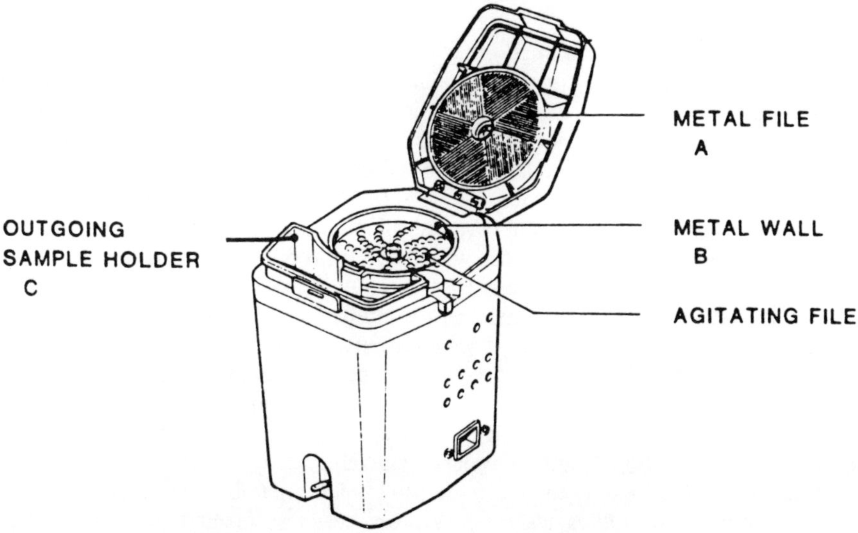

Fig. 12. Principle of the microscale dehuller from Kett Electronic Laboratories (Hallgren et al, 1988; reprinted with permission).

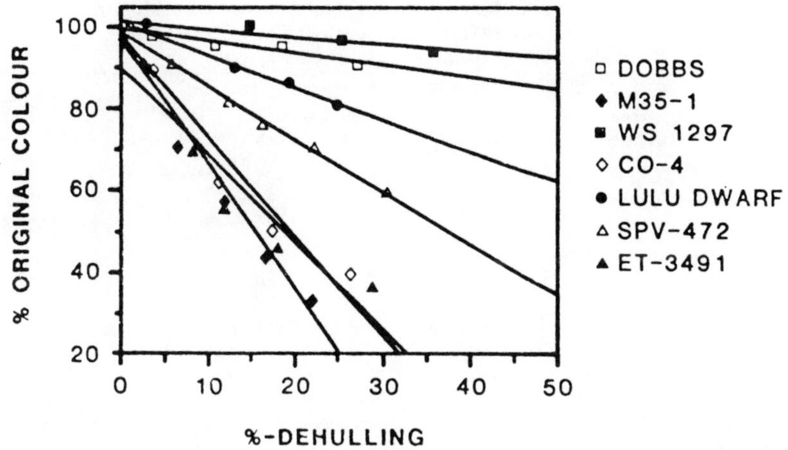

Fig. 13. Curves obtained by dehulling seven sorghum cultivars in the Kett microscale dehuller. Points with the same symbol represent different dehulling times for the same cultivar (see also Table 7). (Hallgren et al, 1988; reprinted with permission).

Sudan, and China. In sorghum decortication, color reduction is of primary importance for consumer acceptance. In the above cited experiment, the different sorghum varieties were, therefore, milled to varying decortication rates, which were related to the decline in color (Agtron green wet value was measured on Udy milled samples of the decorticated seeds) as a percentage of the color of the undecorticated sample (i.e., rise in the Agtron value). From each mill and sorghum variety, the regression lines of color versus extraction rate were used as indices of dehulling as shown in Figure 13 for the Kett micro scale dehuller. The correlation matrix of the results obtained from the three dehullers (dehulling indices) and from the seed hardness and density parameters of the sorghum raw materials is displayed in Table 7.

Table 7 surprisingly shows that the endosperm texture, grinding work, and flour particle size <250 µm are less well correlated with the color index of dehulling for the different decorticators than the density parameters: grams per milliliter of seeds, percent seed floaters, and weight/volume of dry seeds. There are only small differences in dehulling index due to the kind of decorticator, the results of which are highly correlated to each other, $r = 0.91–0.93$. It can be concluded that the inexpensive, simple, and convenient laboratory Kett micro dehuller gives representative decortication results comparable with those from pilot and full-scale industrial dehullers and may, therefore, be used in screening work in plant breeding and for the acquisition of sorghum raw material for industrial sorghum millers. It must be emphasized that there must be a correlation between the laboratory machine and commercial equipment. *For prescreening, simple seed density parameters would be preferred before measurements of endosperm texture, milling energy, and coarseness of milled flour.* Obviously, half-cut sorghum seeds, where the endosperm texture (hard versus soft endosperm) is evaluated by scoring or by the image analyzer, gives a safe guide to quality only *within* comparable seed lots and varieties. In unpublished studies with the Wickers hardness meter (for function, see

TABLE 7
Partial Correlation Matrix for Dehulling Indices with a Laboratory Kett, a Pilot (Schule LUSM), and an Industrial UMS DVA Dehuller and for Hardness-Related Parameters[a]

Dehuller	Endosperm Texture[b]	Density[c]	Floaters in $NaNO_3$[d]	Weight Volume[e]	Grinding Work[f]	Flour[g] (<250 μm)	Dehulling Index[h] Kett	Schule
Kett	−0.49 ns[i]	−0.70**	0.65**	−0.81***	−0.53*	0.62*	1	
Schule LUSM	−0.43 ns	−0.71**	0.66**	−0.70**	−0.50 ns	0.47 ns	0.91***	1
UMS DVA	−0.54*	−0.69**	0.56*	−0.70**	−0.57*	0.57*	0.93***	0.93***

[a]Source: Hallgren et al (1988); used by permission.
[b]Endosperm texture (percent vitreousness of 20 half seeds analyzed under microscope by an image analyzer).
[c]Density (g/ml, by pouring 6 g of kernels into a burette containing ethanol).
[d]Seed floaters (%, in $NaNo_3$ (1,275 g/ml).
[e]Weight/volume of dry seeds (hectoliter weight).
[f]Grinding work by Brabender hardness tester.
[g]Effect of grinding (Flour particle size <250 μm sieved by an Alpine laboratory air-jet sifter from the flour ground in the Brabender hardness tester).
[h]Dehulling index based on green (wet) Agtron color measurements; decorticated product versus raw material.
[i]* = $P < 0.05$, ** = $P < 0.01$, *** = $P < 0.001$, ns = not significant.

Munck et al, 1982a) for evaluation of the specific tissues in a microscope, we found quite a wide range between different varieties in hardness, measuring within the soft as well as within the hard-looking endosperm areas, although there was no overlapping. The density parameters take these variations into account in measuring a total integrated density of the whole seed.

DRY AND SEMI-WET GRAIN MILLING PROCESSES AND PRODUCTS

For a commercially viable process, there are several factors to be considered: a high yield of premium product is of paramount importance. Thus, in a wheat mill, a few percent more of prime product will pay all the running costs of the mill. Other important factors are the availability of raw material and the potential production, as well as by-product markets and product market requirements, which will be discussed at the end of this chapter. Secondly, in contemplating the choices of technology, several scenarios may be discussed, for example: 1) relying on an existing wheat or maize milling factory based on rollers and modifying it to produce sorghum grits and flour; 2) building up a new sorghum mill for a developing country emphasizing flexibility in choice of raw material and final products, simplicity in operation and maintenance and low operation costs including cheap spare parts; or 3) establishing a modern industrial large-scale (sorghum) milling plant in a technically advanced environment for a specific purpose based on a sound business concept that can be financed with optimal capital.

There is now a surplus of milling capacity both in industrialized as well as in developing countries. This could be used for processing sorghum. If a red sorghum hybrid (semisoft, no testa) is used for raw material in a wheat roller mill with normal water conditioning (3–5%) to obtain brewer's grits and a fine sorghum flour (to be used for biscuits and composite flour for breadmaking), there is a serious speck problem even at a low extraction rate (below 70%). This could be partly solved by milling super white, low-tannin varieties. Adding an abrasive carborundum decorticator that works under dry conditions, taking off 5–10% before the conditioning step and the first roller break, may only partly relieve the situation, as would the adding of purifiers to clean the grits. It would then be more interesting to try to scale-up the pilot mill results and the single full-scale trial of Cecil (1986, 1989) of semi-wet milling involving the acquisition of more effective conditioners and steeping and resting tanks as well as, in certain climates, a bran dryer.

Another alternative to be evaluated is to include a disk mill (e.g., a modified UMS MHA 600 with hard metal milling surfaces, and/or apply a decorticator based on attrition, e.g., the UMS DVA) working on semi-wet conditioned sorghum. The decorticated seeds and endosperm chunks recovered from the hull flour by sifting and aspirating would then be further milled dry in the roller process.

In discussing the second alternative to erecting a new sorghum or millet mill that could function in an environment with a low level of technical awareness, the improved UMS MHA disk mill should be evaluated. In a semi-wet process, this machine would break the sorghum berry into large bran flakes and endosperm pieces, which could be separated by aspiration and sifting. The milling surfaces of this machine are based on easily exchangeable hard metal sections, which are easier and less costly to maintain than a roller mill or an attrition dehuller (e.g., UMS DVA) where the metal sieves wear down and must be exchanged more often. The coarse endosperm particles would in this process be reduced to grits and flour by one to two steps of disk grinders of similar construction. Each step would involve grading by sifting. A purifier that could remove specks could be introduced for grits cleaning, probably making the process more difficult to manage.

In approaching the third alternative, the large industrial sorghum mill, a wide range of alternatives is available, but only one will be discussed here as an example of the principles involved, displayed in Figure 3, where hard and soft endosperm is used for different purposes. The decortication experiment (Munck et al, 1982a) applied to sorghum and millet (outlined in Fig. 14) involved moderate conditioning (H_2O levels approximately 5%) with the UMS DVA decorticator (Fig. 7) with three sorghum varieties, one red U.S. hybrid (Dekalb) without testa and two high-tannin sorghums with testa, namely Argentine and IS 4225.

After a thorough cleaning, destoning and tempering, the sorghum (position A1, Fig. 14) is fed through the UMS decorticator, which produces two products, decorticated seeds and screen flour (A2), containing endosperm chunks, bran flakes, and more or less whole embryos. The fine hulls (A3) are separated with a centrifugal sifter from the screen flour (A2), and the overflow (B1) is separated on an air sifter (B1) into coarse hulls (B2) and endosperm fragments (B3), which are joined with the decorticated seeds (C1); these are fed to the milling and sifting sections, consisting of two or three passages in disk mills producing grits (C2 40–65%) and fine flour (C3 20–85%). The UMS DVA decorticator produces, if the raw material is

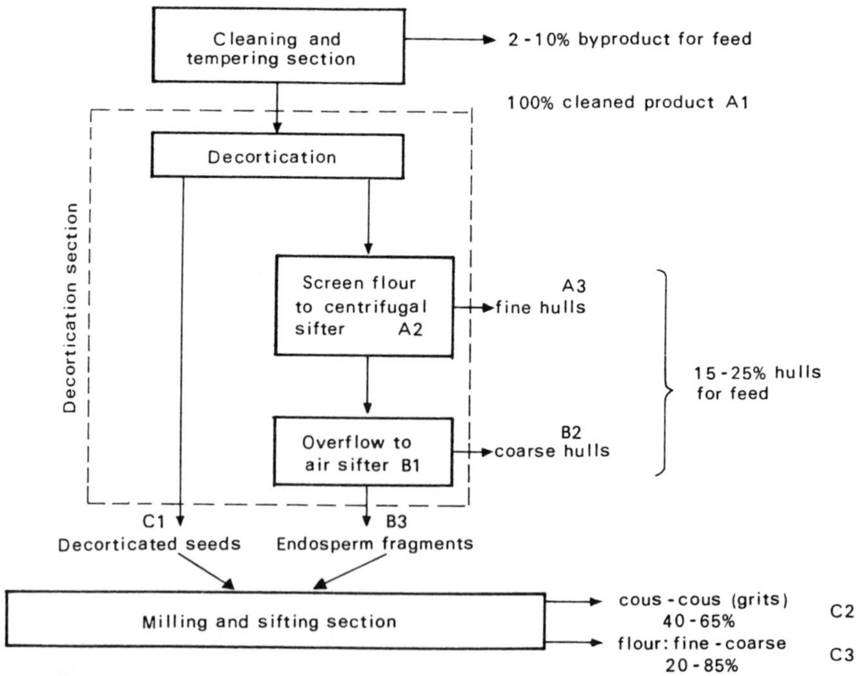

Fig. 14. Principle of a compact industrial mill for sorghum with a capacity of 2 t/hr (United Milling Systems A/S, Copenhagen, Denmark) (Munck et al, 1982a,b; reprinted with permission).

properly conditioned, more or less intact sorghum endosperms with the embryos removed (Fig. 15).

The embryos are included in the bran fraction. Seeds decorticated with abrasive stone mills have a rougher surface and thus appear whiter than seeds from the UMS DVA decorticator, which still have intact smooth endosperm surfaces. Obviously, the rough surface indicates losses of fine particles of endosperm. Breakage of kernels does not affect yield in the milling system employing the UMS DVA decorticator (Fig. 14) as much as in the abrasive polishing technique. The losses are regained (B3 in Fig. 14) by sifting from the screen flour (A2) as long as the endosperm particles are kept sufficiently coarse by the correct choice of hole size of the cylindrical sieve surrounding the decortication chamber of the DVA dehuller (Fig. 7).

In Table 8, the results from the decortication of the three sorghum varieties previously mentioned are presented. The difficulty of decorticating the soft high-tannin Argentine variety is clearly seen. In spite of low extraction (52%), the Agtron reflectance shows extremely low values equal to high color with the blue and the red filters. Ash is, however, surprisingly low (0.53%) indicating that ash should not be used as a criterion of quality in sorghum milling as it is in wheat milling. However, as also shown in Table 8, it is possible to obtain an acceptable flour from the other high-tannin variety, IS 4225, at a yield of 71.4%. The content of fiber, fat, and ash in the flour product from IS 4225 is less than half that of the same variety milled with a Schüle LUSM laboratory vertical carborundum dehuller at a yield of 75% as discussed by Munck et al (1982a).

For comparison, a typical low-tannin U.S. hybrid (Dekalb 3P3866US) with red pericarp is included in the experiment with the Argentine and IS 4225 high-tannin

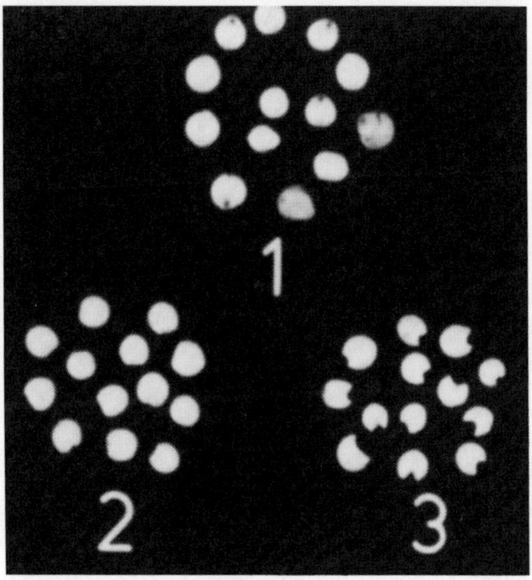

Fig. 15. Sorghum grains (variety Dabar, 1) dehulled on an abrasive decorticator (Fig. 8, 2) and by an attrition dehuller (Fig. 7, 3). Note the removed embryos in the decorticated sample 3 (Munck et al, 1982b; reprinted with permission).

varieties (Table 8). Milling trials with the red hybrid display a total milling yield of 74.7, with no significant losses of broken endosperm in the screen flour (B3). The milled, decorticated products are less colored than those of IS 4225, which, however, are much nearer to the color score of the hybrid variety than the Argentine. Fiber and fat content of the flour is much lower in the hybrid than for either the IS 4225 or the Argentine varieties, while there are less differences in ash between the flours of all three varieties. It is concluded that the described process (Munck et al, 1982a), employing a conditioning system, gives higher yields and better products than conventional roller milling systems. It would, however, be interesting to compare the UMS DVA decortication system with an optimized, semi-wet conditioning procedure for roller milling on a full industrial scale, including both chemical and color analyses based on the technique described by Cecil (1986, 1989). The modified UMS MHA 600 disk mill run under semi-wet conditions should also be included in the comparison.

With regard to milling pearl millet, the process (Fig. 14) in which the grain is conditioned with up to 10% water addition seems to give advantages over conventional carborundum dehullers. To a large extent, the small pearl millet seeds pass through the DVA decorticator sieve but are readily recovered by sifting and aspiration from the screen flour (B3) (Table 8). When comparing the milling results from pearl millet with the DVA dehuller (Table 8) with Cecil's (1986) roller mill trial (Table 6), it is seen that the DVA dehuller gives a higher yield (82 vs. 48.5%) and lower fat (2.2 vs. 3.0%), while Cecil's apparently gives lower fiber (0.15 vs. 1.0%). The levels of ash are similar in the two experiments. It would be interesting to optimize the two milling methods under industrial conditions and compare them using the same raw material.

To summarize, one can conclude that sorghum dry milling on an industrial scale in the future will use systems that start with a high level of conditioning, time, and temperature; this is supported by centuries of experience of local housewives in sorghum-consuming countries. The sorghum to be processed must be carefully cleaned and destoned. If sorghum hardness is not sufficient to gain a reasonable output of prime sorghum products, the seeds could be graded for density on a

TABLE 8
Decortication of Three Sorghum Varieties and One Senegalese Pearl Millet by the Process Described in Figure 14, Decorticated in the UMS DVA Dehuller (C1) and Endosperm Recovered from the Screen Flour (B3)

					Color C1 + B3 (% dm)							
					Agtron Reflectance							
			Percent Yield		Raw Material	C1 + B3						
Sample	Percent Endo-sperm	Total Product	C1	B3	(blue filter)	(blue filter)	(red filter)	Starch	Fiber	Fat	Ash	Protein
Sorghum[a]												
Dekalb	34.7	74.7	74.7	...	25.0	56.4	75.5	86.5	0.5	0.6	0.53	12.1
Argentine	43.4	52.5	36.0	16.5	18.5	27.0	34.5	83.2	1.5	1.1	0.53	11.7
IS 4225	38.7	71.4	51.0	20.0	17.4	46.6	64.0	81.6	0.9	1.1	0.69	11.5
Pearl millet[b]	...	82	56	26	...	...	...	82	1.0	2.2	0.73	...

[a] Data from Munck et al (1982a,b).
[b] Munck (*unpublished data*).

gravity table. This procedure would only be profitable if the soft grains could be sold as feed or for other purposes at a reasonable price. The future industrial sorghum mill will depend on how to process the total output, including grits from hard endosperm, flour from soft endosperm, bran, and germ, by finding the most profitable combination of product applications that will justify the commercial advantages of each fraction (Fig. 3). The decorticating step is here of paramount importance and in the future may be transformed into a breaking-up step similar to that of dry maize milling by optimizing the previously described improved disk mills. It seems less practical because of high equipment and energy costs to improve flour technology with an air-classification step that would separate starch and protein the dry way.

WET MILLING OF SORGHUM

Industrial wet milling of sorghum grains for starch production was developed in the United States during the Second World War as an alternative to maize starch production (Watson, 1970). In 1948, Corn Products Co. built a modern sorghum wet-milling plant in Texas (20,000 bu of grain per day) which ran until the 1970s on an enlarged capacity as the only one in United States. In 1966, sorghum contributed about 3.5% to the total sorghum-maize starch production of about 850,000 t/year (Watson, 1970). In 1975, the Texas wet-milling factory closed down because sorghum prices increased to the same level as maize (Rooney and Serna-Saldivar, 1991). In 1987, Caransa and Bakker reported the establishment of a new sorghum starch plant in Sudan (capacity 150 t/day). Watson's article on sorghum wet milling (Watson, 1970) is the major source of information in this review. Watson's contribution is still outstanding in the field. His data are mainly built on pilot plant trials rather than full-scale process information because of industrial secrecy (a more recent review is given in Watson, 1984).

Wet milling of sorghum greatly resembles maize wet milling, and the two starches cannot be microscopically differentiated from each other, even if the mean starch granule diameter is a little higher in sorghum (about 16 μm) than maize (10 μm). Gelatinization temperature as judged by the birefringence endpoint in the polarization microscope is about 2°C higher for sorghum than maize, ranging in normal sorghum starch from 64 to 74°C, where 98% of all granules have lost birefringence. Gelatinization temperature for waxy starch is 66–76°C. Sorghum starches, as maize starches, are highly purified and contain only 0.3–0.4% proteins and 0.65% of fatty acids (waxy starch only 0.1–0.2%). Sorghum and maize starch are wet milled by steeping the whole grain, whereas wheat starch is produced by wetting dry-milled wheat flour. The hard endosperms of maize, and especially sorghum, are extremely resistant to water penetration, and steeping time of the whole seeds is therefore prolonged (one to two days) under relatively high temperatures and combined with lactic acid fermentation, which loosens the hydrophobic matrix proteins surrounding the starch granules. In addition, sulfur dioxide–treated water (0.1–0.16%) is introduced at the end of the steeping process to reduce the highly cross-linked kafirin proteins and to allow complete water saturation of starch. Strict consumer laws in many countries now exclude sulfur dioxide–treated products from many food applications, thus favoring wet-milling techniques like that for wheat gluten, which does not need SO_2. As indicated in Figure 2, the soft parts of the sorghum grain may be isolated from the hard parts by dry milling, thereby obtaining a raw material for sorghum starch manufacture that should have similar properties as wheat flours with regard to the ease by which the starch granules can take up water and be freed from the protein matrix.

The elaborate sorghum wet milling process is outlined in Figure 16. After the steeping process (step 3 in Fig. 16), the solids (7.0–7.5% of dry grain substance) from the steep water (3B) are recovered in the vacuum evaporators (3D). They are then dried and mixed with sorghum gluten (8C) and sold as sorghum gluten meal, a feed of comparable quality to the corresponding maize wet-milling product. The

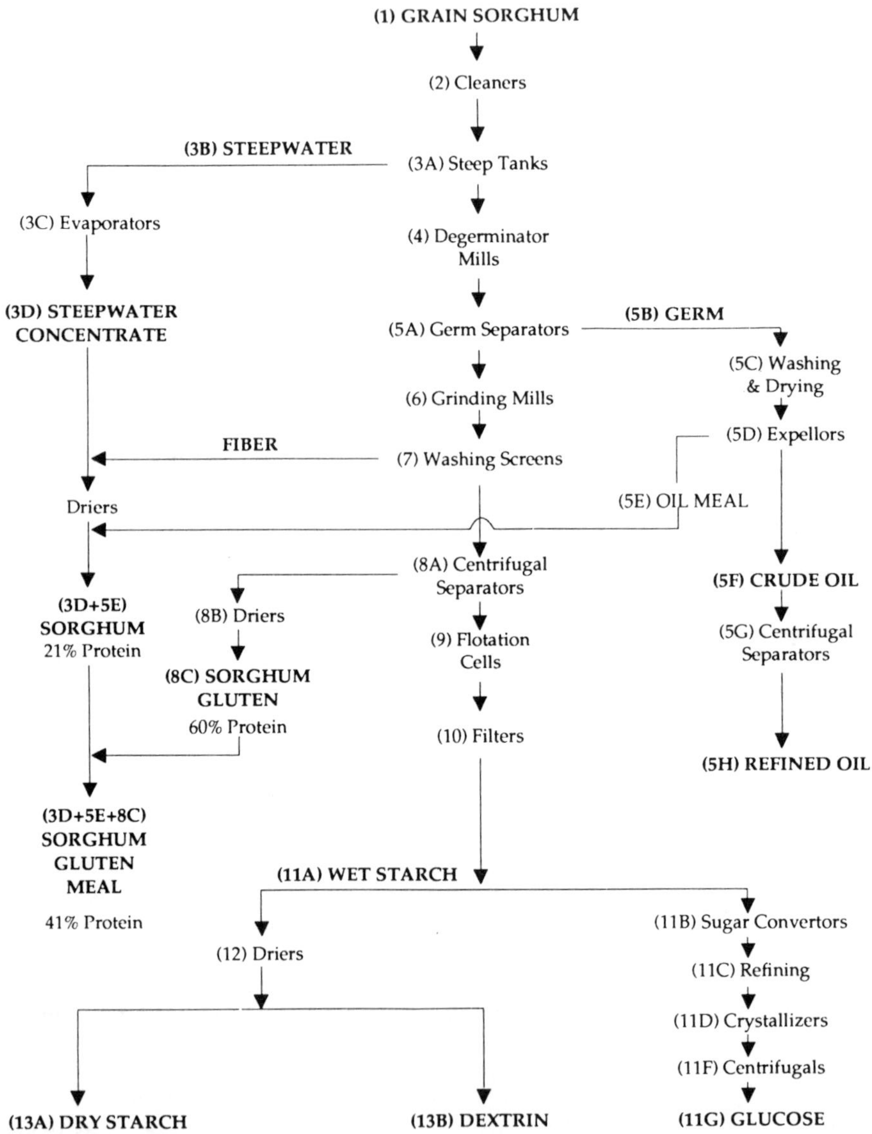

Fig. 16. Flow sheet of grain sorghum wet milling (after Watson, 1970).

softened, steeped seeds are gently milled in degerminator mills (4) in a water slurry to loosen germ, starch, gluten, and fibers but keep them as intact as possible. The sorghum germ, which is much smaller and more difficult to separate whole than the maize germ, is separated from the starch milk together with the gluten by the overflow in liquid cyclones (5A). The purified germs (5B) are dried (5C), and the oil (5F) is pressed out in an expeller machine (5D), whereas the resulting protein is blended into the sorghum gluten meal. The starch slurry from the germ separators (5A) is further milled at high speed in an impact mill (6), and fibers are removed by washing screens (7) and joined into the feed fraction. The purified starch "milk" is finally separated from gluten impurities in continuous starch centrifuges (8). Grain sorghum starch granules have a density of 1.5 compared to 1.1 for gluten particles, which constitutes the physical basis for the centrifugal separation. After removing most of the water on starch filters (10), the wet starch (11) is air-dried (12) to a sufficiently low temperature to avoid gelatinization. The very pure, fin-

TABLE 9
Laboratory Wet Milling Results for Three Types of Grain Sorghum[a]

Fractions	Regular Red (commercial)	Waxy (white)	Yellow Endosperm[b]
Whole grain analysis			
Moisture	14.5	13.5	14.0
Starch, % db	73.1	71.9	74.8
Protein	12.1	11.4	10.3
Fat, total[c]	3.6	4.1	3.3
Solubles			
Yield,[d] % db	6.9	7.6	7.4
Protein	48.2	45.8	46.2
Table starch			
Yield	64.2	61.3	67.9
Protein,[e] % db	0.3	0.2	0.4
Fat[f] (total)	0.7	0.1	...
Germ			
Yield	6.2	5.9	4.0
Starch	18.6	10.8	16.6
Protein	11.8	10.3	15.0
Fat	38.8	42.6	43.4
Oil yield[d]	2.4	2.5	1.7
Fiber			
Yield	8.2	9.0	7.7
Starch	30.6	35.4	22.2
Protein	23.2	21.3	15.7
Fat	2.4	3.7	7.0
Table gluten			
Yield	10.6	11.0	9.2
Starch	42.8	41.9	31.6
Protein	46.7	40.9	50.3
Fat	7.9	8.7	9.9
Middlings, yield	1.2	2.0	1.6
Total dry substance recovery	97.0	96.8	97.8

[a]Source: Watson (1970).
[b]DeKalb hybrid G600.
[c]Includes wax.
[d]Percent of original grain, dm basis.
[e]Percent of fraction, dm basis.
[f]Determined by acid hydrolysis.

ished starch (13A) may be further processed into dextrins (13B) by heating under acid conditions or into glucose (11G) by acid or enzymatic hydrolysis.

The peculiar sorghum pigments may affect the whiteness of the sorghum starch. Low-tannin sorghum without testa is preferred before high-tannin types such as Feterita. Also pericarp and glume pigments may discolor the final starch product. Thus, there are noncarotenoid, colorless pigments in the pericarp of white sorghum that become bright yellow when pH is raised from acid or neutral pH to pH 9–11. Several yellow endosperm hybrids give a superior separation of starch and gluten and an excellent white starch product. A typical pilot trial (Watson, 1970) from the wet milling of one red, one waxy white, and one yellow endosperm sorghum including composition of raw materials, yield, and composition of the different fractions (solubles, starch, germ, fiber, gluten, and middlings) is outlined in Table 9.

WHOLE SORGHUM PLANT MILLING AND SEPARATION

Traditionally, the subsistence farmer uses the whole plant including the seeds, the dried stalk, and the leaves, e.g., for roofing, cattle feed, and cooking. One of the more specialized uses is for broomcorn, a special sorghum species with a long slender relaxed panicle with small, dark high-tannin seeds on which the glumes adhere. The name *broomcorn* is related to the preferred use of the threshed plant for brooms, i.e., long-handled domestic sweeping brushes now used in North America exported from South America and China.

In the early industrial revolution, agricultural products such as (straw) cellulose and starch were extensively utilized in the paper and textile industry, respectively. Agricultural raw materials were also used in a wide range of other nonfood uses from glue to energy (for animal-driven transport) (Gomez and Munck, 1993). About 100 years ago, Swedish engineers developed processes for removing the lignin in wood; wood contains generally much more lignin than straw and, therefore, new processes involving intense chemical treatment have to be developed to make wood competitive with straw as a source for paper pulp. When using trees for cellulose, the leaves (and needles) were removed, and only the stem (wood) was utilized, differing from utilization straw, which was used whole, combining the botanical constituents leaves, nodes, internodes, and internode pit. While unseparated straw of cereals contains only about 33–37% α-cellulose compared to 40–45% in wood, separated internodes from straw have the same level of α-cellulose as wood (Bjørn Pedersen and Munck, 1993). Hydrated silica (SiO_2) is a problem in the straw paper pulp industry, because it burns and isolates the evaporation boilers, preventing a recirculation of the extraction liquors. Leaves are much higher in ash and silica than internodes and contain less lignin and more protein, making them more nutritious as cattle feed (Ramanzin et al, 1986). It is well known (Rexen and Bach Knudsen, 1984) that the digestibility of cereal straw can be improved considerably by spraying with a 4% solution of saturated NaOH. Traditionally, high-tannin sorghum grains have been processed with alkali (potash), and this treatment is known to reduce the nutritional drawbacks caused by polyphenols. Consequently, the nutritional utilization of sorghum straw for cattle feed would be improved considerably by isolating the leaf fraction and treating this fraction with alkali (see also chapter 11). The stem (internode) fraction is ideal for other purposes such as paper pulp and building boards. The highly lignified thick stems from certain landraces of sorghum are used as an ideal fuel for cooking (e.g., in upper Egypt). At the Carlsberg Research Center, a straw separation system based on the United Milling Systems disk mill (Rexen and Munck, 1984; Bjørn Pedersen and Munck, 1993) was developed in the beginning of the 1980s (disk mill illustrated in Fig. 10) involving a stationary and a rotary disk equipped with hard metal milling surfaces in the periphery (Andersen, 1987). Dried (10–12% H_2O) chopped straw is ground in the mill with a distance between the disks of 0.9 mm. By this treatment the brittle leaves and nodes are milled to a fine powder, while the more elastic internodes are split up as coarse chips and cleaned from adhering pith, which is added to the meal fraction. The fractions are obtained by sieving on a 1,000-mesh screen. The internode fraction is optimized for fiber board or paper pulp manufacturing, while the meal fraction (especially after sodium hydroxide treatment) pelleted is ideal as a cattle feed. In order to process the internode chips for particle boards, a second milling step involving a carborundum mill is needed to scratch the wax layer of the internode particles, which make them adhere better to the glue in the particle board process. In a few experiments with sorghum and maize straws, the internode fraction of the entire straw constituted 56 and 36%, respectively, compared with 44–51% in barley and wheat straws and 56–57% in oats and rye. At Carlsberg, the sorghum internode fractions for particle board were studied in pilot scale in cooperation with the SADCC/ICRISAT Sorghum Center in Bulawayo, Zimbabwe (Hallgren et al, 1992) (Table 10).

The results show that the particle boards obtained from sorghum internodes had a similar strength to that of the control board from wood. The difference in tensile strength was not statistically different at $P = 0.05$. In a preliminary experiment, it was shown that it is possible to produce a carbohydrate-modified phenol-formaldehyde glue based on either glucose, wheat starch, wheat flour, or maize starch (Frandsen, 1987). Sorghum starch and even dry-milled sorghum flour should also be suitable. By introducing the cereal carbohydrate, both the phenol and formaldehyde levels needed could be lowered, diminishing the residual values of these chemical substances considerably, and thus producing a product with a higher environmental acceptance. Industrial adhesives have been made from condensed tannins of sources other than sorghum (Ayla, 1984; Hemmingway and Kreibich, 1984; Pizzi and Roux, 1978). It would be highly interesting to explore further the positive use of the wide range of very high-tannin (bird-resistant) sorghums in Africa, which are presently used for brewing and as (poor quality) animal feed.

In Figure 1, several other nonfood uses of sorghum are suggested, also discussed by Subramanian and Jambunathan (1988). This research in alternative uses of plants was stimulated during the Second World War and again after the first energy crisis in 1973, especially by incentives at the United States Department of

TABLE 10
Comparison Between Particle Boards of Sorghum Internode Chips and Wood Chips[a,b]

Board Material	Modulus of Elasticity (N/mm^2)	Modulus of Rupture (N/mm^2)	Tensile Strength (N/mm^2)
Sorghum internode chips	3,237 ± 229	22.0 ± 1.8	0.30 ± 0.99
Wood chips	3,257 ± 112	21.9 ± 1.8	0.39 ± 0.11

[a]Source: Hallgren et al (1992); used by permission.
[b]Results are means and standard deviation from six determinations from the same board.

Agriculture and its Northern Regional Laboratory in Peoria (Doane, 1981) and were later investigated for European conditions in the 1980s (Rexen and Munck, 1984, Munck et al, 1989, Munck, 1993) and by the interest in the "biomass utilization" research for nonfood and energy purposes worldwide published extensively.

UTILIZATION OF THE MILLED PRODUCTS

With regard to the industrial utilization of milled sorghum products, there are two entirely different panoramas depending on the technical and economic situations in industrialized (such as the United States) and developing countries; these will be described from a research and development point of view as well as from a market standpoint.

Industrialized Countries

RESEARCH AND DEVELOPMENT

Industrial sorghum milling technology obviously reached its peak in the USA at the end of the 1960s as documented in the chapters on dry and wet sorghum milling by Hahn (1970) and by Watson (1970). Both authors were deeply engaged in sorghum milling technology employed by large companies that invested in sorghum technology alongside the maize alternative. Much of their published quantitative and technical data are from pilot experiments, but if we read these articles between the lines there is much valid information to be considered for the future. In this review, an attempt is made to fill in the gaps in our knowledge of sorghum milling technology. Much unpublished information may be hidden in industrial archives, and it would be of general interest to document this sorghum experience before it has vanished in order to prepare for the next wave of development of the sorghum milling industry. In spite of sorghum's success in the United States as a cereal that has the highest expansion rate in agriculture in the 1970s, most of the industrial milling operations, including all wet milling and most of the dry milling units, have now (1994) been closed down. In the following we will try to find out why.

THE MARKET

It is simplistic to argue for using maize instead of sorghum because maize technology is the outstanding market leader. More and Majors reported in 1970 that at the end of the 1960s utilization of sorghum grain for food and industrial purposes was increasing in absolute terms, but due to increased sorghum growing, these uses of sorghum were declining as a percentage of the total usage. At that time the industrial sorghum consumption was only 2% of the total production; 40% of the industrial sorghum grains were used by wet millers, another 40% for fermentation (e.g., brewing and alcohol production), and 20% by the dry milling industry. Hahn (1970) lists a range of nonfood uses of dry milled sorghum, e.g., in the gypsum and plywood wallboard industry, as foundry binders in mineral processing, in hard coal briquettes, and in oil drilling starch. Watson (1970) indicated that in 1966, more than 100 million pounds of sorghum starches and 2.75 billion pounds of corn starches were manufactured in the United States. Approximately equal quantities were converted to sugar, syrups, and dextrins. The sorghum and maize starches had the same customers, e.g., the paper (45%), food (25%), and textile (17%) industries, and to manufacturers of building materials, adhesives, explosives, and for miscellaneous industrial uses (9%). The rest was exported.

As a further explanation for the negative development since 1970, we could list the pros and cons of utilization of sorghum for milling in competition with maize.

Advantages of Sorghum

- More productive than maize when moisture stressed and can better stand extensive heat.
- Hybrid technology is as well developed as that of maize.
- Price 90% or less than that of maize.
- Sorghum is highly competitive with maize on the feed market.
- Sorghum, being more nearly spherical, can be more tightly packed in silos than maize, which saves storage volume.

Disadvantages of Sorghum

- In general, sorghum has a poor image and is considered as feed grain, which traditionally is reflected in less restrictive rules with regard to grading (e.g., in the United States), resulting in variable quality and less material for processing left after cleaning (see chapter 13).
- Sorghum requires more energy in milling (e.g., in a hammer mill) than maize.
- Sorghum quality is more variable than maize quality.
- Sorghum may contain polyphenols as well as other deleterious pigments, while compared to maize useful pigments such as xanthophylls are much lower or absent.
- Sorghum contains less useful oils than maize, and the much smaller germs are difficult to separate without breakage, leading to less income from oil extraction in sorghum wet milling.
- In a market with overproduction of both maize and sorghum, the use of sorghum for industrial purposes is less profitable.

Developing Countries

RESEARCH AND DEVELOPMENT

During the 1960s, there was a rising general concern about the population explosion and the need for increased food production with regard to tonnage and quality. Here, protein content became a major issue, in that cereal commodities and especially sorghum were extremely low in the essential amino acid lysine, which is of prime importance for growing children and the lack of which results in nutritionally induced diseases. The debate urged the politicians and international bodies to launch several grand research and development programs in plant breeding, nonwheat cereal food processing, and cereal nutritional quality. The largest and most successful of those programs is the international plant breeding program of the CGIAR (Consultative Group on International Agricultural Research)—the IARCS (International Agricultural Research Centers), including ICRISAT, with headquarters at Hyderabad in India, the center of sorghum and millet research. The prime target of these programs was to improve cultivars and plant husbandry practices to increase availability of food for the small and middle-sized farmer. In the more fertile areas of the developing countries where wheat and rice were grown, the efforts from the IARC institutes resulted in a great increase in total production—the so-called green revolution, which produced new varieties to gratify the absorption of external inputs based on fossil fuels like fertilizers, pesticides, and machinery. This development was not so successful in the arid sorghum- and millet-growing regions, e.g., in Africa, due to the harsh environment and deep socioeconomic constraints.

In South Africa, extensive research from the 1950s to the 1980s on sorghum malt and grits (Novellie, 1993; Olatunji, 1993) stimulated production and marketing of sorghum flours and decorticated grains for porridge and rice-like products, which may compete with imported yellow maize during a situation as in 1992 when production of the preferred white maize was low due to drought (Sooliman, 1993; van Vuuren, 1993). Thus, maize is marketed as "the maize giant" (Counie, 1993) in Africa even though a more balanced production of both maize and the locally adapted sorghum would give a considerable boost to food security.

During the 1960s and 1970s, a range of research and development programs for sorghum and millets utilization for developing countries was started. The UNDP/FAO started a sorghum and millet program in Senegal (1970–1974) and in Sudan (1975–1980) involving small-scale and industrial-scale equipment (Perten, 1984) as previously discussed. Since 1970, the IDRC (International Development Research Center) of Ottawa, Canada, has been involved in sorghum and millet milling research (Reichert, 1982; Young, 1991) mostly with regard to village mills, with a major effort in Botswana (Bassey and Schmidt, 1989; Schmidt, 1992; Dinat, 1992). During 1977–1991, United Milling Systems A/S of Denmark in cooperation with the Carlsberg Research Center was one of the few—if not the only—commercial company in the world that devoted time and effort to sorghum and millet milling development on an industrial scale (Munck et al, 1982a,b).

In 1964, Piere Chatelanat initiated the FAO "Composite Flour Technology Program" also involving the Tropical Products Institute (now the NRI) in the United Kingdom, Kansas State University, United States, and the Institute for Cereals, Flour and Bread at the TNO, Wageningen, Holland (Asselbergs, 1970). The program was built on the FAO motto: *Fiat Panis*—let there be bread—underlining the wholesomeness and sensory attractiveness of a leavened, white, textured wheat bread, including its availability and association with modern life.

Boosted by export of wheat and wheat mills from the industrialized countries in Europe and America to the tropical countries, bread and wheat consumption in the those countries accelerated. Thus, for example, Nigeria found itself in a financial dilemma—a wheat trap (Andrae and Beckmann, 1985)—of increasing wheat imports from 56,000 t in 1965 to 1,670,000 t in 1984, an increase from 0.4% of the total carbohydrate diet to 13.3% (Dendy and Trotter, 1988). Even a country like Nigeria, with natural assets such as mineral oil, could not stand this drain of hard currency. The government thus in 1988 took the brave decision to stop import of grains, leaving the bread-demanding population in the cities to Nigeria's own production (45,000 t of wheat in 1985, 2.7% of the total wheat consumption) diluted by local flours from cassava, maize, and sorghum.

For the first time, the bread producers and consumers had a strong incentive for implementing breads from composite flours, and even nonwheat bread (Dendy and Trotter, 1988) as well as other industrial uses (Bogunjoko, 1992; Olatunji, 1993) as the increased demand fulfilled the positive effect on the national economy that had previously been calculated (Dendy and Trotter, 1988).

From 1970 to 1991, more than 1,200 publications on composite flours and bread technology were published from various laboratories in the world, with little success with regard to implementation in practice (see reviews by Dendy and Trotter, 1988; Dendy and James, 1991). In composite flours with millets (Dendy, 1991) to produce leavened breads, smaller amounts (5–20%) in the mix give an acceptable product than with low-tannin sorghum, where up to 30% may be acceptable. In pearl millet, the strong taste (Hoseney et al, 1992) might affect acceptability in communities where pearl millet no longer is the staple food. Although the minor millets are considered the least important class of cereals, with annual production less than 2% of the world's grain, they are of great local importance as staples and reserve crops in marginal areas, because many of them are productive under arid conditions even more so than sorghum (Dendy, 1991; Malleshi, 1988b). For obvious reasons, very little development has been made on an industrial scale with regard to millet food products. The research made in this direction has been summarized by Hulse et al (1980), Malleshi (1988a,b), and Dendy (1991). A major effect of mixing wheat flour with 50% sorghum flour is a drastic reduction in loaf volume, which could be partly overcome by single and combined additives such as glyceryl monostearate, lecithin, methyl cellulose, pentosans (e.g., from rye), and pregelatinized starch materials (Satin, 1988). If wheat (and wheat gluten) is totally omitted (wheatless bread), natural gums such as guar and acacia as well as pregelatinized starch can be used to provide crumb structure by a "custard process." It is important to develop locally produced alternatives to these additives to improve the practical and economical feasibility of the new technology. However, even after using the above-mentioned additives, the sensory quality of a 50-50% sorghum-wheat loaf is quite changed from that of 100% wheat bread, mainly because the sorghum bread has a hard crumb structure and rapid staling, which decreases shelf life and makes consumption within a few hours after baking necessary. Dendy et al (1990) stress the compatibility of the composite flour product with consumers' end-use and state that "it would seem that for most commodities, however well prepared for incorporation, there is a limit of 30%. Beyond that the product ceases to resemble bread though it may have selling points of its own—a point that deserves more emphasis."

TABLE 11
Milling Yield, Damaged Starch, and Proximate Composition of Sorghum Grains and Flours[a,b]

Sorghum Sample	Milling Yield (%)	Damaged Starch (%)	Ash (%)	Protein (%, N × 6.25)	Starch (%)	Fat (%)	Fiber (%)
296B raw material	100	...	1.37	11.0	71.6	3.0	2.3
296B dehulled grain	82.9	...	ND[c]	9.7	ND	0.7	ND
296B hard flour	22.9	10.2	0.71	11.9	85.5	1.1	0.4
296B soft flour	19.6	4.1	0.36	5.8	90.3	0.6	0.6
2219 raw material	100	...	1.45	12.6	71.2	3.8	2.0
2219B dehulled grain	77.0	...	ND	10.7	ND	1.2	ND
2219B hard flour	37.8	14.8	0.42	11.2	84.8	0.6	0.8
2219B soft flour	11.6	8.2	0.82	8.3	85.9	2.0	0.7

[a]Source: Hallgren (1984); used by permission.
[b]All proximate analyses are means of duplicate determinations.
[c]Not determined.

Surprisingly, the possibility of differential milling of sorghum (Hallgren, 1984, 1985) separating the hard endosperm, optimal for brewer's grits (see chapter 9) from the soft endosperm ideally suited for baking (Hallgren, 1984) has not been generally acknowledged. Hard and soft endosperm were separated by dehulling, milling, and sieving from the two ICRISAT sorghum varieties: 2219B (vitreousity 72%) and 296B (vitreousity 37%). All sorghum flour fractions, either hard or soft, were milled to approximately the same particle size distribution. Starch damage ranged from 8.2 to 4.1% in the soft and from 14.8 to 10.2% in the hard fraction (Table 11).

The gelatinization data (Table 12) show that suspensions from the soft endosperm fractions have much higher Brabender amylograph viscosity than the hard fractions in spite of their lower starch damage. Baking experiments in pans were performed with the additives of ascorbic acid (50 ppm) and a 99% soybean lecithin solution (2%) to a 50% sorghum-wheat flour.

The baking results are shown in Table 12. It is evident that the hard sorghum flour needs more water addition in the Brabender Farinograph to obtain the same consistency as the soft flour and that the soft flours give a higher specific volume of bread.

It is clearly seen from compression studies of bread (Fig. 17) that the addition of the soybean lecithin greatly improved softness and counteracted staling during a 48 hr storage period. Crude lecithin could easily be produced from locally produced soybean, if available. As with composite sorghum flours for leavened bread, similar flours preferably from white pearled sorghum could be combined with (durum) wheat (at least 25% wheat) and would give a pasta of reasonable eating quality (see review by Hulse et al, 1980 and Faure, 1992). However, the color still remains a problem. The experiments of Hallgren (1984) show the usefulness of fractional milling where the different fractions should be sold to the purpose which gives the highest technological and economical advantage.

For a long time, gluten-free wheatless breads have been available in industrial countries to be consumed by people who are allergic to wheat gluten (celiac disease). From a sensory quality point of view, it makes much more sense to produce biscuits and crackers from sorghum flour than leavened bread. In Table 13, an ex-

TABLE 12
Amylograph Data and Baking Results from Sorghum and Composite Flours[a]

Sample	Sorghum Flour Amylograph Peak Viscosity (BU)[b]	Composite Flour[c] Farinograph Dough Consistency Maximum (BU)	Composite Flour[c] Farinograph Dough Consistency 5 min (BU)	Water Added (%)	Bread Specific Volume (ml/g)
Soft variety 296B					
Hard flour	500	275	240	62.5	2.8
Soft flour	945	270	245	58.7	3.0
Hard variety 2219B					
Hard flour	570	275	235	67.9	2.6
Soft flour	770	270	230	62.2	3.0
Dark northern spring wheat	190	...	...	...	...

[a]Source: Hallgren (1984); used by permission.
[b]Brabender units.
[c]Sorghum/wheat, 50:50.

periment by Grøndal (1988) is outlined comparing the sensory characteristics of high-fat sorghum cookies with a corresponding wheat recipe. There are significant differences with respect to appearance, crispness, sandiness, and mouthfeel, whereas the differences in taste and total impression were not significant. Taste had the highest correlation to total impression (r = 0.71). Addition of 25% extruded sorghum flour to an untreated sorghum flour significantly improved crispness and gave less sandiness and a better mouthfeel. A similar effect could be obtained if soft endosperm flour from a suitable sorghum variety were carefully milled and separated from the hard endosperm particles, which cause the feeling of sandiness.

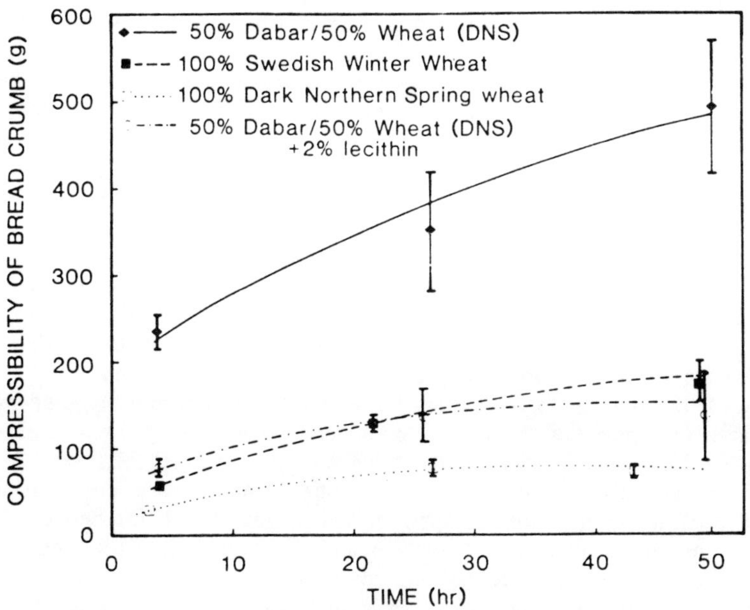

Fig. 17. The hardness and staling of the bread crumb in wheat and wheat-sorghum composite flour breads with and without lecithin addition (Hallgren, 1984; reprinted with permission).

TABLE 13
Mean Rating of Sorghum Cookies by a Taste Panel of 10 Persons[a,b]

Fat	I Sorghum 20%	II Sorghum 35%	III[c] Sorghum 35%	IV Wheat 35%
Appearance	5.5 a	3.6 a	3.4 ab	3.7 a
Crispness	3.4 bc	3.0 c	4.3 a	3.8 ab
Sandiness	2.1 cd	1.8 d	3.0 ab	3.3 a
Mouthfeel	1.7 b	2.2 b	3.6 a	3.5 a
Taste	3.1 a	2.9 a	3.1 a	2.4 a
Total impression	2.8 a	2.6 a	3.3 a	2.9 a

[a]Source: Grøndal (1988).
[b]Treatments marked with the same letter are not significantly different.
[c]As in treatment II, but with 25% extruded sorghum flour in the mix.

Obviously, one of the major problems in sorghum food technology and nutrition is to open the highly cross-linked and hydrophobic kafirin protein matrix, which is encapsulating the starch granules, hindering them from taking up water and swelling, so that they cannot gelatinize properly. A project from Colorado State University, United States, introduced low-cost Brady extruders to make baby food in Sri Lanka and Tanzania (cited by Hallgren, 1985), combining the extrusion of maize grits (65%), dehulled full-fat soybean (29%), and mineral mix (1.8%). This technique also works with sorghum (Chigumira, 1992), implying full utilization of the energy in sorghum starch for children, at the same time giving the necessary reduction in viscosity when cooked as a gruel in order to obtain a high caloric density. The soybean additive provides valuable fat and more importantly supplements lysine to the sorghum protein, which is extremely low in this essential amino acid, thus also making it possible for the children to utilize the sorghum protein effectively.

MacLean et al (1982), in clinical feeding studies with small children, had a positive result with an extruded product made from 20% dehulled sorghum and 80% wheat-soybean mixture. However, unextruded sorghum fed as cooked gruel for one week as the only protein source (with a blend of soybean and cotton-seed oil providing 30% of the calories fortified with minerals and vitamins) gave very low gains in weight, low apparent absorption and nitrogen retention, large stool weight, and abundant loss of calories (starch) in the feces. In comparison with a similar maize diet, there was no difference between high-lysine and normal sorghums. One of the reasons of the poor performance of children on these diets could be that apparently no enzymes were used to decrease viscosity of the very viscous sorghum gruels to increase the caloric density and ease bottle feeding.

It seems meaningless to reascertain the fact that the sorghum level of lysine (about 1/3–1/4 of the dietary need) given to toddlers is limiting for maintenance and growth. This condition is so extreme that it may even affect the production of the necessary digestive enzymes, partly explaining the high level of energy (starch) in the feces. It is important in attempting to understand the relation between food technology and human nutrition of sorghum and millets to include the expertise of the local people who have lived and thrived on these cereals as their major staple food for a thousand years. Thus, in a later study made by the same research group from Peru (Graham et al, 1986), more positive results for nitrogen and energy balances much closer to those of the control diets were obtained in the study of a traditional Sudanese fermented sorghum weaning food, *nasha*. Lysine addition gave here a significantly positive response. An instant powder version of *nasha* was developed in cooperation between the Food Research Center (L. Monowar) and Purdue University (A.W. Kirleis), which had approximately three times the caloric density of traditional *nasha* and an improved lysine content due to fortification (Kirleis, 1989). At the Central Food Technological Research Center (CFTRI) in Mysore, India, a range of weaning foods has been developed, utilizing sprouted finger millet and otherwise treated roller-dried sorghum and finger millet products in combination with legume seeds such as gram and cowpea (Malleshi, 1988a; Malleshi et al, 1989). In all cultures traditionally depending on cereals, a range of treatments of *the whole seed* before milling and sifting has been applied (Vogel and Graham, 1978; Pushpamma and Rao, 1981; Pushpamma and Geervani, 1981; Malleshi, 1988a,b; Young et al, 1990). The treatments involve one or several of the following procedures: steeping, fermentation, malting, alkali or acid treatment, popping, roasting (dry or wet), parboiling, and drying. The effects of the whole seed treatments are as follows:

1) Improvement of dehulling and (hand) milling of the treated, dried, whole seed. Steeping and/or steaming followed by (partial) drying is especially efficient in treatment of the small-grained millets such as finger millet to facilitate the separation of bran and endosperm by milling and sieving (see Malleshi, 1988a).

2) Improved storage quality of dry products.

3) Improved flavor and acceptability of the resulting cooked products.

4) Highly reduced viscosity of the resulting gruels, which improves their caloric density and acceptance in feeding as gruels.

5) Improved digestibility of the products as well as improved utilization of micro-elements like minerals and vitamins.

6) Phytic acid, which depresses uptake of important minerals such as calcium, zinc, and iron is broken down by the appearance of phytases during steeping and germination.

For further details see reviews by Hulse et al (1980), Pushpamma and Rao (1981), Eggum et al (1983), Malleshi and Desikachar (1985), Malleshi (1988a,b), Hansen et al (1989), Pedersen et al (1989), and Serna-Saldivar et al (1987).

Remarkably, very few of the traditional processes of whole-grain processing have survived in westernized, urban cultures in the form of industrialized processes. It seems urgent to document the world heritage of whole-grain processing for ethnographic, food technology, and nutritional reference before it is too late. We have just approached the surface of understanding what these processes mean from a food satisfaction and nutritional point of view. One might introduce attractive sorghum foods such as sorghum rice—cooked pearled sorghum—as health foods in the westernized kitchen, comparable to oats and barley in giving a physiological "dietary fiber effect" and a slow release of carbohydrates during digestion.

It is interesting to note that the concept of dietary fiber, which now is one of the main nutritional quality arguments in affluent societies, emerges from Burkitt's observations (Burkitt and Trowell, 1975) from the sorghum-eating parts of East Africa. Burkitt explains the large stool volume of individuals (which he connects with a low incidence of colon cancer) with a high intake of indigestible fiber. The strong effect of a pure sorghum diet on the stool volume in children was confirmed by MacLean et al (1982). Since 1975, chemical "dietary fiber" analyses (Asp et al, 1983) have been developed to forecast the amount of indigestible fiber that passes through to the colon in the digestive system.

However, as shown by Bach Knudsen and Munck (1985), uncooked sorghum contains less dietary fiber, according to the method of Asp et al, than all other uncooked cereals (Fig. 18A). However, the protein content of the total dietary fiber fraction in the analysis is 38.9% for low-tannin, untreated, sorghum compared with 2–3% for the other cereals (Fig. 18B). In the internationally certified analysis of dietary fiber (Asp et al, 1983), the associated protein is measured, but the figure is deducted from the calculation of the dietary fiber figure and not further presented. Although it is unlikely that such a high percentage of sorghum protein as shown in the dietary fiber determination (Fig. 18B) is lost to the colon in feeding uncooked sorghum, it is most likely that a large part of the increased stool volume from East African individuals observed by Burkitt (and confirmed by MacLean et al [1982]) is due to undigested protein and starch from sorghum.

It has been shown that cooking, especially in combination with a high level of polyphenols, increases the dietary fiber fraction and the associated protein and decreases in vivo digestibility of protein and energy in rat trials (Bach Knudsen and Munck, 1985; Bach Knudsen et al, 1987, 1988). It is likely that the polyphenols bind to the proline residues in the kafirin (prolamin) and render these proteins and enclosed starch (enzyme-resistant starch) unavailable. An acidification and fermentation may decrease these negative effects, which also prevail with low-polyphenol varieties. This effect of cooking can be counteracted by adding a reducing agent, as demonstrated in vitro by Kirleis (1992), who pointed out that the digestibility problem in sorghum, apart from the polyphenols, is probably due to the high-cysteine gamma-kafirins, which contribute to a strong cross-linking of the protein matrix enclosing the starch granules.

Large negative effects on protein digestibility in low-tannin sorghum due to cooking were demonstrated in vivo in feeding trials with children (MacLean et al, 1982) and in vitro with pepsin digestion (Axtell et al, 1981; Axtell and Mertz, 1992). The effects in rat tests were much less (Eggum et al, 1982; Bach Knudsen et al, 1987, 1988) as discussed earlier. Fermentation, which should stimulate a reductive environment in Kirleis's sense, counteracted the decrease in protein digestibility in cooking low-tannin sorghum as demonstrated by Eggum et al (1983).

It can thus be concluded that the crude chemical dietary fiber analysis should be complemented with new analyses to better reflect the total indigestible material that reaches the human colon, including indigestible protein and resistant starch, which in toto are likely to be of significant physiological importance and which also are influencing the stool weight. In this respect, sorghum food products seem to have unique dietary advantages in human nutrition in line with those of oats and barley and should be further evaluated. The issue of nutritional evaluation is further discussed in chapter 5 in this book.

As earlier discussed, a lasting benefit of the past 20 years of sorghum and millet research is the realization, in both technological and nutritional senses, of the connection between the characteristics of the raw materials and the traditional rural cooking regimes. But very little of this important research has resulted in industrial developments of direct impact on sorghum-eating habits in the cities of the developing world. As well as FAO claims *Fiat Panis,* the international community should extend this limited concept to *Fiat Kisra, Fiat Roti, Fiat Ugali,* etc., in order to keep the connection between rural and urban areas fundamental to the development process of society.

In our work on sorghum and milling with United Milling Systems A/S, we started with an inventory of the local practices (Eggum et al, 1982) out of respect for the accumulated knowledge of the people who since ancient times have survived on sorghum and millets. On the basis of this knowledge, we designed technologies for producing convenient sorghum products on an industrial scale, aimed at ensuring that the rural people would maintain their sorghum-eating habits when

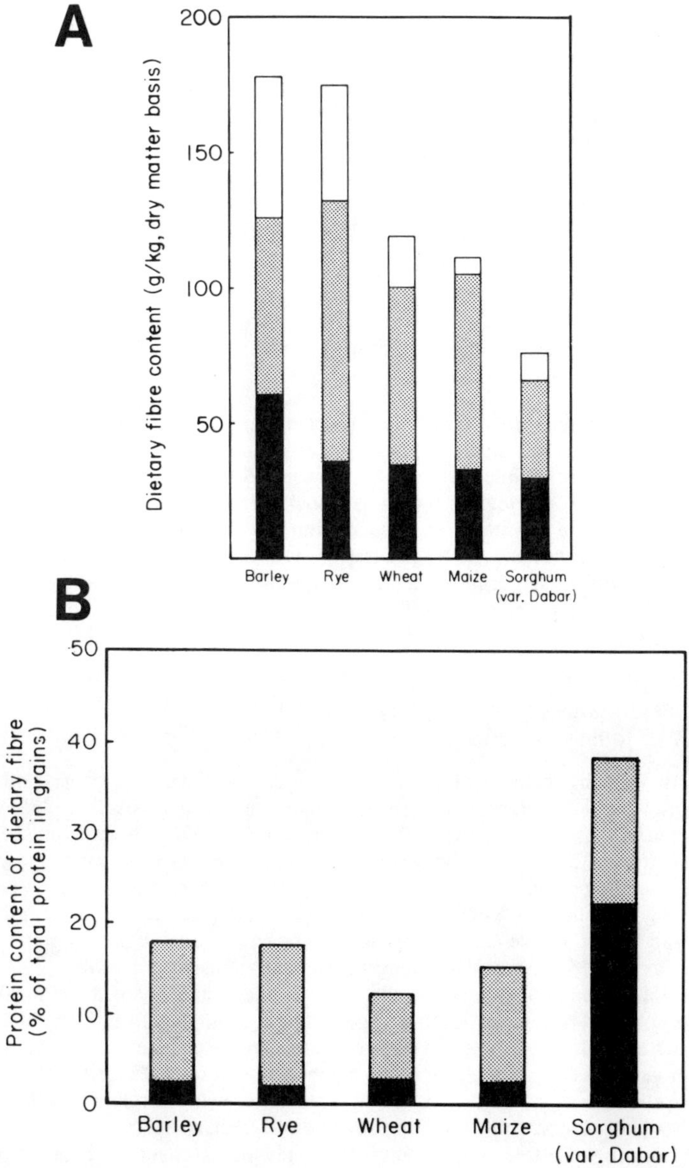

Fig. 18. A, Dietary fiber composition of uncooked whole grain barley (with husk), rye, wheat, maize and sorghum (Dabar). B, Protein associated with total dietary fiber and with acid detergent fiber. Solid bars = acid detergent fiber; open bars = soluble dietary fiber; solid + shaded bars = insoluble dietary fiber; all three together = total dietary fiber. (Bach Knudsen and Munck, 1985; reprinted with permission).

moving to the cities. Thus, a link between urban and rural areas should be established by the demand for sorghum deliveries to the mills. We developed equipment and processes for sorghum flour and grits for porridge (*ugali*), flat bread (*kisra*), and mixes for instant *ogi* as well as for baby food, bread, and cookies. A few industries were established in Africa that will be described in the following.

In order to make a realistic and just diagnosis, we have to understand both the cultures in the developing world and the industrialization process in general, including its association with fossil fuels, which has formed the foundation for the prosperity of western countries and which is looked upon as an ideal by the rapidly growing populations in Africa, Asia, and South America.

THE MARKET

Gomez (1987) described the situation in central Africa (Zimbabwe) in the following terms:

> Traditionally, farming relied on drought-resistant, small-grain millets and sorghum. When maize became the staple food (porridge) in the 1930s, it was associated with the sphere of women, while sorghum was reduced to use for beer and threshed outside the homestead as a communal activity associated with the sphere of men. Peasant producers who were producing for consumption and for cash found themselves getting further and further into the "maize trap" of producing maize in largely unsuitable ecological conditions requiring higher inputs of fertilizer, labor, and (where available) pesticides for intensive cultivation. Sorghum was not generally a grain traded by the marketing boards

For example, in Zimbabwe, sorghum (for food, not beer) and pearl millet were not designated grains for bread purchase until 1986 (Dendy, 1992, *personal communication*). Gomez continues,

> The early breeding work in central and eastern Africa, therefore, focused on high-tannin brewing varieties (such as Serena), introducing the short-statured plant with a compact head that had been one of the most successful yielders in the United States. Also low-tannin, compact, white-seeded food types such as Lulu were introduced. Recognizing the potential of sorghum especially in periods of drier climate, in 1984–1985 the government of Zimbabwe introduced a new price policy for sorghum in parity with maize. The scheme had an instant impact and quadrupled sorghum deliveries to the marketing board. However, this amount represented in 1985 only 15% of the communal sorghum production that was produced and retained for local consumption. A well-developed local storage technology is, therefore, of great importance, and losses due to storage damage are estimated to be high.... Traditional hard sorghum varieties store much better than the modern soft ones. Unfortunately, the increased production was only from red high-tannin brewing sorghum, while whole, white, low-tannin food sorghum deliveries to the marketing board declined from 9.5 to 1.9%, leading to a total surplus in 1987 of brewing sorghum of about 195,000 t. This is far above the need for sorghum in the opaque beer industry, and the large surplus of high-tannin sorghum thus had to be sold cheaply as feed, because this sorghum is not of the highest nutritional quality.

Gomez (1987) concluded that there is a need to develop a whole production processing and utilization technology package for white food sorghum. Shumba (1993) more recently described the situation regarding small grains in Zimbabwe, where the maize crop failed in 1992 due to drought and where food security would have been strongly improved if food grades of sorghum had been available for sowing and production. There is also a definite need for technologies that look beyond the traditional uses of sorghum in beer and porridge to more diverse applications that offer greater economic, nutritional, and technological advantages, such as in flour composited with wheat for bread, as a rice substitute (pearled sorghum), and for industry, e.g., brewer's grits (Gomez and Munck, 1993).

Imported sorghum hybrids from the United States and the previously mentioned new varieties Serena and Lulu with shorter straw (from 4–6 m in local African food varieties to 1.5 m in the new ones) are not well adapted to African conditions (Eggum et al, 1982). The compact head weathers poorly and often ripens in the wet period under East African conditions, thus supporting head molds, which may produce very toxic products (see chapter 6). The introduced varieties are often as a rule much softer (30–60% soft endosperm) than the harder local varieties (5–15% soft endosperm). The latter mill much better, both in traditional and industrial milling, as discussed before. In 1980, when the writer asked a school girl from the sorghum district Dodoma in Tanzania about what she knew about cereals, she mentioned pearl millet, maize, and sorghum, and then she added, after thinking for a while, Lulu and Serena, as if they were separate cereal species.

In rural Tanzania, pearl millet and sorghum are generally preferred to maize, while the choice is reversed in the cities. When sorghum is eaten for breakfast, the farmer can stay replete while working a whole day in the fields, which are often remote from the village. Satisfaction from eating maize is far less. In Denmark, people prefer rye bread to wheat bread much for the same reason. At the International Sorghum and Millets Workshop in Bulawayo, Zimbabwe, in 1988 (Gomez et al, 1992), Deputy Secretary of Agriculture S. C. Machena, in his opening speech, commented on the dominance of maize over sorghum and millets. "A national market test for small grain meal in Zimbabwe revealed that 45% of respondents preferred pearl millet and 31% preferred sorghum meal, while only 6% preferred maize meal. The remaining 18% had no preference. It is thus apparent that factors other than palatability lead to the dominance of maize in areas of marginal rainfall. Probably most important is the drudgery involved in the processing of grains."

ICRISAT in Hyderabad, India, cooperating with the SADCC/ICRISAT Sorghum and Millet Improvement Program in Bulawayo, Zimbabwe, has worked hard to supply the local market with sorghum varieties that will be recognized as useful for milling in the African consumer sense, both by hand and industrial mills (Murty, 1992). Until such varieties are widely grown, there is no sorghum available in most African countries of a defined reasonable quality and that can be purchased in large quantities, other than that for opaque beer in Zimbabwe and South Africa (see chapter 10). Bearing in mind that sorghum is the cereal species with the largest variability in nutritional quality and acceptability, in most African countries it is almost impossible to set up an industrial sorghum mill with a reliable flow of raw material. This is why the technically well-functioning industrial mill set up for brewer's grits in Northern Nigeria by United Milling Systems A/S, Copenhagen, with two separate lines—one for sorghum and one for maize—has now converted the sorghum line to maize, because it is available as a consistent commodity. While it is quite possible to introduce milling technology units to existing well-founded industries such as breweries in Africa, this is much more difficult if the experience of the local owner is limited to trade, and if he does not have experience regarding maintenance of machines and purchase and storage of raw materials. In many countries where there is no tradition for interest on capital and where the sorghum mills and food factories would be bought for local money and (a sizable amount of) foreign hard currency, the local entrepreneur does not have a feeling for the need of foreign capital to earn interest, and the unit may, therefore, stay inactive for months or years. Its value for the owner is in this case the same whether it runs or not. Gomez (1987) concluded with regard to the sorghum milling situation in Zimbabwe:

- Village mills delivered by foreign aid seem to be under-utilized, and the technical, economic and social reasons for this need to be investigated and evaluated.
- Commercial dehulling output (e.g., for composite flour) needs to be expanded. In Zimbabwe, there are 12 large-scale mills with a total capacity of 240,000 t/year of wheat and 1.2 million of maize. Three of these mills are experimenting with sorghum milling involving commercially and conventional roller mills as well as sorghum dehullers. However, the problem of raw material quality is limiting the further expansion of sorghum milling. There is thus need for an intense improvement program to upgrade the sorghum market with regard to quality. This is especially true for building an infrastructure from state and private institutions, including the seed industry, and trade standards that could support sorghum and millets as trade commodities.

The IDRC village mill project has been a success in Botswana, where even a Mill Owners Association has been formed to lobby the government on policies related to sorghum (Schmidt, 1992; Dinat, 1992). However, the IDRC programs in other African countries do not seem to have taken off yet. However, in Kenya a joint pilot project (NRI/KIRDI) has overcome problems of sorghum supply, and a factory has now been built in western Kenya to manufacture pearled sorghum and sorghum meal. The large milling companies are also taking an interest (Dendy, 1992, *personal communication*).

The SADCC/ICRISAT sorghum and pearl millet development program (Gomez, 1992, *personal communication*) was the first CGIAR institute to adopt a total approach to grain utilization involving postharvest semi-industrial and industrial technology. The bottom line of this concept is not merely a sustainable agriculture but sustainable agro-industrial systems that ensure not only food security but also simultaneous economic growth.

In Table 14, priorities proposed by the grain quality and crop utilization unit in SADCC/ICRISAT are outlined. Food, feed, and nonfood (including alcohol, adhesives, paper pulp, and fiber board) semi-industrial and industrial uses are included that could become of paramount importance to farmers in countries where there is little or no incentive to produce and sell sorghum due to subventions of imported grains and government policies that, until recently, often rendered sorghum and millets more expensive than maize and wheat.

In the industrial sector in Zimbabwe (Gomez, 1987), there is a yearly need for about 5,000 t of starch for food and nonfood industries, about 3,900 t of glucose syrup and 15,000 t of lactic acid for the brewing industry, 400 t of citric acid for the soft drink industry, and 20,000 t of rice; these are examples of possible areas where sorghum could be substituted for imports. These amounts seem small compared to those in industrialized countries but nevertheless represent a starting basis for initiatives to build up local industries of the greatest importance for development of a sustainable economy including employment, just as was done 100

TABLE 14
Processes and Products and Their Priority Identified in Two Crop Utilization Conferences[a]

Process or Product	Use	Priority[b] (SADCC)
	Food: Primary Processing	
Immature white grain	Roasting, boiling, popping, steaming	5
Whole grain composite	Sorghum + legumes	1
Whole dry grain	Roasted and ground	5
Use of whole grain treated with: acid, alkali, ash, formaldehyde, detoxification of tanning for feed use	...	1
Dehulling	Traditional and mechanical	1
Dry milling	Hammer and roller mills	2
Semi-wet milling	Traditional, commercial	2
Co-milling of grains from different crops; micronization and flaking; extrusion of whole grain or dehulled; whole grain puffing by steam gun	...	4
Starch extraction	Wet milling	5
Malting	Opaque beer	3
	Food malt	1
	Food: Secondary Processing	
Composition	Blending for bread, biscuits, pastas	1
Use of non-wheat flour	Bread and bakery products	
100% sorghum		5
100% millet flour		3
Acid fermentation	For weaning food	1
	For beverages	2
Glucose	Dextrins from starch, modified starches	5
Cereal binders	Extenders for meat products (sausage)	3
	Feed Uses	
Green forage and hay	...	2
Use of crop residue	...	1
Silage	...	5
Use of bran	...	1
Milling of stem	Pelleting	4
Flaking, reconstituting, popping etc.	...	4
Alkali treatment	Tannin detoxification	2
Feeding whole grain	Low tannin detoxification or detoxification of tannin	2
	Industrial Uses	
Sweet stemmed sorghum for alcohol	...	2
Dry milling	Adhesives, core binding or refining etc., role of waxy sorghums	2
Cellulose	Paper pulp, particle board, pellets	2

[a]Source: Project Proposal for a Grain Quality and Crop Utilization Unit in SADCC/ICRISAT SMIP, SADCC/ICRISAT Bulawayo, Zimbabwe, 1988.
[b]Priority, 1 = high, 5 = low.

years ago in Europe and the United States. Unfortunately, the SADCC/ICRISAT research on quality of sorghum and pearl millet at the center in Bulawayo was closed down in 1993 due to lack of funds.

Dendy and Trotter (1988), Dendy et al (1990), and Dendy and James (1991) comment that in the seemingly applied research on composite flours "One remains amazed at how little has been published on the economics of composite flour" and "It is extraordinary that so much technical research should have been carried out with no prior demonstration of a genuine need, though, of course, this has now been demonstrated." Obviously, the authors here refer to the decision by governments, e.g., that of Nigeria in 1987–1988 to ban imports of wheat, maize, and barley malt. In a short time this ban created an incentive for local entrepreneurs not only to explore the composite flour concept but also to produce traditional food in new ways or even to invent new foods. Dendy and James (1991) summarize, "Composite flour and wheat-less products may succeed if they are *acceptable foods*, at the *right price,* and given *good marketing*". As described in this chapter, this apparently simple statement hangs together on a whole series of circumstances.

Foods. Acceptable foods from sorghum means availability of high-quality raw material, which puts demands on breeders, farmers, and transportation systems. In 1988 in Tanzania, transportation of sorghum from Dodoma to Dar cost 50% of the price of imported wheat at harbor in Dar according to Dendy and James (1991). Acceptable foods mean also educated small- and large-scale industrial entrepreneurs with reasonable equipment, as well as spare parts, and reliable communal services like electricity and transport (Bagachwa, 1991). However, as Young (1991) rightly points out, "Breeding for the most appropriate quality in crops as raw material for processed products is more economical than improving properties through processing."

Price. In addition to sound international market competition (for those countries with an infrastructure that enables them to compete), Dendy and James (1991) state that the right price depends on whether the local government's price policy is allowed by the World Bank and the international community to aim for self-reliance, e.g., to ban imports or put up large import levies in an international political climate that praises the virtues of free trade. They emphasize, likewise, that a true free market has no subsidies for overproducing Western exports and no false valuation of currencies.

The local government's influence on competition between maize and sorghum, for example, can be exemplified by the case of South Africa, where maize lobbyists convinced the government to abandon the value added tax (VAT) on maize but not on other food products such as sorghum. This is partly the reason why about 30,000 t of sorghum was consumed (in 1991–1992) compared with 2.8 million tons of maize. Due to drought in 1992, sorghum consumption rose to 54,000 t in 1992–1993, and maize consumption dropped to 2.2 million tons (Sooliman, 1993).

Marketing. Good marketing depends on motivated local entrepreneurs who will exist only when the whole production chain, except the one part which they can fill themselves, hangs together. Such a production chain involves motivated farmers who produce the right grain and efficient and cheap transportation of raw material to well-functioning mills in cities with enough consumers to support a range of bakeries where flour is made into bread and local food products. These products should be so attractive that they sell themselves and give their producers cash for further investments. Later, when the demand is filled and competition increases, marketing in the western sense might be introduced. Obviously, alternatives to wheat and maize flours have a great future (Satin, 1988) if production and marketing are well coordinated. Most people who grew up in an industrialized society do not appreciate the complexity of the society they live in. Thus, they underestimate the importance of the education of the consumer that is necessary for the market to function in a foreseeable way, which is taken for granted by economists.

THE FUTURE OF SORGHUM AND MILLET TECHNOLOGY AS A VEHICLE FOR AGRO-INDUSTRIAL DEVELOPMENT

Like the pioneers in dry (Hahn, 1970) and wet (Watson, 1970) milling previously cited, the writer is optimistic for the long-range future of sorghum technology. An improved symbiosis should be created between the sorghum plant and man, where the material produced becomes a vehicle for human development mediated by agriculture and technology, both in industrialized and in developing countries. Exploitation of the drought-resistant millet species will be of great significance in local extreme environments. The recent developments, both with the focus on research funds and on technological compatibility of sorghum with maize, have been quite negative due to the low prestige of the sorghum crop, low world market prices, the apparent grain surplus, and the depression in agriculture. We still have 5–10 years, very difficult years, to survive with our know-how, in preparing for the long-term future. With the increased environmental pressure from rapidly growing populations and with the global marginalization of soil, climate, and water resources (a further example of which occurred in southern Africa in 1992), crops adaptable to extreme conditions, such as sorghum, millets, and barley (Munck, 1992a), will be able to compete with maize and wheat. The future of sorghum and millets depends on how far we can develop cooperation between individuals locally and globally, combining the industrial and the green revolutions in sustainable agro-industrial systems. The main component of these systems is people (Young, 1991). In order to succeed, we have to understand and appreciate the limitations of the human mind, including that of scientists, in understanding and directing complex systems and dynamic processes and to develop special methods, e.g., aided by interdisciplinary research and computer programs. The human species is an experimental animal. The plant breeder, the technologist, and the housewife all select extreme items from their environment to combine and test in order to get experience for an optimal life (Munck, 1991). In the 1950s, the technologists were in favor; in the 1960s it was the marketing specialists, in the 1970s the economists, and in the 1980s the lawyers. In the 1960s, the issues of protein malnutrition and farmers' food security led to starting up the IARCs and to research on high-lysine cereals (Axtell et al, 1982; Munck, 1992b) and composite flours (Dendy and James, 1991). In the 1980s, these priorities were almost forgotten, and today biotechnology and environment are in favor as single-question issues. The real needs of developing countries have frequently been overlooked in this top-down approach. It thus seems that our modern society has a split approach to reality, where the ruling paradigm for research funds is not realistically connected to the market of real life. We have to confront the limitations in research funds and markets to guide our development. Thus, money leads to specialization, so that many people lose their empirical overview when moving from rural to urban areas and from everyday life to science. As pointed out by Young (1991), commodities must span time and space, and the empirical local knowledge that was successful before in a static society must now be changed towards a scientifically generalized knowledge that can cope with the dynamics of specialized urban areas. Obviously, in each step of the production chain, money sums up a highly visible selection criterion which, however, has to be supplemented by quantification of amount and quality to be meaningful and operative (Munck, 1991). This is why the Japanese have introduced the concept of "total quality management and control" (Ishikawa, 1985), which is behind the success of much of Japanese industry. This approach emphasizes a holistic approach to production, including environmental and ethical aspects, implemented by prevention of failures through certification of an optimal process rather than by post-factum analysis of products and evaluation of end results. In the 1980s, both the IARCs (e.g., Gomez, 1987) and their partners in governmental bodies in industrialized countries (e.g., Young, 1991) started focusing on multidisciplinary research in developing countries, involving the postharvest technology through small and medium-sized companies. Thus, agricultural products, e.g., sorghum, are used to produce working places and to initiate development and purchasing power. Since then, this fragile start of an essential development has lost impetus—demonstrated by diminishing funding for the postharvest quality aspect of sorghum and millets from local governments and international organizations.

The lasting lesson from the green revolution is that farmers are extremely responsive if they have incentives for developing production—this is also true for local industrialists. The governments of the industrialized and developing countries must cooperate and create new incentives for small, large, and multinational industries to invest in sustainable industrial development based on local agriculture and on a long-term perspective in developing countries.

The multinationals should appreciate that such development, where short-term profit is renounced by accepting a selective intermediate protection of the vulnerable local industry, is a successful way of improving the infrastructure in developing countries over a longer perspective, creating new, significant, stable international markets, which will benefit international trade and secure national and global stability.

We must realize that in this development, food and improved nutrition can never be the end-objective as it was thought of in the 1960s. The ultimate aim should be to use agricultural production to generate work and purchasing power, e.g., for food and a stable, culturally rewarding sustainable development.

A liberal market economy is at present held in high esteem due to the failure of the planned economy in the former communist countries. We all agree that we should not exert guardianship with regard to the decisions of the consumer within the framework of the market, but we have difficulties in understanding why it takes so long for the eastern countries to adapt to the market economy. The reasons seem to be the same in the developing countries—lack of a homogenous tradition and education that can give a consensus within the market, making it *foreseeable*, so that it is possible to plan and invest in this market for the future.

It should be remembered that a considerable part of the North American and European Economic Community economies, as with agriculture, is, de facto, planned. The harmful effect of subsidizing has been treated above. In industrialized countries, taxes are highly biased towards making human work more expensive instead of taxing the consumption of limited natural resources. In this respect, there is now a slow shift in emphasis towards the philosophy of sustainable development (United Nations, 1987; Munck, 1993), which in practice, however, has hardly started.

In the future, one should try to differentiate between those regulations in the market framework that paralyze the positive, self-regulating ability of the market and those that give the individual the necessary incentives for a change—for example, the decision of the Nigerian government in 1987, which overnight created a market outlet for research on composite flours, in place since 1965 (Dendy and James, 1991), and for other industrial uses (Bogunjoko, 1992; Olatunji, 1993). These authors favor the market approach to starting development by questioning the consumers. While this seems to be sound if the economic incentives, e.g., for composite flour, are present, it will give little feedback of value if not. Then the government and politicians should be asked if they can improve the incentives (Dendy, 1992b, 1993) of the free market within a new market framework—that is, an enlightened planned economy. Thus, the future is not extreme solutions but a careful balancing between regulation and self-regulation of the market forces, with a clear sense of what realistically can be regulated and self-regulated. Thus, reverting to the Nigerian case, subventions to support the local sorghum, millet, and maize producers would have had little effect compared to the ban on importing grain, which made local production and consumption take off.

In the future, massive research should be devoted to a better understanding of the agro-industrial development process, comparing the historical perspective of some of the industrialized countries in the West and East with representative cases from the ongoing process in developing countries. Thus, with a parallel to sorghum, we may conclude that rye and rye bread consumption in Denmark today would have been nonexistent if industrial milling and baking technology had not been introduced (Gomez and Munck, 1993). This development may seem small and insignificant as a part of the Danish industrial revolution, but it was just one of the many agro-industrial building blocks that laid the foundation for the present Danish affluence.

Research efforts should involve concerted teams featuring historians, ethnographers, sociologists, economists, biologists, agronomists, and resource technologists. We must admit we are still in the learning phase of understanding how agriculture has boosted our agro-industrial culture to the present state of glory and confusion. In the Western countries, we badly need such a realistic perspective, including that *a certain degree of liberalism may be efficient at the end of the agro-industrial development process and that selective protectionism is essential in the beginning.* Thus in the future, we in the industrialized world should both be able to revitalize our own depressed agriculture (Gomez and Munck, 1993), including sorghum as a durable basis in our lives (Munck, 1993), and to provide the developing nations a better incentive for agro-industrial development, including sorghum and millets for a sustainable agro-industrial production, than the present aid, to avoid chaos and distress in the global society.

ACKNOWLEDGMENTS

When traveling in the sorghum and millet areas in Africa, Asia, and the Americas, I met a great number of farmers, housewives, businessmen, industrialists, scientists and officials who not only taught me about the essentials of these grains but also opened my eyes for a better understanding of the advantages and drawbacks of the past and the ongoing development process in my own country and the industrialized world.

I am most grateful to Bo Löfqvist and co-workers at the United Milling Systems A/S and to the management of the Carlsberg Research Laboratory, Copenhagen, for establishing such a generous foundation, spiritually, technically, and financially, for our sorghum research during 1976–1989. I am most grateful to my former co-workers Kasper Hagemann, Michael Andersen (milling); John Mundy (biochemistry); Knud Erik Bach Knudsen, Birthe Pedersen (nutrition); Lars Hallgren, Jan Grøndal (food technology); Jane Tinning Johansen, Mette Høj, Maj Britt Rask (analyses); and Finn Rexen and Pernille Bjørn Petersen (nonfood products) for their great professional share of this work, partly reviewed in this paper. Warm thanks are also due to John Axtell, Alan Kirleis, and Loyd Rooney, United States, and to Lee House and Manel Gomez, SADCC/ICRISAT, Zimbabwe and India, for inspiring introduction and support. Finally, I am indebted to AACC, David Dendy, and my former secretary Britha Jensen for their helpful advice and for making the publishing of this review possible.

LITERATURE CITED

ANDERSEN, M. 1987. Method of milling and apparatus for carrying out the method. U.S. patent 4,667,888.

ANDERSON, R. A. 1969. Producing quality sorghum flour on wheat milling equipment. Northwest. Miller 276:10-12, 14-15.

ANDRAE, G., and BECKMAN, B. 1985. The Wheat Trap: Bread and Underdevelopment in Nigeria. Scandinavian Institute of African Studies and Zed Press, London.

ASP, N.-G., JOHANSSON, C. G., HALLAUER, H., and SILJESTRÖM, M. 1983. Rapid enzymatic assay of insoluble and soluble dietary fiber. J. Agric. Food Chem. 31:476-482.

ASSELBERGS, E. 1970. Introduction. Page 1 in: Proc. Symp. Use of Non-Wheat Flour in Bread and Baked Goods Manufacture. Tropical Research Institute London.

AXTELL, J. D., KIRLEIS, A. W., HANSEN, M. M., D'CROZ MASON, N., MERTZ, E. T., and MUNCK, L. 1981. Digestibility of sorghum proteins. Proc. Natl. Acad. Sci. USA 78:1333-1335.

AXTELL, J. D., EJETA, G., and MUNCK, L. 1982. Sorghum nutritional quality progress and prospects. Pages 589-604 in: Sorghum in the Eighties. L. R. House, L. K. Mughogho, and J. M. Peacock, eds. Proc. Int. Symp. on Sorghum. Int. Crops Res. Inst. Semi-Arid Tropics (ICRISAT), Patancheru, A. P., India.

AXTELL, J., and MERTZ, E. 1992. Historical perspectives of sorghum protein digestibility. Pages 164-167 in: Proc. Int. Conf. Sorghum Nutritional Quality, 1990. G. Ejeta, E. T. Mertz, L. Rooney, R. Schaffert, and Y. Yoke, eds. Purdue University, West Lafayette, IN.

AYLA, C. 1984. *Pinus brutia* tannin adhesives. Wood adhesives: Present and future. J. Appl. Polym. Sci. 40:69-78.

BACH KNUDSEN, K. E., and MUNCK, L. 1985. Dietary fibre contents and compositions of sorghum and sorghum-based foods. J. Cereal Sci. 3:153-164.

BACH KNUDSEN, K. E., MUNCK, L., and EGGUM, B. O. 1987. Effect of cooking pH and polyphenol level on carbohydrate composition and nutritional quality of a sorghum (*Sorghum bicolor* (L.) Moench.) food, *ugali.* Br. J. Nutr. 59:341-47.

BACH KNUDSEN, K. E., KIRLEIS, A. W., EGGUM, B. O., and MUNCK, L. 1988. Carbohydrate composition and nutritional quality for rats of sorghum *tô* prepared from decorticated white and whole grain red flour. , Pages 588-597 in: Carbohydrate and Fiber. American Institute of Nutrition, Bethesda, MD.

BAGACHWA, M. S. D. 1991. Choice of technology in Industry—The economics of grain milling in Tanzania. Rep. IDRC 279e. International Research Centre, Ottawa, Ontario, Canada.

BASSEY, M. W., and SCHMIDT, O. G. 1989. Abrasive disk dehullers in Africa. International Development Research Centre, Ottawa, Ontario, Canada.

BJØRN PEDERSEN, P., and MUNCK, L. 1993. Whole-crop utilization of barley, including potential new uses. Pages 437-474 in: Barley: Chemistry and Technology. A. W. MacGregor and R. S. Bhatty, eds. Am. Assoc. Cereal Chem., St. Paul. MN.

BOGUNJOKO, J. S. T. 1992. Industrial uses of sorghum in Nigeria. Pages 115-120 in: Utilization of Sorghum and Millets. Proc. Int. Workshop, Bulawayo, Zimbabwe, 1988. M. I. Gomez, L. R. House, L. W. Rooney, and D. A.

V. Dendy, eds. Int. Crops Res. Inst. Semi-Arid Tropics, Patancheru, A. P., India.

BURKITT, D. P., and TROWELL, H. C. 1975. Refined Carbohydrate Foods and Disease—Some Implications of Dietary Fiber. Academic Press, London.

CARANSA, A., and BAKKER, W. G. M. 1987. Modern process for the production of sorghum starch. Starch/Staerke 39 (11):381.

CECIL, J. E. 1986. Roller milling sorghum and millet grain using a semi-wet process. Rep. L74. Tropical Development and Research Institute, London.

CECIL, J. E. 1989. Sorghum and millet decortication technology and food security in SADCC countries. SADCC Post Production Systems Newsl. 12. SADCC/ICRISAT, P. O. Box 776, Bulawayo, Zimbabwe.

CHIGUMIRA, P. 1992. The potential for extruded sorghum food products in Zimbabwe and regional markets. Pages 89-94 in: Utilization of Sorghum and Millets. Proc. Int. Workshop, Bulawayo, Zimbabwe, 1988. M. I. Gomez, L. R. House, L. W. Rooney, and D. A. V. Dendy, eds. Int. Crops Res. Inst. Semi-Arid Tropics, Patancheru, A. P., India.

CRABTREE, J., and DENDY, D. A. V. 1979. Comilling of four species of millet with wheat to produce composite flours. Cereal Foods World 24:103-107.

DADA, L. O., and DENDY, D. A. V. 1987. A preliminary study of broomcorn (*Sorghum bicolor*) grain milling. Rep. L77. Tropical Development and Research Institute, London.

DE FRANCISCO, A., SHEPHERD, A. D., HOSENEY, R. C., and VARRIANO-MARSTON, E. 1982. Decorticating pearl millet and grain sorghum in a laboratory abrasive mill. Cereal Chem. 59:1-5.

DENDY, D. A. V., ed. 1976. Proc. ICC Symp. Sorghum and Millets for Human Food, Vienna, 1976. Tropical Products Institute, London.

DENDY, D. A. V., ed. 1984. Symp. Processing of Sorghum and Millets. ICC Congr 11th. International Association for Cereal Science and Technology, Vienna.

DENDY, D. A. V., ed. 1988. Proc. Quad. Symp. Sorghum Millets 4th, Int. Cereal Bread Congr. 8th. ICC, Lausanne, Switzerland.

DENDY, D. A. V. 1991. Opportunities for traditional uses of the minor millets. SADCC/ICRISAT, P. O. Box 776, Bulawayo, Zimbabwe.

DENDY, D. A. V., ed. 1992a. Proc. Quad. Symp. Sorghum Millets, 5th, Paris. International Association for Cereal Science and Technology, Vienna.

DENDY, D. A. V. 1992b. Composite flour—Past, present and future: A review with special emphasis on the place of composite flour in semi-arid zones. Pages 67-64 in: Utilization of Sorghum and Millets. Proc. Int. Workshop, Bulawayo, Zimbabwe, 1988. M. I. Gomez, L. R. House, L. W. Rooney, and D. A. V. Dendy, eds. Int. Crops Res. Inst. Semi-Arid Tropics, Patancheru, A. P., India.

DENDY, D. A. V. 1993. Drought and urbanization in Africa: The future role of sorghum and millets. Pages 299-318 in: Cereal Science and Technology—Impact on a Changing Africa. Proc. ICC Int. Symp. 1993. J. R. N. Taylor, P. G. Randall and J. H. Viljoen, eds. Council for Scientific and Industrial Research (CSIR), Pretoria, South Africa.

DENDY, D. A. V., and TROTTER, B. W. 1988. Wheatless and composite breads—Technologies awaiting adoption. Entwickl. Ländl. Raum 6:13-18.

DENDY, D. A. V., JAMES, A. W., and OLUGBEMI, L. B. 1990. The use of sorghum in composite and blended products. SADCC/ICRISAT, P. O. Box 776, Bulawayo, Zimbabwe.

DENDY, D. A. V., and JAMES, A. W. 1991. New directions for alternative flour products. World Congress of Food Science and Technology, 8th.

DESIKACHAR, H. S. R. 1982. Pearling and milling studies on sorghum. Pages 194-199 in: Proc. Int. Symp. Sorghum Grain Quality. L. W. Rooney an D. S. Murty, eds. Int. Crops Res. Ins. Semi-Arid Tropics (ICRISAT), Patancheru, A. P., India

DINAT, S. 1992. Status of sorghum milling technology at rural industries innovation centre. Botswana. Pages 33-40 in: Utilization of Sorghum and Millets. Proc. Int. Workshop, Bulawayo, Zimbabwe, 1988. M. I. Gomez, L. R. House, L. W. Rooney, and D. A. V. Dendy, eds. Int. Crops Res. Inst. Semi-Arid Tropics, Patancheru, A. P., India.

DOANE, W. M. 1981. Starch industrial raw material. Pages 265-290 in: Cereals, A Renewable Resource—Theory and Practice. Y. Pomeranz, and L. Munck, eds. Am. Assoc. Cereal Chem., St. Paul, MN.

DOGGETT, H. 1982. A look back on the 70's. Pages 15-24 in: Sorghum in the Eighties. L. R. House, L. K. Mughogho, and J. M. Peacock, eds. Proc. Int. Symp. on Sorghum. Int. Crops Res. Inst. Semi-Arid Tropics (ICRISAT), Patancheru, A. P., India.

EASTMAN, P. 1980. An end to pounding. IDRC-152e. International Development Research Centre, Ottawa, Canada.

EGGUM, B. O., KNUDSEN, K. E. B., MUNCK, L., AXTELL, J., and MUKURU, S. Z. 1982. Milling and nutritional value of sorghum in Tanzania. Pages 211-225 in: Proc. Int. Symp. Sorghum Grain Quality. L. W. Rooney and D.

S. Murty, eds. Int. Crops Res. Ins. Semi-Arid Tropics (ICRISAT), Patancheru, A. P., India.

EGGUM, B. O., MONOWAR, L., KNUDSEN, K. E. B., MUNCK, L., and AXTELL, J. 1983. Nutritional quality of sorghum and sorghum foods from Sudan. J. Cereal Sci. 1:127-137.

EJETA, G. 1982. Kisra quality: Testing new sorghum varieties and hybrids. Pages 67-72 in: Proc. Int. Symp. Sorghum Grain Quality. L. W. Rooney and D. S. Murty, eds. Int. Crops Res. Ins. Semi-Arid Tropics (ICRISAT), Patancheru, A. P., India.

FAURE, J., and M'BENQUE, M. Y. 1991. The dehulling and milling of sorghum. SADCC/ICRISAT, P. O. Box 776, Bulawayo, Zimbabwe.

FAURE, J. 1992. Sorghum and maize pasta and extruded products. Pages 75-82 in: Utilization of Sorghum and Millets. Proc. Int. Workshop, Bulawayo, Zimbabwe, 1988. M. I. Gomez, L. R. House, L. W. Rooney, and D. A. V. Dendy, eds. Int. Crops Res. Inst. Semi-Arid Tropics, Patancheru, A. P., India.

FRANDSEN, E. 1987. M.Sc. thesis. (In Danish.) The Technical University of Denmark and Carlsberg Research Laboratory, Copenhagen.

GOMEZ, M. I. 1987. The technological status of sorghum and food grain in Zimbabwe. SADCC/ICRISAT, P. O. Box 776, Bulawayo, Zimbabwe.

GOMEZ, M. I. 1993. Comparative evaluation and optimization of a milling system for small grains. Pages 463 in: Cereal Science and Technology—Impact on a Changing Africa. Proc. ICC Int. Symp. 1993. J. R. N. Taylor, P. G. Randall, and J. H. Viljoen, eds. Council for Scientific and Industrial Research (CSIR), Pretoria, South Africa.

GOMEZ, M. I., and MUNCK, L. 1993. Whole cereal crop utilization—Towards a sustainable agroindustrial revolution. Pages 653-668 in: Cereal Science and Technology—Impact on a Changing Africa. Proc. ICC Int. Symp. 1993. J. R. N. Taylor, P. G. Randall, and J. H. Viljoen, eds. Council for Scientific and Industrial Research (CSIR), Pretoria, South Africa.

GOMEZ, M. I., HOUSE, L. R., ROONEY, L. W., and DENDY, D. A. V., eds. 1992. Utilization of sorghum and millets. Proc. Int. Workshop Bulawayo, Zimbabwe, 1988. Int. Crops Res. Inst. Semi-Arid Tropics (ICRISAT), Patancheru, A. P., India.

GRAHAM, G. C., MacLEAN, W. C., MORALES, E., HAMAKER, B. R., KIRLEIS, A. W. MERTZ, E. T., and AXTELL, J. D. 1986. Digestibility and utilization of protein and energy from *nasha,* a traditional Sudanese fermental sorghum weaning food. J. Nutr. 116:978-984.

GRAM, L. E., MUNCK, L., and ANDERSEN, M. P. 1989. A short milling process for wheat. Pages 445-455 in: Wheat is Unique. Y. Pomeranz, ed. Am. Assoc. Cereal Chem., St. Paul, MN.

GRØNDAL, J. 1988. Industriel fremstilling af glutenfri bagte cerealieprodukter til brug for det afrikanske marked. M.Sc. thesis. The Royal Veterinary and Agricultural University, Copenhagen, and the Carlsberg Research Laboratory, Copenhagen.

HAHN, R. R. 1970. Dry milling and products of grain sorghum. Pages 573-595 in: Sorghum Production and Utilization. J. S. Wall and W. M. Ross, eds. Avi Publ. Co., Westport, CT.

HALLGREN, L. 1984. Functional properties of sorghum and wheat in breadbaking. Pages 16-23 in: Symp. Processing of Sorghum and Millets. ICC Congr 11th. International Association for Cereal Science and Technology, Vienna..

HALLGREN, L. 1985. Physical and structural properties of cereals, sorghum in particular, in relation to milling methods and product use. Ph.D. thesis no. 96. Technical Academy of Denmark, Lyngby.

HALLGREN, L., and MURTY, D. S. 1983. A screening test for grain hardness in sorghum employing density grading in sodium nitrate solution. J. Cereal Sci. 1:265-274.

HALLGREN, L., RØGE, E., MUNCK, L., and YI, S. 1988. Defining milling quality of sorghum cultivars. Pages 96-122 in: Proc. Quad. Symp. Sorghum Millets 4th, Int. Cereal Bread Congr. 8th. D. A. V. Dendy, ed. ICC, Lausanne, Switzerland.

HALLGREN, L., REXEN, F., BJØRN PETERSEN, P., and MUNCK, L. 1992. Industrial utilization of whole crop sorghum for food and industry. Pages 121-130 in: Utilization of Sorghum and Millets. Proc. Int. Workshop, Bulawayo, Zimbabwe, 1988. M. I. Gomez, L. R. House, L. W. Rooney, and D. A. V. Dendy, eds. Int. Crops Res. Inst. Semi-Arid Tropics, Patancheru, A. P., India.

HANSEN, M., PEDERSEN, B., MUNCK, L., and EGGUM, B. O. 1989. Weaning foods with improved energy and nutrient density prepared from germinated cereals. 1. Preparation and dietary bulk of gruels based on barley. Food Nutr. Bull. U.N. Univ. 11(2):40-45.

HEMMINGWAY, R. W., and KREIBICH, R. E. 1984. Condensed tannin: Resorcinol adducts and their use in wood laminating adhesives—An exploratory study. In: Wood Adhesives. Present and Future. J. Appl. Polym. Sci. 40:79-90.

HOSENEY, R. C., FAUBION, J. M., and REDDY, V. P. 1992. Organoleptic inplications of milled pearl millet. Pages 27-32 in: Utilization of Sorghum and Millets. Proc. Int. Workshop, Bulawayo, Zimbabwe, 1988. M. I.

Gomez, L. R. House, L. W. Rooney, and D. A. V. Dendy, eds. Int. Crops Res. Inst. Semi-Arid Tropics, Patancheru, A. P., India.

HOUSE, L. R., MUGHOGHO, L. K., and PEACOCK, J. M., eds. 1982. Sorghum in the Eighties. Vols. 1 and 2. Proc. Int. Symp. Sorghum, 1981. Int. Crops Res. Inst. Semi-Arid Tropics (ICRISAT) Patancheru, A. P., India.

HULSE, J. H., LAING, E. M., and PEARSON, D. E. 1980. Sorghum and the Millets: Their Composition and Nutritive Value. Academic Press, London.

ISHIKAWA, K. 1985. What is Total Quality Control? The Japanese Way. Prentice-Hall, Englewood Cliffs, NJ.

KIRLEIS, A. W. 1989. Chemical and physical aspects of food and nutritional quality of sorghum. Pages 147-152 in: INTSORMIL Annu. Rep. 241. University of Nebraska, Lincoln.

KIRLEIS, A. W. 1992. The prolamins of sorghum: Their role in protein digestibility. Pages 168-176 in: in: Proc. Int. Conf. Sorghum Nutritional Quality, 1990. G. Ejeta, E. T. Mertz, L. Rooney, R. Schaffert, and J. Yohe, eds. Purdue University, West Lafayette, IN.

KLOPFENSTEIN, C. F., LEIPOLD, H. W., and CECIL, J. E. 1991. Semi-wet milling of pearl millet for reduced goitrogenicity. Cereal Chem. 68:177-179.

MacLEAN, W. C., LOPEZ DE ROMANA, J. R., JR., PLACKO, R. P., and GRAHAM, G. G. 1982. Nutritional value of sorghum in prescool children: Digestibility utilization and plasma free amino acids. Pages 377-382 in: Proc. Int. Symp. Sorghum Grain Quality. L. W. Rooney and D. S. Murty, eds. Int. Crops Res. Ins. Semi-Arid Tropics (ICRISAT), Patancheru, A. P., India.

MALLESHI, N. G. 1988a. Weaning Foods. Central Food Technological Research Institute, Myosore, India. RESC Sci. Ser. 8.

MALLESHI, N. G. 1988b. Processing small millets for food and industrial uses. Pages 325-339 in: Proc. Small Millets Workshop, 1st, Bangalore, 1986. Oxford and IBH Publishing, New Delhi.

MALLESHI, N. G., and DESIKACHAR, H. S. R. 1985. Milling, popping and malting characteristics of some minor millets. J. Food Sci. Technol. 22:400-403.

MALLESHI, N. G., DAODU, M. A., and CHANDRASEKHAR, A. 1989. Development of weaning food formulations based on malting and roller drying of sorghum and cowpea. J. Food Sci. Technol. 24:511-519.

MORE, C. A., and MAJORS, K. R. 1970. Economic considerations. In: Sorghum Production and Utilization. J. S. Wall and W. M. Ross, eds. Avi Publ. Co., Westport, CT.

MUNCK, L. 1991. Man as selector—a Darwinian boomerang striking through natural selection. Pages 211-227 in: Environmental Concerns—An Interdiciplinary Exercise. J. A. Hansen, ed. Elsevier Applied Science Publ., London.

MUNCK, L. 1993. On the utilization of the renewable plant resources. Pages 500-522 in: Plant Breeding: Principles and Prospects. M. D. Hayware, N. O. Bosemark, and I. Romagosa, eds. Chapman and Hall, London, England.

MUNCK, L. 1992a. The contribution of barley to agriculture today and in the future. Proc. Int. Barley Genet. Symp. 6th, 1991, Helsingborg, Sweden.

MUNCK, L. 1992b. The case of high lysine barley. Pages 573-602 in: Barley Genetics Biochemistry, Molecular Biology and Biotechnology. P. R. Shewry, ed. C.A.B. International, Wallingford, Oxon, UK.

MUNCK, L., BACH KNUDSEN, K. E., and AXTELL, J. D. 1982a. Milling processes and products as related to kernel morphology. Pages 200-210 in: Proc. Int. Symp. Sorghum Grain Quality. L. W. Rooney an D. S. Murty, eds. Int. Crops Res. Ins. Semi-Arid Tropics (ICRISAT), Patancheru, A. P., India.

MUNCK, L., BACH KNUDSEN, K. E., and AXTELL, J. D. 1982b. Industrial milling of sorghum for the 1980's. Pages 565-570 in: Sorghum in the Eighties. Vol. 2. Proc. Int. Symp. Sorghum, 1981. Int. Crops Res. Inst. Semi-Arid Tropics (ICRISAT), Patancheru, A. P., India.

MUNCK, L., REXEN, F., PETERSEN, B., and BJØRN PETERSEN, P. 1989. A new agricultural system for Europe? Pages 49-70 in: Wheat Is Unique. Y. Pomeranz, ed. Am. Assoc. Cereal Chem., St. Paul, MN.

MURTY, D. S. 1992. The breeders role in sorghum utilization: A perspective. Pages 157-167 in: Utilization of Sorghum and Millets. Proc. Int. Workshop, Bulawayo, Zimbabwe, 1988. M. I. Gomez, L. R. House, L. W. Rooney, and D. A. V. Dendy, eds. Int. Crops Res. Inst. Semi-Arid Tropics, Patancheru, A. P., India.

MURTY, D. S., and SUBRAMANIAN, V. 1982. Sorghum roti: Traditional methods of consumption and standard procedures for evaluation. Pages 73-78 in: Proc. Int. Symp. Sorghum Grain Quality. L. W. Rooney an D. S. Murty, eds. Int. Crops Res. Ins. Semi-Arid Tropics (ICRISAT), Patancheru, A. P., India.

NOVELLIE, L. W. 1993. Sorghum research perspective. Pages 525-540 in: Cereal Science and Technology—Impact on a Changing Africa. Proc. ICC Int. Symp. 1993. J. R. N. Taylor, P. G. Randall, and J. H. Viljoen, eds. Council for Scientific and Industrial Research (CSIR), Pretoria, South Africa.

OLATUNJI, O. 1993. Developed technologies for

substitution of locally grown cereals in selected food products. Pages 509-524 in: Cereal Science and Technology—Impact on a Changing Africa. Proc. ICC Int. Symp. 1993. J. R. N. Taylor, P. G. Randall, and J. H. Viljoen, eds. Council for Scientific and Industrial Research (CSIR), Pretoria, South Africa.

OOMAH, B. D., REICHERT, R. D., and YOUNGS, C. D. 1981. A novel, multisample tangential abrasive dehulling device (TADD). Cereal Chem. 56:392-395.

PEDERSEN, B., HANSEN, M., MUNCK, L., and EGGUM, B. O. 1989. Weaning foods with improved energy and nutritional density prepared from germinated cereals 2. Nutritional evaluation of gruels based on barley. Food Nutr. Bull. U.N. Univ. 11(2)46-51.

PERTEN, H. A. 1976. UNDP/FAO sorghum processing project in Sudan. In: Proc. ICC Symp. Sorghum and Millets for Human Food, Vienna, 1976. D. A. V. Dendy, ed. Tropical Products Institute, London.

PERTEN, H. 1984. Industrial processing of millet and sorghum. Pages 52-55 in: Symp. Processing of Sorghum and Millets. ICC Congr 11th. D. A. V. Dendy, ed. International Association for Cereal Science and Technology, Vienna.

PIZZI, A., and ROUX, D. G. 1978. The chemistry and development of tannin based weather and boil proof cold-setting and fast-setting adhesives of wood. J. Appl. Polym. Sci. 22: 1945-1954.

PUSHPAMMA, P., RATNAKUMARI, A., and GEERVANI, D. L. 1979. Nutritional quality of sorghum and legume based food mixtures for infants and prescool children. Nutr. Rep. Int. 19:643-648.

PUSHPAMMA, P., and GEERVANI, P. 1981. Nutritive value of traditional recipes of Andhra Pradesh. College of Home Science, Andhra Pradesh Agricultural University, Hyderabad, India.

PUSHPAMMA, P., and RAO, C. K. 1981. Varietal Preference Marketing, Storage, Processing and Utilization of Sorghum and Millets in Andra Pradesh. Andhra Pradesh Agricultural University, Hyderabad, India.

RAMANZIN, M., ØRSKOV, E. R., and TUAK, A. K. 1986. Feed value of straw components. Animal Production 43:271-278.

REICHERT, R. D. 1982. Sorghum dry milling. 1982. Pages 547-564 in: Sorghum in the Eighties. Vol. 9. Proc. Int. Symp. Sorghum, 1981. L. R. House, L. K. Mughogho, and J. M. Peacock, eds. Int. Crops Res. Inst. Semi-Arid Tropics (ICRISAT) Patancheru, A. P., India.

REXEN, F., and MUNCK, L. 1984. Cereal Crops for Industrial Use in Europe, EUR 9617 EN. Rep. Comm. European Communities.

REXEN, F., and BACH KNUDSEN, K. E. 1984. Industrial scale dry scale dry treatment with sodium hydroxyde. Pages 127-161 in: Straw and Fibrous By-Products as Feed. F. Sundstøl and E. Owen, eds. Elsevier Publ., New York.

ROONEY, L. W., and SULLINS, R. D. 1969. A laboratory method for milling small samples of sorghum grain. Cereal Chem. 46:486-490.

ROONEY, L. W., FRYAR, W. B., and CATER, C. M. 1972. Protein and amino acid contents of successive layers removed by abrasive milling of sorghum grain. Cereal Chem. 49:399-406.

ROONEY, L. W., and MILLER, F. R. 1982. Variation in the structure and kernel characteristics of sorghum.Pages in: Proc. Int. Symp. Sorghum Grain Quality. L. W. Rooney an D. S. Murty, eds. Int. Crops Res. Ins. Semi-Arid Tropics (ICRISAT), Patancheru, A. P., India., p. 143-162.

ROONEY, L. W., and MURTY, D. S., eds. 1982. Proc. Int. Symp. Sorghum Grain Quality. Int. Crops Res. Inst. Semi-Arid Tropics (ICRISAT), Patancheru, A. P. India.

ROONEY, L., and SERNA-SALDIVAR, S. O. 1991. Sorghum. In: Handbook of Cereal Science and Technology. K. J. Lorenz and K. Kulp, eds. Marcel Dekker, New York.

SATIN, M. 1988. Bread without wheat. New Scientist. April 28, pp. 56-9.

SCHMIDT, O. G. 1992. An analysis of progress and achievements in regional dehulling projects. In: Utilization of Sorghum and Millets. Proc. Int. Workshop, Bulawayo, Zimbabwe, 1988. M. I. Gomez, L. R. House, L. W. Rooney, and D. A. V. Dendy, eds. Int. Crops Res. Inst. Semi-Arid Tropics, Patancheru, A. P., India.

SERNA-SALDIVAR, S. O., KNABE, D. A., ROONEY, C. W., and TANKSLY, T. D., JR. 1987. Effects of lime cooking on energy and protein digestibilities of maize and sorghum. Cereal Chem. 64:247-252.

SHEPHERD, A. D. 1979. Laboratory abrasive decortication mill for small grains. Cereal Chem. 56:517-519.

SHUMBA, E. M. 1993. Impact of technology and related support services on smallholder cereal production in Zimbabwe. Pages 173-180 in: Cereal Science and Technology—Impact on a Changing Africa. Proc. ICC Int. Symp. 1993. J. R. N. Taylor, P. G. Randall, and J. H. Viljoen, eds. Council for Scientific and Industrial Research (CSIR), Pretoria, South Africa.

SOOLIMAN, A. O. G. 1993. Sorghum processing at the crossroads in South Africa. Pages 497-509 in: Cereal Science and Technology—Impact on a Changing Africa. Proc. ICC Int. Symp. 1993. J. R. N. Taylor, P. G. Randall, and J. H. Viljoen, eds. Council for Scientific and Industrial Research (CSIR), Pretoria, South

Africa.

SUBRAMANIAN, V., and JAMBUNATHAN, R. 1988. Potential alternative uses of sorghum and pearl millet. Pages 46-66 in: Proc. Int. Workshop, Bulawayo, Zimbabwe, 1988. M. I. Gomez, L. R. House, L. W. Rooney, and D. A. V. Dendy, eds. Int. Crops. Res. Inst. Semi-Arid Tropics (ICRISAT), Patancheru, A. P., India.

SUBRAMANIAN, V., SURYAPRAKASH, S., JAMBUNATHAN R., and MURTY, D. S. 1988. Dehulling and milling characteristics of some sorghum cultivars. Pages 126-140 in: Proc. Int. Workshop, Bulawayo, Zimbabwe, 1988. M. I. Gomez, L. R. House, L. W. Rooney, and D. A. V. Dendy, eds. Int. Crops. Res. Inst. Semi-Arid Tropics (ICRISAT), Patancheru, A. P., India.

SUBRAMANIAN, V., and JAMBUNATHAN, R. 1992. Laboratory procedures for evaluating grain and food quality of sorghum and pearl millet: Problems and prospects. Pages 143-150 in: Utilization of Sorghum and Millets. Proc. Int. Workshop, Bulawayo, Zimbabwe, 1988. M. I. Gomez, L. R. House, L. W. Rooney, and D. A. V. Dendy, eds. Int. Crops Res. Inst. Semi-Arid Tropics, Patancheru, A. P., India.

TAYLOR, J. R. N., RANDALL, P. G., and VILJOEN, J. H., eds. 1993. Cereal Science and Technology—Impact on a Changing Africa. Proc. ICC Int. Symp. Council for Scientific and Industrial Research (CSIR), Pretoria, South Africa.

UNITED NATIONS. 1987. U. N. World Commission Report on Environment and Development. Oxford University Press, Oxford, England.

VAN VUUREN, J. J., and SKINNER, P. R. 1993. Marketing opportunities for grain sorghum (a new generation of sorghum consumers). Pages 489-497 in: Cereal Science and Technology—Impact on a Changing Africa. Proc. ICC Int. Symp. 1993. J. R. N. Taylor, P. G. Randall, and J. H. Viljoen, eds. Council for Scientific and Industrial Research (CSIR), Pretoria, South Africa.

VOGEL, S., and GRAHAM, M. 1978. Sorghum and millet: Food production and use. Report of a workshop held in Nairobi, Kenya, 4-7 July, 1978. International Development Research Centre, Ottawa, Ontario, Canada.

WATSON, S. A. 1970. Wet-milling and products. In: Sorghum Production and Utilization. J. S. Wall and W. M. Ross, eds. Avi Publ. Co., Westport, CT.

WATSON, S. A. 1984. Corn and sorghum starches: Production. In: Starch Chemistry and Technology, 2nd ed. Academic Press, Orlando, FL.

WILLS, R. B. H., and ALI, M. R. 1983. Effect of grain size on dehulling of sorghum. Cereal Chem. 60:12-14.

YOUNG, R., HAIDARA, M., ROONEY, L. W., and WANISKA, R. D. 1990. Parboiled sorghum: Development of a novel decorticated product. J. Cereal. Sci. 11:277-289.

YOUNG, R. H. 1991. Addressing the human dimension in Nutrition Sciences, Agro-Industries and International Agricultural Research. Rep. IDRC-285e. International Research Centre, Ottawa, Ontario, Canada.

CHAPTER 9

LAGER BEERS FROM SORGHUM

L. Hallgren
Carlsberg Brewery A/S
Copenhagen, Denmark

INTRODUCTION

Barley is the traditional material for producing lager beer (Narziss, 1976). Since ancient times, brewers have depended on malted barley grain. This cereal, however, is primarily grown in temperate regions of the world and does not thrive well in tropical and subtropical climatic conditions. In contrast to barley, sorghum and millets are indigenous to and grow well in the semiarid tropics. These crops are the traditional brewing materials for African opaque beer (Novellie, 1968; Vogel and Graham, 1979). For centuries the indigenous sorghum beers have occupied an important place in the diets of many African people, and the industrialization of this kind of beer started more than 50 years ago (Schwartz, 1956; Novellie, 1968; Haggblade and Holzapfel, 1989).

Due to an increasing demand for Western clear beer, research work on the use of locally grown sorghum and auxiliary enzymes for producing conventional European beer in Nigeria was carried out in the 1970s and early 1980s (Skinner, 1976; Okafor and Aniche, 1980). This work, however, was not recognized by the breweries until January 1988, when the import of barley malt into Nigeria was banned. This was part of a program to encourage local agriculture and to deal with the economic crisis by cutting down on expensive imports. The Nigerian government's intention was to force breweries to use local crops such as sorghum and corn. As a consequence, Nigeria's 23 breweries, with an annual production of around 10 million hectoliters (Malomo, 1989), had to learn to brew beer with reduced amounts or even without the raw material that is considered vital for making good beer—barley malt. Clear beers made in a way that was never thought possible only a few years ago are now being made successfully in Nigeria (Fig. 1).

THE LAGER BREWING PROCESS

Conventional lager beer brewing involves four major steps: malting of barley grain, wort production, fermentation, and finishing (Hough et al, 1971). The purpose of the brewing process is to extract the soluble components of the malt and

Present affiliation: Novo Nordisk A/S, Bagsvaerd, Denmark.

Fig. 1. Typical sorghum beers of Nigeria. (Source: Novo Nordisk A/S, 1991; used by permission.)

the hops. During malting, the barley grain is induced to germinate in order to solubilize cell walls in the barley endosperm and to produce enzymes essential for the brewing process. The most important enzymes formed are amylases (α- and β-amylases) and proteases. Specific mashing temperature programs have been developed for barley beer brewing in order to ensure the formation of an optimal soluble substrate for the yeast fermentation. These programs influence the quality and the storability of the finished beer. Some breweries may add grits or starch from unmalted, gelatinized cereal adjuncts like corn and rice. This is possible due to the fact that barley during malting produces more enzymatic activity than needed for the solubilization of the malt itself. Adjuncts are a cheaper extract source than malt. However, the major reason for their use in breweries in industrialized countries is the lighter drinking quality of the beers, which has resulted in a continuous increase in their consumption. Today, adjuncts are used in different proportions in more than 90% of the world's beer (Canales, 1979). In brewing, the soluble extract (wort) is separated from the solids (spent grains). This separation may involve filtration in a lauter tun, a vessel with a bottom plate with many holes on which the barley husk will sediment and act as a filter aid. Then the wort is boiled with hops, filtered and cooled. The cold wort is pitched with brewer's yeast to initiate the fermentation process. After an initial primary fermentation, the beer is allowed to mature in a secondary fermentation. Then the beer is clarified through filtration, carbonated and bottled.

SORGHUM IN LAGER BEER BREWING

Sorghum is used in the lager brewing process as malt or as adjunct. However, differences in functional properties of the sorghum grain for brewing compared to barley must be taken into account when designing industrial brewing processes for grists with low levels of barley malt. In the processing industry, the technology of handling raw materials depends mainly on the physical and chemical properties of the raw material. For years, barley malt has been the major raw material for brewing of lager beers, and therefore, the brewery equipment and procedures are designed on the basis of the functional properties of the barley grain (Arri, 1989).

SORGHUM AS BREWING ADJUNCT

Although brewing trials have been done on extruded, dehulled sorghum (Dale et al, 1989; Delcour et al, 1989), the primary use of this cereal as adjunct in lager beer brewing is in the form of brewer's grits. Sorghum brewer's grits are obtained from dry milling of the dehulled grain (Hallgren et al, 1992; and see Chapter 8). The relative production of flour and brewer's grits are highly correlated with the hardness of the sorghum grain (Hallgren and Murty, 1983). A high proportion of vitreous to floury endosperm will favor the yield of grits. From a study in Nigeria in 1987 (*unpublished*), we found an average yield of approximately 50% brewer's grits from an industrial milling process of a relatively hard sorghum cultivar (SK 5912). The result was obtained from an industrial 5-t sorghum mill (United Milling Systems A/S, Denmark), which included both dehulling and grinding. The fine flour from the mill that was a by-product of producing brewer's grits was sold to local biscuit manufacturers. The beer quality brewed from grits of this mill was found to be as good as when using conventional corn grits as adjunct. In a similar study Canales (1979), who investigated the production and use of sorghum adjunct in Mexico, found a much lower yield (35% of sorghum grits). In this case, the flour fraction of 12% was mixed with the grits fraction, thus resulting in a total adjunct yield of 47%. As we shall see this fine flour can give technological and quality problems in the brewing process.

As with corn grits, the quality of sorghum grits may vary depending on the quality of the sorghum grain, the milling method involved, the oil and crude fiber content, the extract yield, etc.

Laboratory Testing of Sorghum Brewer's Grits

The test for quality of brewer's grits from sorghum includes both physical and chemical characterization.

PARTICLE SIZE DISTRIBUTION

A sifting test of the brewer's grits is performed in order to determine the particle size distribution. A relatively well-defined particle size distribution of the grits is important to ensure good brewing performance. Addition of the fine flour fraction as described by Canales (1979) may therefore give problems during brewing, as fine mash particles tend to block the fine pores in the filter cake of the lauter tun. Consequently, the fine sorghum flour may produce a mash with a relatively long run-off time in conventional lauter tuns. For corn, the following specifications for particle size distribution are given for Scandinavian breweries (sifting on three sieves): no particles larger than 3,000 μm; a maximum of 3.5% larger than 1,270 μm and a maximum of 1.5% through the 253–μm sieve (Canales, 1979). The same specifications should be recommended for sorghum brewer's grits.

PROXIMATE ANALYSIS

Kernels of corn, traditionally used as adjunct in lager beer brewing, are very similar to sorghum grain regarding content of starch, protein, and fiber (Canales and Sierra, 1976). Sorghum contains slightly less oil and slightly more protein than corn. The starch and protein of the two cereals are practically identical for industrial uses.

A laboratory analysis of sorghum brewer's grits should at least include an analysis for oil (fat) content. The oil content of the sorghum grits should be as low as possible (<1%), as oil reduces the storability of the grits due to oxidation and produces a final beer with poor foam stability. During the rough sorghum grinding process, parts of the soft and brittle germ often separate with the fine flour from the central endosperm. Consequently, fine sorghum flour by-products in grits milling tend to have a high fat content. Other proximate analyses could include analysis for protein, crude fiber, and ash content, which are negatively correlated with extract yield. Fiber and ash are indicators of the performance of the milling system used. The starch content, which is positively correlated to extract yield, should be high, giving an extract (on dry basis) of approximately 90%. The international Malt Conversion Method suggested by the ASBC (American Society of Brewing Chemists, 1976) may be used to determine extract yield in sorghum brewer's grits.

Grits with adhering colored pigments from brown-seeded sorghums may contain high levels of tannins and should be avoided as adjunct for lager beers (Canales and Sierra, 1976).

TESTS FOR INDUSTRIAL LAUTERING PERFORMANCE

One of the most serious problems identified in breweries using sorghum adjunct is variation in the run-off time of the wort from the lauter tun. This is mainly caused by insufficient separation in the milling system, which may influence the lautering performance due to a high proportion of soft endosperm in the raw material that may result from e.g., using the wrong variety, bad weathering, or poor storage and handling. In a study involving industrial samples, we developed a laboratory test for prediction of lautering performance.

a) *Density Grading.* Density of sorghum brewer's grits was investigated in relation to the run-off time in full-scale breweries. A screening test for grain hardness in sorghum employing density grading sorghum grain was developed (Hallgren and Murty, 1983). The principle of this method is based on density differences between vitreous (heavy) and floury (light) endosperm. The percentage of floating kernels in a sodium nitrate solution of 1.315 g/ml was highly positively correlated with grain hardness and therefore to grits yield from the milling process. Consequently, the sodium nitrate test can be used at the grain elevator to predict the yield of brewer's grits. In a later paper (Pomeranz, 1986), four other screening methods for sorghum grain hardness were evaluated. All these methods and other methods can be used as indirect measurements of endosperm vitreousity. However, very few methods are as simple as the $NaNO_3$ flotation method.

The floatation test for whole grains with sodium nitrate solution could not be used on brewer's grits due to viscosity and surface tension problems. Instead we found that a mixture of organic solvents, such as, 100 ml of trichloroethylene and 3 ml of petroleum ether (boiling point 40–60°C) was useful. This solvent mixture had a density of 1.44 g/ml at 20°C. Depending on the particle size of the grits sample, it may be necessary to adjust the density of the mixture accordingly. The solvents should be carefully handled in a fume hood due to their toxicity. A 5-g sample of sorghum grits was placed in a normal test tube and 10 ml of the solvent mixture added. After covering and shaking, the suspension was left to settle for 30 min. Then the thickness of the upper and the lower layer was measured (in millimeters), and the ratio between these two layers (upper/lower) was defined as the floatation performance ratio (FPR). With this modified flotation method, we were able to group a range of industrial sorghum grits samples into a group representing long run-off times (high FPR) and one group with normal lautering performance (low FPR).

Our results indicated that endosperm vitreousness (hardness) of sorghum brewer's grits is a very important parameter to measure both on the grain level in selection of the raw material for the production of grits and on the final grits received at the brewery.

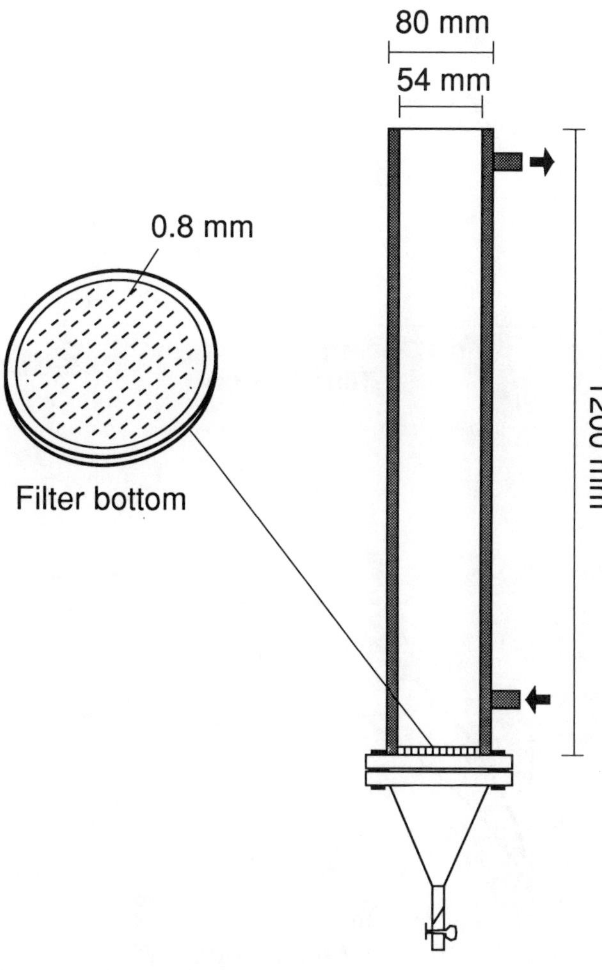

Fig. 2. Laboratory lautering column with specially designed filter bottom. The arrows indicate the inlet and outlet of the water jacket, thermostated at 76°C.

b) *Laboratory Lautering Studies.* A laboratory column was set up in order to make small-scale experiments on lautering performance of sorghum grits. The laboratory lauter column was a cylindrical water-jacketed vessel made of Plexiglas with a conical funnel outlet. The dimensions are shown in Figure 2.

The water jacket was connected to a water bath thermostated at 76°C. The filter bottom (sieve in Fig. 2) was made from the same stainless steel material as used in full-scale lauter tuns (Steinecker bottom).

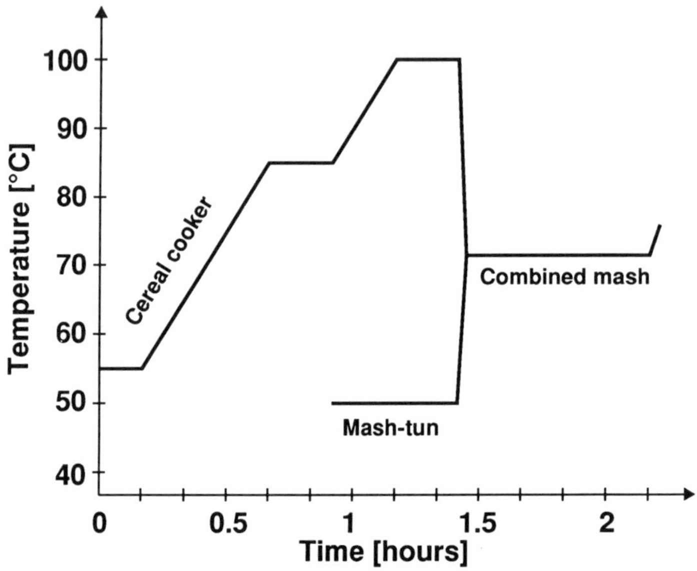

Fig. 3. Mashing cycle used in the laboratory test for lautering performance.

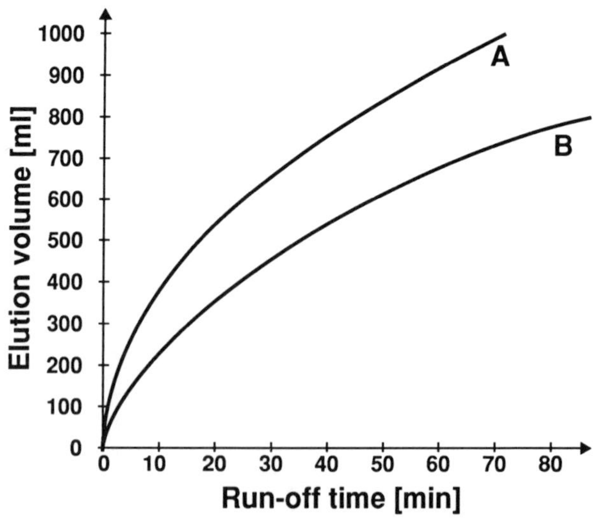

Fig. 4. Wort volume versus run-off time in the laboratory lautering column. The upper (A) and the lower (B) curves are samples of normal and poor industrial lautering performance.

The laboratory lauter tun was filled with 76°C water up to 3 cm over the filter bottom. The mash sample (76°C) was then carefully added to the column, batchwise in small portions. The time of packing the column to a total volume of 2,000 ml should take about 5 min. After a resting period of 15 min, the wort was drained through the bottom outlet. Three portions of 200-ml filtrate were collected and recycled prior to the starting point of registration of run-off time performance.

The mash for the laboratory lautering experiments was prepared from the following recipe: barley malt 248 g, sorghum grits 258 g, water 910 g (cereal cooker), water (mash-tun) 884 g, and 0.8 g of commercial thermostable α-amylase (Termamyl 120L, Novo Nordisk A/S, Denmark). The brewing program is shown in Figure 3.

Figure 4 shows the curves obtained from average results of testing sorghum grits samples with good (normal) and poor industrial lautering performance. The correlation between run-off time for 600 ml of wort in the laboratory column and the FPR values of all six industrial samples was $r = 0.76$.

c) *Sedimentation of Spent Grain.* The rate of sedimentation of spent grain during lautering was expected to be of importance for a proper formation of the filter-bed and therefore expected to influence run-off time. From the laboratory lautering experiments the sedimentation rate was followed and measured as the height of clear wort in the top of the column after 15 min resting time, prior to the elution. The results from the two sets of commercial sorghum grits samples again split into two separate groups indicating that the density of the sorghum grits reflected the density of the spent grain. This was further confirmed from the correlation between FPR value of the sorghum grits samples and the corresponding sedimentation of the spent grain which was as high as $r = -0.91$.

COOKING PROPERTIES

The gelatinization temperature (GT) of sorghum starch can be followed by the loss of birefringence in the hot stage microscope. In one study (Hallgren, 1984), GT's between 68 and 76°C were found for sorghum starch. From these and other findings, it can be concluded that on average sorghum starch has a slightly higher GT (2–6°C) than corn starch (Canales and Sierra, 1976; Canales, 1979). Comparing starches isolated from the vitreous and the floury part of the endosperm of the same variety showed that the vitreous endosperm starches exhibited a slightly higher GT than the floury endosperm starches (Cagampang, 1983; Hallgren, 1984). This relatively high GT is important to consider in relation to the extra energy and longer time needed during adjunct cooking of sorghum grits compared with brewer's grits from other cereals.

Performance of sorghum starch and grits particles in the adjunct cooker can be tested in a Brabender Viscoamylograph (Cantu et al, 1984; Hallgren, 1985). Changes in paste viscosity for sorghum endosperm material isolated from the vitreous and floury endosperm of three varieties was investigated (Hallgren, 1984). Suspensions of similar particle sizes were heated from 25 to 95°C. After a holding period of 30 min, the suspension was cooled to 50°C. The rate of heating and cooling was 1.5°C/min. The curves obtained are shown in Figure 5.

We found a much higher hot paste viscosity for the soft endosperm flours than for the hard endosperm flours of almost equal particle size distribution. No peak was obtained during cooking of the hard flour fractions, where all three varieties displayed increasing viscosity during the whole heating (and cooling) program. A

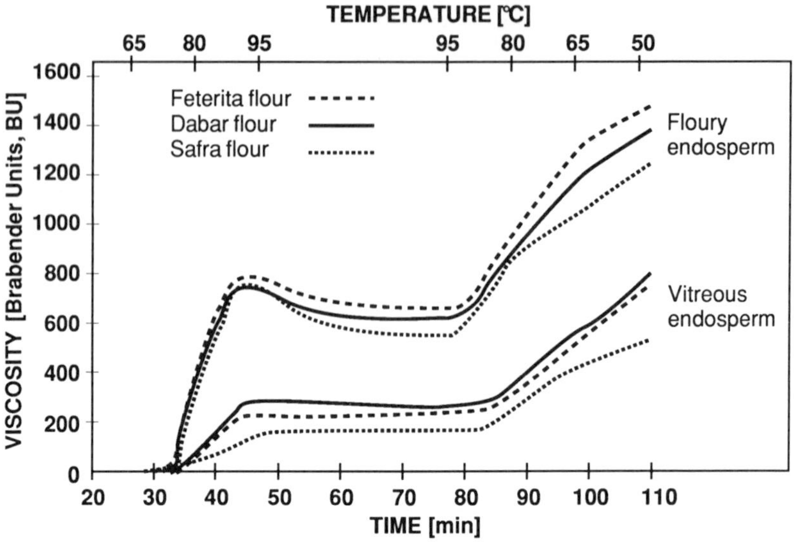

Fig. 5. Viscoamylograph cooking properties of flour from floury and vitreous endosperm sorghum (Hallgren, 1985).

study of differences in particle size showed that fine flour and coarse grits from the same cultivar produced similar pasting curves. The only differences in viscosity were obtained during the initial swelling and gelatinization, indicating differences in hydration rates. These results agree well with those obtained by Cantu et al (1984). They used the Viscoamylograph for optimization of the temperature-time mashing program, the amount of thermostable α-amylase, pH etc. for brews with sorghum adjunct. When no α-amylase was added, they found similar curves as shown in Figure 5 for vitreous endosperm flour. Enzyme addition (α-amylase) lowered the GT of the starch and the viscosity of the slurry during gelatinization, resulting in a peak on the curve during the heating stage of the Viscoamylograph program.

Industrial Brewing with Sorghum Adjunct

It is well documented that brewer's grits from sorghum are used routinely in the United States, Mexico, and Nigeria (Stewart and Hahn, 1965; Canales, 1979; A. O. Aisien, Nigerian Breweries Plc, Lagos, 1991, *personal communication*). Due to improved milling performance and better separation methods, sorghum brewer's grits are today comparable to the best corn and rice grits, giving beers of high quality (Canales and Sierra, 1976). The vitreous endosperm of sorghum is very similar to that of corn, and therefore sorghum adjunct can be used in the same way as corn adjunct. However, because of the slightly higher GT of sorghum starch, some technical modifications of the brewing program may be necessary. In order to increase capacity and to save energy costs, the high viscosity during gelatinization of the sorghum starch can be reduced by heating the sorghum grits to 85°C in presence of a thermostable bacterial α-amylase like Thermamyl 120L (Novo Nordisk A/S). A proper liquefaction step resulting in a low-viscosity mash with complete starch gelatinization is suggested, independent of the proportion of sorghum adjunct to barley malt in the grists (Anonymous, 1988; Lisbjerg and Nielsen, 1991; MacFadden and Clayton, 1989). Three examples of industrial use of sorghum adjunct in the form of brewer's grits are cited below.

BREWING WITH UP TO 50% SORGHUM GRITS WITH BARLEY MALT

Apart from the technical considerations regarding the capacity of the cooker and the liquefaction step involving addition of thermostable α-amylase, beer can be brewed with sorghum adjuncts with the same limitations known from the conventional use of corn. Commercial productions have confirmed that beers from grists made with up to 50% sorghum grits adjunct can be produced without significant changes in quality level compared with beers from 50% corn adjunct. If the sorghum adjunct is of a good quality, there will be no run-off problems in the lauter tun as the husk associated with 50% barley malt is sufficient to ensure the formation of the necessary filter bed in the lauter tun.

BREWING WITH 50 TO 100% SORGHUM GRITS WITH SORGHUM OR BARLEY MALT

Increasing the amount of sorghum means decreasing the proportion of barley or sorghum malt in the grists. With limited malt available, there will be a lower enzyme level for the hydrolysis of starch and protein in the production of the wort with sufficient fermentable sugars and a correct composition of free amino nitrogen (FAN). This is important in order to ensure good yeast growth, flavor development, and foaming properties. Consequently, in this case it becomes necessary

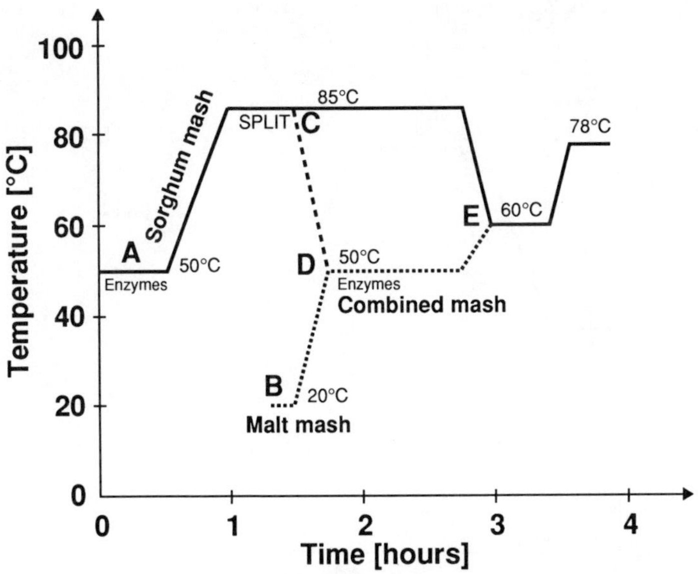

Fig. 6. Mashing cycle for brewing with 50 to 100% sorghum grits with sorghum or barley malt. A is mashing-in of 50°C, B is the malt-mash at 20°C, C is the point where the sorghum-mash eventually is divided into to parts, D is the combined mashes at 50°C and E is the point of pH adjustment to 5.5 at 60°C. The points of enzyme additions are indicated. (Source: Lisbjerg and Nielsen, 1991; used by permission.)

to add auxiliary proteolytic enzymes during the brewing process. Regarding the FAN, it is general experience that yeast needs a minimum of 130–150 mg/L assimilable nitrogen in order to perform well (O'Connor-Cox and Ingledew, 1989) and, as an alternative or as a supplement to the protease addition, yeast nutrient mixtures (Hug et al, 1991) or yeast autolysate products (Lisbjerg and Nielsen, 1991) may be used as assimilable nitrogen source.

Figure 6 shows a typical mashing cycle for a brewery using 50 to 100% sorghum adjunct in the grist (Anonymous, 1988; Lisbjerg and Nielsen, 1991). The actual time and temperature program varies between breweries depending on the ratio between sorghum and malt, grain quality, enzyme quality, and process equipment available. Mashing-in (A in Fig. 6) is performed at 50°C using. At this temperature, sorghum grits are mixed with water and pH is adjusted to the optimum for the enzymes involved (6.5–7.0) with the addition of $Ca(OH)_2$ to a Ca^{++} level of 50–150 mg/L. Then a mixture of enzymes is added including a thermostable α-amylase (for the liquefaction), a neutral protease like Neutrase (Novo Nordisk A/S) for production of a part of the necessary FAN, and a β-glucanase or cellulase like Celluclast (Novo Nordisk A/S) to open up the endosperm cell walls. After 30 min at 50°C (the protein rest period), the temperature is slowly increased to 85°C and maintained at this temperature for 30 min in order to liquefy the starch.

The malt (barley or sorghum) can be prepared from water at 20°C (Anonymous, 1988; Lisbjerg and Nielsen, 1991; B in Fig. 6). After 15 min, the cold malt mash is combined with the hot liquefied sorghum mash or part of this (D in Fig. 6) and maintained at 50°C for 60 min.

More enzymes are now added for starch saccharification. At this temperature (50°C), a liquid fungal α-amylase like Fungamyl 800 L (Novo Nordisk A/S) hydrolyzes α-1,4 linkages of starch and dextrins (from the action of the thermostable

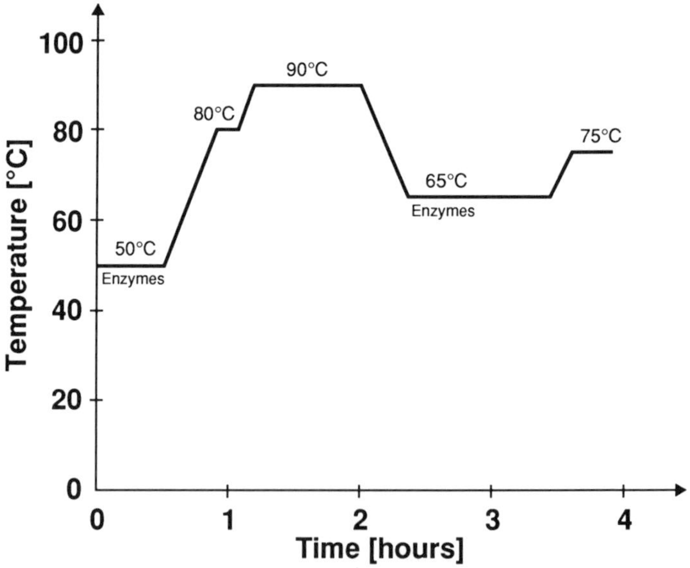

Fig. 7. Typical mashing cycle for brewing with 100 % sorghum (MacFadden and Clayton, 1989; A. O. Aisien, 1991, *personal communication*).

α-amylase), producing maltotriose, oligosaccharides, and large amounts of maltose. At the same time, a new enzyme mixture of neutral protease, cellulase, and amyloglucosidase is added.

After 60 min at 50°C, the combined mash is mixed with the rest of the 85°C sorghum mash, and the temperature is slowly raised to 60°C (E in Fig. 6); at this point the pH is adjusted to 5.5. The temperature is maintained at 60°C for 30 min. The temperature is then raised to 75°C, and after 20 min at this temperature, the mash is transferred to the mash filter.

BREWING WITH 100% SORGHUM GRITS (NO MALT)

Brewing with no malt (barley or sorghum) is possible. However, unmalted sorghum has no enzymes, and the brew-house extraction process requires the use of heat-stable external enzymes. A typical mashing program for 100% sorghum grits is shown in Figure 7 (MacFadden and Clayton, 1989; A. O. Aisien, Nigerian Breweries Plc, Lagos, *personal communication*). The procedure for mashing-in, pH adjustment, and enzyme addition is similar to that described for mashing 50 to 100% sorghum grits with sorghum or barley malt. However, instead of heating the sorghum mash to 85°C, it has been suggested that there be a rest period for 10 min at 80°C and then slowly raise the temperature to 90°C or even 100°C. This temperature is held for 50 min before the mash is cooled to 65°C and saccharified with heat-stable fungal amylase at pH 5.5. After 60 min at 65°C, the temperature of the mash is raised to 75°C and held at this temperature for 20 min prior to filtering.

MALTING OF SORGHUM

The malting of sorghum and millets for use in the brewing of African opaque beer is highly industrialized, and it has expanded in parallel to the brewing industry (Daiber, 1988). The major asset of a good sorghum malt for this kind of beer is a high diastatic power, since the mash consists mainly of adjunct, either corn or sorghum, which has to be converted (Nout and Davies, 1982). In the brewing of clear lager beer, however, other quality parameters are also important, and the use of sorghum or millet malt may not be suitable for this kind of beer production. Nout and Davies (1982) concluded in a study on malting characteristics of sorghum and finger millet that as extenders for barley malt in lager beer brewing, finger millet and sorghum do not appear adequate, mainly because of their low extract, poor filterability (for sorghum), and their dark color (finger millet). As a result of the Nigerian ban on cereal imports, however, much more research is now allocated into malting of sorghum and the use of sorghum malt in clear lager beer brewing with the use of more or fewer external enzymes (Koleoso and Olatunji, 1992; Aniche and Palmer, 1990). A process is now being perfected by the Nigerian lager brewing industry. The substitution of barley malt (which is not grown locally) with sorghum malt (which could be locally available) is expected to save the country about U.S.$100 million annually (Obilana, 1990).

The diastatic power (α- and β-amylase) developed in the sorghum malt is very low compared with that in barley malt and varies a lot due to cultivar and processing differences. The saccharification power of malted sorghum is restricted mainly because of limited β-amylase (Nout and Davies, 1982). Some sorghum cultivars, however, have been reported to develop significant β-amylase (Palmer, 1992). Malting losses for sorghum between 20 and 25% have been reported after 4–5 days germination (Ilori et al, 1990), which are much higher than those for barley malt.

The microbiological quality of sorghum grain material must be carefully tested because of the frequent occurrence of potentially very toxic fungi, e.g., *Aspergillus flavus,* which can produce highly toxic secondary metabolites. The elevated temperature for sorghum germination (25°C) compared with that of barley (16°C) makes it difficult to control microbial (mold) growth in sorghum malt, which may result in mycotoxin formation, problems with flavor, and possible beer gushing (Ilori et al, 1991). In order to reduce the microbial infections of the steeped grains of sorghum, an addition of 0.05–0.1% of formaldehyde may be added during the steeping process (Palmer, 1992; Ilori et al, 1990). These factors combined with the lack of sorghum malting capacity presently in Nigeria favors the use of sorghum as an unmalted adjunct in combination with the necessary exogenous enzymes.

It is important to recognize the fact that brewing with sorghum in whatever form (i.e., malted or unmalted) will require the addition of enzymes. Although there have been reports made from laboratory-scale or pilot-scale experiments that acceptable lager beer can be produced from sorghum malt without the need to supplement endogenous enzymes with external heat-stable enzymes (Skinner, 1976; Okafor and Aniche, 1987; Koleoso and Olatunji, 1992), there is yet no evidence available for the existence of a commercial production plant based on such a process of lager beer brewing (A. O. Aisien, Nigerian Breweries Plc, Lagos, *personal communication*). The use of locally produced sorghum may require less supplementation of external enzymes, but against this must be weighed the cost of sorghum malt, which in Nigeria presently is twice the cost of the raw sorghum grain.

MASH FILTRATION

The use of a conventional lauter tun in wort separation relies on the husk material from the barley to provide a filter bed for satisfactory performance. Because of this, conveyer and elevator systems used in conventional lager beer production are designed to minimize any damage to the barley husk, and the roller milling technology used for grinding the barley malt is now changing from dry to wet milling in an attempt to preserve the barley husk, which during mashing provides a natural filter bed for the mash. Sorghum, on the other hand, is a naked grain without husk. Consequently, a sorghum mash cannot be handled in a lauter tun with good result. A mash having a relatively low barley content (<50%) may, however, be separated in an artificial mash filter (Arri, 1989; Obilana, 1990). Due to reduced production capacities on lauter tun systems when brewing with a low malt content, many breweries in Nigeria have switched to conventional mash filters or invested in a more sophisticated thin-bed filter (B. Eloy, Nordon & Cie, France, 1990, *personal communication*). According to the vendor, the thin-bed filter is a more efficient filter for sorghum mashes than the conventional mash filter. In the thin-bed filter, the mash is introduced at full speed in vertical frames in order to form an evenly spread filter bed. Sparging time is reduced from 60–90 min to below 35 min due to the special construction that allows inside sparging. This is in contrast to the conventional mash filter where the sparging water has to run through a double filter bed, which means a higher counter pressure and lower production capacity than the thin-bed filter (Waesberghe, 1990).

THE FUTURE

The intensive research during the last decades has revealed new technology to the industrial sorghum brewer in the form of mashing programs and relatively cheap auxiliary enzymes for lager beer brewing. Many breweries in Africa have been forced to make changes in their brewing programs originally designed for the use of a high proportion of barley malt in order to make beers from 100% locally grown sorghum. Sorghum malt is an interesting substitute for barley malt. However, improved sorghum varieties suitable for malting (higher β-amylase) combined with sufficient malting facilities are necessary in order to compete with and reduce the quantity of brewing enzymes needed today. In Nigeria, only two commercial sorghum malting plants are in operation, providing raw materials for a few breweries for lager beer production (A. O. Aisien, Nigerian Breweries Plc, Lagos, *personal communication*), but there is a strong indication that the utilization of malted sorghum for lager beer production may increase in the future.

There is an urgent need for breeding sorghum varieties that fulfill the needs of the industrial lager beer brewer. Today only a few cultivars are grown under commercial production and the same sorghum cultivar is often used by the brewing industry for both grits and malt as well as for several other purposes by the food industry. However, for grits the brewer needs hard endosperm, low-fat types, and for malting he needs grain with medium or soft endosperm, fast water uptake, good germination, low malting loss, low polyphenol content, low production of rootlets, and free of molds and mycotoxins (Murty et al, 1990).

ACKNOWLEDGMENT

Thanks are due to the management of the Carlsberg Research Laboratory, where I worked from 1980 to 1988. Also thanks to D. S. Murty and M. I. Gomez of ICRISAT for excellent cooperation. Lars Munck of the Royal Danish Veterinary and Agricultural University of Copenhagen is acknowledged for his assistance in preparing this manuscript.

LITERATURE CITED

AMERICAN SOCIETY OF BREWING CHEMISTS. 1976. Methods of Analysis of the ASBC. Adjunct Materials, Cereals-5, Extract, Method A. Malt Conversion (International Method). The Society, St. Paul, MN.

ANICHE, G. N., and PALMER, G. H. 1990. Development of amylolytic activities in sorghum and barley malt. J. Inst. Brew. 96:377-379.

ANONYMOUS. 1988. U-Brew. Brewing with up to 100% unmalted grain and enzymes. Novo Nordisk A/S, Bagsvaerd, Denmark.

ARRI, B. K. 1989. Problems associated with the use of sorghum for lager beer production. Industrial Utilization of Sorghum: Symp. Current Status and Potential of Industrial Uses of Sorghum in Nigeria, December, 1989. International Crops Research Institute for the Semi-Arid Tropics (ICRISAT), Kano, Nigeria.

CAGAMPANG, G. B. 1983. Studies on the elucidation of the physicochemical basis of sorghum grain quality. Ph.D. thesis. Purdue University, West Lafayette, IN.

CANALES, A. M. 1979. Unmalted grains in brewing. Pages 233-278 in: Brewing Science, Vol. 1. J. R. A. Pollock, ed. Academic Press, London.

CANALES, A. M., and SIERRA, J. A. 1976. Use of sorghum. Tech. Q. Master Brew. Assoc. Am. 13:114-116.

CANTU, R. G., GARZA-ULLOA, H., and CANALES-GAJA, A. M. 1984. Selection of the optimal working conditions in the cooker using a Viscograph. Tech. Q. Master Brew. Assoc. Am. 21:160-165.

DAIBER, K. H. 1988. Malted and treated sorghum for the industrial production of beer, feed and food. Pages 15-30 in: Proc. Quad. Symp. Sorghum Millets 4th, Int. Cereal Bread Congr. 8th. ICC, Lausanne, Switzerland.

DALE, C. J., YOUNG, T. W., and MAKINDE, A. 1989. Extruded sorghum as a brewing material. J. Inst. Brew. 95:157-164.

DELCOUR, J. A., HENNEBERT, M. E., VANCRAENENBROECK, R., and MOERMAN, E. 1989. Unmalted cereal products for beer brewing. Part I. The use of high percentages of extruded or regular corn starch and sorghum. J. Inst. Brew. 95:271-276.

HAGGBLADE, S., and HOLZAPFEL, W. H. 1989. Industrialization of Indigenous Fermented Foods. K. H. Steinkraus, ed. Marcel Dekker, New York.

HALLGREN, L. 1984. Functional properties of sorghum and wheat in bread baking. Pages 16-23 in: Proc. ICC Symp. Processing Sorghum and Millets for Human Food, Vienna, 1984.

HALLGREN, L. 1985. Physical and structural properties of cereals, sorghum in particular, in relation to milling methods and product use. Vol. 1. Ph.D. thesis. The Technical University of Denmark, Carlsberg Research Laboratory, Copenhagen.

HALLGREN, L., and MURTY, D. S. 1983. A screening test for grain hardness in sorghum employing density grading in sodium nitrate solution. J. Cereal Sci. 1:265-274.

HALLGREN, L., REXEN, F., PETERSEN, P. B., and MUNCK, L. 1992. Industrial utilization of whole crop sorghum for food and industry. Pages 121-130 in: Utilization of Sorghum and Millets. M. I. Gomez, L. R. House, L. W. Rooney, and D. A. V. Dendy, eds. International Crops Research Institute for the Semi-Arid Tropics (ICRISAT), Patancheru, A. P., India.

HOUGH, J. S., BRIGGS, D. E., and STEVENS, R. 1971. Malting and Brewing Science. Chapman and Hall Ltd, London.

HUG, H., STROZYK, F., and BOROWCZAK, M. P. 1991. Production of beers based on 100% raw grain, using a yeast nutrient mixture. Brauwelt 2:114-20.

ILORI, M. O., AKINGBALA, J. O., OGUNTIMEIN, G. B., and OGUNDIWIN, J. O. 1990. Effect of grain bed thickness, duration of steeping and germination on the malting properties of improved Nigerian sorghum varieties. Lebensm. Wiss. Technol. 23:505-512.

ILORI, M. O., FESSEHATZION, B., OLAJUYIGBE, O. A., BABALOLA, G. O., and OGUNDIWIN, J. O. 1991. Effect of malting and brewing processes on the microorganisms associated with sorghum grains and malt. Tech. Q. Master Brew. Assoc. Am. 28:45-49.

KOLEOSO, O. A., and OLATUNJI, O. 1992. Sorghum malt/adjunct replacement in clear (lager) beer: Policy and practice in Nigeria. Pages 41-45 in: Utilization of Sorghum and Millets. M. I. Gomez, L. R. House, L. W. Rooney, and D. A. V. Dendy, eds. International Crops Research Institute for the Semi-Arid Tropics (ICRISAT), Patancheru, A. P., India.

LISBJERG, S., and NIELSEN, H. 1991. Brewing with 80% or more unmalted cereal. Proc. Sci. Tech. Conv. Inst. Brew., 3rd, Central and Southern African Section, March 1991, Victoria Falls, Zimbabwe.

MacFADDEN, D. P., and CLAYTON, M. 1989. Brewing with sorghum—Use of exogenous enzymes. Brew. Bev. Ind. Int. 1:77-81.

MALOMO, O. 1989. Sorghum in lager beer production: Progress and prospects. Presented at Industrial Utilization of Sorghum: Symp. Current Status and Potential of Industrial Uses of Sorghum in Nigeria. International Crops Research Institute for the Semi-Arid Tropics (ICRISAT), Kano, Nigeria.

MURTY, D. S., ALAYI, O., TABO, R., OLUGBEMI, L. B., NWASIKE, C. C., and MISARI, S. M. 1990. Recommendations. Industrial utilization of sorghum: Summary Proceedings. Industrial Utilization of Sorghum: Symp. Current Status and Potential of Industrial Uses of Sorghum in Nigeria. International Crops Research Institute for the Semi-Arid Tropics (ICRISAT), Kano, Nigeria.

NARZISS, L. 1976. Die Technologie der Malzbereitung. Bierbrauerei, 6th ed. Enke, Stuttgart.

NOUT, M. J. R., and DAVIES, B. J. 1982. Malting characteristics of finger millet, sorghum and barley. J. Inst. Brew. 88:157-163.

NOVELLIE, L. 1968. Kaffir beer brewing, ancient art and modern industry. Wallerstein Lab. Commun. 31:17-32.

OBILANA, A. T. 1990. Sorghum seed production and grain utilization in the brewing industries: The Nigerian experience. Presented at the Second Workshop on Seed Production and Utilization, Lusaka, Zambia.

O'CONNOR-COX, E., and INGLEDEW, W. M. 1989. Wort nitrogenous sources—Their use by brewing yeast: A review. J. Am. Soc. Brew. Chem. 47:102.

OKAFOR, N., and ANICHE, G. N. 1980. Brewing a lager beer from Nigerian sorghum. Brew. Distil. Int. 10:32-35.

PALMER, G. H. 1992. Sorghum—Food, Beverage and Brewing Potentials. Process Biochem.. 27:145-153.

POMERANZ, Y. 1986. Comparison of screening methods for indirect determination of sorghum hardness. Cereal Chem. 63:36-38.

SCHWARTZ, H. M. 1956. Kaffircorn malting and brewing studies. I. The kaffir beer brewing industry in South Africa. J Sci. Food Agric. 7:101-105.

SKINNER, R. 1976. Tropical lager brewing with sorghum malt. Brew. Distil. Int. 6:26-27.

STEWART, E. D., and HAHN, R. H. 1965.

Sorghum brewer's grits in the brewing of lager beer. Am. Brew. 7:21-26.

VOGEL, S., and GRAHAM, M. 1979. Sorghum and millet: Food production and use. Rep. Workshop, Nairobi, Kenya, July 1978. International Development Research Centre (IDRC), Ottawa, Canada

WAESBERGHE, J. W. M. 1990. Sorghum mash separation on thin bed filter. Brauwelt Int. 2:125-126.

CHAPTER 10

OPAQUE BEERS

Klaus H. Daiber
Division of Food Science and Technology
CSIR
Pretoria, Republic of South Africa

John R. N. Taylor
Cereal Foods Research Unit
Department of Food Science
University of Pretoria
Pretoria, Republic of South Africa

Terminology. The term opaque beer comprises the synonyms sorghum beer, Bantu beer, kaffir beer, African beer, sorghum-maize-millet beers, and a number of vernacular terms such as *utshwala, joala, busaa, dolo,* and other local terms. Other beers will be termed according to the cereal name of the malt, thus either barley beer, wheat or rye beers.

INTRODUCTION

The origin of the art of brewing opaque beer is shrouded in the beginning of history, but it is arguably of common origin with that of the barley beer in the Fertile Crescent of the Middle East. All the first explorers of Africa from the pioneering Portuguese to David Livingstone report and comment on the widespread existence of the gratifying and nutritious beverage.

Detailed reviews of the ancient art and the modern industry of opaque beers have been written (Novellie, 1968; Novellie and De Schaepdrijver, 1986; Haggblade and Holzapfel, 1989).

INGREDIENTS

In traditional home brewing the opaque beers are commonly made from a gruel of cooked, gelatinized starchy adjunct, which is subsequently liquefied and saccharified by the enzymes of a malted indigenous cereal. In industrial brewing in South Africa, the same ingredients are used today.

The success of brewing is critically dependent on the quality of the malt, particularly its enzyme composition.

Malt is made by germinating under well-controlled conditions the cereals available in the semiarid to humid conditions of subtropical Africa. These cereals have been sorghum and millets; they have been selected over the centuries for their suitability in malting.

Sorghum Malts

Malts made from grain sorghum have primarily been used for the brewing of opaque beer and have received foremost attention in research. Sorghum suitable for malting must meet more stringent and specific requirements than sorghum used for feed, food, or adjunct.

GRAIN REQUIREMENTS

The grain must be viable, that is, germinate under favorable conditions rapidly, within 48 hr, to at least 92 to 95% and be free of parasitic fungi and bacteria. Growth, harvest, and storage conditions of the grain require special precautions to maintain and improve viability and hygiene of the grain.

Inherent or genetic properties of sorghum have been found to play a critical role in the suitability of sorghum for malting and brewing (Novellie, 1962b; Daiber, 1978; Jayatissa et al, 1980).

The malt of the bird-resistant group of cultivars has been found unsuitable for brewing. Responsible are the tannins or condensed polyphenols deposited under genetic control in varying concentration in the peripheral tissue of the testa (Rooney et al, 1980). The enzymes produced during malting are complexed in the aqueous solution of the mash to become insoluble and inactive. Only when the causative tannins are neutralized in the intact grain prior to malting (Daiber, 1975) are malts of these cultivars suitable for brewing.

Inherent grain properties also determine the enzyme concentration, the extract, the composition of soluble solids in the malt, and they determine the susceptibility of grain to microbial infection. It is thus essential to select and market the consistently most suitable sorghum cultivar for successful malting. Efforts over the last decade (Daiber, 1988) have defined objective selection procedures, identified components of malting quality, and guided sorghum breeders to improve the malting quality of new agronomically better sorghum hybrids by active, selective breeding. In South Africa (Novellie, 1984) and Zimbabwe (House, 1988), identification and marketing of sorghum cultivars suitable for malting are actively pursued. These efforts correspond to those for the improvement of barley for malting. They must continue to meet better-defined requirements for agronomic performance and for brewing opaque beer, but also those for additional specifications for brewing clear, hopped beers with sorghum malts to replace expensive imported barley malts.

Sorghum malting quality is also affected by conditions during growth, production, and storage of the crop. Soil fertility, particularly available nitrogen, improves the protein content of the grain, enzyme content, and soluble amino nitrogen in the malt (Daiber, 1978). All measures and conditions, such as late and narrow planting and drought conditions, which tend to increase the proportion of the embryo improve enzyme formation and concentration as well as the content of free amino nitrogen (FAN) but decrease yield of grain and extract of malt.

Harvest and storage conditions are crucial for the preservation of sorghum viability and hygiene. Sorghum for malting must be harvested at full ripeness, at low grain moisture, with particular care to prevent breaking or cracking the grain by threshing. Prior to storage, grain must be dried to a safe moisture level of below 12% moisture, either pneumatically at temperatures below 50°C with high air circulation or in thin layers with frequent turning. Dust, loose husks, broken grain, and foreign substances must be removed. Size selection by sieving should retain small sound grains, which produce high diastatic activity although lower extract than larger-sized grain. Extract in opaque beer brewing is, however, more economically supplied by unmalted cereals used as adjunct.

The cleaned sorghum must be stored protected against storage insects. If insecticides must be used, they must be free of residue, and not reduce the viability of the seed.

MALTING TECHNOLOGY

The malting process of sorghum, in principle similar to that of barley, is initiated by steeping (soaking) the grain to hydrate the embryo for active growth and germination. The germinating grain is subsequently maintained at optimal levels of moisture, temperature, and aeration to support the synthesis of all enzyme systems, and the solubilization of the reserve materials in the endosperm at a rate to keep respiration losses to the minimum necessary to achieve the quality of malt for brewing. At that stage, malting is arrested by drying to preserve the quality and the enzymes during storage (Novellie, 1962a; Pathirana et al, 1983).

Sorghum is still malted outdoors in the traditional way in relatively thin layers on open or covered floors. In southern Africa, it is increasingly malted indoors in modern pneumatic industrial installations of hygienic design under controlled conditions of moisture, temperature, and aeration (Daiber, 1988). Sorghum malting differs in certain specific aspects from the well-documented malting of barley (Hough et al, 1982; Narziss, 1976).

Steeping. The uptake of water during steeping is controlled by physical absorption and diffusion to about 35% (wet basis) moisture. Further hydration is dependent on the metabolic activity of the germinating embryo. Inhibition of metabolism by temperatures below 15°C, by high CO_2 concentration, and other causes restricts water absorption to below the level of about 37% moisture essential for germination. Steeping after the physical absorption phase of about 6 to 8 hr requires temperatures above 18°C and sufficient aeration of the steep water or periods of air rest adjusted to the physiological requirements of the germinating sorghum (Hofmeyr, 1970).

The application of gibberellic acid during or at the end of steeping has failed to trigger enzyme synthesis in the embryo or aleurone (Daiber and Novellie, 1968; Aisien et al, 1983; Mundy et al, 1983) as in the case of barley.

To prevent the subsequent inhibitory effects of tannins during brewing, bird-resistant sorghum is steeped in a dilute (0.04–0.08%, w/w) formaldehyde solution for the first 4 hr of steeping. The formaldehyde diffusing radially into the grain complexes the inhibiting polyphenol deposited in the outer, peripheral layer of such sorghum (Daiber, 1975). The fungicidal properties of the formaldehyde cause in addition a degree of surface sterilization, although subcutaneous infection is not or incompletely controlled. The phytotoxic formaldehyde must be precisely dispensed, drained, and rinsed after 4 hr.

Germination. The steeped grain is cast onto a suitable support. This can be an open or covered, slightly sloping floor for beds of up to 0.3-m depth. For beds of up to 1.5-m depth, it must be a perforated floor on top of a ventilation duct. On the floor, the steeped grain is maintained at optimum temperature, moisture, and aeration for the desired rate of germination. Morrall et al (1986) obtained the highest malt quality attributes, namely, diastatic power, FAN, and extract at malt temperatures of 24°C, slightly, but not significantly, better than results at 28°C. Malting performed under frequent watering reached maximum quality after 4.5 to 5.5 days on the floor. Higher malting temperatures and lower moisture of the malt reduced all quality attributes significantly. Previously, Novellie (1962a) had similar results for diastatic power, which were significantly lower for malting temperatures at and below 20°C.

In floor malting the maintenance of the germinating grain at the optimum conditions is achieved by turning , by controlling the depth of the malt bed, and by frequent watering. In industrial malting, temperature maintenance is achieved by forced aeration with humidified air of the required temperature. At the optimum range of germination temperature, more heat is created than in barley malting. Rates of air flow are, therefore, higher than usual for barley maltings at 1,000 to 1,200 m^3/hr per metric ton of grain (Joustra and Field, 1978). Optimum moisture maintenance requires copious and frequent watering during the turning operation.

Turning and loosening the germinating sorghum is essential to prevent matting of the rapidly growing seedling, to reduce the resistance to aeration, and to assist the uniform watering of the green malt. In germinating sorghum, both the root and the shoot protrude from the grain berry. The shoot is not protected by the husk as in barley, and in addition the seedling in sorghum grows faster and more profusely. The seedling is thus far more tender and prone to injury by abrasive turning. Injury to the seedling and the tissue of the soft grain give rise to unproductive regrowth of the seedling and give access to parasitic, potentially toxin-producing fungi. Furthermore, the loss of the roots and shoots cause losses of FAN essential for the opaque brewing process.

Malting of sorghum under ideal conditions requires four to six days of active growth. Malt quality increases rapidly during the initial phase of malting, then remains static or even decreases with prolonged malting (Novellie, 1962a; Dyer and Novellie, 1966). Loss of grain substance by metabolic processes and formation of roots and shoots increase in direct linear relation to malting intensity and time. The malting time should be kept to the minimum required by the specified quality of the malt and the relative importance of the critical component of the quality.

Microbial Infection. The conditions of high temperature and moisture in combination with high microbial load, and the incidence of damage to the seedling and the epidermis of the grain, favor the contamination of the malting grain by potentially toxigenic fungi. Although floor and industrially malted sorghum are frequently contaminated by molds (Rabie and Thiel, 1985), only about 10% of industrial malts were found to contain more than 5 ppb, the legal limit of aflatoxin B_1, over a period of five years of intensive sampling. No aflatoxin was found in the opaque beers (Trinder, 1988). Nevertheless, strict hygiene and selection of clean sorghum of cultivars least susceptible to mold contamination are essential. Surface-sterilization of the grain (e.g., formaldehyde and hypochlorite) and strict control of the malting process can reduce mold infection and proliferation.

Drying. At the end of malting, the malt is dried to 12% or less of moisture.

Drying should be rapid to avoid further metabolic losses and to arrest microbial growth. Traditionally, the green malt is dried in thin layers in the sun. Industrially, the malt is dried in forced-draft dryers with high volumes of air heated to not more than 50°C. Higher temperatures, particularly during the initial stages of the process, cause losses of enzymes (Novellie, 1962a; Okon and Uwaifo, 1985).

The dried malt is not polished, as the roots and shoots are an important source of FAN (Taylor, 1983) and flavor. It is milled, usually in hammer mills, to a fineness specified by the brewer, usually to pass an 0.5-mm sieve to at least 95%.

MALT PROPERTIES

Sorghum malt is a composite material of partially solubilized, nutritionally and technologically beneficiated solids containing a spectrum of enzymes, flavor substances, microorganisms, buffers, and other components essential for brewing. Although industrially produced microbial enzymes and soluble adjuncts can be used for brewing (Olesen, 1985; MacFadden, 1989), there is no economic benefit for African countries to replace malts produced locally from indigenous cereals by these substitutes.

Modification. The structural, enzyme-mediated changes of the hard grain to the soft, friable, and porous malt is termed modification. Grain modification is linearly related to time and intensity of malting. It is dependent on the concentration and interaction of carbohydrases and proteases. Modification of sorghum can be quantitatively analyzed (Daiber et al, 1973).

Carbohydrates. Starch, the predominant (70–75% of grain) carbohydrate in the grain, is hydrolyzed during malting by the combined action of the diastatic enzymes α- and β-amylase. The hydrolysis starts at the interface of the scutellum and floury endosperm and progresses into the horny endosperm. The starch granules are pitted and are hollowed out from within (Glennie et al, 1982). The cell walls persist (Glennie, 1984). Von Holdt and Brand (1960) found only 60 and 54% of starch remaining in 5- and 7-day malted grain. Considering the malting loss, 30 and 44% of the original starch of the grain was metabolized during malting. This starch is used for energy and to synthesize the seedling. In malt germinated for seven days, Von Holdt and Brand (1960) found in increasing order fructose, sucrose, glucose, maltose, and oligosaccharides, at a total level of 16% of malt mass. By a more accurate and specific procedure, levels of 0.6% fructose, 1.1% maltose, 3.1% sucrose, and 3.6% glucose were found in sorghum malted seven days (Anonymous, 1971). Glennie and Wight (1986) found only traces of dextrins.

Carbohydrases. The hydrolysis of carbohydrates is brought about by five carbohydrases identified so far. The best researched and technologically most important are the α- and β-amylases, which liquefy and saccharify the gelatinized starch during mashing. Both enzymes are synthesized de novo during germination in the embryo and scutellum (Dyer and Novellie, 1966; Daiber and Novellie 1968; Aisien, 1982; Mundy, 1982) and progress radially into the endosperm. The enzymes are analyzed as combined or diastatic activity (Novellie, 1959) and are expressed as sorghum diastatic units (SDU/g dry malt). One SDU is roughly equivalent to 0.5°Lintner. The activity of one amylase is determined by the inactivation of the other amylase, formerly by the Preece method (Dyer and Novellie, 1966), more recently by the Taylor and Von Benecke (1990) procedure or with enzyme-specific substrates (Mundy, 1982).

The amylases have been prepared in high degree of purity and studied (Botes et al, 1967a,b; Mundy, 1982). Two α-amylase isozymes and a homogeneous β-amylase were identified with broad ranges of pH optima at pH 4.5–5.0 and pH 5.2–5.5 for α- and β-amylase, respectively, similar in properties to other cereal amylases.

In sorghum malt, α-amylase constitutes 60–80% (barley 18–50%) of the total diastatic activity (Dyer and Novellie, 1966; Okon and Uwaifo, 1985). This composition ensures proper liquefaction of starch, but in relation to barley malt limits formation of fermentable sugars.

Maltase (α-glucosidase) has been found (Watson and Novellie, 1974) preformed in the grain of sorghum; its level is not increased during malting (Watson and Novellie, 1975). As in barley, maltase in sorghum grain and malt is insoluble in water, but active and present at a 10-fold level and is more acid and extractant resistant.

Limit dextrinase, which hydrolyzes the branched component of starch, the α- and β-limit dextrins, was identified by Hardie et al (1976) in malted sorghum. Aisien (1982) found no preformed activity in the grain, a rapid increase of activity between 24 and 48 hr of germination (mainly in the endosperm), and lower activity of limit dextrinase in the embryo.

Another carbohydrase, β-glucanase-hydrolyzing cell wall material, was found by Aisien (1982) and by Nout and Davies (1982) in the ungerminated grain. The activity increased during germination by about 40%, however, only in the endosperm. Glennie et al (1983) found little microscopic evidence of cell wall hydrolysis in malted sorghum.

Proteins and Proteases. The structural and reserve proteins of sorghum (Taylor, 1983) are hydrolyzed during malting by proteolytic enzyme systems yielding soluble peptides and FAN. The products of hydrolysis are translocated from the endosperm; they resynthesize in the seedling and also accumulate in the malt berry (Taylor, 1983).

The presence, properties, and development of proteolytic enzymes in sorghum grain were investigated by Garg and Virupaksha (1970), Aisien et al (1983), and Evans and Taylor (1990b). The latter developed specific assays for sorghum with an improved extractant and purified kafirin as substrate (Evans and Taylor, 1990a).

The endoprotease activity of sorghum was defined by Garg and Virupaksha (1970) as acid protease with optimum activity at pH 3.6 at 50°C. Aisien et al (1983) determined the activity that they termed endoprotease, as reduction of viscosity of gelatin, whereas Evans and Taylor (1990a) measured the release of total soluble nitrogen from kafirin as the criterion for the activity they termed proteinase. The enzyme showed optimum activity at pH 4.0 to 4.25 at temperatures between 45 and 50°C. The enzyme(s) was found preformed in the ungerminated grain by Garg and Virupaksha (1970) and Evans and Taylor (1990b). It declined during early germination and steeping, respectively, and was found by Garg and Virupaksha (1970) to increase threefold after six days of germination, whereas Evans and Taylor (1990b) observed only a twofold increase after steeping, with a maximum after two to four days and a decline with further malting to six days. Aisien et al (1983) found no activity preformed in the grain. In their study, it increased considerably over three days of germination, to decline sharply during the fourth day of germination. Evans and Taylor (1990b) found significant differences in proteinase activity between cultivars malted identically over six days, which might also apply to the pattern of development of the enzyme.

The enzyme yielding FAN, termed carboxypeptidase by Evans and Taylor (1990a,b) or protease by Aisien et al (1983), increased more than fourfold to a maximum after four days of germination. Aisien et al (1983) found the enzyme confined to the embryo. Optimum activity of the enzyme was found (Evans and Taylor, 1990a) over a wide range of pH 4.0–5.3 at temperatures between 45 and 50°C. Genetic differences between cultivars and malting conditions affected the level of exopeptidase.

Lipids. During malting, the lipid content of sorghum (2.0–4.1%) decreased by 20 to 50% (Aucamp et al, 1961; Glennie et al, 1985). In the process the content of free fatty acids doubled, and the content of saturated fatty acids increased (Du Plessis and Burger, 1980). Lipase activity during malting is thus indicated but has not yet been studied.

Phenols. The wide spectrum of monomeric to highly polymeric phenols (reviewed by Gupta and Haslam, 1980) preformed in sorghum (Rooney et al, 1980) and synthesized in the seedling (Stafford, 1965) is present in the malt and has been studied by McGrath et al (1982) and by Glennie (1983). Polymeric condensed tannins have been found to inhibit enzymes in the malt of bird-resistant sorghum (Daiber, 1975). These tannins are synthesized and deposited in the testa of bird-resistant sorghum (Glennie, 1983) and have been identified as a distinct, protein-precipitating fraction in the extract of grain and of malt (McGrath et al, 1982). Steeping bird-resistant grain in dilute formaldehyde reduced the condensed tannins in direct relation to increasing concentrations of formaldehyde (McGrath et al, 1982) and reduced the inhibition of enzyme in malt of treated grain.

Soluble monomeric to oligomeric flavonoids are present in the grain, identified as the first eluting fraction (F_1) of the sorghum extract. The flavonoids are not affected and are not reduced by the formaldehyde treatment (McGrath et al, 1982). In addition, intensely red colored cyanidins are synthesized in the developing root (Stafford, 1965) to increase the flavonoid content and the color of the malt (Glennie, 1983).

The phenolic glycoside dhurrin is synthesized in the seedling (Panasiuk and Bills, 1984). Upon disruption of the seedling, dhurrin is enzymatically hydrolyzed with the release of hydrocyanic acid (HCN). Glennie (1983) found, however, only low concentration of HCN during various stages of malting and brewing. Further processing, even by traditional methods, removes the HCN (Dada and Dendy, 1987).

Minerals. Minerals are important as buffering substances and trace elements in enzyme synthesis, in souring, mashing, and in fermentation (Narziss, 1976). The mineral composition of sorghum malt as reported by Glennie et al (1985) and Van Heerden (1985) shows large variation between samples but only small changes as a result of malting. The composition of minerals in the grain is obviously determined by the concentration of available minerals in the soil. During malting, only the concentration of calcium is increased (Glennie et al, 1985). Up to 80% of minerals are components of organic compounds of distinct biochemical properties. The minerals are determined after ashing, thus usually obscuring biochemically relevant changes during malting. Such changes were investigated in the case of phosphates. Phosphates are between 73 and 90% deposited in the form of phytates (Doherty et al, 1982), a metal organic salt of calcium, magnesium, potassium, and zinc. During malting 25–100% of phytate (Van Heerden, 1985; Glennie et al, 1985) is hydrolyzed. The constituent minerals are released and contribute to the changes in the chemical and biochemical properties of the grain by malting.

Vitamins. Malt contains the vitamins of the grain, notably the B vitamins thiamin, riboflavin, and niacin. While the level of thiamin and niacin is maintained or slightly reduced, the level of riboflavin increases by more than 60% (reviewed in Haggblade and Holzapfel, 1989).

Microbial Aspects. Malts carry a mixed flora of microorganisms and their metabolites. The composition of this flora and the overall microbial load depend on the contamination remaining on the steeped grain, the state of hygiene in the malting plant, malting conditions, and the susceptibility of the sorghum to support microbial growth. The microorganisms include species of yeast, bacteria, and molds (Van der Walt, 1956), which are responsible for both lactic acid and alcoholic fermentation, but also for acetic acid and mycotoxins. As a foodstuff, the mycotoxin level in the malt as well as in the grain to be used for malting has to be analytically assured to be below the legal limit.

Millet Malts

Under arid conditions and high insect predation, conditions marginal for the production of sorghum, millets still produce a small, though more reliable crop (Rachie, 1975). In the semiarid tropics of Africa, India, and China, millets are produced as a crop of last resort for food and in malted form as supplement for weaned children (Desikachar, 1980; Malleshi et al, 1986). In times of shortage of sorghum, malt of pearl and finger millet is regionally used in Africa for the brewing of opaque beers (Nout and Davies, 1982; Shambe et al, 1989; Gillott, 1990; Harris, 1990). In East Africa (Kenya), finger millet malt is used to brew the traditional *busaa* opaque beer (Nout, 1980a).

GRAIN PROPERTIES

Structure, properties, and composition of millet grain are extensively reviewed by Hoseney et al (1981). The properties of the two millets used for traditional malting differ in the following aspects from the more commonly malted sorghum.

Pearl millet (*Pennisetum glaucum,* also identified as *P. americanum* and *P. typhoides*) known as *babala, nyati* (Zimbabwe), and *bajra* (India) is considerably smaller, about one-third the kernel weight of sorghum, and the embryo is proportionally larger. Pearl millet has a higher protein but lower starch content. The grain has no inhibitory tannins; it is reported to contain preformed α-amylase (Hoseney et al, 1981).

Finger millet (*Eleusine coracana*), known as *rapoko* (Zimbabwe) and *ragi* (India), is a small, generally red grain, about one-tenth the kernel weight of sorghum. The grain has a high lysine but low lipid content. While Ramachandra et al (1977) found evidence of inhibitory tannins in finger millet, Daiber (Anonymous, 1974) found no inhibition of diastatic enzymes in malted samples of finger nor of pearl millet from Zimbabwe.

Malting Technology. While most malting studies were done on petri dish scale, two (Nout and Davies, 1982; Anonymous, 1974) were malted under pilot plant conditions, and R. Harris (1990, *personal communication*) reported experience on industrial-scale malting. In industrial malting, the small-grained millets tend to pass or block the slotted malting floor, sized for sorghum grain. Rapid and intense seedling growth of the germinating small-sized millets tends to block the passage of air through the densely packed and highly matted malt bed. By obstructing sufficient aeration, the control of the temperature of the intensely respiring millet is impaired, oxygen supply becomes limiting, and turning is complicated. In Zimbabwe (T. Gillott, 1990, *personal communication*; R. Harris, 1990, *personal communication*), pearl and finger millets are, therefore, mixed with sorghum to be steeped and malted together. The cereals seem to complement each other for more rapid germination. It is reported that the mixed malt has better souring ability, and its taste is locally preferred. Drying of finger millet malt was investigated by Nout and Davies (1982). Kilning temperatures above 50°C reduce diastatic activity, particularly β-amylase activity, whereas the proteolytic activity was less temperature sensitive.

MALT PROPERTIES

Pearl and finger millets produce diastatic enzymes at similar magnitude, though at a slightly faster rate than sorghum (Table 1) (Anonymous, 1974; Novellie, 1976). Nout and Davies (1982) found in germinating finger millet pronounced production of the liquefying α-amylase, comparable to that of barley and higher than in sorghum. The enzyme developed at a higher rate at 30°C but reached maximum production after five days at 20°C. In the same malt β-amylase concentration was particularly low. Gibberellic acid at 0.2 ppm concentration apparently promoted diastatic and β-amylase activity up to about five days of malting. The composition of carbohydrates in finger, pearl, and foxtail millets malted for two and four days was analyzed by Malleshi et al (1986). In decreasing order of importance they found maltose, glucose, fructose, and sucrose increasing up to a total of 5.9%, while the mass of grain and its starch content decreased.

Proteolysis. Nout and Davies (1982) determined the solubilization of the endogenous proteins of finger millet during malting. The results suggest a high level of preformed proteolytic activity, which increased in the course of malting to about

TABLE 1
Some Grain and Malt Properties of Barley, Sorghum, and Millets

		Sorghum		Millets	
Properties	**Barley**	**Normal**	**BR**[a]	**Pearl**	**Finger**
Grain properties					
Starch gelatinization temp. range (°C)	51–60	68–78	68–78	62–78	62–72
Lipids (%) av.	2.9	3.3	3.3	5.4	1.5
Tannins	Absent	Absent	High	Absent	cv. dependent
Malt properties					
Opt. malting temp. (°C)	14–18	24–28	24–28	(25–30)[b]	(25–30)
Malting loss (%)	7	10–20	10–20	High	High
Diastatic power					
SDU[c]/g malt	150–200	20–60	20–60	(20–60)	(20–60)
α-Amylase (% of DP[d])	18–50	60–80	60–80	(60–80)	(60–80)
Extract at 60°C	High	Medium	Low	NI[e]	NI
Extract at 45–70°C	High	High	Medium	NI	NI
Effect of GA[f]	High	Absent	Absent	(Present)	(Present)

[a] BR = Bird-resistant, high-tannin sorghum.
[b] Figures in parentheses are estimates.
[c] SDU = Sorghum diastatic units.
[d] DP = Degree of polymerization.
[e] NI = No information.
[f] GA = Gibberellic acid.

the same magnitude as found in sorghum and barley. In contrast to barley, gibberellic acid had no significant effect on the activity.

Adjuncts

In traditional and industrial brewing of opaque beer, unmalted cereal adjunct is an important ingredient to supplement the expensive malt with a cheaper source of starch and protein and give the beer its characteristic viscous body (Novellie and de Schaepdrijver, 1986). In home brewing unrefined sorghum and millets and in industrial brewing refined maize grits are the major sources of cereal adjunct, while cassava and bananas are also used in Zaire and Uganda (Haggblade and Holzapfel, 1989). The requirements for an adjunct are availability, competitive price, low fat content, and, lately, freedom from mycotoxins.

SORGHUM ADJUNCT

Traditionally, whole, coarsely ground sorghum, probably grain less suitable for malting but not bird-resistant, has been used most commonly as adjunct. In the unrefined sorghum, the nutrients, in particular the vitamins, are retained and enrich the beer (Novellie, 1968). With development of industrial brewing, problems due to the variable and sometimes high fat content of sorghum developed. Lipids reduced the desirable foam of the beer and became in milled grain rancid during storage. Concurrently, refined, low-fat maize grits became abundantly available at a competitive price and increasingly replaced the frequently more expensive and more variable sorghum (Novellie, 1968). The nutritional quality of opaque beers decreased when the cereal adjuncts became more refined as a result of selective milling (Novellie, 1966).

As a result of developments in the milling technology of sorghum (Hahn, 1969; Schweigart and Vlietstra, 1971; Reichert and Youngs, 1977; Eastman, 1980) and a more favorable price, sorghum is again used in industrial brewing in the following, often experimental forms: 1) whole grain, dry-milled just before cooking; 2) whole grain, wet-milled by partial homogenization; 3) abraded, refined grits; 4) abraded, refined, and extruded grits (Joustra and Jansen, 1988).

Anthocyanins in unrefined whole sorghum adjunct intensify the reddish color and the flavor of opaque beer to locally well-accepted levels. The gelatinization of starch and denaturation of protein to facilitate enzymatic hydrolysis are similar to that of the more commonly used maize grits. However, the quality and composition of whole and also of refined, processed adjunct sorghum is not yet sufficiently stable due to variation in raw material and processing (Joustra and Jansen, 1988).

MILLET ADJUNCT

Haggblade and Holzapfel (1989) listed millets as sources of starch. Apart from the contribution of the residual starch in malted millet, it is assumed that millets are also used as adjuncts. The low starch and high lipid content of pearl millet (Hoseney et al, 1981) mitigate against profitable and technologically acceptable use of pearl millet as an adjunct. In contrast, the particularly low lipid content (1.8%) of finger millet (Kurien and Desikachar, 1966) makes the finger millet potentially more suitable.

MAIZE ADJUNCT

Maize is the principal adjunct ingredient of opaque beer produced industrially and locally, and also of traditionally produced beer. Highly refined maize grits, a product of the sophisticated dry-milling industry, are used as a competitively priced material of guaranteed low fat (1–1.5%) and high starch content (85–87%). The loss of nutrients by selective milling has been mentioned. Maize grits are gelatinized by cooking or by extrusion (Joustra and Jansen, 1988).

BREWING TECHNOLOGY

The opaque beer brewing process comprises a number of steps: souring (lactic acid fermentation), cereal cooking (starch gelatinization), mashing, straining (spent grain separation), and alcoholic fermentation. These steps are illustrated clearly in the Reef process (Fig. 1), one of the processes used to brew sorghum beer industrially in South Africa. The order of the steps varies between different brewing processes. In the *iJuba* process, used in the Kwazulu-Natal Province of South Africa, souring takes place after mashing (Novellie and De Schaepdrijver, 1986). In industrial brewing in Bulawayo, Zimbabwe, straining takes place during fermentation (Anonymous, 1983). Two or more steps may take place simultaneously. This happens particularly in home brewing, where strict microbiological hygiene is not practiced. Home brewing processes involving simultaneous mashing, souring, and fermentation have been described (Faparusi, 1970; Novellie and

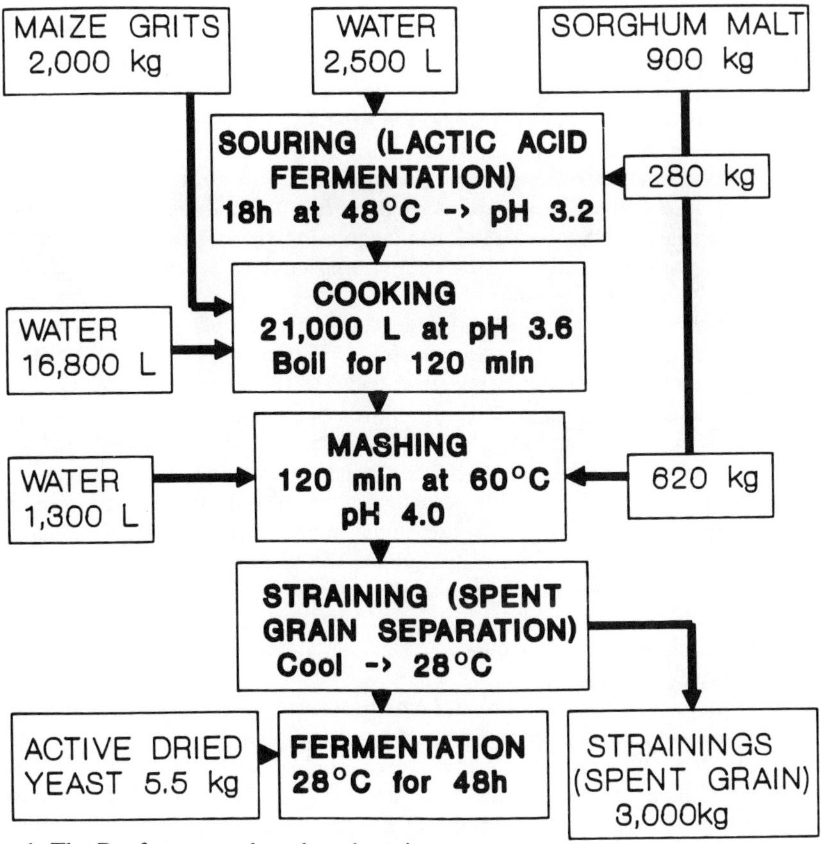

Fig. 1. The Reef-type sorghum beer brewing process.

De Schaepdrijver, 1986). However, for the sake of clarity the brewing steps will be discussed separately.

Souring

Souring, the lactic acid fermentation, performs three major functions in opaque beer brewing: the lactic acid produced imparts the beer with its characteristic sour taste; it also lowers the pH of the beer, which slows down the rate of microbial spoilage and inhibits the growth of pathogenic organisms; and, where the mash is acidified, the low pH helps to prevent the complete hydrolysis of starch into sugars. The resulting residual starch in the beer is primarily responsible for its opaque, viscous character.

Where souring is practiced as a separate step, it involves the growth of lactic acid bacteria on a slurry of sorghum malt (approximately 8–10% malt). Two types of souring can be distinguished; spontaneous and inoculated. Spontaneous souring is a rather hit or miss process, whose success is dependent on the presence of the lactic acid bacteria which form part of the natural microflora of the sorghum malt. Lactic acid production varies greatly between different batches of malt (Novellie and De Schaepdrijver, 1986). Inoculated souring is the norm in industrial brewing and in many home brewing processes. In its simplest form, a portion of the previous sour, normally about 10% by volume, which contains a high concentration of viable lactic acid bacteria, is used to inoculate the new sour. Although apparently crude, the process if properly carried out, can be highly effective and rapid, and consistent lactic acid production takes place. It appears that over many inoculation cycles, strains of lactic acid bacteria are selected that are well adapted to growing in that environment. The Council for Scientific and Industrial Research (CSIR), Pretoria, has also isolated pure strains of *Lactobacillus* and freeze-dried cultures have been prepared which are used for inoculation by one of the breweries in South Africa. Pure cultures of *Lactobacilli* are reportedly also used for inoculation in Zimbabwe (Anonymous, 1983).

The conditions of temperature and time of souring are highly variable. In home brewing, temperature is not strictly controlled. This can result in the growth of mesophilic, heterofermentative lactic acid bacteria (Van der Walt, 1956). These organisms produce acetic acid, ethanol, and carbon dioxide in addition to lactic acid (Kandler, 1983) as well as minor by-products such as formic acid and glycerin. The flavor imparted to home-brewed beer by compounds other than lactic acid may or may not be acceptable according to local taste. In industrial brewing, the temperature of souring is strictly controlled in the range of 48–50°C. This temperature prevents the growth of mesophilic organisms and favors the growth of the thermophilic, homofermentative *L. delbrueckii* (*leichmanni*), which produces solely lactic acid (Van der Walt, 1956). In industrial practice, souring is carried out for up to two days. The final concentration of lactic acid falls in the range of approximately 1–2% and pH 3.0–3.2.

The sorghum malt provides the nutrients necessary for the growth of the lactic acid bacteria. Since souring is an extended period of all malt mashing at moderate temperature, the malt enzymes are also active. Hydrolysis of malt starch into sugars by the diastatic enzymes appears to be limited, since the temperature of souring is below the gelatinization temperature of the starch, and the low pH for most of souring inactivates the enzymes (Taylor, 1989). In fact, recent research found that under conditions of very high lactic acid production, sugars can be exhausted during souring. Production of amino acids and peptides (free amino nitrogen or FAN) during souring by the action of the malt proteolytic enzymes on malt protein is, however, considerable (J. R. N. Taylor, *unpublished*). Further, the amount of FAN taken up by the bacteria is negligible in comparison with that produced.

Cooking

All opaque brewing processes involve a cooking stage. Its purpose is to gelatinize the starch in the unmalted cereal adjunct, thus rendering it readily hydrolyzable during mashing by the malt diastatic enzymes. Ungelatinized (raw) starch is only slowly attacked by the malt α-amylase enzyme and not at all by the β-amylase. A separate cooking stage is necessary, because the starch in the cereals used as adjuncts in opaque beer brewing—maize, sorghum, and millets—has a high gelatinization temperature, in the range 62–75°C (Briggs et al, 1981). Above 70°C, the malt diastatic enzymes are normally rapidly inactivated (Novellie, 1966; Taylor, 1989). Thus, simultaneous gelatinization and hydrolysis of the adjunct starch during mashing is problematical.

A wide variety of cooking conditions are used. In home brewing, a slurry of the adjunct is boiled for about 2 hr, but sometimes for as long as 7 hr (Haggblade and Holzapfel, 1989). The high altitude of most of southern Africa (1,000–2,000 meters) lowers the boiling point of water and necessitates these apparently long cooking times. In industrial brewing, boiling at atmospheric pressure for 1.5–3 hr is normal. Cooking under pressure up to 110°C for reduced times is also employed to save time and energy. In the Reef process (Novellie and De Schaepdrijver, 1986) (Fig. 1) and in many home-brewing processes (Haggblade and Holzapfel, 1989), the sour is cooked with the unmalted cereal adjunct. The lactic acid has the effect of softening the endosperm protein, which tightly encloses the starch granules. Softening the protein allows more rapid water uptake by the starch granules and thus speeds up gelatinization (Novellie, 1968). The lactic acid steep in corn wet-milling performs a similar purpose.

A novel development is the use of high-temperature, short-time extrusion cooking to "dry cook" the cereal adjunct (Joustra and Jansen, 1988). At the Kwaggakop brewery in South Africa, maize grits have been cooked using a single-screw extruder. The extruded grits are milled, then added directly to the mash without further treatment. The resulting beer is apparently indistinguishable from that brewed with conventionally "wet cooked" maize grits, and yield of extract was increased. Extrusion cooking was far more energy efficient than conventional wet cooking.

Mashing

The mashing process involves the incubation at optimal temperatures of milled sorghum malt with the cooked adjunct in aqueous suspension with the objective of physically and enzymically solubilizing components of the malt and adjunct to produce a fermentable wort.

STARCH HYDROLYSIS

The most important solubilization process during mashing is the hydrolysis of gelatinized adjunct starch, and to a lesser extent ungelatinized malt starch, into fermentable sugars and water-soluble dextrins by the malt diastatic enzymes. Starch hydrolysis during mashing is brought about mainly by the joint action of the sorghum malt α- and β-amylase enzymes (Taylor, 1989). Alpha-amylase attacks α-(1→4) glucosidic bonds at random in the starch molecules to produce a variety of dextrins and fermentable sugars. This enzyme rapidly reduces mash viscosity and provides substrate for β-amylase. Beta-amylase attacks gelatinized starch and dextrin molecules at their non-reducing ends, hydrolyzing the penultimate α-(1→4) glucosidic bond to release maltose. Sorghum malt contains other diastatic enzymes: limit dextrinase (debranching enzyme) (Hardie et al, 1976), which hydrolyzes the α-(1→6) bond in amylopectin, and α-glucosidase (maltase) (Watson and Novellie, 1974), which hydrolyzes maltotriose and maltose into glucose. However, the importance of these two enzymes during opaque beer mashing is not known.

Unlike barley beer brewing, the aim of mashing in opaque beer brewing is not to hydrolyze all the starch. A proportion of the gelatinized adjunct starch must remain at the end of the mashing (Novellie, 1966). This residual starch is primarily responsible for the opaque character and high viscosity of the beer. A number of factors play a role in limiting starch hydrolysis during mashing. Sorghum malt intrinsically has low diastatic activity in comparison with barley malt (Novellie, 1959). The low diastatic activity of sorghum malt is compounded by the fact that a high ratio of adjunct to malt is used in the mash. In industrial brewing in South Africa, a ratio of adjunct to malt in the mash of approximately 4:1 is normal (Taylor, 1989). Starch hydrolysis is commonly also limited by mashing at low pH, particularly in those brewing processes where the sour is cooked with the adjunct, as in the Reef process (Fig. 1). In this process, the mash pH is normally in the range pH 3.8–4.0 (Novellie, 1966). This low pH is below the optimum for sorghum α- and β-amylase, pH 4.5–5.0 (Botes et al, 1967a) and pH 5.2–5.5 (Botes et al, 1967b), respectively. This retards the rate of enzymic action (Novellie, 1966; Taylor, 1989) and accelerates enzyme destruction (Taylor and Daiber, 1988). In South African industrial brewing, mashing is carried out for 1½ to 2 hr at constant temperature, between 50 and 60°C (Novellie, 1968). The relatively low mashing temperature is necessary, because higher mashing temperatures at the low pH employed would limit sugar production due to enzyme inactivation (Novellie, 1966). The low mash temperature in comparison with barley malt mashing is, however, below the gelatinization temperature of sorghum malt starch 64–68°C (Taylor, 1989), which severely limits its hydrolysis. In some industrial brewing processes such as the iJuba process, the mash is cooked at 80–100°C at the end of the mashing period (Novellie and De Schaepdrijver, 1986). This cooking period gelatinizes the sorghum malt starch and thus provides additional residual starch in the beer.

At the end of the mashing period, the mash in a Reef brewing process contains 6–7% fermentable sugars in the approximate ratio of 1:3:1 glucose to maltose to maltotriose; 3% water-soluble dextrins, of which virtually all are of degrees of polymerization 10 glucose units or higher and somewhat more than 1% gelatinized starch and approximately 2% ungelatinized starch (Glennie and Wight, 1986; Taylor, 1989; J. R. N. Taylor, *unpublished*). In processes where the mash is cooked at the end of mashing, the quantity of ungelatinized starch is reduced to virtually zero. In such processes, it is normal that after this second cooking period a second short mashing period is carried out with a small addition of malt or commercial amylase of microbial origin. The objective of this second mashing period is to reduce the viscosity of the mash to the desired level by the action of amylase (Novellie and De Schaepdrijver, 1986).

PROTEOLYSIS

During opaque beer mashing a certain amount of proteolysis takes place due to the action of the sorghum malt proteolytic enzymes on the malt and adjunct proteins (Taylor and Boyd, 1986). It appears that it is specifically the prolamin and glutelin storage proteins which are hydrolyzed during mashing. The products of proteolysis, free amino acids and small peptides, are collectively referred to as free amino nitrogen or FAN. The optimum mashing conditions for FAN production have been found to be around 51°C and pH 4.6. These conditions are close to the optimum temperature and pH for the sorghum malt proteinase and carboxypeptidase enzymes, 45°C and pH 4.25–4.50 and pH 4.00–5.25, respectively (Evans and Taylor, 1990a). The actual quantity of FAN produced during mashing is relatively small, accounting for some 30% of wort FAN. The remaining 70% comes from FAN preformed in the malt and that formed during souring. FAN is of importance because it forms the nitrogen source for yeast growth during fermentation (Taylor et al, 1985). Analysis of the FAN at the end of mashing shows it to be of good quality with regard to yeast nutrition. The two most abundant amino acids in the FAN are aspartic acid and glutamic acid, both of which are classified as group A amino acids, those which are taken up immediately by yeasts during fermentation (Jones and Pierce, 1964). Group A amino acids account for some 50% of the amino acids in opaque beer wort. In contrast, the group D amino acid proline, which is only slightly absorbed by yeast under the anaerobic conditions of fermentation, accounts for only some 10% of the total. This differs from the situation in barley malt wort where proline is the predominant amino acid and constitutes up to one-third of the FAN (Jones and Pierce, 1964). The difference is probably a result of the fact that the composition of the prolamin protein from which the FAN is derived differs between sorghum and barley. Sorghum prolamin is only 10% proline (Taylor and Schüssler, 1986) as against the 22–23% in barley prolamin (Miflin and Shewry, 1979).

LIPIDS, VITAMINS, AND POLYPHENOLS

Other substances in the malt and adjunct such as lipids, vitamins, and minerals are physically solubilized or hydrolyzed during mashing. Of particular importance in opaque beer brewing is the solubilization of the reddish anthocyanin pigments occurring in sorghum malt (Glennie, 1983). These polyphenolic compounds are solubilized during souring and mashing and are responsible for the characteristic pinkish-brown color of opaque beer. The color density of anthocyanins is affected by pH (Van Buren et al, 1974), and their reddish color is enhanced at the low pH of opaque beer.

Straining

The objective of straining is to remove coarse particles of cereal such as malt pericarp from the mash or beer. In traditional home brewing, the fermented beer was strained through a bag of woven grass, prior to consumption (Novellie, 1968). Today many home brewers use metal screens of appropriate mesh size to strain the beer (Haggblade and Holzapfel, 1989). In industrial brewing in South Africa,

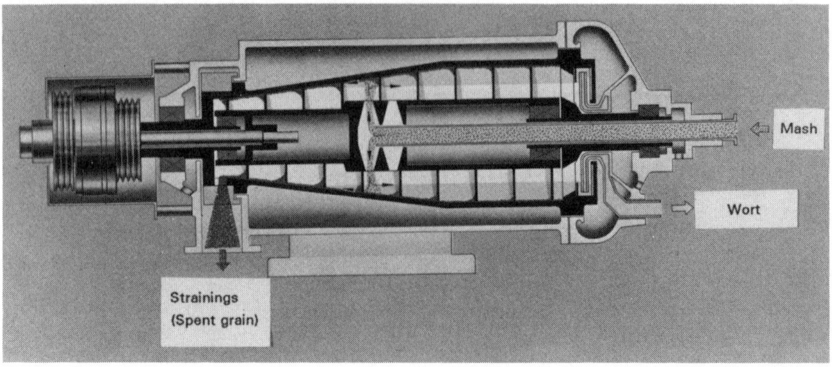

Fig. 2. Mode of operation of a solid bowl centrifuge, which is used to separate the spent grain (strainings) from the sorghum beer wort.

straining is carried out directly after mashing (prior to fermentation). Solid bowl centrifuges (decanters) (Fig. 2) are used to strain the mash (Novellie and De Schaepdrijver, 1986). The specifications for decanter performance are that they should produce a wort that essentially contains no particles larger than 250-μm diameter. A characteristic of their operation is that separation is primarily according to density. This results in the by-product of straining, the spent grains (referred to as strainings) being enriched in the dense, ungelatinized starch and insoluble protein. On a dry basis, strainings from industrial brewing typically contain some 46% starch and 25% protein (Van Heerden, 1987). The remainder is virtually all carbohydrate in the form of sugars, dextrins, and fiber.

In contrast, the less dense gelatinized starch remains in the wort. On a dry basis, a wort typically contains some 60% fermentable sugars, 25% dextrins, 10% gelatinized starch, and the remainder is mainly protein (J. R. N. Taylor, *unpublished*). The strainings from industrial brewing are sold for animal feed. Research has, however, shown the feasibility of saccharifying the starch in recooked strainings using industrial amylase enzymes to produce a second fermentable wort (Taylor and Schüssler, 1988).

Fermentation

In traditional home brewing of opaque beer, alcoholic fermentation is brought about by the growth of wild yeasts in the wort. In industrial brewing, commercially produced active dried yeast is used to inoculate the wort (Novellie and De Schaepdrijver, 1986). Strains of the top fermenting yeast *Saccharomyces cerevisiae,* which rapidly ferment opaque beer wort at ambient temperature, were specially selected by the CSIR and have been in general use in South Africa for many years.

In industrial brewing in South Africa, the beer is either packed shortly after pitching with yeast or fermented at approximately 25°C in batch-sized fermentors for at least 24 hr. Unlike barley beer, the fermented opaque beer is not filtered, nor is the yeast removed from it. After fermentation, this draft beer is transported in road tankers to its place of consumption. There the beer is dispensed by "Buckmeter" beer pumps into beakers known as *skaals,* of 2- or 4-L capacity, from which the beer is consumed.

As regards the packaged beer, in South Africa a wide range of different types of containers are used (Fig. 3). These include 1-L cardboard, 1- and 2-L low-density polypropylene, nonreturnable milk-type containers, and 1–50-L high-density propylene returnable containers. In all containers, however, there is a vent (small hole or slit) in the top to allow the escape of carbon dioxide. This is necessary to prevent the container from bursting due to excess pressure, as the beer ferments during transport from the brewery to its place of sale. Opaque beer is normally always consumed while it is still actively fermenting.

The basic biochemistry of opaque beer fermentation is identical to that of barley beer fermentation and will not be discussed here. Two subjects regarding the fermentation are, however, worthy of note. First, because such a low proportion of malt is used in the grist, yeast nutrients, particularly FAN, may limit yeast growth during fermentation (Taylor et al, 1985; Taylor and Boyd, 1986). Pickerell (1986b) found a linear relationship between the sugar content of opaque beer wort and the quantity of FAN in the wort required to rapidly ferment the sugars to alcohol. A minimum requirement for FAN in wort of 100 mg/L was proposed.

Secondly, opaque beers contain a relatively high level of fusel oils in comparison with lager beers. O'Donovan and Novellie (1966) found an average level of 227 ppm, compared with an average of 110 ppm in barley beer. Fusel oils are potentially aromatic compounds and exert a considerable influence on the flavor and aroma of alcoholic beverages (Hough et al, 1982). They are also potentially toxic substances. However, rat feeding trials using the combination of ethanol and fusel oils found in opaque beer did not produce any evidence of toxicity (Johannsen and Purchase, 1969). Fusel oil formation during fermentation is linked to yeast amino acid biosynthesis (Hough et al, 1982). The high levels of fusel oils in opaque beer are related at least in part to the relatively high fermentation temperature (20–30°C) in comparison with barley beer (10–20°C) (Pickerell, 1986a).

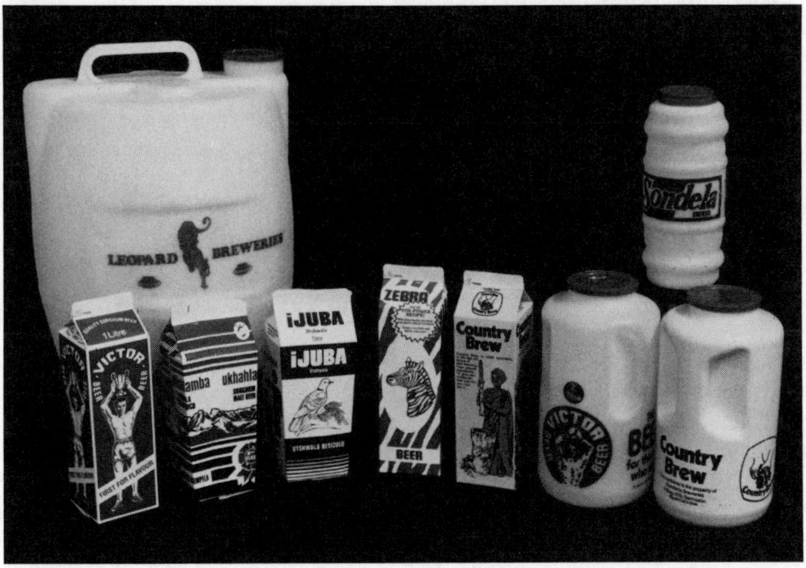

Fig. 3. A selection of sorghum beer containers. One-liter cardboard nonreturnable cartons and 1-, 2-, and 16-L returnable high-density polyethylene containers.

Because opaque beer is not pasteurized and is fermented at relatively high temperature, it is subject to microbial spoilage despite its low pH. The life of the beer is in the range of only one to five days depending on how hygienically it was prepared and fermented. Spoilage is apparently primarily due to the growth of mesophilic lactic acid bacteria (Haggblade and Holzapfel, 1989). The beer may also be infected with acetic acid bacteria and wild yeasts (Van der Walt, 1956).

Millet Beers

The use of malted millets to brew opaque beer either as the sole source of malt or in combination with sorghum is or was widespread throughout Africa (Haggblade and Holzapfel, 1989). The practice of using mixed millet and sorghum malts has been adopted in industrial opaque brewing in central Africa, Zambia, and Zimbabwe (Young, 1949; Anonymous, 1983; Gillott, 1990; Harris, 1990). A mixture of both finger millet (*E. coracana*) and pearl millet (*P. glaucum*) malts with sorghum malt and barley malt may be used in souring and finger millet malt and barley malt in mashing (Harris, 1990). This variety of different malts apparently provides extra nutrients for *Lactobacillus* growth during souring and produces a beer of very acceptable taste to the consumer.

As regards opaque beer brewed with solely millet malt, Nout (1980a) reported on the traditional brewing process for *busaa*, a Kenyan opaque beer brewed with finger millet malt. The process is similar to that described earlier for industrial brewing of sorghum malt opaque beer, with two exceptions. Firstly, it is the maize grits adjunct which is soured and not the malt. Despite the apparent lack of nutrients in maize endosperm for the lactic acid bacteria, a relatively high concentration of lactic acid (>2%) is obtained in the sour. Secondly, the maize grits adjunct is cooked by roasting on a steel sheet, as opposed to boiling water. In general terms, the composition of the final beer; 0.5–1% lactic acid and 2–4% alcohol, is similar to that of sorghum malt–based opaque beer. A characteristic of opaque beers brewed with millet malt is that they are buff in color in contrast to the pinkish-brown of sorghum malt–based opaque beer.

A rather different type of millet beer is the Nigerian *oyokpo* beer brewed with pearl millet malt (Iwuagwu and Izuagbe, 1985). This product appears to be essentially a nonhopped, clear beer. The brewing process differs fundamentally from an opaque brewing process in that there is no souring step, no starchy adjunct is employed, and the beer is clarified.

BEER QUALITY

In home-brewed opaque beer, specific statements about quality requirements are virtually meaningless, since beer recipes differ across the continent of Africa, and variations occur even between brewers in the same village (Novellie and De Schaepdrijver, 1986; Haggblade and Holzapfel, 1989). However, some generalizations can be made. Firstly, the consumer frequently regards the sourness of the beer due to lactic acid as being more important than its alcoholic content (Novellie, 1968). A second important property to the consumer is the body or viscosity of the beer (Novellie, 1968), which results from the presence of the semisuspended gelatinized starch and other particulate matter.

Quality Specifications

For industrial brewing, more quantitative quality specifications are required in order to produce a consistent product. In South Africa there is a standard specification for the production of opaque beer that describes a number of criteria to which the beer should conform (South African Bureau of Standards, 1971). The specification sets out minimum values for total solids, protein, and lactic acid content of the beer, its pH, and maximum values for alcohol and volatile acids. There are also microbiological requirements that the beer should be free from *Escherichia coli* I and *Acetobacter* species. In addition, individual breweries have quality specifications with respect to beer characteristics such as viscosity, foam, and settling out of the beer, as well as evaluation by taste panel.

Nutritive Value

The early literature on opaque beer laid strong emphasis on its apparently excellent nutritive value; see reviews by Novellie and De Schaepdrijver (1986) and Haggblade and Holzapfel (1989). However, the nutritive value of opaque beer or any food product is dependent on the ingredients from which it is made, how it is prepared, and the biological availability of its nutrients. As stated, traditional home-brewed opaque beers vary greatly. As with beer quality requirements, this means that it is difficult to make specific statements about actual nutritive value. Nevertheless, it can be said that opaque beer is an important source of food energy,

TABLE 2
Nutritive Content of Industrially Brewed Opaque Beer[a]

Nutrient	Mean Content per Liter	RDA[b]	% Contribution to RDA by 1 L of Beer
Alcohol (g)	25.4	NR[c]	...
Fat (g)	Trace	96[d]	...
Crude fibre (g)	1	NR	...
Protein (g)	5.4	56	9.6
Ash (g)	1.1	NR	...
Carbohydrate (g)	47.6	446[d]	10.7[e]
Starch (g)	29.7	NR	...
Food energy (kJ)	1,651	12,600	13.1[e]
Vitamin B-1 (thiamin) (mg)	0.2	1.5	16.0[e]
Vitamin B-2 (riboflavin) (mg)	0.4	1.8	21.7[e]
Niacin (mg)	2.9	20	14.7[e]
Copper (mg)	0.2	2	10.0[e]
Iron (mg)	2.0	18	11.1[e]
Zinc (mg)	1.8	15	12.0[e]
Manganese (mg)	1.5	3.8	39.5[e]
Calcium (mg)	53	800	6.6
Magnesium (mg)	140	400	35.0[e]
Potassium (mg)	276	4,000	6.9
Sodium (mg)	9	$<$3,000	0.3
Phosphorus (mg)	218	800	27.3[e]

[a] Adapted from Van Heerden (1989a).
[b] RDA = Recommended dietary allowance.
[c] NR = no recommendation.
[d] Suggested daily intake.
[e] Significant contribution to RDA ($>$10%/L).

B vitamins, and protein, and thus can constitute a valuable supplement to the diet (Haggblade and Holzapfel, 1989).

An accurate picture has, however, been obtained of the nutritive value of industrially brewed opaque beer through the comprehensive research of Van Heerden and co-workers (Zammit, 1980; Van Heerden, 1985, 1988, 1989a,b; Van Heerden and Glennie, 1987; Van Heerden et al, 1987). Some of her data are summarized in Table 2. Opaque beer is, as would be expected, rich in carbohydrate. The consumption of 1-L of beer can provide 13% of an adult's daily food energy requirements. The high starch content of opaque beer together with its moderate alcohol content is also believed to be responsible for preventing the occurrence of reactive hypoglycemia when it is consumed (Joffe et al, 1981). Opaque beer is also rich in protein. One liter can supply nearly 10% of the recommended daily dietary allowance. However, because much of the protein is of cereal origin it is somewhat deficient in the essential amino acid lysine (Van Heerden et al, 1987). From the Table 2 it would appear that opaque beer is also an excellent source of the B vitamins thiamin, riboflavin, and niacin. Research, however, has shown that many of these vitamins are contained within the yeast cells in the beer and are probably unavailable to the consumer (Van Heerden and Glennie, 1987). B vitamin availability in opaque beer was considerable improved by thermal treatment of the beer to disrupt the yeast cells, as would occur if the beer was pasteurized. As regards minerals, opaque beer is a rich source of copper, iron, zinc, manganese, magnesium, and phosphorus. The use of sorghum rather than maize grits as a starchy adjunct increases the nutritional value of opaque beer with respect to vitamins and minerals (Van Heerden, 1989b). In comparison with barley beer, opaque beer has a considerably higher protein, thiamin, riboflavin, and mineral content, but a lower content of alcohol and niacin (Van Heerden, 1988).

NEW DEVELOPMENTS IN OPAQUE BEER

A major challenge facing commercial opaque brewing is to develop products with an extended shelf life of up to several weeks. Because opaque beer is currently sold in a microbially active state, its short shelf life of a maximum of five to seven days prevents the product being distributed any great distance from its place of manufacture. Also, the short shelf life, coupled with the fact that the beer has to be marketed in vented containers, makes the product inconvenient to handle in comparison with, for example, bottled barley beer.

Research, particularly by the CSIR in South Africa, has resulted in a number of product concepts designed to solve these problems in different sectors of the opaque beer market (De Schaepdrijver, 1983). A pasteurized, bottled, carbonated opaque beer has been developed (De Schaepdrijver et al, 1980; Novellie and De Schaepdrijver, 1986). After fermentation, the beer is pasteurized and bottled with a precisely dosed mixture of sugar and yeast. A second fermentation takes place in the bottle and carbonates the beer.

Similarly, Nout (1980b) experimentally produced a bottled, pasteurized Kenyan opaque beer (*busaa*). Bottling in combination with chemical preservation using sodium benzoate has been successfully applied to Nigerian millet beer (*oyokpo*) on an experimental basis (Iwuagwu and Izuagbe, 1985). A different type of long shelf life opaque beer that has been developed by the CSIR is a still (noncarbonated) product. The manufacturing process involves pasteurizing the beer and then aseptically packing it in 1-L Tetra Brik aseptic, lined cardboard cartons (De Schaepdrijver and Joustra, 1985; Novellie and De Schaepdrijver, 1986). As stated earlier, pasteurization as well as extending shelf life has the additional advantage of improving the availability to the consumer of the B vitamins in the beer (Van Heerden and Glennie, 1987).

Another development by the CSIR involves the production of "prebrewed" shelf-stable worts. An opaque beer wort is produced in a brewery in the normal way. The wort is then concentrated to a syrup of high solids content and consequently low water activity or further spray-dried into a powder (Joustra, 1975; De Schaepdrijver, 1987). Not only are these wort products shelf-stable, being concentrated they can be transported much more cheaply than beer. The wort products can be either sold directly to the retail consumer who ferments them into beer for home consumption, or to a satellite brewery that ferments, packs, and sells the product. This second process is ideal for communities that are distant from the major towns and cities where the breweries are presently situated. It has been used successfully in Namibia. In either case, the beer produced from the wort concentrates should be of much more consistent quality than can be produced by traditional home brewing.

A third shelf-stable product which is growing in popularity in South Africa is the so-called "instant" beer powder. This dry powder product consists essentially of finely milled sorghum malt, a precooked starchy adjunct, and active dried yeast (Novellie, 1986). The small brewer or retail consumer merely mixes the powder with warm water. The suspension is maintained at appropriate temperature to allow simultaneous mashing and lactic acid fermentation. Alcoholic fermentation then takes place in due course. The product is ready for consumption after approximately one day. As would be expected, the resulting beer can be of very variable quality depending on the success of the three brewing processes.

Looking to the future, there is no doubt that opaque beer will continue to be a significant beverage in Africa. The relative importance of opaque beer versus beverages originating in Europe and America such as barley beer and carbonated soft drinks will depend to a large extent on innovative research and development to meet the demands of this rapidly changing continent.

LITERATURE CITED

AISIEN, A. O. 1982. Enzymic modification of sorghum endosperm during seedling growth and malting. J. Sci. Food Agric. 33:754-759.

AISIEN, A. O., PALMER, G. H., and STARK, J. R. 1983. The development of enzymes during germination and seedlng growth in Nigerian sorghum. Starch/Staerke 35:316-320.

ANONYMOUS. 1971. Changes in the carbohydrates of kaffircorn during germination. Pages 14-19 in: CHEM 175. Bantu Beer Unit Annu. Rep. 1971, Council for Scientific and Industrial Research (CSIR), Pretoria, South Africa.

ANONYMOUS. 1974. Grain and malting quality of rapoko and babala. Pages 13-14 in: BB 48. Annu. Res. Rep. 1974. Bantu Beer Unit, Council for Scientific and Industrial Research (CSIR), Pretoria, South Africa.

ANONYMOUS. 1983. The Bulawayo municipal brewery. The Brewer 69:153-155.

AUCAMP, M. C., GRIEFF, J. T., NOVELLIE, L., PAPENDICK, B., SCHWARTZ, H. M., and STEER, A. D. 1961. Kaffircorn malting and brewing studies. VIII. Nutritive value of some kaffircorn products. J. Sci. Food Agric. 12:449-456.

BOTES, D. P., JOUBERT, F. J., and NOVELLIE, L. 1967a. Kaffircorn malting and brewing studies. XVII. Purification and properties of sorghum malt α-amylase. J. Sci. Food Agric. 18:409-415.

BOTES, D. P., JOUBERT, F. J., and NOVELLIE, L. 1967b. Kaffircorn malting and brewing studies. XVIII. Purification and properties of sorghum malt β-amylase. J. Sci. Food Agric.

18:415-419.

BRIGGS, D. E., HOUGH, J. S., STEVENS, R., and YOUNG, T. W. 1981. Malting and Brewing Science, Vol. I. Chapman and Hall, London.

DADA, L. O., and DENDY, D. A. V. 1987. The effect of various processing techniques on the cyanide content of processed germinated sorghum. Trop. Sci. 27:101-104.

DAIBER, K. H. 1975. Treatment of cereal grain. S. Afr. patent 75/4957.

DAIBER, K. Habiliment. 1978. BB 118. The selection of sorghum for malting. Sorghum Beer Unit, Council for Scientific and Industrial Research (CSIR), Pretoria, South Africa.

DAIBER, K. H. 1988. Malted and treated sorghum for industrial production of beer, feed and food. Pages 15-30 in: Proc. Quad. Symp. Sorghum Millets 4th, Int. Cereal Bread Congr. 8th. ICC, Lausanne, Switzerland.

DAIBER, K. H., and NOVELLIE, L. 1968. Gibberellic acid and amylase formation in kaffircorn. J. Sci. Food Agric. 19:87-90.

DAIBER, K. H., MALHERBE, L., and NOVELLIE, L. 1973. The modification of sorghum malt. Brauwissenschaft 26:220-225, 248-251.

DE SCHAEPDRIJVER, P. 1983. Development of new generation sorghum beers by the Sorghum Beer Unit, CSIR. S. Afr. Wine Bev. 34(4):9-11.

DE SCHAEPDRIJVER, P. 1987. Better beer brewing. S. Afr. Food Rev. 14 (2 Suppl.):9,11.

DE SCHAEPDRIJVER, P., and JOUSTRA, S. M. 1985. New process for long life sorghum beer. Food Ind. S. Afr. 12(4):11,13.

DE SCHAEPDRIJVER, P., WATSON, T. G., and JOUSTRA, S. M. 1980. Method of making sorghum beer. S. Afr. patent 80/1636. U.K. patent GB 2072218B.

DESIKACHAR, H. S. R. 1980. Development of weaning foods with high caloric density and low hot-paste viscosity using traditional technologies. Food Nutr. Bull. 2:21-23.

DOHERTY, C., ROONEY, L. W., and FAUBION, J. M. 1982. Phytin content of sorghum and sorghum products. Pages 328-333 in: Proc. Int. Symp. on Sorghum Grain Quality. L. W. Rooney and D. S. Murty, eds. Int. Crops Res. Inst. Semi-Arid Tropics (ICRISAT), Patancheru, A. P., India.

DYER, T. A., and NOVELLIE, L. 1966. Kaffircorn malting and brewing studies. XVI. The distribution and activity of α- and β-amylases in germinating kaffircorn. J. Sci. Food Agric. 17:449-456.

EASTMAN, P. 1980. An end to pounding: A new mechanical flour milling system in use in Africa. International Development Research Centre (IDRC), Ottawa, Canada.

EVANS, D. J., and TAYLOR, J. R. N. 1990a. Extraction and assay of proteolytic activities in sorghum malt. J. Inst. Brew. 96:201-207.

EVANS, D. J., and TAYLOR, J. R. N. 1990b. Influence of cultivar and germination conditions on proteolytic activies in sorghum malt. J. Inst. Brew. 96:399-402.

FAPARUSI, S. I. 1970. Sugar changes during the preparation of Barakutu beer. J. Sci. Food Agric. 21:79-81.

GARG, G. K., and VIRUPAKSHA, T. K. 1970. Acid protease from germinated sorghum. Eur. J. Biochem. 17:4-12.

GLENNIE, C. W. 1983. Polyphenol changes in sorghum grain during malting. J. Agric. Food Chem. 31:1295-1299.

GLENNIE, C. W. 1984. Endosperm cell wall modification in sorghum grain during germination. Cereal Chem. 61:285-289.

GLENNIE, C. W., and WIGHT, A. W. 1986. Dextrins in sorghum beer. J. Inst. Brew. 92:384-386.

GLENNIE, C. W., HARRIS, J., and LIEBENBERG, N. V. D. W. 1983. Endosperm modification in germinating sorghum grain. Cereal Chem. 60:27-31.

GLENNIE, C. W., CILLIERS, J. J. L., and GEYER, H. L. 1985. Changes in phytate and related compounds in grain sorghum during germination. Nutr. Rep. Int. 32:349-355.

GUPTA, R. K., and HASLAM, E. 1980. Vegetable tannins—Structure and biosynthesis. Pages 15-24 in: Polyphenols in Cereals and Legumes. J. H. Hulse, ed. International Development Research Centre (IDRC), Ottawa, Canada.

HAGGBLADE, S., and HOLZAPFEL, W. H. 1989. Industrialization of Africa's indigenous beer brewing. Pages 191-283 in: Industrialization of Indigenous Fermented Foods. K. H. Steinkraus, ed. Marcel Dekker, New York.

HAHN, R. R. 1969. Dry milling of grain sorghum. Cereal Sci. Today 14:234-237.

HARDIE, D. G., MANNERS, D. J., and YELLOWLEES, D. 1976. The limit dextrinase from malted sorghum (*Sorghum vulgare*). Carbohydr. Res. 50:75-85.

HOFMEYR, J. F. 1970. CHEM 127. Moisture uptake by sorghum grain during the steep stage of malting. Council for Scientific and Industrial Research (CSIR), Pretoria, South Africa.

HOSENEY, R. C., VARRIANO-MARSTON, E., and DENDY, D. A. V. 1981. Sorghum and Millets. Pages 71-144 in: Advances in Cereal Science and Technology, Vol. 4. Y. Pomeranz, ed. American Association of Cereal Chemists, St. Paul, MN.

HOUGH, J. S., BRIGGS, D. E., STEVENS, R., and YOUNG T. W. 1982. Malting and Brewing Science, Vol. 2. Chapman and Hall, London.

HOUSE, L. R. 1988. An overview of ICRISAT interest in sorghum utilization. Pages 9-14 in: Proc. Quad. Symp. Sorghum Millets 4th, Int. Cereal Bread Congr. 8th. ICC, Lausanne, Switzerland.

IWUAGWU, Y. O. U., and IZUAGBE, Y. S. 1985. Studies on the preservation and bottling of Oyokpo—A Nigerian beer from millet (*Pennisetum typhoideum*). J. Appl. Bacteriol. 59:487-492.

JAYATISSA, P. M., PATHIRANA, R. A., and SIVAYOGASUNDERAM, K. 1980. Malting quality of Sri Lankan varieties of sorghum. J. Inst. Brew. 86:18-20.

JOFFE, B. I., ROACH, L., BAKER, S., SHIRES, R., SANDLER, M., and SEFTEL, H. C. 1981. Failure to induce reactive hypoglycaemia by drinking a starch-base alcohol beverage (sorghum beer). Ann. Clin. Biochem. 18:22-24.

JOHANNSEN, E., and PURCHASE, I. F. H. 1969. Kaffircorn malting and brewing studies. XXI. The effect of the fusel oils of bantu beer on rat liver. S. Afr. Med. J. 43:326-328.

JONES, M., and PIERCE, J. S. 1964. Absorption of amino acids from wort by yeasts. J. Inst. Brew. 70:307-315.

JOUSTRA, S. M. 1975. Method of producing a product for use in the brewing of bantu beer. S. Afr. patent 75/0958.

JOUSTRA, S. M., and FIELD, A. 1978. Technological factors which influence malt quality. Pages 1-20 in: BB 118. Sorghum Beer Unit Silver Jubilee Congress. Sorghum Beer Unit, Council for Scientific and Industrial Research (CSIR), Pretoria, South Africa.

JOUSTRA, S. M., and JANSEN, S. B. J. 1988. Extruders for cooking adjunct for sorghum beer. S. Afr. Food. Rev. 15(1 Suppl.):73-76.

KANDLER, O. 1983. Carbohydrate metabolism in lactic acid bacteria. Antonie van Leeuwenhoek 49:209-224.

KURIEN, P. P., and DESIKACHAR, H. S. R. 1966. Preparation of a refined white flour from ragi. (*Eleusine coracana*) using a laboratory mill. J. Food Sci. Technol. 3:56-58.

MacFADDEN, D. P. 1989. Use of exogenous enzymes in sorghum brewing. Pages 306-322 in: Proc. Sci. Tech. Conv., 2nd. Institute of Brewing Central and Southern African Section, Johannesburg.

MALLESHI, N. G., DESIKACHAR, H. S. R., and THARANATHAN, R. N. 1986. Free sugars and non-starchy polysaccharides of finger millet (*Eleucine coracana*), pearl millet (*Pennisetum typoideum*), foxtail millet (*Setaria italica*) and their malts. Food Chem. 20:253-261.

McGRATH, R. M., KALUZA, W. Z., DAIBER, K. H., VAN DER RIET, W. B., and GLENNIE, C. W. 1982. Polyphenols of sorghum grain, their changes during malting and their inhibitory nature. J. Agric. Food Chem. 30:450-456.

MIFLIN, B. J., and SHEWRY, P. R. 1979. The biology and biochemistry of cereal seed prolamins. Pages 137-158 in: Seed Protein Improvement in Cereals and Grain Legumes, Vol. 1. International Atomic Energy Agency, Vienna.

MORRALL, P., BOYD, H. K., TAYLOR, J. R. N., and VAN DER WALT, W. H. 1986. Effect of germination time, temperature and moisture on malting of sorghum. J. Inst. Brew. 92:439-445.

MUNDY, J. 1982. Isolation and characterization of two immunologically distinct forms of α-amylase and a β-amylase from seeds of germinated *Sorghum bicolor* (L.) Moench. Carlsberg Res. Commun. 47:263-274.

MUNDY, J., GIBBONS, G. C., and MUNCK, L. 1983. Sorghum and barley malt amylases—A comparison. Proc. Congr. Eur Brew. Conv. 19:39-46.

NARZISS, L. 1976. Die Technologie der Malzbereitung, F. Enke, Stuttgart.

NOUT, M. J. R. 1980a. Microbiological aspects of the traditional manufacture of Busaa, a Kenyan opaque maize beer. Chem. Mikrobiol. Technol. Lebensm. 6:137-142.

NOUT, M. J. R. 1980b. Process development and preservation of Busaa, a Kenyan traditional opaque maize beer. Chem. Mikrobiol. Technol. Lebensm. 6:175-182.

NOUT, M. J. R., and DAVIES, B. J. 1982. Malting characteristics of finger millet, sorghum and barley. J. Inst. Brew. 88:157-163.

NOVELLIE, L. 1959. Kaffircorn malting and brewing studies. III. Determination of amylases in kaffircorn malts. J. Sci. Food Agric. 10:441-449.

NOVELLIE, L. 1962a. Kaffircorn malting and brewing studies. XI. Effect of malting conditions on the diastatic power of kaffircorn malt. J. Sci. Food Agric. 13:115-120.

NOVELLIE, L. 1962b. Kaffircorn malting and brewing studies. XII. Variation of diastatic power with variety, season, maturity and age of grain. J. Sci. Food Agric. 13:124-126.

NOVELLIE, L. 1966. Kaffircorn malting and brewing studies. XIV. Mashing with kaffircorn malt: Factors affecting sugar production. J. Sci. Food Agric. 17:354-361.

NOVELLIE, L. 1968. Kaffir beer brewing ancient art and modern industry. Wallerstein Lab. Commun. 31(104):17-32.

NOVELLIE, L. 1976. Beverages from sorghum and millets. Pages 73-77 in: Proc. ICC Symp. Sorghum and Millets for Human Food, Vienna, 1976. D. A. V. Dendy, ed. Tropical Products Institute, London.

NOVELLIE, L. 1984. Sorghum quality for malting and brewing. Pages 59-63 in: Symp. Process. Sorghum and Millets: Criteria for Quality of Grains and Products for Human Food. International Association for Cereal Science and Technology(ICC), Vienna.

NOVELLIE, L. 1986. Sorghum beer and related fermentations in southern Africa. Pages 219-235 in: Indigenous Fermented Food of Non-Western Origin. C. W. Hesseltine and H. L. Wang, eds. J. Cramer, Berlin.

NOVELLIE, L., and DE SCHAEPDRIJVER, P. 1986. Modern developments in traditional African beers. Pages 73-157 in: Progress in Industrial Microbiology, Vol. 23. M. R. Adams, ed. Elsevier, Amsterdam.

O'DONOVAN, M. B., and NOVELLIE, L. 1966. Kaffircorn malting and brewing studies. XV. The fusel oils of kaffir beer. J. Sci. Food Agric. 17:262-265.

OKON, E. U., and UWAIFO, A. O. 1985. Evaluation of malting sorghums. Brew. Dig. 60(12):24-29.

OLESEN, T. 1985. Microbial enzymes for production of African and European types of beer. Pages 267-294 in: Proc. Sci. Tech. Conv., 1st, Institute of Brewing Central and Southern African Section, Johannesburg.

PANASIUK, O., and BILLS, D. D. 1984. Cyanide content of sorghum sprouts. J. Food Sci. 49:791-793.

PATHIRANA, R. A., SIVAYOGASUNDARAM, K., and JAYATISSA, P. M. 1983. Optimization of conditions for malting of sorghum. J. Food Sci. Technol. 20:108-111.

PICKERELL, A. T. W. 1986a. Factors affecting fermentations of sorghum beer. M.Sc. thesis. University of the Witwatersrand, Johannesburg.

PICKERELL, A. T. W. 1986b. The influence of free alpha-amino nitrogen in sorghum beer fermentations. J. Inst. Brew. 92:568-571.

RABIE, C. J., and THIEL, P. E. 1985. Toxigenic fungi and mycotoxins in sorghum malt. Pages 252-265 in: Proc. Sci. Tech. Conv., 1st, Institute of Brewing Central and Southern African Section, Johannesburg.

RACHIE, K. O. 1975. The millets: Importance, utilization and outlook. Tech. Bull. Int. Crops. Res. Inst. Semi-Arid Tropics (ICRISAT), Hyderabad, India.

RAMACHANDRA, G., VIRUPAKSHA, T. K., and SHADAKSHARASWAMY, M. 1977. Relationship between tannin levels and in vitro protein digestibility in finger millet (*Eleucine coracana* Gaertn.) J. Agric. Food. Chem. 25:1101-1104.

REICHERT, R. D., and YOUNGS, C. G. 1977. Chemical compositions of mechanically and traditionally dehulled sorghum and millet. Cereal Chem. 54:174-178.

ROONEY, L. W., BLAKELY, M. E., MILLER, F. R., and ROSENOW, D. T. 1980. Factors affecting the polyphenols of sorghum and their development and location in the sorghum kernel. Pages 25-35 in: Polyphenols in Cereals and Legumes. J. H. Hulse, ed. International Development Research Centre (IDRC), Ottawa, Canada.

SCHWEIGART, F., and VLIETSTRA, H. 1971. Die herstellung von Kafferkorngrits. Mühle 108:538-539, 551-552, 586-589.

SHAMBE, T., VONCIR, N., and GAMBO, E. 1989. Enzyme and acid hydrolysis of malted millet (*Pennisetum typhoides*) and sorghum (*Sorghum bicolor*). J. Inst. Brew. 95:13-16.

SOUTH AFRICAN BUREAU OF STANDARDS. 1971. SABS 183-1971. Standard specification for the production of bantu beer. Council of the South African Bureau of Standards, Pretoria.

STAFFORD, H. A. 1965. Flavanoids and related phenolic compounds produced in the first internode of *Sorghum vulgare* in darkness and in light. Plant Physiol. 40:130-138.

TAYLOR, J. R. N. 1983. Effect of malting on the protein and free amino nitrogen composition of sorghum. J. Sci. Food Agric. 34:885-892.

TAYLOR, J. R. N. 1989. The biochemistry of sorghum beer brewing. Pages 275-292 in: Proc. Sci. Tech. Conv., 2nd, Institute of Brewing Central and Southern African Section, Johannesburg.

TAYLOR, J. R. N., and BOYD, H. K. 1986. Free α-amino nitrogen production in sorghum beer mashing. J. Sci. Food Agric. 37:1109-1117.

TAYLOR, J. R. N., and DAIBER, K. H. 1988. Effect of calcium ions in sorghum beer mashing. J. Inst. Brew. 94:68-70.

TAYLOR, J. R. N., and SCHÜSSLER, L. 1986. The protein compositions of the different anatomical parts of sorghum grain. J. Cereal Sci. 4:361-369.

TAYLOR, J. R. N., and SCHÜSSLER, L. 1988. Production of fermentable sugars from sorghum beer strainings (brewer's grains) by industrial enzymes. S. Afr. Food Rev. 15(1 Suppl.):77-79.

TAYLOR, J. R. N., and VON BENECKE, R. 1990. Development of a simple assay for the direct determination of β-amylase in sorghum malt. Pages 344-347 in: Proc. Aviemore Conf. Malt. Brew. Distil. 3rd. I. Campbell, ed. Institute of Brewing, London.

TAYLOR, J. R. N., BOYD, H. K., and PICKERELL, A. T. W. 1985. The role of free alpha-amino nitrogen (FAN) in the brewing of sorghum beer. S. Afr. Food Rev. 12(4):82-83.

TRINDER, D. W. 1988. A survey of aflatoxins in industrially brewed South African sorghum beer and beer strainings. J. Inst. Brew. 95:307-309.

VAN BUREN, J. P., HRAZDINA, G., and ROBINSON, W. B. 1974. Color of anthocyanin solutions expressed in lightness and chromaticity terms. Effect of pH and type of anthocyanin. J. Food Sci. 39:325-328.

VAN DER WALT, J. P. 1956. Kaffircorn malting and brewing studies. II. Studies on the microbiology of kaffir beer. J. Sci. Food Agric. 7:105-113.

VAN HEERDEN, I. V. 1985. Sorghum beer: An analysis of the raw materials, intermediates and final products. S. Afr. J. Sci. 81:587-589.

VAN HEERDEN, I. V. 1987. Nutrient content of sorghum beer strainings. S. Afr. J. Anim. Sci. 17:171-175.

VAN HEERDEN, I. V. 1988. In defence of sorghum beer. S. Afr. Food Rev. 15(1 Suppl.): 83-84.

VAN HEERDEN, I. V. 1989a. Sorghum beer—A decade of nutritional research. Pages 293-305 in: Proc. Sci. Tech. Conv., 2nd, Institute of Brewing Central and Southern African Section, Johannesburg.

VAN HEERDEN, I. V. 1989b. The nutritive content of African beers brewed with maize grits or sorghum adjunct. J. Inst. Brew. 95:17-20.

VAN HEERDEN, I. V., and GLENNIE, C. W. 1987. Availability of B vitamins in sorghum beer. Nutr. Rep. Int. 35:147-155.

VAN HEERDEN, I. V., TAYLOR, J. R. N., and GLENNIE, C. W. 1987. The contribution of minerals, trace elements, phytin, starch and amino acids to the nutritional quality of sorghum beer. S. Afr. J. Sci. 83:5-7.

VON HOLDT, M. M., and BRAND, T. C. 1960. Kaffircorn malting and brewing studies. VII. Changes in the carbohydrates of kaffircorn during malting. J. Sci. Food Agric. 11:467-471.

WATSON, T. G., and NOVELLIE, L. 1974. Extraction of *Sorghum vulgare* and *Hordeum vulgare* α-glucosidase. Phytochemistry 13:1037-1041.

WATSON, T. G., and NOVELLIE, L. 1975. The development of amylase and maltase during the malting of *Sorghum vulgare*. Agrochemophysica 7:61-64.

YOUNG, R. S. 1949. Brewing of kaffir beer in Northern Rhodesia. J. Inst. Brew. 55:371-372.

ZAMMIT, I. V. 1980. The nutritive value of sorghum beer. S. Afr. Food Rev. 7(1 Suppl.):73-75.

CHAPTER 11

SORGHUM AND MILLETS FOR FORAGE AND FEED

P. J. Bramel-Cox
Sorghum Geneticist
Department of Agronomy
Kansas State University
Manhattan, Kansas

K. A. Kumar
Pearl Millet Breeder and Team Leader
Pearl Millet Improvement Program
ICRISAT Sahelien Center
Niamey, Niger

J. D. Hancock
Animal Nutritionist
Department of Animal Sciences and Industry
Kansas State University
Manhattan, Kansas

D. J. Andrews
Sorghum and Pearl Millet Breeder
Department of Agronomy
University of Nebraska
Lincoln, Nebraska

INTRODUCTION

Sorghum (*Sorghum bicolor* (L.) Moench) and pearl millet (*Pennisetum glaucum* (L.) R. Br.) were both domesticated in tropical Africa 3,000–5,000 years ago and subsequently spread into the Indian subcontinent, with sorghum further becoming established in China. Despite the introduction of maize into the Old World, these two crops still remain staple cereals for millions of people living in drought-prone regions of Africa and the Indian subcontinent. A great diversity of types was developed in both species, and natural introgression with their wild progenitors has continued in Africa, resulting in persistent weedy populations. These wild relatives and their weedy introgressions represent a large, untapped genetic pool for improvement in the future.

Although domestication was primarily for food grains (and in sorghum for beer and sweet stems in Africa and for brooms in China), crop residues have been valued as animal fodder, building materials, and fuel. Farmer-consumer selection over the centuries in food grain types has been mainly for traits such as appearance of grain and food, ease of processing, taste, and only indirectly for nutritional value.

Joint contribution of the Kansas Agricultural Experiment Station, contribution No. 91-461-B, International Crop Research Institute for the Semi-Arid Tropics, Journal Article No. 1142, and Department of Agronomy, University of Nebraska.

White or cream grain colors in sorghum and gray-blue or light brown in pearl millet with hard endosperms have been preferred for human food. Sorghum beer types were selected for quite different criteria, including high tannin levels, malting ability, and easily hydrolyzable starch. Beer types are often earlier maturing because of the superior mold resistance of their grain. Today, the total world area planted to sorghum is about 47 million hectares, of which about 4 million are grown for animal feed and fodder in the United States. Some 26 million hectares of pearl millet are grown almost exclusively for human food in Africa and the Indian subcontinent. Possibly 0.4 million hectares of pearl millet forage is grown in the United States.

Sorghum and pearl millet have been grown for forage in the United States for about 140 years, the first sorghum forage introductions being attributed to Leanard Wray in 1851 (Vinall et al, 1936). Sorghum was also cultivated for grain, syrup, and brooms. Although pearl millet essentially has remained a forage crop in the United States, sorghum has become a major feed grain crop in the Great Plains where moisture stress makes maize production risky. This was made possible by breeding of semidwarf (2-dwarf) varieties and use of cytoplasmic male sterility, which allowed development of fully dwarf (3-dwarf) hybrids that could be combine-harvested. Many early grain sorghum hybrids had brown pericarps with high tannin levels, which kept the grain moldfree in humid ripening conditions and reduced bird damage. When these brown-grained sorghums were found to have low feeding values, new hybrids were developed with red, white-cream, or translucent pericarps and white or yellow endosperms with little or no tannins. These newer hybrids have substantially better feed values, but, unfortunately, sorghum has remained characterized by the poor reputation of the earlier brown-grained types. Newer hybrids have 95% of the feed value of maize but are priced 10–15% lower.

Pearl millet is now being bred as a feed grain crop in the United States because the grain has excellent nutritional value. The development of early-maturing, drought-resistant hybrids appears possible for areas marginal for sorghum production. A number of other small-grain cereals unrelated to pearl millet and termed minor millets are grown principally for human consumption in various parts of the world. Of these only two, proso millet (*Panicum miliaceum* (L.)) and foxtail millet (*Setaria italica* (L.) Beauv.) are grown on a small scale in the United States for bird seed and forage.

SORGHUM

Sorghum Forage

The various uses of *Sorghum* species as forage have included silage, fodder, green chop, pasture, and hay for livestock. Owen and Moline (1970) proposed a general classification of the forage sorghums based upon their use: 1) pasture using Columbus grass (*S. almum*); 2) pasture and green chop or hay using Johnsongrass (*S. halapense*), sudangrass, or sudangrass hybrids; 3) green chop or hay using sorghum-sudangrass hybrids; 4) silage using sorgos and canes; and 5) dual-purpose (i.e., for production of grain and forage) using grain sorghums or 2-dwarf sorghums.

Forage sorghums have been grown primarily in the Great Plains of the United States where moisture is insufficient for maize production. Most forage sorghums grown in the semiarid tropics or subtropics can be classified as dual-purpose, with the grain used for human consumption and the stover for other purposes. The growth requirements for these multipurpose crops and their management have been very well discussed by Owen and Moline (1970), Coleman (1970), Weibel (1970), and more recently by Doggett (1988). All of these reviews concluded that these sorghums are well adapted to environments with little rainfall, high temperatures and low soil fertility, where they produce greater quantities of high-quality forage than alternative crops such as maize. The general constraints to sorghum use and production have included high levels of hydrocyanic acid precursors in early growth or regrowth, accumulation of high levels of nitrates under certain environmental conditions, and a generally lower feeding value than forage options such as maize. The effects of various management alternatives and genetic factors, such as plant height, maturity, grain yield, and brown midrib, are known to have major effects on the nutritional value of this crop.

CROP MANAGEMENT

Owen and Moline (1970) described three basic crop management factors known to influence the quantity of forage produced. These were planting date, fertility of the soil, and plant spacing both between and within the row. Sorghums for forage produce greater quantities of biomass when planted later than types for grain production and with adequate levels of nitrogen, phosphorous, and potassium provided in split applications. Nelson (1953) found that high levels of nitrogen applied at one time resulted in dangerous nitrate concentrations and high levels of HCN content if the phosphorus level was low. These problems could be accentuated under moisture stress conditions. The use of adequate nitrogen fertilization, however, is important for both acceptable protein levels and an increased quantity of biomass. Koller and Scholl (1968) concluded that sudangrass produced a higher quality and a greater quantity of forage if planted in narrower spacing with a higher seeding rate. This was especially pronounced in the first cutting. Doggett (1988) reviewed specific crop requirements for the various forage types.

Owen and Moline (1970) concluded that the stage of maturity was the most important factor influencing the quality and quantity of forage produced. Generally, from heading to the ripe-seed stage, forage sorghums declined in protein content, crude fiber, and ash, while they increased in nitrogen-free extract. In grain sorghums, the changes associated with maturity were an increase in lignin content and variable changes in the sugar content. The sudangrasses and sorghum-sudangrass hybrids were most affected by maturity, with protein content being the most dramatically reduced. This general reduction in digestibility was associated with an increase in cellulose and cell wall constituents. Although advances in maturity reduced the nutrient value of the various types on a dry matter basis, the dry matter increased more on a per hectare basis, so the actual nutrients harvested per hectare increased.

Other authors have reported that digestibility of nutrients decreases with stage of maturity in the grass- and forage-type sorghums, whereas the high-grain types were relatively more constant, especially from milk to hard-dough stage (Helm and Leighton, 1960; Anthony et al, 1961; Owen and Kuhlman, 1967). Browning and Lusk (1967) found that total crude protein and crude protein digestibility decreased with maturity, but these changes had little influence on the energy digestibility. Thurman et al (1960) noted that the increased yield of the forage types associated with maturity was due to an increased stalk dry weight, whereas percentage heads, blades, and sheaths declined.

Browning and Lusk (1967) and Owen and Moline (1970) reported that the most important effect of maturity on silage quality was related to the moisture level. Stage of maturity affects the constituents produced during ensiling, especially those associated with the increased fermentation acid found under the higher moisture conditions in the early harvested forages. Owen (1962) reported that dry matter consumption in dairy cattle increased with stage of maturity in forage. For high-grain sorghums, Marshall et al (1966) and Browning and Lusk (1967) found that, although maturity reduced feed efficiency of dairy cattle, silage yield and total milk yield were optimized when plants were harvested past the milk stage.

BREEDING AND GENETICS

Sorghums used for forages and feed constitute a very diverse group, including various races within *Sorghum bicolor* subspecies *bicolor* and two other species, *S. almum* and *S. halapense*. Comparisons between types or within types for nutritional composition and quantity have been reported by a number of authors. Burns and Kimbrough (1981) found no differences in the in vitro dry matter digestibility (IVDMD) between forage sorghums, grain sorghums, sorghum-sudangrass hybrids, and pearl millet. A number of other studies have reported differences. Ritchie et al (1972) reported a 0.14-kg advantage in average daily gain for cattle fed grain sorghum silage compared with a forage sorghum. Ward and Smith (1968) found that the major difference between a sterile forage sorghum and a better grain-producing hybrid was in the higher consumption of silage from the latter. Bolsen et al (1983) also found that silage from a grain-producing forage sorghum resulted in better cattle gains due to better feed efficiency and higher dry matter yield per hectare compared with a sterile forage sorghum. Nordquist and Rumesy (1967) found that silage from a low-grain forage sorghum produced lower fat-corrected milk yields than intermediate- or high-grain types. Ritchie et al (1972) reported similar advantages for silages from grain-type forages when fed to breed heifers.

When sorghums with low grain-to-stalk ratios were compared with dual-purpose or high-grain varieties, weight gains by steers and calves were very similar or better with the grain types. Meyers et al (1959) and Bolsen et al (1983) found very similar gains with forage or grain sorghums, but the relative feeding value of a nonheading forage was much lower. Browning and Lusk (1967) concluded that grain-type sorghums generally had a lower cellulose or crude fiber content than forage sorghums.

Great variation exists among varieties or hybrids within a type for a number of nutritional quality parameters. Arora et al (1975) reported a wide variability in structural carbohydrates and IVDMD among grain sorghums; acid detergent fiber (ADF) ranged from 31 to 38%, cellulose content ranged from 21 to 32%, and IVDMD ranged from 45 to 66%.

Goldy et al (1986) studied 188 grain sorghum hybrids for IVDMD, crude protein content, and calcium and phosphorus levels at 12 locations in Kansas. They found only slight differences among hybrids within a location but large differences between sorghums from different locations. Comparisons among forage or silage sorghums have also shown wide variation in nutritional value. White et al (1988) compared 80 forage sorghum cultivars produced one year and 60 cultivars from the next for a number of agronomic and silage quality traits. They reported a wide range between the cultivars for silage yield, grain yield, maturity, height, IVDMD, crude protein, neutral detergent fiber (NDF), and ADF. The late maturity, nearly nonheading types tended to have the lowest digestibility, lowest protein content, and the highest fiber content. Most of these cultivars also had considerable tannin concentrations. Similar diversity among quality components was reported by Gourley and Lusk (1978), Dickerson et al (1986), Minor et al (1987), Honeycutt (1988), Kalton (1988), and Raymer et al (1988).

These differences among cultivars could be due to various quantitatively inherited traits, as well as a few major genetic factors. Those identified as quantitative traits were quality factors such as IVDMD, crude protein, NDF, ADF, and Brix. A few studies addressed the genetic nature of these traits in forage sorghum hybrids (Ross et al, 1979) and sorghum-sudangrass hybrids (Pederson et al, 1982, 1983). The quality components IVDMD and Brix had a high general combining ability for both types of hybrids (Ross et al, 1979; Pederson et al, 1983).

A number of more simply inherited factors have been evaluated for their influence on forage quality or quantity. In sorghum, there are four basic height genes that are related to the dwarfing of the plant (Quinby and Karper, 1954). Gourley and Lusk (1978) compared sorghums with one, two, or three of these dwarfing genes. They concluded that leaf percentage was the same in all three height groups, but the percentage stalk versus head was altered. The 1-dwarf type had 75% stalks and 7% heads, the 2-dwarf type had 56% stalks and 23% heads, and the 3-dwarf type had 36% stalks and 43% heads. Another factor that is known to be fairly simply inherited is maturity measured as anthesis date (Quinby, 1967), the effect of which on forage quality was discussed in the previous section.

Cummins (1981) concurred that silage sorghums with a high percentage of heads but a low stalk ADF and high stalk IVDMD were still important because stalks made up 50% or more of the total forage. The induction of the brown midrib trait in sorghum has made possible a major increase in the stalk or leaf IVDMD. Porter et al (1978) identified and isolated 19 different brown midrib plants (*bmr*-1 to *bmr*-19). The reduction in permanganate lignin concentrate ranged from 1 to 33% in stems and 2 to 18% in the leaves of forage sorghum. Fritz et al (1981) found a similar expression when these genes were transferred to sudangrass. Lusk et al (1984) reported no difference between dry matter digestibility of *bmr*-12 sorghum and maize silage; thus, the *bmr*-12 gene in sorghum was credited with increased dry matter digestion and consumption. Wedig et al (1987) found no difference in dry matter intake and cell wall digestion between *bmr*-12 and normal sorghums. Hanna et al (1981) reported greater IVDMD of young vegetative forage with this brown midrib trait because of lower lignin concentration and no differences in total sugars.

Bucholtz et al (1980) determined that the lignin of *bmr*-6 had about three times more aldehyde groups than lignin of normal sorghum, and the activity of cinnamyl alcohol/NADPH oxidoreductase, an enzyme responsible for reduction of the aldehyde lignin intermediate, was significantly reduced. Jung et al (1983) described the lignin of *bmr* types as less polymerized with less para-coumaric acid (PCA), parahydroxybenzaldehyde, and syringealdehyde. Akin et al (1986b), Cherney et al (1986), Wedig et al (1987), and Fritz et al (1990) also reported reduced PCA concentration in the mutant *bmr*. Wedig et al (1989) found that the digestibilities of hemicellulosic monosacharides, arabinose, xylose, and uronic acids were greater in *bmr*-12 silage. This was probably due to fewer bonds between the lignin and carbohydrates; thus, the cell wall fractions were more digestible.

Fritz et al (1981) reported that *bmr* mutation did not change the rate of *in situ* digestion or rate of ruminal digestion flow, but lignin concentration was decreased and apparent digestibility was increased. Fritz et al (1990) concluded that one of the major alterations in lignin chemistry in the *bmr* mutant was a decrease in the amount of PCA esterified to the lignin core. This reduction in the PCA–ferulic acid ratio and changes in the linear xylan content were given credit for changes in the rate and extent of NDF digestion. Akin et al (1986a), compared leaf blade anatomy, histological reactions, and degradation of specific tissues, and reported increased availability of degradable to marginally degradable tissues was responsible for the increased digestibility of the *bmr*-12 plants. Akin et al (1986b), in a further evaluation of these two types, concluded that *bmr*-12 mutation affected the less complex or less rigidly bound phenolic compounds within the cell wall and that removal of these phenolic compounds resulted in a more pronounced degradation of tissues.

Two fairly simply inherited traits known to cause reduced digestibility are the antinutritional and toxic properties of sorghum. These are tannins and prussic (hydrocyanic) acid, respectively. Montgomery et al (1988) compared bird-resistant and bird-susceptible types to determine the influence of tannins on quality components. They concluded that a negative relationship exists between IVDMD and the tannin content of sorghum grain and whole plant samples. These results agreed with those of Harris et al (1970), Cummins (1971), and Johnson et al (1971). Because tannin level was correlated negatively with crude protein content and IVDMD, but positively associated with the fibrous constituents, high-quality silage from bird-resistant sorghums was considered unlikely. Pund (1970) and Maxson et al (1973) also concluded that bird-resistant sorghums generally gave lower feed conversions and produced less total gain than low-tannin sorghums.

The concentration of cyanogens and toxic hydrocyanic acid (HCN) is fairly simply inherited. Earlier studies by Nass (1972) showed that several loci were involved and the realized heritability in a sudangrass population was 0.4 (Gorz et al, 1986). Lamb et al (1987) concluded that additive gene effects were most important for seedling HCN, but Gorz et al (1986) reported that at least one line, KS 8, had a major gene with partial dominance for high HCN.

Although a few genes have major effects, such as brown midrib or HCN concentration, the majority of characteristics known to improve the nutritional value of forage sorghum are not well understood. Kalton (1988) emphasized that development of improved forages in the future, through manipulation of genetics, has a tremendous potential. He described a number of breeding objectives for the various types, including disease and insect resistance, improved standability, better regrowth potential, and better adaptation to stress environments, but concluded that greatest improvement will be possible with increased incorporation of the brown midrib trait into all types of forage sorghums.

Mohammad (1989) found a wide range of responses among a diversity of sorghum types to variable dryland environments. He concluded that the most stable performance for the various quality parameters was generally found within the 2-dwarf types that had higher grain yielding ability. This concurs with results indicating better performance by animals fed high-yielding grain types, as reported in the previous section of this chapter. Thus, in the future more emphasis should be put on developing 2-dwarf forage sorghums with greater grain yield potential and stability of performance.

The genetic base of improvement within forage sorghums has been very limited, depending upon eight early introductions. The genetic base for sudangrass is one introduction made in 1909. The development of sweet sorghums was based upon the introduction of cv. Collyer in 1881 (Duncan et al, 1991). Although the incorporation of a single trait, such as the brown midrib gene, offers potential for progress in the short term, continued progress of sorghum forages will depend on the use of greater amounts of the genetic diversity available in the world, especially from areas such as China and Russia.

OTHER USES

Sorghum has other uses besides forage and grain. The management for and utilization of sorghum for syrup has been discussed by Coleman (1970), Doggett (1988), and in chapter 12 of this book. This use is limited, as is use for the production of brooms. Weibel (1970), Froment (1986), and Doggett (1988) concluded that little research has been done on broomcorn, and its production has declined except as a small specialty crop outside the United States. Doggett (1988) discussed other miscellaneous uses of sorghum, including as construction material, fuel, and dye.

Sorghum Grain

The value of sorghum grain for livestock feeding is different for ruminants (e.g., cattle, sheep, and goats) than for nonruminants (e.g., swine, poultry, and fish). In ruminants, microflora of the rumen can upgrade poor-quality proteins and nonprotein nitrogen to the protein quality of the microflora itself. Therefore, ruminant nutritionists view sorghum and other cereal grains primarily as sources of carbohydrates (e.g., starch). In nonruminants, sorghum is viewed as an energy source, but the quality and quantity of protein is important, because proteins from cereal grains in cereal grain based diets can contribute more than one-third of the dietary crude protein for chicks and more than one-half of the dietary crude protein for growing-finishing pigs. In diets for nonruminants, a cereal grain serves as the base ingredient, to which other ingredients are added to supplement its nutrient composition. In Table 1, the nutrient contents of maize and sorghum (NRC, 1982, 1988) are expressed relative to published nutrient requirements (NRC, 1988) for growing pigs.

Crude protein contents of maize and sorghum are only 57–59% of the pig's requirement, but of even greater significance is the very low concentration of lysine (i.e., 33 and 31% of the pig's requirement in maize and sorghum, respectively) in the cereal grains. Human nutritionists have similar concerns about low concentrations of protein and limiting amino acids in cereal grains used for human diets, thus selection for improved amino acid content would be of benefit. Such advances have been made in maize (Mertz et al, 1964, 1965), but the 10% loss of yield typical to high-lysine maize has prevented commercial acceptance. Although this yield loss has been largely overcome (Hancock et al, 1990a), high-lysine maize is still subject to a certain prejudice. Thus, even though sorghums with greater concentrations of lysine than presently available in commercial hybrids have been identified, sorghum breeders must incorporate these into hybrids without sacrificing the present high yields if high-lysine sorghums are to be accepted.

TABLE 1
Nutrient Content of Maize and Sorghum as a Percentage of the Nutrient Requirements for the Growing Pig (20 to 50 kg live weight)[a]

Nutrient	Requirement (% or amount/kg of diet)	Percentage of Requirement	
		Maize	Sorghum
Protein and amino acids, %			
Protein	15	57	59
Arginine	0.25	172	148
Histidine	0.22	123	109
Isoleucine	0.46	76	96
Leucine	0.60	198	220
Lysine	0.75	33	31
Methionine + cystine	0.41	98	71
Phenylkalanine + tyrosine	0.66	127	130
Threonine	0.48	75	56
Tryptophan	0.12	75	83
Valine	0.48	100	110
Linoleic acid, %	0.1	2,200	1,130
Mineral elements			
Calcium, %	0.60	5	5
Phosphorus, %	0.50	56	56
Sodium, %	0.10	10	40
Chloride, %	0.08	63	100
Magnesium, %	0.04	275	325
Potassium, %	0.23	143	135
Copper, mg	4	88	250
Iodine, mg	0.14	21	29
Iron, mg	60	55	80
Manganese, mg	2	285	790
Selenium, mg	0.15	47	133
Zinc, mg	60	32	28
Vitamins			
A, IU[b] (assuming 150 IU/mg of β-carotene)	1,300	23	12
D, IU (assuming equal activity for D-2 and D-3)	150	...[c]	17
E, IU	11	190	110
K, mg of menadione	0.5	40	40
Biotin, mg	0.05	140	460
Choline, mg	300	168	213
Folic acid, mg	0.3	100	67
Niacin, mg of available	10	...	...
Pantothenic acid, mg	8	64	138
Riboflavin, mg	2.5	44	44
Thiamin, mg		370	410
B-6, mg	10	...	...
B-12, μg	10	...	...

[a] Adapted from NRC (1982) and NRC (1988).
[b] International units.
[c] Blank spaces indicate no reported value or concentrations and bioavailability so low as to be considered of no biological significance.

Vitamin concentrations in maize and sorghum are similar, with slightly greater concentrations of biotin, choline, pantothenic acid, and thiamin in sorghum, and slightly greater concentrations of vitamins A, E, B-6, and folic acid in maize. Concentrations of most minerals are greater in sorghum than maize. However, the variability in vitamin and mineral contents of maize and sorghum caused by growing conditions, soil fertility, and harvesting and storage techniques has made these differences in nutrient content of minimal use to animal nutritionists. The amount of vitamin and mineral supplementation required for a cereal-based diet can add $3–4 (U.S.) to the cost of a metric ton of feed but has been accepted as necessary.

Digestible energy, metabolizable energy (ME), and net energy are used to express the energy value of feedstuffs, but the ultimate measure of suitability for livestock feeding is growth performance. A compilation of university experiments comparing sorghum and maize for pig feeding is given in Table 2. The feeding value (i.e., efficiency of gain) of sorghum relative to maize ranged from 91 to 99%, with an average of 95%. Despite the extreme differences in anatomy of their digestive tracts, this average value in swine is the same as the relative feeding value of sorghum for beef cattle, as proposed in a review article by Riley (1985).

TABLE 2
Nutritional Value of Sorghum Relative to Corn

Animal Studied	Investigator	Response Criteria Measured	Relative Value, Percent of Corn
Nursery pigs	White et al (1983)		
		Rate of gain	90
		Efficiency of gain	94
Nursery pigs	Combs et al (1980)		
		Rate of gain	98
		Efficiency of gain	91
Nursery pigs	Hamilton et al (1980)		
		Metabolizable energy	97
		N retention	98
Nursery pigs	Hamilton et al (1979)		
		Rate of gain	106
		Efficiency of gain	99
Nursery pigs	Hines and Allee (1971)		
		Rate of gain	92
		Efficiency of gain	95
Growing-finishing pigs	Cromwell et al (1985)		
		Rate of gain	99
		Efficiency of gain	96
Growing-finishing pigs	Meyer (1982)		
		Rate of gain	106
		Efficiency of gain	95
Growing-finishing pigs	Danielson et al (1978)		
		Rate of gain	97
		Efficiency of gain	95
Growing-finishing pigs	Tribble et al (1975)		
		Rate of gain	95
		Efficiency of gain	96
		N digestibility	89
		Energy digestibility	94
Lactating sows	Louis et al (1989)		
		Metabolizable energy	99
		N digestibility	88

Although this average feeding value has merit for use in decision making about sorghum in diets for livestock, perhaps of greater importance to plant breeders and sorghum producers is the range in these values. An understanding of factors contributing to this variability (e.g., environment, genetics, and processing), that could lead to higher, more consistent feeding values, would improve the status of sorghum as a feed grain.

CROP MANAGEMENT

The general effects of growing conditions on yield of sorghum have been documented. Increased available water and nitrogen results in increased yield of dry matter and protein. However, little information is available on the effects of available water and nitrogen, or other growing conditions, on nutritional value of the sorghum grain. An experiment designed to test the effects of early frost on the nutritional value of sorghum grain for growing-finishing pigs was reported by Tribble et al (1987). Average daily gain and feed efficiency of pigs fed diets with 32 lb/bu (412 g/L) test weight sorghum were inferior to those of pigs fed diets with 48 and 56 lb/bu (611 and 714 g/L) test weight sorghums, which were similar. Hansen et al (1990) reported similar results. Hancock et al (1991a) reported that broiler chicks were more sensitive to differences in test weight than growing-finishing pigs, with a linear decrease in rate of gain, efficiency of gain, and ME as test weight decreased from 55 to 35 lb/bu (708 to 451 g/L). An additional observation from this experiment was that the common practice of blending normal and light test weight grains (i.e., a 55/35 blend) did not give a product of equal feeding value or ME content to the intermediate test weight sorghum (i.e., 45 lb/bu or 579 g/L).

TABLE 3
Effects of Agronomics on Yield of Utilizable Nutrients from Maize and Sorghum[a]

	Maize[c]				Bronze Sorghum				Yellow Sorghum			
	Irr		Low-Irr		Irr		Low-Irr		Irr		Low-Irr	
Item[b]	+N	−N	+N	−N	+N	−N	+N	−N	+N	−N	+N	−N
Experiment 1												
Yield, kg/ha	7,601	7,272	5.687	5.197	7,937	6,597	5,790	5,229	8.427	7,802	4,827	4,865
G/F	0.710	0.712	0.716	0.693	0.723	0.741	0.725	0.716	0.668	0.693	0.705	0.697
UN, kg/ha	5,397	5,178	4,072	3,602	5,738	4,888	4,198	3,744	5,629	5,407	3,403	3,391
Experiment 2												
CP yield, kg/ha	671	601	508	481	821	609	609	544	752	665	462	468
G/F	0.645	0.646	0.651	0.658	0.635	0.638	0.607	0.633	0.661	0.652	0.647	0.654
UP, kg/ha	433	388	331	317	521	389	370	344	497	434	299	306
Experiment 3												
GE yield, Gcal/ha	31.5	30.7	23.6	21.7	32.8	27.3	24.0	21.9	34.6	32.4	19.7	19.7
ME, kcal/kg	3,593	3,282	3,137	3,314	3,699	3,323	3,187	3,484	3,315	3,610	3,111	2,977
UE, Gcal/ha	27.3	23.9	17.8	17.2	29.4	21.9	18.5	18.2	27.9	28.2	15.0	15.5

[a] Adapted from Richert et al (1991).
[b] Item abbreviations are G/F = gain to feed ratio, UN = utilizable nutrients (G/F × yield), CP = crude protein, UP = utilizable protein (G/F × protein yield), GE = gross energy, ME = metabolizable energy, and UE = utilizable energy (ME × yield).
[c] Treatment abbreviations are Irr = irrigated, Low-Irr = minimal irrigation, +N = with nitrogen application, −N = without nitrogen application.

In contrast with the preceding discussion, reports of nutritional evaluation of grains produced with well-defined growing conditions, and at the same location, are difficult to find. Of the few reported experiments, Cromwell et al (1983) and Hancock et al (1990a) indicated that irrigation and nitrogen fertilization increased the yield of utilizable nutrients in normal and high-lysine maize. Richert et al (1991) reported similar results with sorghums grown in an arid location (Table 3). Broiler chicks were used to evaluate the nutritional value of maize, a bronze pericarp sorghum (with hetero-yellow endosperm), and a translucent pericarp sorghum (with homozygous yellow endosperm), produced with full irrigation, minimal irrigation, nitrogen applied, and without nitrogen applied. Nitrogen was applied at the same rate for maize and the sorghums, but additional water was applied to maize at tasseling to ensure enough grain to conduct feeding experiments. Full irrigation increased yield by 44% compared with minimal irrigation, whereas the application of nitrogen only increased yield by 9% compared with no nitrogen application. When data were pooled across all agronomic conditions, the bronze pericarp sorghum yielded the highest total utilizable nutrients and utilizable protein even with the extra water given to maize. Interactions between grain source, irrigation level, and nitrogen application suggested that water stress resulted in greater loss of nutrient yield in the translucent pericarp, yellow endosperm sorghum than in the bronze pericarp sorghum. Also, these data confirm that in geographic locations where rainfall and soil fertility can be limiting, sorghum will yield more nutrients per unit of land area.

BREEDING AND GENETICS

Of factors contributing to nutritional value of a feedstuff, genotype, the interaction of genotype with growing conditions, and feed processing methods will be of major importance. Undoubtedly, the combined effects of genotype and growing conditions resulted in the extreme range in crude protein (10.9–16.5%) and lysine as a percentage of that protein (1.9–2.3%) reported from a 1988 survey of sorghums produced throughout the state of Oklahoma (Luce et al, 1988).

The nutritional consequences of various fairly simply inherited traits such as pericarp color, and endosperm color, type, and texture have been investigated. Grabouski et al (1987) compared sorghums with bronze, cream, and translucent pericarp (yellow endosperm) colors with maize for feeding nursery and growing-finishing pigs. Nursery pigs gained 5 to 8% more slowly with all sorghum types but with a similar efficiency to pigs fed maize. In the growing-finishing experiment, pigs fed the sorghums gained 4% more slowly and were 4 to 9% less efficient than pigs fed maize. Combining the nursery and growing-finishing results, maize generally supported improved gains, which were related very closely to differences in feed intake. In contrast, efficiency of gain was similar in nursery pigs fed maize or sorghums, but finishing pigs consumed more of the diets with sorghum than maize and converted feed to weight gain with less efficiency. The authors argued that reduced palatability of sorghums may be more of a problem in very young animals and reduced energy value more of a problem in older animals, but there were no consistent differences due to pericarp color.

Noland et al (1977) reported that sorghums with a translucent pericarp and yellow endosperm were better utilized by nursery pigs (fed from 10 to 20 kg body weight) than sorghums with brown pericarp, but the latter had high tannin content (i.e., 0.67 vs. 0.22% for the brown and yellow endosperm sorghums). Simple correlations coefficients were –0.68 and –0.58 between tannin content and digestible energy and digestible protein, respectively. These negative relations were also reported by Nelson et al (1975) in growing chicks. However, these experiments did not address the issue of differences in nutritional value between sorghums with different pericarp colors but similar tannin content. Richert et al (1991) reported data from chick growth assays comparing the nutritional value of sorghums with low tannin content but bronze and yellow pericarp color. The bronze sorghum had 5% more ME, but rats fed the yellow sorghum were 4% greater in protein efficiency. Thus, these differences were not associated directly with pericarp color, because it would have affected both energy and protein value in a similar manner (Noland et al, 1977).

In contrast to pericarp color, differences in endosperm characteristics have been suggested to affect nutritional value. There are reports of higher nutritional value for sorghums with yellow versus white endosperm color (McCullough et al, 1972; Noland et al, 1977; Hibberd et al, 1978), suggesting a positive relationship between yellow pigment in the endosperm and nutritional value. However, it is most difficult to rationalize why digestive enzymes might prefer endosperm with yellow pigmentation. More probable is that early experiments reporting differences among sorghums with different endosperm colors actually may have been comparisons of differences in endosperm type and texture. Hibberd et al (1978), using an in vitro procedure with strained rumen fluid, reported that dry matter disappearance for waxy sorghums was approximately 4% greater than disappearance for normal endosperm sorghums. A bird-resistant sorghum had the poorest digestibility, and maize was superior to all of the sorghums.

Lichtenwalner et al (1978) added incremental dosages of the waxy gene to sorghum to create nonwaxy (*WxWxWx*), heterowaxy (*WxWxwx, Wxwxwx*), and waxy (*wxwxwx*) genotypes. Effects of the waxy gene on in vitro starch hydrolysis and in situ dry matter digestibility indicated that, as the dosage of the waxy gene was increased, starch hydrolysis and dry matter digestibility of the ground grain increased. To separate the effects of protein composition and starch waxiness of the grain, isolated starch was also evaluated for digestibility. The isolated starch was more susceptible to hydrolysis than the ground grain at all dosages of the waxy gene, indicating that both waxiness and protein composition of the endosperm have major effects on the observed differences in digestibility. Reports from feedlot experiments indicate that cattle fed waxy versus normal endosperm sorghums had greater rates of gain, and were 9 to 17% more efficient (Brethour and Duitsman, 1965; McCullough et al, 1972).

Pig feeding experiments have been less encouraging. Cohen and Tanksley (1973) reported no differences in dry matter, energy, or protein digestibility for waxy versus normal endosperm types. Furthermore, comparison of floury, intermediate, and corneous endosperm types indicated that sorghum with intermediate-type endosperm was superior to sorghum with floury-type endosperm for both dry matter and energy digestibility. Sorghums with floury and corneous endosperm types had similar digestibilities. Noland et al (1977) reported that digestibilities of energy and protein were slightly greater in growing pigs fed sorghums with intermediate endosperm texture than sorghums with corneous endosperm texture. Myer and Gorbet (1983) compared sorghum grain with waxy and normal starch types and low-, medium-, and high-tannin content in nursery pigs. As tannin content increased, rate and efficiency of gain decreased, but sorghum with the waxy starch type was no better nutritionally than sorghum with normal starch type.

The lack of improvement in energy digestibility and utilization in swine fed floury and waxy endosperm sorghums and the realization that very soft or floury sorghums have reduced yield and poor weathering ability (Sullins and Rooney, 1974) stimulated research into the relationship of yield and digestibility. Bramel-Cox et al (1990) made direct selection in a sorghum population, KP7B, for improved in vitro protein digestibility (digestibility in pepsin) using 100 S_1 families, which were grown at five locations for two years. There were significant location and year effects, suggesting that growing conditions influenced sorghum digestibility. Higher protein digestibility was associated with lower yield and later maturity. In view of these undesirable relationships, a rank summation procedure was developed including the rank for yield plus the rank for digestibility minus the rank for bloom date, with selection restricted to families in the top 50% for yield and protein digestibility. Heritability for this index was 38%, which identified 20 families with improved protein digestibility, better grain yield, and an earlier maturity. This moderate heritability indicated that further improvements could be made but would require time and patience. From this population, two lines were selected with consistently low in vitro digestibility and two lines with consistently high in vitro digestibility. In chick feeding assays, these lines, respectively, supported rates of gains that were 95, 96, 98, and 100% that of the maize control and efficiencies of gain that were 95, 97, 98, and 100% that of the maize control (Hancock et al, 1990b). Overall, the authors concluded that use of in vitro protein digestibility, in conjunction with yield and maturity date, has potential to genetically improve grain sorghum as a feed grain for livestock, even though the environment will still have a major effect on nutritional quality. Unlike selection for floury and waxy endosperm textures, this selection index would result in genetic material for use in hybrids with acceptable yield, maturity, and weathering ability, and improved nutritional value.

PROCESSING

A third area of considerable research effort in sorghum grain utilization is the effects of handling and processing on nutritional value. A novel approach to grain handling, i.e., germination before feeding (Elmalik et al, 1986), resulted in an 8 to 26% loss of dry matter, but increased percentage lysine and other essential amino acids (except leucine) when expressed as a percentage of crude protein. In vitro protein and starch digestibility were improved, and in a rat feeding experiment germination improved efficiency of gain by 1 to 12%. However, the loss of dry matter would negate the gains in feed efficiency. Additionally, a major limitation to the application of germination before feeding is that the grain must be dried after sprouting to prevent molding, thus greatly increasing handling and storage costs.

The potential for molding is also a problem associated with a second alternative to storage and processing dried grain, i.e., feeding high-moisture sorghum. In a commercial feeding situation, high-moisture sorghum, harvested and stored at 23 to 30% moisture, can be preserved with organic acids or oxygen-limiting storage, both of which add considerable expense and require careful management. Results from swine feeding experiments with high-moisture sorghum (Meyer et al, 1982) are similar to results using other grains such as maize (Hancock et al, 1989), with no differences in rate of gain or efficiency of gain when expressed as weight gain per unit of dry matter intake.

Processing techniques have included grinding, crushing, steaming, steam flaking, popping, and extruding. Zhuge et al (1990) reported an experiment designed to compare the effects of fine and coarse grinding (hammermill vs. roller mill), steam flaking, and extruding at two temperatures on nutritional value to growing chicks. Fineness of grind, mill type, steam flaking or extrusion at 100–110°C did not affect performance, and extruding at 130–135°C reduced growth performance. The authors concluded that hammermilling and roller milling were equally efficient and that additional heat processing of sorghum grain (steam flaking or extruding) did not improve performance in growing chicks. Hancock et al (1991b) reported that extrusion processing of sorghum grain improved nutrient digestibility and growth performance in finishing pigs (Table 4). These data are in partial agreement with those of Herkelman et al (1990), who reported that energy utilization was improved when maize was extruded prior to feeding, but nitrogen and amino acid digestibilities were not affected. Owsley et al (1981) reported a linear improvement in ileal and total tract energy digestibility as particle size of sorghum was reduced from approximately 1,300 to 600 μm. Ohh et al (1983) and Wu (1984) reported similar findings, in that nitrogen and dry matter digestibilities and efficiency of gain for pigs fed maize and sorghum improved linearly as particle size was reduced from greater than 1,000 μm to below 700 μm.

In a direct comparison of the effects of coarse grinding and fine grinding on different cereal grains (maize, yellow sorghum, and bronze sorghum), Giesemann et al (1990) reported that fine grinding improved gain-to-feed ratio in finishing pigs fed maize or yellow sorghum, and that dry matter and nitrogen digestibilities were improved more by fine grinding of sorghum than maize (Table 5). Healy et al (1991) found that power usage was three to four times greater for maize than sorghum to roller mill grain to particle sizes ranging from 900 to 500 μm (Table 6). When these grains were fed to broiler chicks, rate of gain was optimized at 700–500 μm, and efficiency of gain was optimized at 500–300 μm. An important observation was that at 900 μm, sorghum had a feeding value of 95% of maize, which was increased to 99% with grinding to the optimum particle size for each grain. Thus, much of the previously reported difference in feeding values between maize

TABLE 4
Effect of Extrusion Processing on the Nutritional Value of Sorghum and Soybeans for Finishing Pigs[a]

Item	Ground Sorghum and Soybean Meal-Oil Control	Ground Sorghum and Extruded Soybeans	Extruded Sorghum and Soybean Meal-Oil	Extruded Sorghum and Extruded Soybeans
Growth performance, kg				
Av. daily gain	1.02	1.03	0.98	1.00
Daily feed intake	3.33	3.20	3.05	2.97
Gain/feed	0.306	0.322	0.321	0.337
Apparent digestibility, %				
Dry matter	81.2	83.2	88.0	86.4
Nitrogen	68.5	73.1	84.2	84.6

[a]Adapted from Hancock et al (1991b).

TABLE 5
Effect of Particle Size of Maize and Sorghum Grain on Growth Performance and Nutrient Digestibility in Growing-Finishing Pigs[a]

Item	Maize		Yellow Sorghum		Bronze Sorghum	
	Coarse	Fine	Coarse	Fine	Coarse	Fine
Particle size, μm	1,506	641	1,122	613	1,056	616
Growth performance, kg						
Av. daily gain	686	719	696	699	716	680
Gain/feed	0.257	0.279	0.259	0.272	0.284	0.278
Apparent digestibility, %						
Dry matter	80.5	81.6	75.1	83.2	78.2	84.7
Gross energy	79.2	81.0	73.6	82.3	76.5	83.8
Nitrogen	70.3	74.8	63.4	74.6	65.8	74.1

[a]Adapted from Giesemann et al (1990).

TABLE 6
Power Requirements to Roller Mill Maize, Hard and Soft Sorghums, and Comparative Nutritional Values for Broiler Chicks[a]

Item	Roller Milled 900 μm			Roller Milled 700 μm			Roller Milled 500 μm			Hammer Milled 300 μm		
	Soft	Hard	Maize	Soft	Hard	Maize	Soft	Hard	Maize	Soft	Hard	Maize
Power req., kW hr/t	1.9	1.7	5.3	2.5	2.4	9.2	4.3	3.8	15.7	...	...	...
Growth, g[b]												
Av. daily gain	31.8	31.0	34.8	34.4	35.9	35.0	35.9	35.4	34.9	35.0	35.0	34.0
Gain/feed	0.624	0.632	0.661	0.653	0.645	0.665	0.671	0.665	0.677	0.655	0.671	0.679

[a]Adapted from Healy et al (1991).
[b]From seven to 28 days of age.

and sorghum was due simply to sorghum being at greater than optimum particle size.

Processing also improves sorghum utilization in high-grain ruminant feeds. Xiong et al (1990a,b) reported improved in vitro starch availability and protein degradation in steam-flaked and reconstituted sorghum grain. Hale (1984) observed that compared to dry grinding, steam flaking (the most widely used method of processing) improved sorghum utilization of beef cattle by 12–15% and returned three to five times the cost of processing. Huber et al (1991) estimated that in a large milk-cow operation, steam flaking sorghum at a cost of 1.1 cents/kg would give a net profit of 37 cents a day per cow and also would increase milk protein levels.

In conclusion, sorghum is an excellent feedstuff for livestock. Although the feeding value of sorghum is on average 3 to 7% less than that of maize, cost of sorghum in many areas is commonly 10% less than cost of maize. In addition to the current economic and environmental incentives for using sorghum, scientists are improving its feeding value through plant breeding, increasing understanding of agronomic practices, and improving milling and processing procedures that will result in optimum nutritional values. With these advances and with the superior sustainability of sorghum production in minimal rainfall, the future of sorghum as a major cereal feed grain is promising.

PEARL MILLET

Mueller (1880) noted that pearl millet was useful for hay and green fodder in the United States and was being grown in several southern states. The plethora of names being used for pearl millet by the turn of the century, such as bulrush, cattail, African, and "Mands wonder millet," and the high prices being charged for seed prompted Ball (1903) to publish a bulletin clarifying description, cultivation, and use. He suggested that pearl millet had reached the United States via South America or Mexico, being originally introduced there by the Spaniards.

Green (1973) reported that pearl millet was tested for silage at the Florida Experiment Station, Lake City, in 1889; however, few data were reported until 1931 when pearl millet, compared to Early Amber sorghum and German (foxtail) millet, gave the highest green fodder yield of 28.7 t/ha when clipped three times. This was the first indication that the rapid recovery rate of pearl millet after cutting or grazing could lead to higher, better quality forage production than other crops. Norman and Begg (1968) reported the use of pearl millet for grazing or standing dry season feed in Australia.

Breeding commenced in the United States in the 1940s for forage productivity and quality at Tifton, Georgia. Many of the cultivars in commercial use today and a considerable body of genetic information about the species originated from that still on-going U.S. Department of Agriculture program. Pearl millet for food, feed, and forage has been reviewed by Andrews and Kumar (1992).

Utilization

GRAZING

Pearl millet is primarily used for grazing and silage for steers and milk cows. Marshall et al (1953) reported a Mississippi experiment in which pearl millet was found superior to Tift sudan or grain sorghum as a summer grazing pasture. In 1951, lactating cows grazed 1,412 kg/ha of total digestible nutrients compared to 930 kg/ha from sorghum. In the following year, the figures were 2,300 and 1,657 kg/ha, respectively. In Tennessee, two millet cultivars provided the same amount of grazing as sudangrass for lactating cows in experiments at several locations (Baxter et al, 1959). Average milk production was 43.4 and 44.7 kg/day on the two millet cultivars (no figure given for sudangrass), but there was a distinct depression in butterfat content. Clark et al (1965) also noted a depression, averaging 0.13% in actual butterfat over three years in cows grazing pearl millet relative to those on sudangrass and sorghum-sudangrass hybrid pastures, which was accentuated by high grain supplement levels. Rumen liquor samples of cows grazing pearl millet tended to be lower in the molar percentage of acetic, butyric, and isovaleric acids and higher in propionic acid (Hemken et al, 1968). Harner et al (1969) found that butterfat depression was influenced by the level of cation fertilization. Bucholtz et al (1969) observed that the lower fat content was associated with a higher degree of fat unsaturation. The molar percentage of rumen butyrate was significantly decreased in cows grazing pearl millet compared to sudangrass. Analysis of plant material showed a higher concentration of oxalic acid in pearl millet than sudangrass.

Norman (1963) investigated wet-season grazing with steers in northern Australia and found that, at 2.5 animals/ha, gains were 102 kg/head in 20–24 weeks, an increase of 51 kg/head over those grazing native pasture. Norman and Stewart (1964) found that cattle grazing mature, standing pearl millet in the dry season had an average live weight gain of 296 kg/ha in 16 weeks during a period when animals grazing native pastures showed a decline in live weight. Norman and Phillips (1968) showed that an early start of grazing, in the mid-wet season, gave the highest live weight gains per head and per hectare. Johnson et al (1978) evaluated the performance of dairy heifers grazing two new hybrids relative to heifers grazing Gahi 1, a widely grown hybrid millet. For two years, pearl millet cvs. Gahi 1, Gahi 3, and Tifleaf 1, after reaching a height of 45–50 cm, were grazed for 56 days by dairy heifers with initial liveweight of 315 kg and 268 kg/animal in the first and second year, respectively, at the stocking rates of 7.2 and 8.8 animals/ha. Daily gain per animal over both years averaged 0.63, 0.76, and 0.84 kg from Gahi 1, Gahi 3, and Tifleaf 1, respectively.

Pearl millet also could be used for grazing by pigs to save on concentrates. Burton (1980) reported that 45.5-kg pigs on full feed of a balanced concentrate and grazing on young Tifleaf 1 for 35 days gained as well as those in dry lots and required less concentrate per kilogram of grain. The concentrate saved by the Tifleaf 1 grazing made the millet crop worth $250/ha.

SILAGE

Pearl millet produces good silage because it has an adequate supply of carbohydrates and a sufficiently high dry matter content to avoid excessive moisture. Johnson and Southwell (1960) compared the performance of lactating Jersey cows fed either pearl millet silage made from cv. Starr or good maize silage. Cows consumed more millet silage and gave 3.3% more milk, but body weights remained

TABLE 7
Yield and Chemical Composition of Silage from Sorghum and Pearl Millet[a]

Item	Pearl Millet	Sorghum
Planting date	July 15	July 15
Harvesting date	Sept. 8	Sept. 8
Seeding rate, kg/ha	11	9
Variety	Tifleaf	Grain Milo 200
Yield, dm, t/ha	4.56	3.62
Digestible dm yield, kg/ha	2.93	2.17
pH	4.6	4.4
ADIN[b]	5.8	5.9
Dry matter, %	38.5	38.8
Crude protein[c] (%)	15.0	16.9
NDF[c,d]	68.7	61.4
ADF[c,e]	40.8	40.8
ADL[c,f]	6.6	6.2
Hemicellulose[c,g]	27.9	23.2
Cellulose[c,h]	34.8	33.1
Ash[c]	14.9	14.2
Calcium[c]	0.37	0.44
Magnesium[c]	0.36	0.40
Potassium[c]	3.47	3.38
Sodium[c]	0.38	0.31

[a]From Jaster et al (1985). Data taken only for pearl millet and sorghum.
[b]ADIN = Acid detergent insoluble nitrogen as a percent of total nitrogen.
[c]Percentage of dry matter (dm).
[d]NDF = Neutral detergent fiber.
[e]ADF = Acid detergent fiber.
[f]ADL = Acid detergent lignin.
[g]Hemicellulose = (NDF − ADF).
[h]Cellulose = (ADF − ADL).

the same. Apparently, the millet silage contained more energy per unit weight. However, Baker (1970), comparing pearl millet and sorghum silages, found that both consumption and milk production were better with sorghum silage.

Bertrand and Dunavin (1973) observed that steers grazing Gahi 1 millet pasture gained faster than comparable steers receiving Gahi 1 millet silage (0.75 and 0.62 kg/head per day, respectively). Though beef production was slightly higher for steers receiving the millet silage (495 vs. 455 kg/ha), the authors concluded that the small increase in amount of beef produced would not compensate for the additional expenditures required for harvesting, storing and feeding millet silage.

Jaster et al (1985) evaluated the nutritive value for dairy heifers of oat, barley-pea, pea, oat-pea, pearl millet, and sorghum as silage grown under a forage double-cropping system. In Table 7, details of the yield and chemical composition of only the silages made from sorghum and pearl millet are shown. Pearl millet had silage characteristics very similar to those of sorghum, a desirable silage crop. The only differences were in the greater yield of the pearl millet, which, given its similar percent dry matter, resulted in a greater dry matter yield than sorghum. Dry matter intakes of pearl millet and sorghum silages were similar, representing approximately 2.4% of the body weight (Table 8), even though pearl millet was 7.3% higher in NDF content than sorghum silage. Pearl millet had the higher apparent digestibilities of all components except protein. The authors concluded that heifers consuming pearl millet and sorghum silages showed higher dry matter intake and dry matter digestibility than those consuming cool-season silages.

Gupta et al (1981) studied the improvements brought about in the nutritional quality of fodder when pearl millet and cowpea (*Vigna unguiculata*) were ensiled together and fed to crossbred (Haryana × Jersey) cattle. Both pearl millet and cowpea were harvested 80 days after sowing and ensiled after wilting for one to two days. Average feed intakes were 6.03 kg/day per animal for pearl millet plus cow-

TABLE 8
Nutrient Intake and Apparent Digestibility of Pearl Millet and Sorghum Experimental Silages Fed to Heifers[a]

Item	Pearl Millet	Sorghum
Intake, % body weight	2.4	2.5
Intake, kg/day		
Dry matter	8.5	8.5
NDF[b]	5.2	5.8
ADF[c]	0.5	0.4
Hemicellulose	2.4	2.0
Cellulose	2.9	2.8
Crude protein	1.3	1.4
Digestibility, %		
Dry matter	64.3	60.1
NDF	69.0	61.1
ADF	65.4	57.0
ADL[d]	33.1	16.4
Hemicellulose	74.2	67.7
Cellulose	71.4	63.4
Crude protein	55.4	65.6

[a]Data from Jaster et al (1985).
[b]NDF = Neutral detergent fiber.
[c]ADF = Acid detergent fiber.
[d]ADL = Acid detergent lignin.

pea and 5.09 kg/day per animal for pearl millet alone. The digestibility coefficients for dry matter and crude protein content, IVDMD, digestible crude protein, and total digestible nutrients were highest in pearl millet–cowpea silage. Similar results were reported from Brazil by Silveira et al (1981) for sheep.

Unlike sorghum, pearl millet does not produce HCN, but high soil nitrogen together with moisture stress at an early growth or regrowth stage can result in high plant nitrate levels. Green (1973) recorded one case of nitrate poisoning from pearl millet grazed in Florida. Rouquette et al (1980) reported that pearl millet growing in apparent drought stress conditions in eastern Texas became unpalatable to grazing cattle. Analysis of the forage showed a potentially toxic level of nitrate and a high level of total alkaloids. Stage of growth and time of grazing or levels of cutting have been known to change the nitrate content of forages. Burger and Hittle (1967) observed that in pearl millet and sudangrass, nitrate accumulation was higher when repeated cuttings were at 7.6 cm compared with 15.2 cm. Lemon and McMurphy (1984) confirmed that the lower parts of the plant contained 2.7 to 3.5 times more nitrate. Nitrate levels increased with higher nitrogen fertilization but decreased at later maturity. Krejsa et al (1984) showed a considerable range of total nitrate (2.4 to 9.8 g/kg) and alkaloid levels (17 to 101 mg/kg) among 11 pearl millet varieties and inbred lines. Although pearl millet has been regarded as a nitrate accumulator, only one inbred line exceeded the potentially toxic level of 9.3 g/kg of total forage nitrate. According to Fales and Wilkinson (1984), caution should be exercised when using quality-enchancing growth regulators, such as mefluidide, in pearl millet; it inhibited growth and increased nitrate to potentially hazardous levels.

GRAIN

Finishing steers fed on pearl millet diets at Hays, Kansas, gained as well as those on sorghum diets (Christensen et al, 1984). When compared with sorghum grain, pearl millet grain had higher levels of both fat and protein, and the protein had a better balance of amino acids. The estimated net energy of pearl millet was 4% higher than that of finely rolled sorghum. When Rumensin was used, millet gave significantly higher gains than sorghum in growing rations for calves (Table 9). The authors concluded that pearl millet is an excellent protein source and growing ration for beef cattle.

Diets based on coarsely ground pearl millet, maize, and sorghum were compared for metabolizable energy value and in a heifer feeding test by Hill and Hanna (1990). The gross energy level of the pearl millet diet was higher, but total digestible nutrients were higher for maize than the other two cereals. Weight gains per day were higher on the maize-based diet, but all three cereals gave comparable feed/gain values. The authors concluded that pearl millet is suitable for use in cattle feeding if diets are adjusted to accomodate the higher protein levels of the grain.

Calder (1955, 1960) reported that, for pig feeding, pearl millet grain should be finely ground so that the risk of internal irritation believed to be caused by the hard grain hull is reduced to a minimum. To determine the value of pearl millet for pig feeding, both for pork and bacon, experimental rations containing pearl millet grain and maize were compared with a control ration of maize. Pigs fed *ad lib.* on diets containing 75 and 50% pearl millet reached the average slaughter weight of 90.8 kg in 10 days less than the maize control group. These experiments established that pearl millet has a high value for pig feeding. In addition, pearl millet promoted the formation of firm white fat, having an influence comparable to that of barley.

French (1948) reported that whole grains of pearl millet fed to sheep were less digested, and a fair percentage passed whole into the feces. He indicated that grinding grain can bring its feeding value nearly equal to that of crushed maize. Martinez (1988) allowed lambs grazing pearl millet (cv. Tifleaf) 4, 6, 9, or 12 kg herbage per 100 kg live weight daily. Average live weight gains per lamb were 65, 68, 90, and 100 g/day, respectively, and corresponding live weight gains were 663, 478, 434, and 374 kg/ha. A 6- to 7-kg allowance gave a reasonable balance between live weight gain/lamb and carrying capacity per hectare.

Studies on formulation of poultry feeds using pearl millet have been carried out to ease competition for energy sources between man and monogastrics and to reduce feed costs. Studies conducted by several workers (French, 1948; Singh and Barsaul, 1976; Sharma et al, 1979) showed that millets compared favorably to maize in poultry diets. French included up to 60% pearl millet or white-seeded finger millet in layers' diets without any reduction in egg production. He concluded that there was little or no difference in feeding value for poultry among pearl millet, white-seeded finger millet, and maize.

Sanford et al (1973) found that, at equal levels, the amino acid profiles, rates of gain in chicks, and efficiencies of utilization of feed were comparable for pearl millet and sorghum grain sources of energy and protein for broiler-strain chicks. Lloyd (1964) observed that broilers fed on millet rations were heavier and had better feed conversion than those fed on maize rations. No significant differences could be found between the diets in slaughter weights and yields, and both were equivalent in the production of good-quality carcasses. However, the carcasses of broilers fed on millet had a slightly higher pigmentation than those fed on maize.

Abate and Gomez (1983) partly substituted pearl millet and finger millet for maize in both broiler starter and finisher diets. The chicks fed pearl millet diets had

TABLE 9
Rolled Pearl Millet Compared to Rolled Sorghum in High-Silage Growing Rations[a]

	Treatment			
	Control		Rumensin[b]	
Item	Sorghum	Millet	Sorghum	Millet
Number of cattle	30	29	30	30
Average initial weight, kg	230.1	231.2	230.1	229.7
Average final weight, kg	365.6	363.1	360.1	372.7
Average gain, kg	135.4	131.8	129.9	142.8
Average daily gain, kg	1.11	1.08	1.07	1.18
Average daily ration, kg				
Sorghum silage	18.9	18.0	17.8	17.5
Rolled sorghum	1.8	...	1.8	...
Rolled millet	...	2.1	...	2.1
Soybean meal	0.56	0.26	0.56	0.26
Premix	0.14	0.14	0.18	0.18
Air-dry total	8.8	8.6	8.5	8.4
kg feed 45.4 kg/gain	368.0	359.0	364.0	337.0

[a] Adapted from Christensen et al (1984).
[b] No endorsement is intended, nor is any criticism implied of similar products not mentioned.

the highest overall body weight gain, whereas the gain of chicks fed finger millet was comparable to the gain of chicks fed maize. Experiments conducted to determine the value of using pearl millet as an energy source and as a partial substitute for protein supplements in broiler diets indicated that pearl millet could effectively replace part of the vegetable protein supplement, provided the diet was supplemented with up to 0.3% lysine.

Smith et al (1989) conducted trials with pearl millet, grain sorghum, and triticale, substituting each grain for 50 or 100% of the maize in the control diet. Pearl millet and grain sorghum replaced maize in the diet of chicks without adversely affecting gain or feed efficiency. The authors concluded that an economic assessment of dietary treatments would be premature, because no fair market price or established production practices exist for pearl millet in the United States.

Sullivan et al (1990) reported two chick feeding tests conducted in Nebraska. In both cases, chick growth rates were similar on diets with pearl millet, maize, and low-tannin sorghum. High-tannin sorghum gave lower growth rates. Pearl millet had higher metabolizable energy values. In two trials conducted at Kansas State University in 1989 and 1990 by J. Hancock and P. J. Bramel-Cox, the nutritional and growth parameters of diets made from maize, four sorghums, and pearl millet were compared when fed to broilers (Table 10). The four sorghum types had previously been selected on the basis of in vitro digestibility values using a pepsin digestion technique (Bramel-Cox et al, 1990). Pearl millet was generally equal to both maize and the best sorghum in total gain and gain-to-feed ratio but had a higher ME. When an approximate comparison was made of the economic cost per kilogram of gain using maize, sorghum, or pearl millet in broiler diets, the best sorghum line and the pearl millet cost nearly 1 cent less per kilogram of gain than maize. These estimates used a differential of 10% between the price of maize and sorghum, which has been a market difference, and assumed that millet would cost the same as sorghum.

Dehydrated and pelleted pearl millet forage has been evaluated in the United States (Wilkinson et al, 1968). Wilkinson and Barbee (1968) reported that dehydrated products of coastal bermudagrass and pearl millet compared favorably in metabolizable energy values to dehydrated alfalfa and may be used as a source of supplementary xanthophyll in poultry feeds. Butler et al (1969) reported that fresh

TABLE 10
Nutrient Content of Millet, Sorghum, and Maize, and Growth Performance of Broiler Chicks[a]

		Sorghum				
Item	**Corn**	**Low dig. 1**	**Low dig. 2**	**High dig. 1**	**High dig. 2**	**Pearl Millet**
Crude protein, %	11.2	12.9	12.0	11.8	12.1	11.4
Lysine, %	0.33	0.30	0.30	0.32	0.27	0.39
Gross energy, kcal/kg	4,335	4,440	4,379	4,515	4,422	4,588
Gain, g	479	456	461	471	481	475
Feed intake, g	732	730	723	735	737	723
Gain/feed	0.654	0.623	0.636	0.641	0.653	0.656
ME, kcal/kg	3,270	3,345	3,352	3,350	3,360	3,455
Cost of gain, cents/kg	29.13	29.13	28.93	28.75	28.16	28.23

[a]J. Hancock and P. J. Bramel-Cox (1991, *unpublished data*), Kansas State University.

pearl millet was high in carotene (302–444 mg/kg), xanthophyll (483–737 mg/kg), and moisture (83–87%).

The results thus far reported indicate that pearl millet can successfully be used in poultry feeds. Feed costs may be lower, because the generally higher grain nitrogen content in pearl millet means that fewer expensive, high-nitrogen additives are needed than for sorghum- or maize-based diets. Additionally, weight gains per unit of millet ration have been good.

Crop Management

Research on management practices for pearl millet forage has focused mainly on the effects of timing, number of cuts, and height of cutting on quality and regrowth potential. The stage of development at harvest can greatly influence yield and regrowth habit. Beaty et al (1965), Begg (1965), Fribourg (1966), Clapp and Chamblee (1970), and Stephenson and Posler (1984) noted that as stubble height was raised, regrowth from terminal buds increased but axillary and basal tillering remained constant. Mays et al (1966) also observed that cutting to a 10-cm stubble gave 7% more regrowth yield than cutting to 20 cm, confirming the results of Broyles and Fribourg (1959). Mays and Washko (1962) showed that more structural reserve carbohydrates were used in regrowth when pearl millet was cut to 5-cm rather than a 15-cm stubble. Beaty et al (1965) found over a three-year period that harvesting at five-week intervals increased forage production 46% compared with harvesting every two weeks. Harvesting seven-eighths of the plant at five-week intervals increased production by 18% over harvesting one-third of the available height.

Management influences not only forage yields but also forage quality. Hoveland and McCloud (1957) noted that, although longer intervals between cuttings produced higher yields, these also had the lowest protein content. Rusoff et al (1961) found that lignin content progressively increased with plant maturity. Burton et al (1964) observed that young leaf blades contained higher levels of crude protein and true protein but lower levels of lignin than older leaves taken from the same culms. Burger and Hittle (1967) reported that frequent and severe cutting improved protein content. According to Westphalen and Jacques (1978), the yield of crude protein increased with delay in cutting, delay in sowing, and decreased stubble height.

A good quality forage should have a high leaf-to-stem ratio, be high in protein and digestible nutrients, and be low in fiber and lignin. Fertility of the soil, soil-moisture availability, and cutting and grazing management are important production factors. Hart (1967) reported that the dry matter digestibility (DMD) of stems and whole plants of eight genotypes was positively and significantly correlated with leafiness from about boot stage to maturity. Lignin content was an excellent predictor of DMD (correlation coefficients were –0.91 [$P < 0.01$] for leaves and –0.68 [$P < 0.01$] for stems). DMD of leaves was positively and significantly correlated with crude protein (0.57 [$P < 0.01$]) and crude fiber content (–0.68 [$P < 0.01$]), but these were much poorer predictors of digestibility than lignin. Achacoso et al (1960) observed that, as dry matter increased, increases occurred in both crude fiber and lignin contents.

Breeding and Genetics

Pearl millet is a cross-pollinated, diploid, warm-season, tillering cereal. Varieties from tropical germ plasm tend to be tall and vegetatively vigorous and to flower in response to short days—an advantage in temperate regions where prolonged, leafy, vegetative growth in summer is desired. Although pearl millet is noted for its drought tolerance, under favorable conditions its growth rate is as high as that of any cereal.

The breeding of successive cultivars of forage pearl millet at Tifton, Georgia, has been a classic example of genetic discovery and use of genetic theory and germ plasm. A widely grown, but early-maturing and "stemmy," open-pollinated variety in the United States called Common millet was first replaced by a later-maturing, more leafy synthetic called Starr (Burton and DeVane, 1951), which when fed to cattle produced 23% more beef per hectare. A higher level of heterosis was then employed in cv. Gahi 1 (Burton, 1962), seed for which was produced from a mixture of four inbred lines. In the resulting crop, any inbred plants were crowded out and did not affect the yield of the mixture of the six possible hybrids (Burton 1948, 1989). Gahi 2, a shorter, more leafy version, was subsequently produced by breeding the same dwarfing genes into each of the parental lines. Then true, single-cross hybrids, in which the maximum level of heterosis could be utilized, became possible with the discovery of cytoplasmic male sterility (Burton, 1965). Gahi 3 was the first of such hybrids, which additionally had some disease and nematode resistance (Burton, 1977). The dwarf hybrid, Tifleaf 1, that followed was easier to manage and had a higher leaf-to-stem ratio. The relative values of Gahi 1, Gahi 2, and Tifleaf 1 were investigated by Johnson et al (1978), as noted in the discussion of utilization for grazing. Finally, Tifleaf 2 was bred, it resembled Tifleaf 1 but incorporated rust and blast resistance obtained from African germ plasm (Hanna et al, 1988).

Napier or elephant grass (*Pennisetum purpureum* Shumacher), a tetraploid, or its interspecific cross with pearl millet are used as forages in warmer climates under low-management conditions, although not in the United States. Napier grass gives high dry matter production, but forage quality is low. Breeding in the interspecific hybrid, which is usually propagated by cuttings, emphasizes quality, yield, and palatability differences between hybrid clones (Gupta, 1975). The pearl millet × napier grass hybrid has other potentials for pearl millet improvement. It may be possible to extract desirable genes from the napier grass genome itself, but the hybrid can also serve as a genetic bridge in crosses to other more distantly related *Pennisetum* species, giving potential access to traits such as apomixis (Dujardin and Hanna, 1989).

Several major genes that have important effects on forage quality have been discovered in pearl millet. Among a number of recessive dwarfing genes that occur naturally in pearl millet (Burton and Fortson, 1966), the d_2 gene, which reduces stature by about 50% and behaves as a simple recessive, is the most widely used. Dwarfing genes reduce internode length and increase the proportion of the more digestible leaves (Burton and Fortson, 1966; Burton et al, 1969; Johnson et al, 1968). Although total dry matter yield was reduced by 9 to 31%, the increased nutrient content and digestibility resulted in gains among heifers fed dwarf forage equal to those of heifers fed tall forage.

Johnson et al (1968) measured the differences in the composition and digestibility of millet forage that were associated with the dwarf trait. The tall millet was 213 cm high at harvest and produced 7.2 t/ha of dried forage compared with a height of 107 cm and 5.6 t/ha for dwarf millet. However, the proportion of leaves in the dwarf millet was 50% higher, and they contained 15% more protein and 17% less lignin than tall millet leaves. Protein, crude fiber, and nitrogen-free extract of dehydrated dwarf millet averaged 13.6, 30.3, and 42%, respectively; these values for tall millet were 11.4, 36.0, and 40.1%. Apparent coefficients of digestibility for dwarf millet of 59.7, 64.4, and 59.2% for dry matter, crude fiber, and nitrogen-free extract were higher than those for tall millet by 6.8, 8.8, and 12.9%, respectively. Daily consumption of dwarf millet was 21% greater than that of tall millet. Animals fed dwarf millet over a period of 51 days gained 0.52 kg/day, which was 40% more than the gain for those fed tall millet. On the basis of these results, the authors concluded that the dwarf characteristic could be effective in improving the nutritive value of millet forage. This gene has been used in the development of dwarf male-sterile lines to breed improved hybrids for forage and extensively in breeding lines for grain production in the United States (Burton, 1980; Hanna et al, 1988; Andrews et al, 1993).

Powell and Burton (1971) reported the discovery of a trichomeless (*tr*) mutant in the very pubescent pearl millet inbred, Tift 23B. When homozygous, this single recessive gene suppressed all trichomes and stigma branches and imparted a smooth waxlike surface to the cuticle. Comparing nearly isogenic Tift 23S (*trtr*) with pubescent Tift 23H (*TrTr*), Burton et al (1977) found that the *tr* gene reduces transpiration without altering stoma length and frequency. Hanna et al (1977) found that normal leaves with trichomes were digested more quickly in rumen fluid than trichomeless leaves of near-isogenic plants. Hanna and Akin (1978) attributed slow digestion of trichomeless leaves to reduced penetration of rumen microbes due to reductions of 1.3- and threefold, respectively, in the number of cuticle cracks on the upper and lower surfaces of the trichomeless leaves.

The *tr* gene increased palatability of forage for cattle and favored increased intake in the first few days in a green feeding trial (Burton et al, 1977). It did not increase overall livestock gains, because the slower digestion rate of *tr* forage slowed its passage through the gastrointestinal tract and reduced intake, thus countering its greater palatability. The loss of IVDMD associated with the *tr* gene tended to be compensated for by associated drought tolerance and better palatability. Based on results of forage and dry matter production and associated beneficial effects of the *tr* gene, Burton et al (1988) concluded that the *tr* gene could be used to increase forage yields.

Because of the negative relationship between lignin and digestibility, one effective means of increasing forage digestibility is reducing lignin content. Brown-midrib mutants in maize and sorghum have low lignin and increased IVDMD (Colenbrander et al, 1973; Porter et al, 1978). Using the mutagen diethyl sulfate, Cherney et al (1988), induced a similar brown-midrib mutant (*bmr*) in pearl millet. Lignin concentration in the *bmr* mutant was 40 g/kg compared with 50 g/kg for its normal counterpart, and IVDMD was 726 g/kg for the *bmr* mutant compared with 659 g/kg for the normal genotype.

Cherney et al (1990) determined the intake and digestibility of this *bmr* mutant in relation to its normal counterpart in wether sheep. Dry matter, NDF, and ADF digestibilities were consistently higher for the *bmr* genotype than for the normal genotype. Lignin concentration was 10% lower in *bmr* plants than in normal millet plants. In a forage palatability trial (four-week regrowth), lambs spent an average of 2.6 min on plots containing *bmr* pearl millet for every 1 min spent on plots of the normal genotype.

Data gathered thus far indicate that the *bmr* character has a good potential to improve forage quality in pearl millet. The *bmr* trait also has been discovered in a collection of Zimbabwean germ plasm, and crosses to the source reported by Cherney et al (1988) indicate that the genes may be the same (S. C. Gupta, ICRISAT, *personal communication*, 1991). Degenhart et al (1991) also reported that the gene for an orange node trait discovered earlier at Tifton, Georgia, was the same as the *bmr* gene induced by Cherney et al (1988). Further studies are needed on the genetics and pleiotropic effects of this gene. One such effect is weaker stems, which leads to earlier stem lodging in tall varieties.

Numerous diseases affect millet growth, forage quality, and grain yield in Africa and Asia. Rust (*Puccinia substriata* var. *indica*), first reported in several southern states in 1972 (Wells et al, 1973), and blast (*Pyricularia penneseti*) can seriously affect millet forage in the United States. Monson et al (1986) showed that digestible dry matter yields could be reduced 50% in rust-infected plants. Sources of effective dominant rust resistance have been identified (Andrews et al, 1985; Hanna et al, 1985), and blast resistance has been obtained from Africa and utilized (Hanna et al, 1988). Pearl millet has been found to be susceptible to several pests that are of major importance on other crops; however, resistance might be obtained by adequate screening. Leuck et al (1968) reported resistance to the fall armyworm, and Merkle et al (1983) identified lines resistant to cinch bugs.

Although numerous reports are available on the composition and nutrient value of pearl millet grain in human diets (reviewed in Hoseney et al, 1987), few reports exist on genetic effects. Kumar et al (1983) reported a weak negative correlation between grain yield and protein content but a strong negative correlation between protein content and percent lysine in protein. Singh et al (1987) described the selection of three inbred lines with normal endosperm development whose grain protein contents were 4 to 6 percentage points higher than average. Although lysine percent in protein was decreased, samples showed a net increase of lysine, which almost doubled utilizable protein.

Exploitation of the enormous genetic diversity available in the cultivated pearl millet germ plasm pool has been inadequate. There has been no systematic per se or test-cross evaluation for forage use in the World Germplasm Collection, now numbering over 21,000 accessions (ICRISAT, 1990). Sweet-stemmed millet cultivars have been reported (Appa Rao et al, 1982). Vegetative growth rates among the wild progenitors of pearl millet (*P. glaucum* ssp. *monodii*) (Brunken et al, 1977) were found to be about twice those of cultivated varieties (Bramel-Cox et al, 1984), whereas the weedy types (*P. glaucum* ssp. *stenostachyum*) showed intermediate growth rates. The weedy types in pearl millet have other obvious desirable forage attributes, such as extensive tillering with fine stems and outstanding drought tolerance.

MINOR MILLETS

A number of other grassy, dryland cereals collectively called minor millets have been domesticated for human food in Africa, Eurasia, and Asia. These are listed in chapter 1, which shows their common and scientific names and major regions of production. Their role in global agriculture is well summarized in Seetharama et al (1989) and Riley et al (1993) and is discussed in chapter 2. Because they were adapted for human nutrition, they are generally suitable as animal feed, though production for this purpose is relatively limited. Where they have been used in undeveloped agriculture, the bran and other byproducts of grain processing, e.g., from beer, have been useful as animal feed, but the crop residues have been most valued for use as dry fodder. In low-input agriculture, varieties that produce a higher quantity of dry fodder, as well as a consistent grain yield, have been preferred.

Utilization

FORAGE AND SILAGE

Earlier literature on the composition and nutritive value of foxtail millet hay was reviewed by Malm and Rachie (1971). Foxtail millet straw is highly valued in China (Jiaju, 1989). In the United States, foxtail millet is not suitable for pasturing because of its shallow root system, but it is used as a hay crop. The best quality hay is obtained by cutting after 55 to 75 days' growth. Mixtures of foxtail millet and legumes produce hay of better quality (Vinall, 1924).

Badve et al (1983) reported that the dry matter intake and weight gains in calves were similar when they were given concentrates with 10 or 20% sorghum and 10% foxtail millet. Superchi and Sabbioni (1985) and Sabbioni and Superchi (1985) compared the chemical composition and digestibility of foxtail hay and silage (Table 11). Crude protein content was slightly lower in silage, and hay appeared to be slightly better nutritionally.

In East Africa, the stover of finger millet left after grain harvest is often considered a valuable feed for animals (Esele, 1989). It is preserved in stacks, or cattle are allowed to graze the stubble. Finger millet has been reported to contain HCN in

TABLE 11
Chemical Composition and Dry Matter Digestibility of Hay and Silage of Foxtail Millet

Item	Hay[a]	Silage[b]
Composition, %		
Moisture	13.43	79.58
Crude protein	12.51	11.15
Ether extract	2.19	4.11
Crude fiber	35.13	35.95
Nitrogen-free extract	36.16	33.05
Ash	12.03	15.74
Digestibility, %		
Dry matter	67.79	64.75
Crude protein	67.72	67.76
Ether extract	38.95	60.32
Crude fiber	79.52	77.01
Nitrogen-free extract	64.35	62.26
Organic matter	70.23	69.11

[a] Adapted from Sabbioni and Superchi (1985). Hay harvested at the milk-ripe stage, digestibility in trials with Bergamo wethers fed hay for maintenance.

[b] Adapted from Superchi and Sabbioni (1985). Silage made from forage harvested at milk-ripe stage, ensiled for 120 days; pH 5.31. Digestibility in trials with Bergamo wethers fed silage for maintenance.

very small amounts (Brunnich, 1924). The dry straw from mature plants contained around 2% crude protein and 45% carbohydrate (Rachie and Peters, 1977). In feeding tests, Patel et al (1961) found that finger millet straw had a higher proportion of total digestible nutrients (61%) than pearl millet, paspalum (kodo) millet, wheat, rice, and sorghum straws.

Chun and Ryu (1982) reported that the crude fiber content in Japanese barnyard millet increased greatly from 52 to 59 days after sowing, but nitrogen-free extract did not change throughout the vegetative period. IVDMD was high in the early growing stages and decreased with age. Axelsen et al (1980) reported that gains (per head per day and per hectare) of yearling cattle were higher on Japanese barnyard millet than on Sudax and lucerne.

Japanese researchers reported that the quality of silage decreased with advancing maturity in Japanese barnyard millet and Job's tears (Kumai et al, 1982, 1983; Nakui et al, 1985).

Vinall (1924) reported that feeding foxtail millet grain to horses caused injury by a glycoside called setarin. Hanumanthaiah (1982) observed severe but reversible symptoms of acidosis in buffalo calves fed finger millet grain at 20 to 60 g/kg body weight. Ramachandra et al (1977) analyzed 19 Indian and 10 African finger millet varieties for tannins and found two African varieties with values over 3% (catechin equivalents). In Africa, finger millet varieties used in brewing beer characteristically have high tannin levels.

GRAIN

In the United States, grain of proso millet has been shown to be a good feed for cattle, swine, and poultry (Nelson, 1986). The limitation for cattle is that proso should contribute no more than half the grain in the diet. For swine and poultry, it should be supplemented with lysine. When it was fed to livestock, cracking the seed coat was necessary to allow better digestion (Nelson, 1986).

Erickson et al (1963) reported that proso millet could be a major component in rations for growing pigs if the feed contained 16% protein. However, at 13.5% protein in the feed, the addition of lysine was required. Danielson and Pollmann (1980) fed pigs diets with 14 or 16% protein supplied by maize, soybean oil meal, proso millet, or wheat, and with or without lysine supplementation. There were no significant differences among pigs on the diets in mean daily gain and feed efficiency, although gain tended to be less on wheat and millet diets than on the control diet with maize and soybean. Lysine tended to increase the mean daily gain. Prasad and Singh (1982) found that proso millet was palatable to rams and had digestibilities of 85% for dry matter and crude protein.

Luis et al (1982b) reported that proso millet generally contained higher amounts of protein, fat, and ash than maize or sorghum but was lower than maize in gross and actual ME. In feeding trials with broilers and young turkeys, diets with proso were inferior to those with sorghum or maize when protein levels were suboptimal (18%) (Luis et al, 1982a,b). At optimum protein levels for broilers, there were no differences among the cereals, but at 28% protein, turkey poults were significantly heavier at 28 days on proso than maize or sorghum. For laying chickens on diets containing either ground or unground cereal, proso supported equivalent egg production, egg weight, feed intake, and conversion efficiency to sorghum or maize (Luis et al, 1982c).

Calder (1960) observed that finger millet was more suitable for breeding stock than for young growing stock and recommended that it be limited to 50% of the grain in pig feeds. Local brewery waste of finger millet at a ratio of 40 parts of local to 60 parts of concentrate was profitably fed to pigs to obtain excellent meat quality (Bhutia et al, 1982).

Baghel and Netke (1982) observed that a diet of not more than 30% foxtail millet and groundnut oil meal gave better results than one of maize and groundnut oil meal for starter chicks. When foxtail millet replaced 24 or 38% of wheat in diets for laying hens, there were no significant differences after 52 days in laying performance, egg weight, mortality, or feed intake (Trenchi, 1985).

French (1948) included up to 60% pearl millet or white-seeded finger millet in layers' diets without any reduction in egg production. Abate and Gomez (1983) substituted pearl millet and finger millet for maize at 20, 40, and 60% in both starter and finisher feeds for broilers. The chicks fed pearl millet diets had highest overall body weight gain and performance, and those on finger millet and maize diets were comparable. Luis et al (1980) observed that in proso millet and sorghum diets for broiler chickens, methionine and lysine supplementation significantly increased body weight and feed efficiency. Sullivan et al (1981) found proso millet diets to be equal to diets with maize or sorghum for laying chickens. Millet diets gave significantly greater scores for yolk color than sorghum diets but significantly lower scores than yellow maize diets.

Kapoor et al (1987) fed chickens with diets containing 60% kodo or maize with 35% groundnut cake or soybean meal. Growth, feed efficiency, and protein efficiency ratio were not different in chickens given kodo-groundnut cake and maize-groundnut cake diets supplemented with fish meal and essential amino acids.

No differences in growth, feed intake, or feed conversion efficiency were observed between groups of pigs that were fed with all maize, 50% maize and 50% kodo millet or all kodo millet (Bhadauria et al, 1988).

Crop Management

In the United States, comparing autumn and spring tillage and replacing tillage operations with herbicides, Anderson et al (1986) observed that controlling autumn weed growth by tillage alone or with atrazine increased grain yield and protein content of proso millet. Nelson and Fenster (1983) reported that methods of seedbed preparation and planting did not affect proso grain yields and seed size. Nelson (1977) reported that grain yields were highest with 18-cm row spacings and lowest with 71-cm spacing. In India, Tripathi et al (1979) observed that increased nitrogen fertilization rates led to increased yields of fresh fodder, dry matter yield, and crude protein content.

Cutting experiments on foxtail millet in Japan gave a maximum production of 42 t/ha of green fodder, but crude protein content and digestibility decreased in late cuttings (Fukumi et al, 1979). In Argentina, Cangiano (1979) obtained a cumulative production in foxtail millet from six cuts of 8 t/ha dry matter, but crude protein levels fell from 20.6% in the first cut to 8.6% in the sixth cut. Bruno et al (1984) reported that the best stage at which to harvest foxtail millet for high-quality hay was initial heading, when yields ranged from 5 to 7.8 t/ha of dry matter.

Nishikawa et al (1977) reported that maximum yield of dry matter was obtained with finger millet cut at 10 cm. Nitrogen content increased with increasing rates of nitrogen fertilizer but decreased with age and increased cutting height. Kogure

(1982) observed that yield at first cut increased with delay in cutting and increased density, but regrowth yields decreased with delay in first cut. Total yield of finger millet was related to plant density.

Breeding and Genetics

In the United States, emphasis has been given to the development of foxtail and proso varieties that are suitable for caged and wild birds but also can be used for human and animal food. The prices in the wild bird food market tend to be considerably higher than in the feed trade. Proso varieties specifically for hay and forage have been bred in the former U.S.S.R..

Variety Butte of foxtail millet was released in the United States as a bulk selection from a varietal introduction from the former U.S.S.R.. Caged birds preferred heads of this variety in trials (Hinze and Hamilton, 1975). Nelson (1981) described development of the variety Sno-Fox, which was high-yielding and suitable for higher elevations. In India, Harinarayana et al (1984) described four different dwarf genotypes in foxtail millet that could be used in the development of high-yielding, fertilizer-responsive, and lodging-resistant varieties.

In the United States, grain proso varieties have been released for various areas in Colorado, Georgia, Minnesota, and Nebraska (Hinze and Mann, 1975; Hinze et al, 1978; Nelson, 1975, 1976, 1983; Powell et al, 1977; Robinson, 1976). Although the newer varieties are shorter in stature, they do not have the stem strength to stand until maturity and must be swathed to dry down for threshing. Some grain loss inevitably occurs, which then creates a weed problem in subsequent crops. The grain yields of these improved varieties ranges from 0.5 to 1.0 t/ha.

Saprykin (1977) reported release of the first cultivar of fodder proso, Chernosemyannoe-1, which was adapted to Omsk province in western Siberia. It was black-grained and gave grain yields of 1 to 1.5 t/ha, fresh fodder yields of 20 to 25 t/ha, and a dry matter yield of 5.3 t/ha. Popov (1978) described two varieties, Koromovoe-1 and Koromovoe-2, which produced good forage and hay, with the second giving the highest forage yield of 15.8 t/ha. Koromovoe-1, when harvested at heading for hay, contained 11 to 14% crude protein, 3 to 4% ether extract, and fodder meal with 4.19% lysine. Yudina (1982) listed 22 varieties suitable for green fodder production in St. Petersburg Province, Russia.

Lee (1981) evaluated three varieties of Japanese barnyard millet and concluded that two local varieties, Jeju and Jinju, were better adapted to Korean conditions as a summer forage.

The potential for minor millets in the United States appears restricted by their low productivity and consequent inability to compete in the general feed grain markets dominated by maize and sorghum. They are thus limited to the specialty grain markets and as forages in cooler short-season areas.

This is in contrast to proso in southeast Russia and foxtail in northeast China, where substantially larger areas are grown and yields are apparently much higher. Both of these countries have active research programs to improve production. In Russia, proso forms part of the total cereal market and is also grown for fodder. In China, foxtail millet varieties have to be dual-purpose, providing grain for human food and stover as animal forage. Li Yinmei reported (1994, *personal communication*) that experimental foxtail millet hybrids made using a recessive nuclear gene for male sterility are being tested in Hebei Province, China. The expression of this male sterility is slightly incomplete, giving 1–5% seed set under selfing bags, which thus permits the increase of the male sterile line.

CONCLUSION

Three basic avenues exist for improvement of forage and feed quality. These are crop management, genetic manipulation, and processing. Although wider applications of known management techniques and procedures for processing offer improvements overall in the quality and quantity of forage, feed, and products such as silage, their potential for the future is limited. Their major use will be in upgrading poor-quality forages and the consistency of the feedstuffs produced. Therefore, the greatest advances for the future will be from further genetic manipulation of these crops to improve their yield and quality.

The development of techniques utilizing biotechnology methods to analyze the basis of genetic change and the role of genetic diversity offer new alternatives. The application of these new technologies for the identification and transfer of genes will be critical for future gains in both quality and quantity of nutrients produced per area of production. Recent research provides good reason to expect that continued gains also can be made in productivity and quality in sorghum and pearl millet through more traditional approaches to breeding. The new *bmr* genes could result in quantum improvements in forage quality in both crops; however, little is yet known of the relationship of the *bmr* trait with the overriding requirement of total productivity. This relationship must be established and taken into account when breeding *bmr* cultivars. Superior growth rates, regrowth potential, and drought tolerance are known to exist in the wild and weedy relatives of both crops. Indeed, much of the wide range of genetic diversity available in both species has not been examined adequately and should provide profitable areas for research.

The improvements to date in the feeding value of sorghum grain have been primarily a result of reducing tannin levels. Recent assays of a range of low- or non-tannin genotypes have shown a substantial range of digestibilities, largely related to protein quality, indicating that further improvements in grain feeding values are possible. The development of a market for pearl millet grain in the United States is primarily dependent on raising productivity rather than grain quality levels, which are relatively good. However, because breeding for grain production and experience in managing the grain crop are at relatively early stages compared to sorghum, gains in productivity can be expected from current research. Pearl millet grain production will probably be established first in areas where the crop has an adaptational advantage over sorghum, particularly for drought tolerance and earliness.

Future advances will depend upon changes in nutritional quality with continued gains in productivity. The focus by plant breeders in the past has been on total productivity and, given the current market emphasis on yield, any nutritional improvements will have to be made without sacrifice in total productivity. As long as no economic value is given to quality components, the impact of genetic changes in quality will be limited unless incorporated concurrently with higher yield potential.

LITERATURE CITED

ABATE, A. N., and GOMEZ, M. 1983. Substitution of finger millet (*Eleusine coracana*) and bulrush millet (*Pennisetum typhoides*) for maize in broiler feeds. Anim. Feed Sci. Technol. 10:291-299.

ACHACOSO, A. S., MONDART, C. L., JR., BONNER, F. L., and RUSOFF, L. L. 1960. Relationship of lignin to other chemical constituents in sudan and millet forages. J. Dairy Sci. 43:443.

AKIN, D. E., HANNA, W. W., and RIGSBY, L. L. 1986a. Normal-12 and brown midrib-12 sorghum. I. Variations in tissue digestibility. Agron. J. 78:827-832.

AKIN, D. E., HANNA, W. W., SNOOK, M. E., HIMMELSBACH, D. S., BARTON, F. E. II, and WINDHAM, W. R. 1986b. Normal-12 and brown midrib-12 sorghum. II. Chemical variations and digestibility. Agron. J. 78:832-837.

ANDERSON, R. L., SHANAHAN, J. F., and GREB, B. W. 1986. Effect of tillage systems on proso millet production. Agron. J. 78:589-592.

ANDREWS, D. J., and KUMAR, K. A. 1992. Pearl millet for food, feed, and forage. Adv. Agron. 48:89-139.

ANDREWS, D. J., RAI, K. N., and SINGH, S. D. 1985. A single dominant gene for rust resistance in pearl millet. Crop Sci. 25:565-566.

ANDREWS, D. J., RAJEWSKI, J. F., and KUMAR, K. A. 1993. Pearl millet: A new feed grain crop. Pages 198-208 in: New Crops. J. Janick and J. Simon, eds. John Wiley & Sons, New York.

ANTHONY, W. B., SMITH, C. A., BROWN, V. L., and HARRIS, R. R. 1961. Nutritive value of dehydrated sorghum silage. (Abstr.) J. Anim. Sci. 20:958.

APPA RAO, S., MENGESHA, M. H., and SUBRAMANIAN, V. 1982. Collection and preliminary evaluation of sweet-stalk pearl millet (*Pennisetum*). Econ. Bot. 36:286-290.

ARORA, S. K., LUTRA, Y. P., and DAS, B. 1975. Locational variation in chemical composition and in vitro digestability of forage sorghum. J. Agric. Food Chem. 23:543.

AXELSEN, A., DANN, P. R., and COOMBE, J. B. 1980. Animal production from summer forage crops (Abstr.) Page 243 in: Proc. Austr. Agron. Conf. Pathways to Productivity. I. M. Wood, ed. Australian Society of Agronomy, Lawes, Australia.

BADVE, V. C., SOBALE, B. N., JOSHI, A. L., and RANGNEKAR, D. V. 1983. Use of rale seeds (*Setaria italica*) for dairy cattle along with sugarcane. Indian J. Anim. Sci. 53:1049-1052.

BAGHEL, R. P. S., and NETKE, S. P. 1982. Studies on the incorporation of kangni (*Setaria italica*) in starter chick diets. Indian J. Anim. Sci. 52:411-417.

BAKER, F. S., JR. 1970/71. Mimeographed data. Florida Agric. Exp. Stn., AREC, Quincy, FL (quoted by Bertrand and Dunavin, 1973).

BALL, C. R. 1903. Pearl millet. Farmers' Bull. 168. U.S. Dep. Agric., Washington, DC.

BAXTER, H. D., OWEN, J. R., and RATCLIFF, L. 1959. Comparison of two varieties of pearl millet for dairy cattle. Tenn. Farm Home Sci. Prog. Rep. 33:4-5.

BEATY, E. R., SMITH, Y. C., MCCREERY, R. A., ETHREDGE, W. J., and BEASLEY, K. 1965. Effect of cutting height and frequency on forage production of summer annuals. Agron. J. 57:277-279.

BEGG, J. E. 1965. The growth and development of a crop of bulrush millet (*Pennisetum typhoides* S.&H.). J. Agric. Sci. 65:341-349.

BERTRAND, J. E., and DUNAVIN, L. S. 1973. Millet pasture and millet silage supplemented and unsupplemented for beef steers. Proc. Soil Crop Sci. Soc. Fla. 32:23-25.

BHADAURIA, S. S., DHINGRA, M. M., and LAKHANI, G. P. 1988. Effect of incorporation of kodo (*Paspalum scrobiculatum*) in the ration of growing indigenous pigs. Livest. Adv. 13:5-7.

BHUTIA, P. N., KATIYAR, R. D., and PAL, R. N. 1982. Note on the growth on white Yorkshire pigs fed on chhang (local brewery waste of *Eleusine coracana*) in the hills of Sikkim. Indian J. Anim. Sci. 52:470-472.

BOLSEN, K., ILG, H., POPE, R., and HOOVER, J. 1983. Whole plant forage, grain and non-heading sorghum silages, cornlage, and feed flavor supplements for growing calves. In: Cattlemen's Day 1983, Kansas Agric. Exp. Stn. Rep. Prog. 427.

BRAMEL-COX, P. J., ANDREWS, D. J., BIDINGER, F. R., and FREY, K. J. 1984. A rapid method of evaluating growth rate in pearl millet and its weedy and wild relatives. Crop Sci. 24:1187-1191.

BRAMEL-COX, P. J., LAUVER, M. A., and WITT, M. E. 1990. Potential gain from selection in grain sorghum for higher protein digestibility. Crop Sci. 30:521-524.

BRETHOUR, J. R., and DUITSMAN, W. W. 1965. Utilization of feterita-type and waxy endosperm sorghum grains in all concentrate rations fed to yearling steers. In: Fort Hays Roundup, Kansas Exp. Stn. Bull. 482.

BROWNING, C. B., and LUSK, J. W. 1967. Effect of stage of maturity at harvest on nutritive value of combine-type grain sorghum. J. Dairy Sci. 50:81-85.

BROYLES, K. R., and FRIBOURG, H. A. 1959. Nitrogen fertilization and cutting management of sudangrass and millets. Agron. J. 51:277-279.

BRUNKEN, J. N., DE WET, J. M. J., and HARLAN, J. R. 1977. The morphology and domestication of pearl millet. Econ. Bot. 31:163-174.

BRUNNICH, J. C. 1924. Hydrocyanic acid content of millets and sorghums. Pages 31-33 in: Queensl. Dep. Agric. Livest. Annu. Rep. 1923–1924.

BRUNO, O. A., FOSSATI, J. L., CALCHA, N. A., and FENOGLIO, H. F. 1984. Evolution of production and quality forage from *Setaria italica* cultivars. Rev. Argent. Prod. Anim. 4:673-682.

BUCHOLTZ, H. F., DAVIS, C. L., PALINQUIST, D. L., and KENDALL, K. A. 1969. Study of the low-fat milk phenomenon in cows grazing pearl millet pastures. J. Dairy Sci. 52:1385-1394.

BUCHOLTZ, D. L., CANTRELL, R. P., AXTELL, J. D., and LECHTENBERG, V. L. 1980. Lignin biochemistry of normal and brown midrib mutant sorghum. J. Agric. Food Chem. 28:1239-1241.

BURGER, A. W., and HITTLE, C. N. 1967. Yield, protein, nitrate and prussic acid content of sudangrass, sudangrass hybrids and pearl millets harvested at two cutting frequencies and two stubble heights. Agron. J. 59:259-262.

BURNS, J. C., and KIMBROUGH, E. L. 1981. Quality estimates and chemical characterization of fermented and unfermented summer annual forages. Agron. J. 73:921-924.

BURTON, G. W. 1948. The performance of various mixtures of hybrid and parent inbred pearl millet, *Pennisetum glaucum* (L.) R. Br. J. Am. Soc. Agron. 40:908-915.

BURTON, G. W. 1962. Registration of Gahi-1 pearl millet. Crop Sci. 2:355.

BURTON, G. W. 1965. Pearl millet Tift 23A released. Crops Soils 17:19.

BURTON, G. W. 1977. Registration of Gahi 3 pearl millet. Crop Sci. 17:345-346.

BURTON, G. W. 1980. Registration of pearl millet inbred Tift 383 and Tifleaf 1 pearl millet. Crop Sci. 20:292.

BURTON, G. W. 1989. Composition and forage yield of hybrid-inbred mixtures of pearl millet. Crop Sci. 29:252-255.

BURTON, G. W., and DEVANE, E. H. 1951. Starr millet—Synthetic cattail lasts longer and produces more beef per acre. South. Seedsman 14:17.

BURTON, G. W., and FORTSON, J. C. 1966. Inheritance and utilization of five dwarfs in pearl millet (*Pennisetum typhoides*) breeding. Crop Sci. 6:69-72.

BURTON, G. W., KNOX, F. E., and BEARDSLEY, D. W. 1964. Effect of age on the chemical composition, palatability, and digestibility of grass leaves. Agron. J. 56:160-161.

BURTON, G. W., MONSON, W. G., JOHNSON, J. C., JR., LOWERY, R. S., CHAPMAN, H. D., and MARCHANT. 1969. Effect of the d_2 dwarfing gene on the forage yield and quality of pearl millet. Agron. J. 6:607-612.

BURTON, G. W., HANNA, W. W., JOHNSON, J. C., JR., LEUCK, D. B., MONSON, W. G., POWELL, J. B., WELLS, H. D., and WIDSTROM, N. W. 1977. Pleiotropic effects of the *tr* trichomeless gene in pearl millet on transpiration, forage quality, and pest resistance. Crop Sci. 17:613-616.

BURTON, G. W., KVIEN, C. S., and MAW, B. W. 1988. Effect of drought stress on productivity of trichomeless pearl millet. Crop Sci. 28:809-811.

BUTLER, J. L., WILKINSON, W. S., HELLWIG, R. E., BARBEE, C., and KNOX, F. E. 1969. Factors involved in production and storage of high-quality dehydrated coastal bermudagrass and pearl millet. Trans. ASAE 12:552-555.

CALDER, A. 1955. Value of munga (pearl millet) for pig feeding. Rhod. Agric. J. 52:161-170.

CALDER, A. 1960. The value of ropoko (finger millet) for pig feeding. Rhod. Agric. J. 57:116-119.

CANGIANO, C. A. 1979. Production and quality of forage of foxtail millet (*Setaria italica*). (In Spanish.) Inf. Tecnica, Inst. Nacional de Tecnologia Agropecuaria 84.

CHERNEY, J. H., MOORE, K. J., VOLENAC, J. J., and AXTELL, J. D. 1986. Rate and extent of digestion of cell wall components of brown midrib sorghum species. Crop Sci. 26:1055-1059.

CHERNEY, J. H., AXTELL, J. D., HASSEN, M. M., and ANLIKER, K. S. 1988. Forage quality characterization of a chemically induced brown-midrib mutant in pearl millet. Crop Sci. 28:783-787.

CHERNEY, D. J. R., PATTERSON, J. A., and JOHNSON, K. D. 1990. Digestibility and feeding value of pearl millet as influenced by the brown-midrib, low-lignin trait. J. Anim. Sci. 68:4345-4351.

CHRISTENSEN, N. B., PALMER, J. C., PRAEGER, H. A., JR., STEGMEIER, W. D., and VANDERLIP, R. L. 1984. Pearl millet: A potential crop for Kansas. Kans. Agric. Exp. Stn. Keeping up with Research 77.

CHUN, W. B., and RYU, N. H. 1982. Changes in dry matter, feed composition and digestibility of Japanese millet (*Echinochloa crus-galli* var. *frumentacea* (Roxb.) Wight) during the period of vegetation. (In Korean.) Korean J. Anim. Sci. 24: 320-325.

CLAPP, J. G., and CHAMBLEE, D. S. 1970. Influence of different defoliation systems on the regrowth of pearl millet, hybrid sudangrass, and two sorghum-sudangrass hybrids from terminal, axillary and basal buds. Crop Sci. 10:345-349.

CLARK, N. A., HEMKEN, R. W., and VANDERSALL, J. H. 1965. A comparison of pearl millet, sudangrass and sorghum-sudangrass hybrid as pasture for lactating dairy cows. Agron. J. 57:266-269.

COHEN, R. S., and TANKSLEY, T. D., JR. 1973. Energy and protein digestibility of sorghum grains with different endosperm textures and starch types of growing swine. J. Anim. Sci. 37:931-935.

COLEMAN, O. H. 1970. Syrup and sugar from sweet sorghum. Pages 416-440 in: Sorghum Production and Utilization. J. W. Wall and W. M. Ross, eds. AVI Publ. Co., Westport, CT.

COLENBRANDER, V. F., LECHTENBERG, V. L., and BAUMAN, L. F. 1973. Digestibility and feeding value of a brown midrib corn stover silage. J. Anim. Sci. 37:294-295.

COMBS, G. E, COPELIN, J. L., and GORBET, D. W. 1980. Performance of swine fed sorghum diets supplemented with fat. Univ. Fla. Annu. Swine Field Day, 28th . Res. Rep. AL-1980-4.

CROMWELL, G. L., BITZER, M. J., STAHLY, T. S., and JOHNSON, T. H. 1983. Effect of soil nitrogen fertility on the protein and lysine content and nutritional value of normal and opaque-2 corn. J. Anim. Sci. 57:1345-1351.

CROMWELL, G. L., STAHLY, T. S., and RANDOLF, J. R. 1985. Grain sorghum and barley as alternative feed grains for growing-finishing swine. Pages 27-28 in: Ky. Agric. Exp. Stn. Rep. 25-173.

CUMMINS, D. G. 1971. Relationships between tannin content and forage digestibility in sorghum. Agron. J. 63:500-502.

CUMMINS, D. G. 1981. Yield and quality changes with maturity of silage type sorghum fodder. Agron. J. 73:988-990.

DADA, L. O., and DENDY, D. A. V. 1987. A preliminary study of broomcorn (*Sorghum bicolor*) grain milling. Rep. L77. Tropical Development and Research Institute, London.

DANIELSON, D. M., ENGLAND, M. E., and POLLMAN, D. D. 1978. Effect on average daily gain: Changing grain sources. Neb. Swine Res. Rep. EC78-219.

DANIELSON, D. M., and POLLMANN, D. S. 1980. Lysine supplementation of proso millet diets for growing-finishing swine. Nutr. Rep. Int. 21:625-630.

DEGENHART, N. R., WERNER, B. K., and BURTON, G. W. 1991. An orange node trait in pearl millet: Its inheritance and effect on digestibility and herbage yield. (Abstr.) Abstr. Tech. Pap. Annu. Meet. 1991, Am. Soc. Agron., Madison, WI.

DICKERSON, J. BOLSEN, K., HAMMA, S., and HOOVER, J. 1986. Effect of maturity at harvest on yield, composition, and feeding value of hybrid forage sorghum silages. Pages 135-143 in: Cattlemen's Day 1986. Kansas Agric. Exp. Stn. Rep. Prog. 494.

DOGGETT, H. 1988. Pages 411-427 in: Sorghum. Longman Scientific and Technical, Essex, England.

DUJARDIN, M., and HANNA, W. W. 1989. Developing apomictic pearl millet—Characterization of a BC_3 plant. J. Genet. Breed. 43:145-151.

DUNCAN, R. R., BRAMEL-COX, P. J., and MILLER, F. R. 1991. Contributions of introduced sorghum germplasm to hybrid development in the USA. Pages 69-102 in: Use of Plant Introductions in Cultivar Development, Part I. H. L. Shands, ed. CSSA Spec. Publ. 17.

ELMALIK, M., KLOPFENSTEIN, C. F., HOSENEY, R. C., and BATES, L. S. 1986. Effects of germination on the nutritional quality of sorghum grain with contrasting kernel characteristics. Nutr. Rep. Int. 34:941-950.

ERICKSON, D. O., HAUGSE, C. N., DINUSSON, W. E., BOLIN, D. W., and BUCHANAN, M. L. 1963. Proso: Lysine deficient for pigs. North Dakota Farm Res. 22:7.

ESELE, J. P. E. 1989. Processing and utilization of small millets in Uganda. Pages 351-353 in: Small Millets in Global Agriculture. A. Seetharam, K. W. Riley, and G. Harinarayana, eds. Proc. Int. Small Millets Workshop, 1st. Oxford and IBH Publ. Co., New Delhi, India.

FALES, S. L., and WILKINSON, R. E. 1984. Mefluidide-induced nitrate accumulation in pearl millet forage. Agron. J. 76:857-860.

FRENCH, M. H. 1948. Local millets as substitutes for maize in feeding of domestic animals. East Afr. Agric. J. 13:217-220.

FRIBOURG, H. A. 1966. The effect of morphology and defoliation intensity on the tillering, regrowth and leafiness of pearl millet, *Pennisetum typhoides* (Burm.) Stapf & C.E. Hubb. Pages 489-491 in: Proc. Int. Grassland Congr., 9th, Sao Paulo, Brazil, 1965.

FRITZ, J. O., CANTRELL, R. P., LECHTENBERG, V. L., AXTELL, J. D., and HERTEL, J. M. 1981. Brown midrib mutants in sudangrass and grain sorghum. Crop Sci. 21:706-709.

FRITZ, J. O., MOORE, K. J., and JASTER, E. H. 1990. Digestion kinetics and cell wall composition of brown midrib sorghum × sudangrass morphological components. Crop Sci. 30:213-219.

FROMENT, M. A. 1986. Broom sorghum research in Santa Cruz, Bolivia 1983-86: An overview of an experimental program carried out using the FSR methodology. Mimeo. Santa Cruz, Boliva: CIAT/BTAM.

FUKUMI, R., KUMAI, S., TAZI, K., and FUJIHARA, K. 1979. Studies on cultivation and feed utilization of miscellaneous grain crops as fodders. 3. Growth, yield and chemical composition of Italian millet for soiling. (In Japanese.) J. Jpn. Soc. Grassl. Sci. 25:269-271.

GIESEMANN, M. A., LEWIS, A. J., HANCOCK, J. D., and PEO, E. R., JR. 1990. Effect of particle size of corn and grain sorghum on growth and digestibility by growing pigs. (Abstr.) J. Anim. Sci. 90(Suppl. 1):350.

GOLDY, G., RILEY, J., WALTER, T., POSLER, G., and LENSEN, A. 1986. Comparative nutritional composition of selected grain sorghum varieties. Pages 12-13 in: Cattlemen's Day 1986, Kansas Agric. Exp. Stn. Rep. Prog. 494.

GORZ, H. J., HASKINS, F. A., and VOGEL, K. P. 1986. Inheritance of dhurrin content in mature sorghum leaves. Crop Sci. 26:65-67.

GOURLEY, L. M., and LUSK, J. W. 1978 . Genetic parameters related to sorghum silage quality. J. Dairy Sci. 61:1821-1827.

GRABOUSKI, H. A., PEO, E. R., JR., LEWIS, A. J., and HANCOCK, J. D. 1987. New varieties of grain sorghum—Are they better for swine? Neb. Swine Res. Rep. EC 87-219.

GREEN, JR., V. E. 1973. Pearl millet, *Pennisetum typhoides* (Burm.) Stapf and Hubb.: Research and observations in Florida—1888-1968. Proc. Soil Crop Sci. Soc. Fla. 32:25-29.

GUPTA, V. P. 1975. Fodder improvement in *Pennisetum*. Forage Res. 1:54-60.

GUPTA, P. C., SINGH, K., and SHARDA, D. P. 1981. Note on the *in vivo* studies on the nutritive value of pearl millet, pearl millet-cowpea forage mixture and its silage. Indian J. Anim. Sci. 51:1166-1167.

HALE, W. H. 1984. Comparison of wet grain processing methods for finishing cattle. In: Proc. Feed Grains Utilization Symposium, Texas Tech University.

HAMILTON, C. R., ORR, D. E., JR., and TRIBBLE, L. F. 1979. Effects of pelleting and grain source on performance of four-week-old pigs. Pages 68-70 in: Texas Tech. Univ. Agric. Sci. Tech. Rep. T-5-144.

HAMILTON, C. R., ORR, D. E., JR., and TRIBBLE, L. F. 1980. Effect of grain source and physical form of diet on nutrient balance in the weaned pig. Pages 26-31 in: Texas Tech. Univ. Agric. Sci. Tech. Rep. T-5-154.

HANCOCK, J. D., BOLSEN, K. K., HINES, R. H., and JONES, D. B. 1989. Effect of microbial inoculants on the storage characteristics and nutritional value of high moisture corn for finishing pigs. Kansas State Univ. Rep. of Prog. 581. Pages 160-163.

HANCOCK, J. D., PEO, E. R., JR., LEWIS, A. J., KNIEP, K. R., and MASON, S. C. 1990a. Effects of irrigation and nitrogen fertilization of normal and high lysine corn on protein utilization by the growing rat. Nutr. Rep. Int. 38:413-422.

HANCOCK, J. D., BRAMEL-COX, P. J., SMITH, K. R., HEALY, B. J., KLOPFENSTEIN, C. F., and WITT, M. D. 1990b. Selection for increased in vitro digestibility improves the feeding value of sorghum grain. (Abstr.) J. Anim. Sci. 90(Suppl. 1):320.

HANCOCK, J. D., GOODBAND, R. D., and HANSEN, J. A. 1991a. Effect of test weight on feeding value and MEn of sorghum for growing chicks. (Abstr.) J. Anim. Sci. 91(Suppl. 1):315.

HANCOCK, J. D., HINES, R. H., FITZNER, G. E., and GUGLE, T. L. 1991b. Effect of extrusion processing on the nutritional value of sorghum and soybeans for finishing pigs. Page 113 in: Proc. Bienn. Grain Sorghum Res. Utilization Conf., 17th, Lubbock, TX.

HANNA, W. W., and AKIN, D. E. 1978. Microscopic observations on cuticle from trichomeless, *tr*, and normal, *Tr*, pearl millet. Crop Sci. 18:904-905.

HANNA, W. W., MONSON, W. G., and BURTON, G. W. 1977. Morphological and anatomical factors affecting forage quality. Ga. Agric. Res. 18:17-21.

HANNA, W. W., MONSON, W. G., and GAINES, T. P. 1981. IVDMD, total sugars and lignin measurements on normal and brown midrib (*bmr*) sorghums at various stages of development. Agron. J. 73:1050-1052.

HANNA, W. W., WELLS, H. D., and BURTON, G. W. 1985. Dominant gene for rust resistance in pearl millet. J. Hered. 76:134.

HANNA, W. W., WELLS, H. D., BURTON, G. W., HILL, G. M., and MONSON, W. G. 1988. Registration of `Tifleaf 2' pearl millet. Crop Sci. 28:1023.

HANSEN, J. A., GOODBAND, R. D., THALER, R. C., HANCOCK, J. D., NELSSEN, J. L., and HINES, R. H. 1990. Low-test weight sorghum for growing finishing swine. (Abstr.) J. Anim. Sci. 68(Suppl. 1):398.

HANUMANTHAIAH, K. 1982. Experimental rumen acidosis in buffalo calves. M.Sc. thesis. University of Agricultural Science, Bangalore, Karnataka, India.

HARINARAYANA, G., NAVALE, P. A.,AND MUNGSE, H. B. 1984. Dwarf genotypes of foxtail millet. J. Maharashtra Agric. Univ. 9:343-344.

HARNER, J. P., HENKEN, R. W., VANDERSALL, J. H., CLARK, N. A., and SCHNEIDER, B. A. 1969. Relationship of pearl millet to milk fat depression in dairy cows. I. Cation fertilization. J. Dairy Sci. 52:1244-1248.

HARRIS, H. B., CUMMINS, D. G., and BURNS, R. E. 1970. Tannin content and digestibility of sorghum grain as influenced by bagging. Agron. J. 62:633-635.

HART, R. H. 1967. Digestibility, morphology, and chemical composition of pearl millet. Crop Sci. 7:581-584.

HEALY, B. J., HANCOCK, J. D., BRAMEL-COX, P. J., BEHNKE, K. C., and KENNEDY, P. K. 1991. Optimum particle size of corn and hard and soft sorghum grain for nursery pigs and broiler chicks. Pages 56-62 in: Kansas State Univ. Swine Res. Rep. Prog. no. 641.

HELM, R. E., and LEIGHTON, R. E. 1960. Relationship of stage of maturity at harvest to the feeding value of sorghum silage for dairy cows. (Abstr.) J. Dairy Sci. 42:868.

HEMKEN, R. W., HARNER, J. P., VANDERSALL, J. H., CLARK, N. A., and SCHNEIDER, B. A. 1968. Fertilization of pearl millet forage as related to milk fat test. (Abstr.) J. Dairy Sci. 51:982.

HERKELMAN, K. L., RODHOUSE, S. L., VEUM, T. L., and ELLERSIECK, M. R. 1990. Effect of extrusion on the ileal and fecal digestibiltiies of lysine in yellow corn in diets for young pigs. J. Anim. Sci. 68:2414-2424.

HIBBERD, C. A., SCHEMM, R., and WAGNER, D. G. 1978. Influence of endosperm type on the nutritive value of grain sorghum and corn. Pages 77-81 in: Okla. St. Univ. Anim. Sci. Res. Rep.

HILL, G. M., and HANNA, W. W. 1990. Nutritive characteristics of pearl millet grain in beef cattle diets. J. Anim. Sci. 68:2061-2066.

HINE, R. H., and KOCH, B. A. 1971. Rations containing corn, sorghum grain or wheat compared for weaned pigs. Pages 37-39 in: Kansas State Univ. Swine Res. Rep.

HINZE, G., and HAMILTON, J. 1975. Registration of Butte foxtail millet. Crop Sci. 15:886.

HINZE, G., and MANN, H. O. 1975. Registration of Abarr proso millet. Crop Sci. 15:886.

HINZE, G., MANN, H. O., LANGIN, E. J., and FISHER, A. 1978. Registration of Cope proso millet. Crop Sci. 18:1093.

HONEYCUTT, H. 1988. Forage variety performance test 1986-1987. Arkansas Agric. Exp. Stn. Res. Ser. 370.

HOSENEY, R. C., ANDREWS, D. J., and CLARK, H. 1987. Sorghum and pearl millet. Pages 397-456 in: Nutritional Quality of Cereal Grains: Genetic and Agronomic Improvement. R. A. Olson and K. J. Frey, eds. Am. Soc. Agron., Agron. Monogr. 28, Madison, WI.

HOVELAND, C. S., and MCCLOUD, D. E. 1957. Manage millet for better yields. Sunshine State Agric. Res. Rep. 2:14.

HUBER, J. T., THEURER, C. B., SWINGLE, R. S., OLIVEIRA, J., MOORE, J. and POORE, M. 1991. Effect of processing sorghum grain on milk composition and feed efficiency. Pages 116-118 in: Proc. Bienn. Grain Sorghum Res. Utilization Conf., 17th Lubbock, TX.

HULSE, J. H., LAING, E., and PEARSON, D. D. 1980. Sorghum and the Millets: Their Composition and Nutritive Value. Academic Press, New York.

ICRISAT. 1990. Annual Report 1989. Patancheru, A. P., India.

JASTER, E. H., FISHER, C. M., and MILLER, D. A. 1985. Nutritive value of oatlage, barley/pea, pea, oat/pea, pearl millet and sorghum as silage ground under a double cropping forage system for dairy heifers. J. Dairy Sci. 68:2914-2921.

JIAJU, C. 1989. Utilization of small millets in China. Pages 347-350 in: Small Millets in Global Agriculture. A. Seetharam, K. W. Riley, and G. Harinarayana, eds. Proc. Int. Small Millets Workshop, 1st. Oxford and IBH Publ. Co., New Delhi, India.

JOHNSON, J. C., JR., and SOUTHWELL, B. L. 1960. Pasture and silage both from Starr millet. Ga. Agric. Res. 1:6-7.

JOHNSON, J. C., JR., LOWREY, R. S., MONSON, W. G., and BURTON, G. W. 1968. Influence of the dwarf characteristic on composition and feeding value of near-isogenic pearl millets. J. Dairy Sci. 51:1423-1425.

JOHNSON, R. R., DEFARIA, V. P., and MCCLURE, K. E. 1971. Effects of maturity on chemical composition and digestibility of bird resistant sorghum plants when fed to sheep as silages. J. Anim. Sci. 33:1102-1109.

JOHNSON, J. C., JR., MCCORMICK, W. C., BURTON, G. W., and MONSON, W. G. 1978. Weight gains of heifers grazing Gahi 1, Gahi 3, and Tifleaf 1 hybrid pearl millet. Univ. Ga. Res. Rep. 274.

JUNG, H. G., FAHEY, G. C., JR., and GARST, J. E. 1983. Simple phenolic monomers of forages and effects of in vitro fermentation on cell wall phenolics. J. Anim. Sci. 57:1294-1305.

KALTON, R. R. 1988. Overview of the forage sorghums. Proc. Am. Seed Trade Assoc. Annu. Corn Sorghum Res. Conf. 43:1-12.

KAPOOR, P. N., NETKE, S. P., and BAJPAI, L. D. 1987. Kodo (*Paspalum scrobiculatum*) as a substitute for maize in chick diets. Indian J.

Anim. Nutr. 4:83-88.

KOGURE, K. 1982. An examination of the relationship between the time of first cut and the growth and chemical composition of finger millet as a summer forage crop. (In Japanese) Tech. Bull. Fac. Agric., Kagawa Univ. 33:81-88.

KOLLER, H. R., and SCHOLL, J. M. 1968. Effect of row spacing and seeding rate on forage production and chemical composition of two sorghum cultivars harvested at two cutting frequencies. Agron. J. 60:456-458.

KREJSA, B. B., ROUQUETTE, JR., F. M., HOLT, E. C., CAMP, B. J., and NELSON, L. R. 1984. Nitrate and total alkaloid concentration of 11 pearl millet lines. Agron. J. 76:157-158.

KUMAI, S., FUKUMI, R., and TAJI, K. 1982. Some experiments in the Job's-tear millet (*Coix lacryma-jobi* L. var. *frumentacea*) on the quality of fresh stover silage and nutritive values of sun-cured stover and fresh stover silage. (In Japanese.) Mem. Coll. Agric. Ehime Univ. 27:171-179.

KUMAI, S., FUKUMI, R., and TAJI, K. 1983. Studies on Job's tears as a forage crop. II. Silage quality and feeding values of soiling crop and silage in Job's tear millet. (In Japanese.) J. Jap. Soc. Grassl. Sci. 29:66-72.

KUMAR, K. A., GUPTA, S. C., and ANDREWS, D. J. 1983. Relationship between nutritional quality characters and grain yield in pearl millet. Crop Sci. 23:232-235.

LAMB, J. F. S., HASKINS, F. A., GORZ, H. J., and VOGEL, K. P. 1987. Inheritance of seedling hydrocyanic acid potential and seed weight in sorghum-sudangrass crosses. Crop Sci. 27:522-525.

LEE, H. W. 1981. Studies on the production and management of Japanese millet (*Echinochloa crus-galli*) for summer forage. I. Agronomic characteristics and yield performance of three Japanese millet cultivars. (In Korean.) Korean J. Anim. Sci. 23:264-269.

LEMON, M. D., and MCMURPHY, W. E. 1984. Forage NO_3 in Sudangrass and pearl millet. Pages 302-306 in: Forage Systems—Leading U.S. Agriculture into the Future. American Forage and Grassland Council, Lexington, KY.

LEUCK, D. B., TALIAFERRO, BURTON, R. L., BURTON, G. W., and BOWMAN, M. C. 1968. Fall armyworm resistance in pearl millet. J. Econ. Entomol. 61:693-695.

LICHTENWALNER, R. E., ELLIS, E. B., and ROONEY, L. W. 1978. Effect of incremental dosages of the waxy gene of sorghum on digestibility. J. Anim. Sci. 46:1113-1119.

LLOYD, G. L. 1964. Use of munga to replace maize in a broiler ration. Rhod. Agric. J. 61:50-51.

LOUIS, G. F., LEWIS, A. J., and PEO, E. R., JR. 1991. Feeding value of grain sorghum for the lactating sow. J. Anim. Sci. 69:223-229

LUCE, W. G., MAXWELL, C. V., HAWKINS, S. E., WEIBEL, D. W., WOLTMANN, M. D., and BULGERIN, L. E. 1988. Nutrient composition of 15 varieties of of Oklahoma sorghum grain. Pages 241-246 in: Okla. St. Univ. Anim. Sci. Res. Rep. 241.

LUIS, E. S., SULLIVAN, T. W., and NELSON, L. A. 1980. The nutritional value of three varieties of proso millet and bird resistant sorghum for poultry. (Abstr.) Poult. Sci. 59:1632.

LUIS, E. S., SULLIVAN, T. W., and NELSON, L. A. 1982a. Nutrient composition and feeding value of proso millets, sorghum grains and corn in broiler diets. Poult. Sci. 61:311-320.

LUIS, E. S., SULLIVAN, T. W., and NELSON, L. A. 1982b. Nutritional value of proso millets, sorghum grains, and corn in turkey starter diets. Poult. Sci. 61:321-326.

LUIS, E. S., SULLIVAN, T. W., and NELSON, L. A. 1982c. Nutritional value of proso millet in layer diets. Poult. Sci. 61:1176-1182.

LUSK, J. W., KARAW, P. K., BALOGU, D. O., and GOURLEY, L. M. 1984. Brown midrib sorghum or corn silage for milk production. J. Dairy Sci. 67:1739-1744.

MALM, N. R., and RACHIE, K. O. 1971. *Setaria* millets: A review of world literature. Neb. Agric. Exp. Stn. Spec. Bull. 513.

MARSHALL, S. P., NORDEN, A. J., ROSS, I. J., and MEYERS, J. M. 1966. Effect of maturity and ensiling procedure on feeding value of sorghum silage. J. Dairy Sci. 49:448-449.

MARSHALL, S. P., SANCHEZ, A. B., SOMERS, H. L., and ARNOLD, P. T. D. 1953. Value of pearl millet pasture for dairy cattle. Fla. Agric. Exp. Stn. Bull. 527.

MARTINEZ, A. 1988. Performance of lambs grazing pearl millet at four levels of herbage allowance. Diss. Abstr. Int. B (Sci. Eng.) 49:578B.

MAXSON, W. E., SHIRLEY, R. L., BERTRAND, J. E., and PALMER, A. Z. 1973. Energy values of corn, bird-resistant and non-bird-resistant sorghum grain in rations fed to steers. J. Anim. Sci. 37:1451-1457.

MAYS, D. A., and WASHKO, J. B. 1962. Refractometric determination of "sucrose equivalent" levels in the stubble of sudangrass and pearl millet. Crop Sci. 2:81-82.

MAYS, D. A., PETERSON, J. R., and BRYANT, H. T. 1966. A clipping management study of two sudangrass-sorghum hybrids, sudangrass and Gahi millet for forage production. Va. Agric. Exp. Stn. Res. Rep. 113.

McCULLOUGH, R. L., DRAKE, C. L., ROTH, G. M., BRENT, B. E., RILEY, R. G., and SCHALLES, R. R. 1972. Several reports on the nutritive value of hybrid sorghum grains.

Pages 15-17 in: 59th Cattlemen's Day. Kansas Agric. Exp. Stn. Bull. 577.

MERKLE, O. G., STARKS, K. J., and CASADY, A. J. 1983. Registration of pearl millet germplasm lines with cinch bug resistance. Crop Sci. 23:601.

MERTZ, E. T., BATES, L. S., and NELSON, O. E. 1964. Mutant gene that changes protein composition and increases lysine content of maize endosperm. Science 145:279-280.

MERTZ, E. T., VERON, O. A., BATES, L. S., and NELSON, O. E. 1965. Growth of rats fed opaque-2 maize. Science 148:1741-1742.

MEYERS, J. H., LOFGREEN, G. P., and ITTNER, N. R. 1959. Alfalfa and sorghum silages. Calif. Agric. 13(1):4114.

MINOR, H. C., MORRIS, C. G., MASON, H. L., MATTAS, R. E., KNERR, D. R., and LAWMAN, E. E. 1987. Forages. 1987 Missouri Crop Performances. Mo. Agric. Exp. Stn. Rep. 369.

MONSON, W. G., HANNA, W. W., and GAINES, T. P. 1986. Effects of rust on yield and quality of pearl millet forage. Crop Sci. 26:637-639.

MONTGOMERY, C. R., NELSON, B. D., JOOST, R., and MASON, L. F. 1988. Tannin concentration and quality changes in sorghum as affected by maturity and sorghum type. Crop Sci. 26:372-375.

MUHAMMAD, D. 1989. Grain yield, forage yield and forage quality of different sorghum types under irrigated and dryland conditions. Ph.D. thesis. Kansas State University, Manhattan.

MYER, R. O., DURRANCE, K. L., GORBET, D. W., WHITE, C. E., and COMBS, G. E. 1982. High moisture grain sorghum in swine growing and finishing diets. Fla. Agric. Res. Rep. MA-1982-5.

MYER, R. O., and GORBET, D. W. 1983. Waxy vs. normal grain sorghums with varying tannin contents in diets for young pigs. Fla. Agric. Res. Rep. MA-1983-3.

NAKUI, T., YAHARA, N., and TAKAI, S. 1985. The effect of the harvesting time, height and length of cutting on the quality and nutrient yield of Japanese millet silage. Pages 899-901 in: Proc. Int. Grassl. Congr. 15th, Kyoto, Japan.

NASS, H. G. 1972. Cyanogenesis: Its inheritance in *Sorghum bicolor*, *Sorghum sudanese*, *Lotus*, and *Trifolium repens*—A review. Crop Sci. 26:372-375.

NELSON, C. C. 1953. Hydrocyanic acid content of certain sorghums under irrigation as affected by nitrogen fertilization and soil moisture stress. Agron. J. 45:615-618.

NELSON, L. A. 1975. Registration of Cerise proso millet. Crop Sci. 15:98.

NELSON, L. A. 1976. Registration of Dawn proso millet. Crop Sci. 16:739.

NELSON, L. A. 1977. Influence of various row widths on yields and agronomic characteristics of proso millet. Agron. J. 69:351-353.

NELSON, L. A. 1981. Registration of Sno-Fox Italian millet. Crop Sci. 21:350.

NELSON, L. A. 1983. Registration of Rise proso millet. Crop Sci. 24:383.

NELSON, L. A. 1986. Producing proso in western Nebraska. Neb. Coop. Ext. Serv. Leafl. EC 86-111.

NELSON, L. A., and FENSTER, C. R. 1983. Seedbed preparation and planter comparisons for proso millet following wheat. Agron. J. 75:9-13.

NELSON, T. S., STEPHENSON, E. L., BURGOS, A., FLOYD, J., and YORK, J. O. 1975. Effects of tannin content and dry matter digestion on energy utilization and average amino acid availability of hybrid sorghum grains. Poult. Sci. 54:1620-1623.

NISHIKAWA, K., TANGE, M., and FUJINAMI, M. 1977. Studies on dry matter production of herbage grasses. 1. The effects of rate of nitrogen application and cutting height on the growth, yield and three essential nutrient elements of African millet (*Eleusine coracana* (L.) Gaertn.). (In Japanese.) Sci. Rep. Fac. Agric. Kobe Univ. 12:183.

NOLAND, P. R., CAMPBELL, D. R., SHARP, R. N., and JOHNSON, Z. B. 1977. Influence of pericarp and endosperm color and type on digestibility of grain sorghum by pigs. Anim. Feed Sci. Technol. 2:219-224.

NORDQUIST, P. T., and RUMESY, G. A. 1967. Corn and sorghum silage for lactating dairy cows. J. Dairy Sci. 50:1254-1261.

NORMAN, M. J. T. 1963. Wet-season grazing of sown pasture and fodder crop at Katherine. CSIRO, Australia, Div. Land Res. Reg. Surv., Tech. Mem. 22.

NORMAN, M. J. T., and BEGG, J. E. 1968. Bulrush millet (*Pennisetum typhoides* (Burm.)) at Katherine, N. T. J. Aust. Inst. Agric. Sci. 34(2):59-68.

NORMAN, M. J. T., and PHILLIPS, L. J. 1968. The effect of time of grazing on bulrush millet (*Pennisetum typhoides*) at Katherine, N. T. Aust. J. Exp. Agric. Anim. Husb. 8:288-293.

NORMAN, M. J. T., and STEWART, G. A. 1964. Investigations on the feeding of beef cattle in the Katherine region, N. T. J. Aust. Inst. Agric. Sci. 30:39-46.

NRC. 1982. United States–Canadian Tables of Feed Composition, 3rd revision. National Research Council/National Academy Press, Washington, DC.

NRC. 1988. Nutrient Requirements of Swine, 9th rev. ed. National Research Council/National

Academy Press, Washington, DC.

OHH, S. J., ALLEE, G. L., BEHNKE, K. C., and DEYOE, C. W. 1983. Effect of particle size of corn and sorghum grain on performance and digestibility of nutrients for weaned pigs. (Abstr.) J. Anim. Sci. 57(Suppl. 1):260.

OWEN, F. G. 1962. Effect of stage of maturity on the nutritive value of Atlas sorghum silage for lactating dairy cows. J. Dairy Sci. 45:625-628.

OWEN, F. G., and KUHLMAN, J. W. 1967. Effect of stage of maturity on the digestibility of sorghum silages. J. Dairy Sci. 50:527-530.

OWEN, F. G., and MOLINE, W. J. 1970. Sorghum for forage. Pages 382-415 in: Sorghum Production and Utilization. J. S. Wall and W. M. Ross, eds. AVI Publ. Co., Westport, CT.

OWSLEY, W. F., KNABE, D. A., and TANKSLEY, T. D., JR. 1981. Effect of sorghum particle size on digestibility of nutrients at the terminal ileum and over the total digestive tract of growing-finishing pigs. J. Anim. Sci. 52:557-566.

PATEL, B. M., SHAH, B. G., PATEL, P. S., and SHUKLA, P. C. 1961. The digestibility and nutritive value of common straws of Gujarat. Indian J. Dairy Sci. 14:12-19.

PEDERSON, J. F., GORZ, H. J., HASKINS, F. A., and ROSS, W. M. 1982. Variability for quality and agronomic traits in forage sorghum. Crop Sci. 22:853-856.

PEDERSON, J. F., HASKINS, F. A., GORZ, H. J., and BRITTON, R. 1983. Variability for traits used to estimate silage quality in forage sorghum hybrids. Crop Sci. 23:376-379.

POPOV, B. V. 1978. Panicum millet for fodder. (In Russian.) Selektsiya i Semenovodstvo. 2:42-43 (Sorghum and Millet Abstracts, 1978, 3:110, 779).

PORTER, K. S., AXTELL, J. D., LECHTENBERG, V. L., and COLENBRANDER, V. F. 1978. Phenotype, fiber composition, and *in vitro* dry matter disappearance of a chemically induced brown midrib (*bmr*) mutants of sorghum. Crop Sci. 18:205-208.

PUND, W. A. 1970. Finishing yearling steers with high energy sorghum silage. Miss. Agric. Exp. Stn. Bull. 780.

POWELL, J. B., and BURTON, G. W. 1971. Genetic suppression of shoot-trichomes in pearl millet, *Pennisetum typhoides*. Crop Sci. 11:763-765.

POWELL, J. D., BEATY, E. R., and YOUNG, W. C. 1977. Registration of Dove proso millet. Crop Sci. 17:978.

PRASAD, R., and SINGH, G. S. 1982. Chemical composition and nutritive value of cheena grain. Indian J. Dairy Sci. 35:96-98.

QUINBY, J. R. 1967. The maturity genes in sorghum. Adv. Agron. 19:267-304.

QUINBY, J. R., and KARPER, R. E. 1954. Inheritance of height in sorghum. Agron. J. 46:211-216.

RACHIE, K. O. 1975. The Millets: Importance, Utilization and Outlook. Int. Crops Res. Inst. Semi-Arid Tropics (ICRISAT), Hyderabad, India.

RACHIE, K. O., and PETERS, L. V. 1977. The Eleusines: A review of the world literature. Int. Crops Res. Inst. Semi-Arid Tropics (ICRISAT), Patancheru, A. P., India.

RAMACHANDRA, G., VIRUPAKSHA, T. K., and SHADAKSHARASWAMY, M. 1977. Relationship between tannin levels and *in vitro* protein digestibility in finger millet (*Eleusine coracana* Gaertn.). J. Agric. Food Chem. 25:1101-1104.

RAYMER, P. L., DAY, J. L., FISHER, C. D., and HEYERDAHL, R. H. 1988. 1987 Field Crops Performance Tests. Ga. Agric. Exp. Stn. Res. Rep. 551.

RICHERT, B. T., HANCOCK, J. D., BRAMEL-COX, P. J., WITT, M. D., KLOPFENSTEIN, C. F., and HEALY, B. J. 1991. Effect of agronomics on yield of nutrients from corn and sorghum: Results from chick assays. Page 112 in: Proc. Bienn. Grain Sorghum Res. Utilization Conf., 17th Lubbock, TX.

RILEY, J. G. 1985. Comparative feedlot performance of corn, wheat, milo and barley. Coop. Ext. Serv. Bull., Manhattan, KS.

RILEY, K. W., GUPTA, S. C., SEETHARAM, A., and MUSHONGA, J. N. 1993. Advances in Small Millets. Oxford and IBH Publishing Co., New Delhi.

RITCHIE, H. D., WOODBURN, D. A., DINGERSON, R. L., HENDERSON, H. E., WOODY, H. D., and STROHBEHN, D. R. 1972. Wintering pregnant heifers in dry lot on two varieties of sorghum silage and two sources of supplemental nitrogen. Mich. Agric. Exp. Stn. Res. Rep. 174.

ROBINSON, R. G. 1976. Registration of Minco proso millet. Crop Sci. 16:884.

ROSS, W. M., GORZ, H. J., HASKINS, F. A., and KOFOID, K. D. 1979. Combining ability in forage sorghum hybrids. Maydica 24:83-93.

ROUQUETTE, F. M., JR., KEISLING, T. C., CAMP, B. J., and SMITH, K. L. 1980. Characteristics of the occurrence and some factors associated with reduced palatability of pearl millet. Agron. J. 72:173-174.

RUSOFF, L. L., ACHACOSO, A. S., MONDART, JR., C. L., and BONNER, F. L. 1961. Relationship of lignin to other chemical constituents in sudan and millet forages. La. Agric. Exp. Stn. Bull. 542.

SABBIONI, A., and SUPERCHI, P. 1985. Chemical composition and digestibility *in vivo* of foxtail millet (*Setaria italica*) hay. (In

Italian.) Atti Soc. Ital. Sci. Vet. 39:437-440.

SANFORD, P. E., CAMACHO, F., KNAKE, R. P., DEYOE, C. W., and CASADY, A. J. 1973. Performance of meat-strain chicks fed pearl millet as a source of energy and protein. (Abstr.) Poult. Sci. 52:2081.

SAPRYKIN, V. 1977. The first cultivar of fodder proso. (In Russian.) Nauchno-Tekh. Byull. Sibirskii Nauchoissl. Inst. Kormov. No. 9:33-39 (from Sorghum and Millet Abstracts, 1978, 3:43, 307).

SEETHARAMA, A., RILEY, K. W., and HARINARAYANA, G., eds. 1989. Small Millets in Global Agriculture. Proc. Int. Small Millets Workshop, 1st. Oxford and IBH Publ. Co., New Delhi, India.

SHARMA, B. D., SADAGOPAN, V. R., and REDDY, V. R. 1979. Utilization of different cereals in broiler rations. Br. Poult. Sci. 20:371-388.

SILVEIRA, C. A. M., SAIBRO, J. C. DE., MARKUS, R., and FREITAS, E. A. G. DE. 1981. Intake, digestibility and nitrogen balance in sheep of pearl millet (*Pennisetum americanum* (L.) Leeke) silage alone or in mixture with cowpea (*Vigna unguiculata* (L.) Walp.). (In Portuguese.) Rev. Soc. Bras. Zootec. 10:361-380.

SINGH, P., SINGH, V., EGGUM, B. O., KUMAR, K. A., and ANDREWS, D. J. 1987. Nutritional evaluations of high protein genotypes of pearl millet (*Pennisteum americanum* (L.) Leeke). J. Food Sci. 38:41.

SINGH, S. D., and BARSAUL, C. S. 1976. Replacement of maize by coarse grains for growth and production in White Leghorn and Rhode Island Red birds. Indian J. Anim. Sci. 46:96-99.

SMITH, R. L., JENSEN, L. S., HOVELAND, C. S., and HANNA, W. W. 1989. Use of pearl millet, sorghum, and triticale grain in broiler diets. J. Prod. Agric. 2:78-82.

STEPHENSON, R. J., and POSLER, G. L. 1984. Forage yield and regrowth of pearl millet. Trans. Kansas Acad. Sci. 87:91-97.

SULLINS, R. D., and ROONEY, L. W. 1974. Microscopic evaluation of the digestibility of sorghum lines that differ in endosperm characteristics. Cereal Chem. 51:134-142.

SULLIVAN, T. W., LUIS, E. S., and NELSON, L. A. 1981. Value of proso millet in layer diets. (Abstr.) Poult. Sci. 60:1741.

SULLIVAN, T. W., DOUGLAS, J. H., ANDREWS, D. J., BOND, P. L., HANCOCK, J. D., BRAMEL-COX, P. J., STEGMEIER, W. D., and BRETHOUR, J. R. 1990. Nutritional value of pearl millet for food and feed. Pages 83-94 in: Proc. Int. Conf. on Sorghum Nutritional Quality. G. Ejeta, E. T. Mertz, L. Rooney, R. Schaffert, J. Yohe, eds. Purdue University, W. Lafayette, IN.

SUPERCHI, P., and SABBIONI, A. 1985. Chemical composition and digestibility *in vivo* of foxtail millet (*Setaria italica*). (In Italian.) Atti Soc. Ital. Sci. Vet. 39:434-437.

THURMAN, R. L., STALLCUP, O. T., and REAMES, C. E. 1960. Quality factors of sorgho as a silage crop. Arkansas Agric. Exp. Stn. Bull. 632.

TRENCHI, H. 1985. Use of foxtail millet (*Setaria italica*) in layer diets. (In Spanish.) Vet. Argent. 2:661-666.

TRIBBLE, L. F. 1975. Grain comparisons for growing-finishing swine. Texas Tech. Univ. Agric. Sci. Tech. Rep. T-5-111.

TRIBBLE, L. F., STANSBURY, W. F., and MCGLONE, J. J. 1987. Feeding value of low bushel test weight sorghum for growing-finishing swine. Texas Tech Univ. Agric. Sci. Tech. Rep. T-5-233.

TRIPATHI, S. N., SINGH, A. P., and MISHRA, M. N. 1979. Effect of nitrogen levels on forage and crude protein yield of Japanese millet. Indian J. Agron. 24:427-428.

VINALL, H. N. 1924. Foxtail millet: Its culture and utilization in the United States. U.S. Dep. Agric. Farmers' Bull. 793.

VINALL, H. N., STEPHENS, J. C., and MARTIN, J. H. 1936. Identification, history, and distribution of common sorghum varieties. U. S. Dep. Agric. Tech. Bull. 506.

VON MUELLER, F. J. H.. 1880. Select extra-tropical plants readily eligible for industrial culture or naturalization, with indications of their native countries and some of their uses. Office of the Superintendent of Government Printing,Calcutta, India.

WARD, G. M., and SMITH, E. F. 1968. Nutritive value of sorghum silage as influenced by grain content. J. Dairy Sci. 51:1471-1473.

WEDIG, C. L., JASTER, E. H., MOORE, K. J., and MERCHEN, N. R. 1987. Rumen turnover and digestion of normal and brown midrib sorghum × sudangrass hybrid silages in dairy cattle. J. Dairy Sci. 70:1220-1227.

WEDIG, C. L., JASTER, E. H., and MOORE, K. J. 1989. Disappearance of hemicellulose monosaccharides and alkali-soluble phenolic compounds of normal and brown midrib sorghum × sudangrass fed to heifers and sheep. J. Dairy Sci. 72:104-111.

WEIBEL, D. E. 1970. Broomcorn. Pages 441-468 in: Sorghum Production and Utilization. J. S. Wall and W. M. Ross, eds. AVI Publ. Co., Westport, CT.

WELLS, H. D., BURTON, G. W., and HENNEN, J. W. 1973. *Puccinia substriata* var. *indica* on pearl millet in the southeast. Plant Dis. Rep. 57:262.

WESTPHALEN, S. L., and JACQUES, A. V. A.

1978. Effects of date of sowing, stage of growth and height of cut on the dry matter and protein yields of pearl millet. 1. Late-flowering cultivar. (In Portuguese.) Agron. Sulriograndense 14:87-105 (from Sorghum and Millets Abstracts, 1979, 4:131, 1002).

WHITE, C. E., RICH, J. R., CAMPBELL, D. R., and SWISHER, M. E. 1983. Performance of swine fed corn or grain sorghum grown in identical field environments. Fla. Agric. Exp. Stn. Res. Rep. SW-1983-3.

WHITE, J., BOLSEN, K., KIRCH, B, and PFAFF, L. 1988. Selecting forage sorghum cutlivars for silage. Pages 177-182 in: 75th Cattlemen's Day. Kansas Agric. Exp. Stn. Res. Rep. 539.

WILKINSON, W. S., and BARBEE, C. 1968. Metabolizable energy value of dehydrated coastal bermudagrass and pearl millet for the growing chick. Poult. Sci. 47:1901-1905.

WILKINSON, W. S., BARBEE, C., and KNOX, F. E. 1968. Nutrient content of dehydrated coastal bermudagrass and pearl millet. J. Agric. Food Chem. 16:665-668.

WU, J. 1984. Effects of particle size of corn, sorghum grain and wheat on pigs performance and nutrient digestibility. Ph.D. Dissertation, Kansas State University, Manahttan.

XIONG, Y., BARTLE, S. J., and PRESTON, R. L. 1990a. Improved enzymatic method to measure processing effects and starch availability in sorghum grain. J. Anim. Sci. 68:3861-3870.

XIONG, Y., BARTLE, S. J., PRESTON, R. L., and MENG, Q. 1990b. Estimating starch availability and protein degradation of steam-flaked and reconstituted sorghum grain through a gas production technique. J. Anim. Sci. 1990. 68:3880-3885.

YUDINA, E. V. 1982. Growing *Panicum* millet for green fodder in Leningrad province. (In Russian.) Byull. Vses. Ord. Lenina Ord. Druzhby Narodov Inst. Rasteniev. Imeni N.I. Vavilova. 116:21-24 (from Sorghum and Millet Abstracts, 1985, 10:98, 770).

ZHUGE, Q., YAN, Z., KLOPFENSTEIN, C. F., and BEHNKE, K. C. 1990. nutritional value of sorghum grain depends on processing. Feedstuffs 26(9):18.

CHAPTER 12

SWEET SORGHUM SUBSTRATE FOR INDUSTRIAL ALCOHOL

R. E. Schaffert
CNPMS/EMBRAPA
Sete Lagoas, Minas Gerais
Brazil

INTRODUCTION

The utilization of biomass as a raw material for transformation into liquid fuels is primarily a function of economics. The transformation of sugar products into ethanol is a centuries old technology utilized in the preservation of foods. This technology has been perfected to a high degree of efficiency by the sugarcane-ethanol industry of Brazil and other countries. The transformation or saccharification of starch products into simple sugars and subsequent fermentations is also a highly perfected technology with a high level of industrial efficiency. The utilization of these technologies to transform sugar and starch biomass into liquid fuels is primarily a question of the interrelation of the price and availability of crude oil and ethanol and the demand and price for food (sugar and starch) products. The transformation of cellulose and hemicellulose into liquid fuels is a much newer technology with a significantly lower level of industrial efficiency.

Sweet sorghum (*Sorghum bicolor* (L.) Moench), because of its juicy, sweet stalks, has been a subject of scientific investigation for over 150 years (Cowley and Smith 1972). Interest in sweet sorghum has been especially important in areas of the world where other sources of sugar are difficult or impossible to produce. The importance of sweet sorghum has also increased during the years when imported sugar supplies have been threatened or interrupted by political or economic problems. Sweet sorghum has been used for over 100 years to produce a concentrated syrup for food use. Sweet sorghums may differ from grain sorghums by only a few genes, those controlling plant height, the presence of juice in the stem, and the presence of sugar in the juice. Sweet sorghum has also been widely used for forage and production of silage for animal feed.

This chapter appeared in an earlier version in *Utilization of Sorghum and Millets.* M. I. Gomez, L. R. House, L. W. Rooney, and D. A. V. Dendy, eds. International Crops Research Institute for the Semi-Arid Tropics (ICRISAT), Patancheru, A. P., India, 1992, pages 131–137, and is used here with permission.

Some varieties of high sucrose sweet sorghum were developed for crystallized sugar production by the U.S. Department of Agriculture (USDA) and the Texas Agriculture Experiment Station. These varieties were developed with the idea of prolonging the milling period of sugar factories, since the milling periods of sugarcane and sweet sorghum complement each other (Cowley 1969). With the development and release of the varieties Rio, Roma, and Ramada, and the development of a method for crystallizing sugar from sweet sorghum juice in the early 1970s (Smith et al 1970), it became possible to produce crystallized sugar from sweet sorghum juice. A pilot industrial run confirmed the viability of the process in the Rio Grande Valley of Texas in the late 1970s. The following year, however, the world price dropped, and the production of sugar from sweet sorghum has remained unprofitable since then.

The oil crises of 1973 and 1976 renewed interest in commercial production of sweet sorghum for biological transformation into ethyl alcohol for use as a fuel and fuel additives (Schaffert and Gourley 1982). In 1975 Brazil created a national alcohol program (PROALCOOL) that resulted in a dramatic increase in alcohol production, principally from sugarcane (Schaffert et al 1986b.) In 1987, nearly 14 million m^3 of alcohol were produced to operate a fleet of approximately 3 million alcolhol motor automobiles. Interest was also generated to produce alcohol from other biomass courses. Sweet sorghum was considered a good choice because it can be processed in sugar-cane mills and the harvest periods of sweet sorghum and sugarcane complement each other, increasing the potential milling period by 3 or 4 months to a 9–10 month milling period.

The great concentration of large distilleries (120 000 to 2 million L day^{-1} capacity) and unavailable land space initially made the idea of processing sweet sorghum for alcohol impractical. However, the idea of using micro-distilleries and mini-distilleries (1000–20 000 L day^{-1} capacity) in outlying areas for fuel self-sufficiency on large farms, for small cooperatives and outlying small communities, awoke interest in the use of sweet sorghum for alcohol production. In central Brazil, the harvesting period for sweet sorghum is from late February to May. Sugarcane is normally harvested from June to November.

There are several advantages to using sweet sorghum instead of sugarcane, as the biomass source for alcohol production.

- Sweet sorghum may be harvested 4 months after planting (whereas the first cut of sugarcane is 18 months after planting).
- Sweet sorghum production can be completely mechanized.
- The sweet sorghum crop is established from seed.
- Sweet sorghum grain may be used as either food or feed.
- The bagasse from sweet sorghum has a higher biological value than the bagasse from sugarcane when used as a forage for animals.

DEVELOPMENT OF SWEET SORGHUM CULTIVARS IN BRAZIL

Because of the strong influence of the sugarcane industry on policymaking in Brazil and the misconception concerning correct sugar levels that arose, the cultivation of sweet sorghum as an adequate biomass source for alcohol production in Brazil was severely delayed in the 1970s.

Breeding Methods for Sweet Sorghum Improvement

In order to successfully carry out a breeding program for crop improvement, the plant breeder must have a thorough understanding of how the end product will be used. In the case of sweet sorghum, whether for sugar or alcohol production, a complete knowledge of the transformation process and the period of the industrial process must be known. In this paper the improvement of sweet sorghum for transformation into alcohol is considered.

The sweet sorghum breeding improvement program in Brazil (Schaffert et al 1986a) was initiated in 1980. It was determined that the stalks and leaves of sweet sorghum could be milled in micro- and mini-distilleries in central and southern Brazil during the season when sugarcane is not normally milled (February to June). The normal planting season for annual crops begins in October and November with the first rains of the cropping year. A 100-day sweet sorghum cultivar planted at the beginning of November could be harvested in February and the ratoon crop could be harvested in June. Considering tbat day length decreases from December to June, the cultivar used must be nonsensitive to photoperiod daylength. Because the stalks and leaves remain green, the separation of the two components is difficult, whether by manual or mechanical methods. The stalks and leaves are therefore milled together. The panicles and grain are removed for their food or feed value and because of possible interference during the extraction of sugars and the subsequent fermentation and distillation processes.

In micro- and mini-distilleries the sugars are extracted from the plant material by crushing, in a three-roller mill or a tandem of three-roller mills, with the liquid phase of the raw material. The juice is then fermented with the brewer's yeast. The fermented juice is subsequently distilled to a minimum of 92% ethyl alcohol for use as a motor fuel or in the alcohol chemical industry (Schaffert and Gourley 1981). The 92% alcohol may be redistilled to produce absolute alcohol for mixing with gasoline or for industrial purposes.

Minimum Quality Requirements

The energy input required during the distillation process requires that the maximum degree of alcohol in the "beer" be obtained during the fermentation process. The biological limit of this process today is between 8 and 9% alcohol. The ideal conversion factor of simple sugars to alcohol is 0.64755. It is therefore necessary to have a minimum sugar content of 12.35% in the juice to produce an alcohol content of 8% under ideal conditions. Normally the juice is diluted to a sugar content of 14–15% for maximum fermentation efficiency. For the purposes of our breeding program, we consider an inferior limit of 12.5% fermentable sugars in the juice. There is no superior limit.

Culture Management

Normally all distilleries in Brazil operate with sugarcane and those that use sweet sorghum will use it for a period ranging from 60 to 120 days (and possibly longer) when the ratoon crop is considered. A cultivar of sweet sorghum normally reaches a peak sugar content about 30 days after flowering and then declines. The decline of sugar content varies from one cultivar to another and from one year to another. In order to have sweet sorghum with a sugar content in the extracted juice superior to 12.5% every day for the programmed harvest period, it is necessary to know the period for industrial utilization (PIU) for each cultivar. In our program we have established a minimum PIU of 21 days for cultural management. It is possible to program a 60- to 120-day or even longer harvest period, either by using cultivars with different maturities or by sowing the same variety at different sowing dates.

Minimum Alcohol Yield

For the sweet sorghum crop to be economically feasible, it is necessary to establish a minimum alcohol yield per t of stalks and leaves as well as a minimum biomass yield. We have considered a minimum of 40 L alcohol t^{-1} biomass (stalks and leaves) for micro- and mini-distilleries, and a minimum biomass yield of 40 t ha^{-1}. Considering 90% efficiency during the fermentation process and 90% efficiency during the distillation process, the total industrial efficiency is 0.9×0.9 or 81%.

The minimum amount of extractable sugar t^{-1} biomass (stalks and leaves) necessary to give an alcohol yield of 40 L t^{-1} with an overall industrial efficiency of 81% is 76 kg. In our program we have considered a minimum of 80 kg of extractable sugars t^{-1} biomass for determining the minimum values for calculating the PIU. The efficiency of sugar extraction would be better in a large distillery (approximately 90–95% instead of 60–65%).

Determining PIU

Each new cultivar developed in our breeding program is characterized by the PIU. We use a hydraulic press that closely approximates the extraction of a three-roller mill (245 kg cm^{-2} for the duration of 1 min). Replicated field samples of a minimum of 10 stalks with panicles removed are harvested at regular intervals (7 or 14 days). Each stalk and leaf sample is run through a forage chopper for maximum open-call preparation. A 500 g sample is pressed and the percentage of extracted juice is calculated. The juice is then immediately analyzed for brix (soluble solids), reduced sugars, and total sugars. Moisture and fiber content are also determined.

A response curve is determined and plotted for each parameter using the regression technique. The possible variation between cultivars and between years is demonstrated in Figures 1 to 4 for juice extraction, fiber, brix, total sugars (juice) and total sugars extracted. The PIU is for the period that extracted sugars are greater than 8 kg per 100 kg biomass.

The year 1981 was characterized by an excessive amount of rain. The variety Brandes is a syrup-type sweet sorghum with low sucrose. It was developed at Meridian, Mississippi by the USDA and Mississippi State University (Coleman and Broadhead 1968). Brandes is photoperiod sensitive and generally produces a large amount of grain. In some environments, it does not meet the minimum standards of the PIU for even 1 day, whereas in other environments it has a PIU of around 30 days. The variety Wray, a high-sucrose cultivar developed by USDA and the Texas Agricultural Experiment Station, is photoperiod nonsensitive and characterized by a small panicle and low grain production. This cultivar has an excellent PIU that in some years is greater than 60 days. The cultivar Wray is generally less productive in biomass production than the cultivar Brandes in periods of long daylength in central Brazil. In periods of short daylengths the production of Brandes is greatly reduced and resembles a grain sorghum cultivar.

New Cultivars

Two new sweet sorghum cultivars for alcohol production, BR 506 and BR 507, derived from the cross of Brandes × Wray, was developed by CNPMS/EMBRAPA

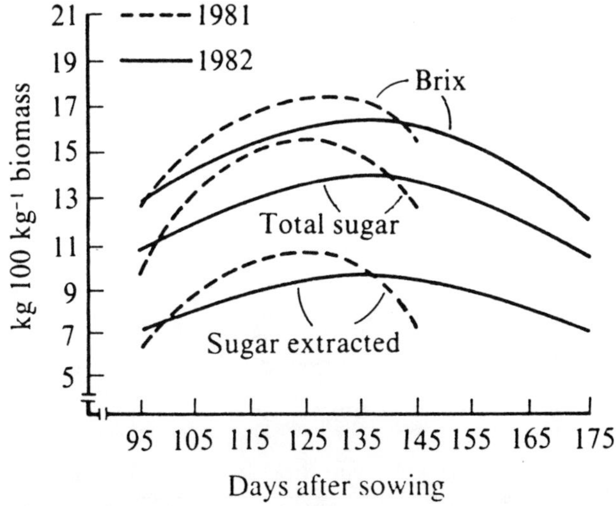

Fig. 1. Brix total sugar and extracted sugars for cultivar Brandes at Araras, CNPMS/EMBRAPA, 1981 and 1982 (Schaffert et al. 1986).

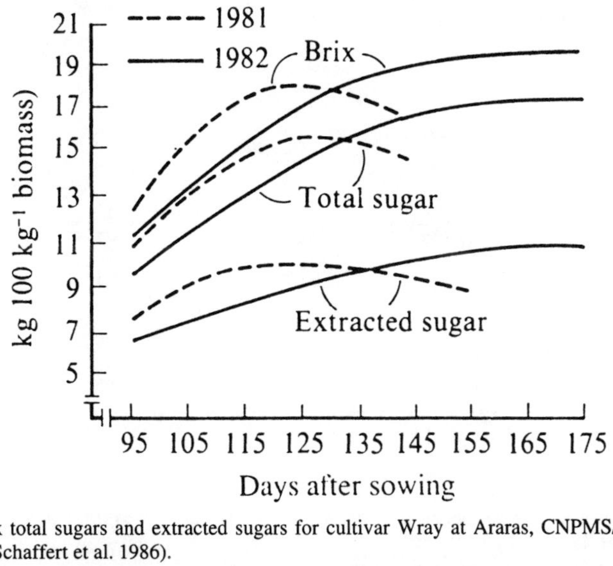

Fig. 2. Brix total sugars and extracted sugars for cultivar Wray at Araras, CNPMS/EMBRAPA, 1981 and 1982 (Schaffert et al. 1986).

for central Brazil. The cultivars, selected for high extractable sugar and a stable PIU (greater than 21 days), were released in 1987. Both varieties are photoperiod nonsensitive, have excellent disease resistance, are intermediate in grain production, and have high biomass production. The PIU of both cultivars is excellent, comparing well with Wray and definitely superior to Brandes. The characteristics are summerized in Table 1 and Figures 5 to 10.

CONCLUSIONS

The emphasis on alcohol production in Brazil and in the USA has been reduced during the past few years, principally because of the cheaper price of crude oil and fuel substitutes, and in the case of Brazil because of over production of alcohol in

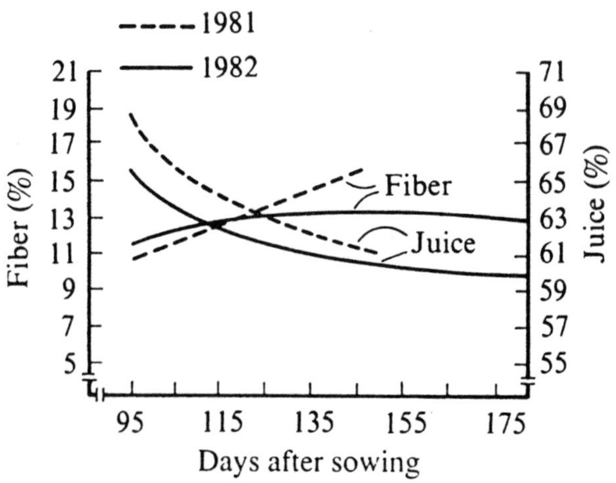

Fig. 3. Percent fiber and juice extraction of cultivar Brandes at Araras, CNPMS/EMBRAPA, 1981 and 1982 (Schaffert et al. 1986).

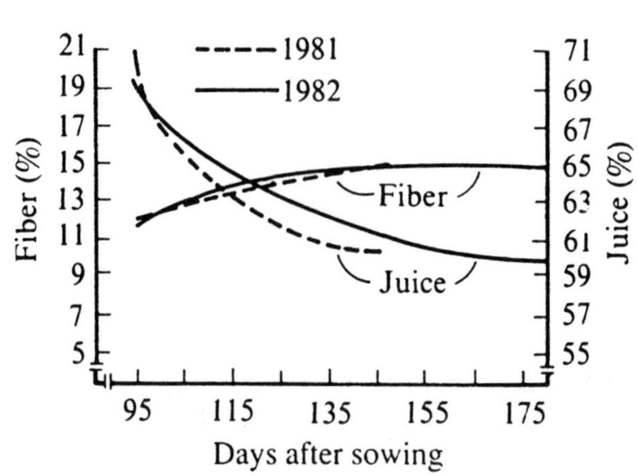

Fig. 4. Percent fiber and juice extraction of cultivar Wray at Araras, CNPMS/EMBRAPA, 1981 and 1982 (Schaffert et al. 1986).

the late 1980s and early 1990s. This in part also reflects reduced demand and price of sugar on the world market. Alcohol can be produced in small- or large-scale distilleries with nearly the same level of efficiency.

EMBRAPA has developed a small-scale horizontal inclined diffusor for extracting residual sugar from bagasse that has been as efficient in test runs in extracting the sugar from biomass as the large-scale distilleries. This in effect reduces the total cost of production as the capital investment of installed liter-per-day capacity is greatly reduced and the transportation distance of the biomass is also reduced.

The production results of the sweet sorghum cultivar Wray, evaluated at the CNPMS/EMBRAPA, indicated that approximately 2500 L ha^{-1} alcohol can be produced from one harvest of sweet sorghum in a small distillery with an 80% transformation efficiency during a 4-month growing period. The ratoon crop may be used either for alcohol production or for animal forage.

In one cycle of sweet sorghum improvement, a gain of 10% has been obtained in alcohol production. In a sweet sorghum improvement program, it is necessary to also select for an adequate PIU besides maximum sugar production. Industrial change from a roller mill to the more efficient technique of diffusion to extract sugars signifies that a change in breeding methodology may be made for selecting for maximum total sugar production.

TABLE 1
Biomass and Alcohol Production of Two New Cultivars, BR 506 and BR 507, Compared to Brandes[a] and Wray

Cultivar	Biomass Production (t ha^{-1})	Extracted Sugar Yield (kg 100 kg^{-1})	Extracted Sugar Yield (t ha^{-1})	Alcohol Yield: Ideal (L ha^{-1})	Alcohol Yield: 81% Efficient (L ha^{-1})	Alcohol Yield: Relative (Wray)
Brandes (BR 501)	47.4	7.2	3.4	2208	1788	75
Wray (BR 505)	44.6	10.9	4.6	2946	2386	100
BR 506	48.8	9.7	4.7	3062	2481	104
BR 507	52.0	9.6	5.0	3231	2617	110

[a]Data from EMBRAPA 1986/87.

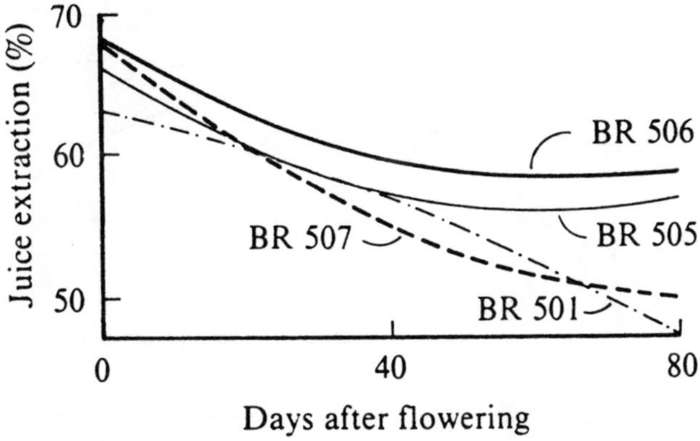

Fig. 5. Percent juice extraction of four sweet sorghum cultivars at CNPMS/EMBRAPA, Sete Lagoas, Brazil, 1986/87. BR 501 = Brandes; BR 505 = Wray.

Most of the earlier varieties evaluated (Brandes, Theis, Rio, Roma, etc.) were inadequate for maximum alcohol production. Adequate cultural practices need to be developed for the nearly released cultivars, including crop ratooning. Photoperiod insensitive cultivars must be used when daylength is declining and during periods of short days. All cultivar development needs to be made within an integrated management system that also utilizes the grain and bagasse.

The policy to explore alcohol production in Brazil using small distilleries (1000-20 000 L day^{-1}) did not develop extensively for several reasons. The early small distillery prototypes were not as efficient as planned, although the later models were greatly improved. The price of crude oil dropped, making other forms of fuel more economical. Also, Brazil actually had a glut of alcohol production when producers worked hard and met the alcohol production targets ahead of schedule. The incentives for establishing micro- and mini-distilleries were consequently reduced.

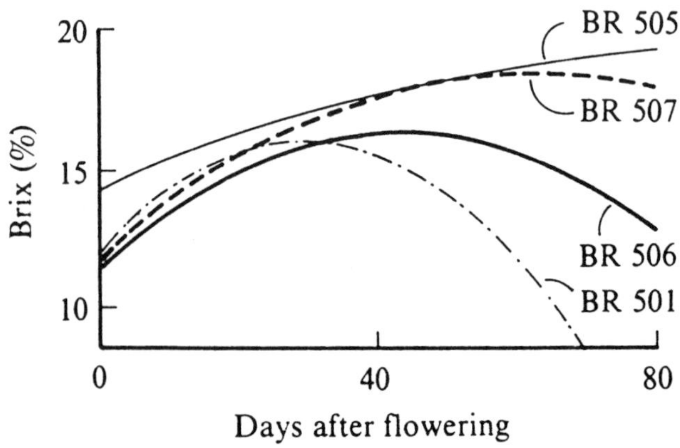

Fig. 6. Percent brix in the juice of four sweet sorghum cultivars at CNPMS/EMBRAPA, Sete Lagoas, Brazil, 1986/87. BR 501 = Brandes; BR 505 = Wray.

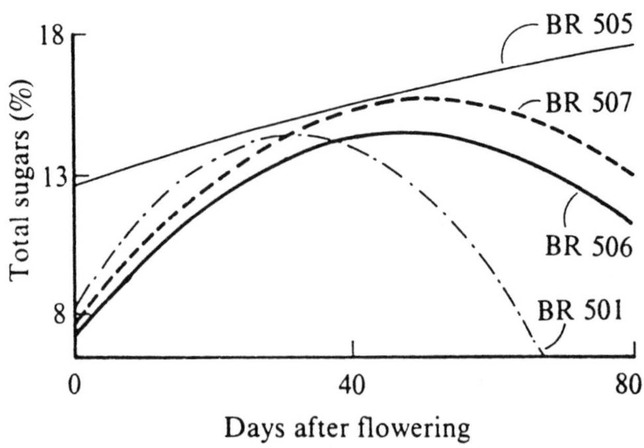

Fig. 7. Percent total sugars in the juice of four sweet sorghum cultivars at CNPMS/EMBRAPA, Sete Lagoas, Brazil, 1986/87. BR 501 = Brandes; BR 505 = Wray.

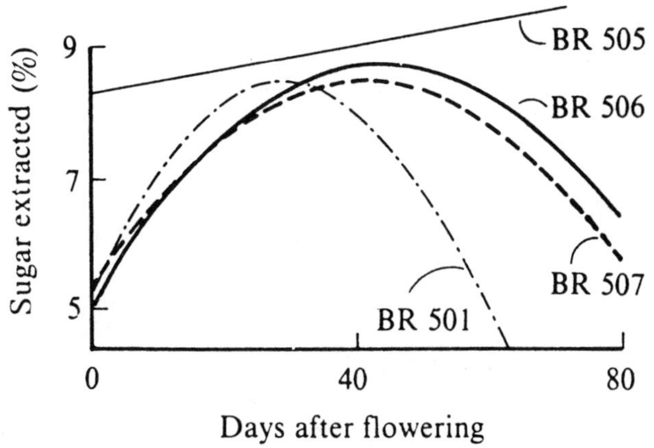

Fig. 8. Sugar extracted (kg 100 kg^{-1} biomass) for four sweet sorghum cultivars at CNPMS/EMBRAPA, Sete Lagoas, Brazil, 1986/87. BR 501 = Brandes; BR 505 = Wray.

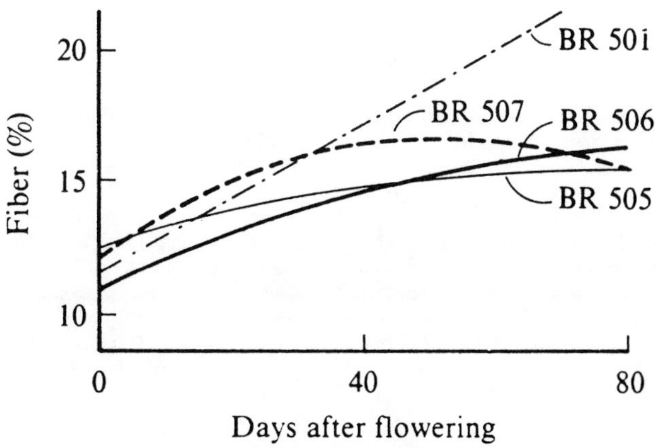

Fig. 9. Percent fiber of fresh biomass of four sweet sorghum cultivars at CNPMS/EMBRAPA, Sete Lagoas, Brazil, 1986/87. BR 501 = Brandes; BR 505 = Wray.

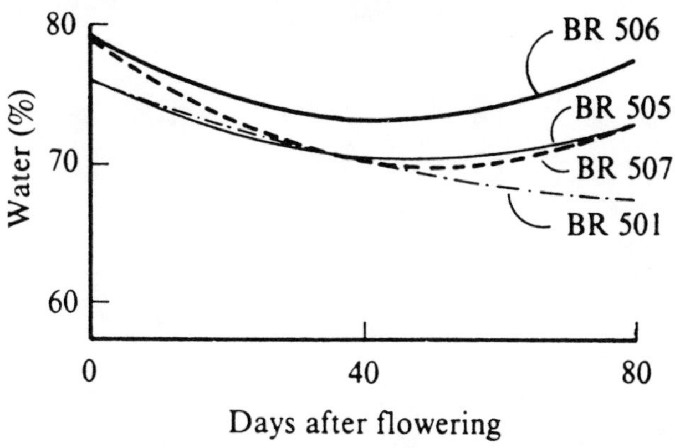

Fig. 10. Percent water in the biomass of four sweet sorghum cultivars at CNPMS/EMBRAPA, Sete Lagoas, Brazil, 1986/87. BR 501 = Brandes; BR 505 = Wray.

RECOMMENDATIONS

The production of alcohol from sweet sorghum, considering only the value of the alcohol at production levels of 2500-3000 L ha^{-1}, will probably never be economical compared to other sources of alcohol products. However, if the value of the subproducts is also considered, the value of sweet sorghum for alcohol production is more viable and sustainable. Normally, 2–3 t of grain and approximately 16 t of bagasse (40 t biomass ha^{-1}) are also produced. In Brazil, in vivo digestibility values of 55% have been obtained in feeding trials of untreated sweet sorghum bagasse. Forms of treatment exist that can improve the digestibility value of bagasse considerably. The bagasse also has value as a soil conditioner in horticultural production and as a source of cellulose.

The total economic value of the various subproducts as well as policy issues must be considered when conducting an economic analysis of the viability of producing alcohol from sweet sorghum.

LITERATURE CITED

COLEMAN, O. H., and BROADHEAD, D. M. 1968. Brandes, a new variety of sweet sorghum for syrup production. Mississippi State University Agriculture Experiment Station Information Sheet 1013. State College, MS.

COWLEY, W. R., and SMITH, B. A. 1972. Sweet sorghum as a potential sugar crop in South Texas. Proc. Congr. ISSCT 14th.

SCHAFFERT, R. E., and GOURLEY, L. M. 1982. Sorghum as an energy source. Pages 605-623 in Sorghum in the eighties: Proceedings of the International Symposium on Sorghum, 2-7 Nov 1981, ICRISAT Center, India. Patancheru, A. P. 502 324, India.

SCHAFFERT, R. E., BORGONOVI, R. A., and CASELA, C. R. 1986a. Metodologia usada para desenvolver novas cultivares de sorgo sacarino. Relatorio Tecnico Anual do Centro Nacional de Pesquisa de Milho e Sorgo. 1980-1884. CNPMS/EMBRAPA, Sete Lagoas, Minas Gerais, Brazil.

SCHAFFERT, R. E., BORGONOVI, R. A., and CASELA, C. R. 1986b. Parametros para selecção de progenies de sorgo sacarino. Relatorio Tecnico Anual do Centro Nacional de Pesquisa de Milho e Sorgo. 1980-1884. EMBRAPA/CNPMS, Sete Lagoas, Minas Gerais, Brazil.

SMITH, B. A., SMITH, R. C., ROMO, R. V., DE LA CRUZ, R. A., and GRIFFTHS, F. P. 1970. Removal of starch from sweet sorghum juices. Sugar J. 32:25-30.

CHAPTER 13

QUALITY EVALUATION AND TRADING STANDARDS

David R. Shipman
Federal Grain Inspection Service
U.S. Department of Agriculture
Washington, D.C.

Michael H. Eustrom
Federal Grain Inspection Service, USDA
Quality Assurance Research Division
Kansas City, Missouri

INTRODUCTION

The quality of grain sorghum is best defined on the basis of its intended end use. Those using grain sorghum for feed have a variety of quality requirements depending on whether they are feeding livestock or poultry. These quality requirements differ further for those relying on grain sorghum as a food ingredient. The need to store grain sorghum prior to feed or food use introduces further quality considerations. However, regardless of how or when grain sorghum is used, its postharvest quality does not improve. Harvesting as well as subsequent handling, drying, transportation, and storage represent potential causes for quality deterioration.

Seed selection, agronomic practices, and the environment influence the physical condition and composition of grain sorghum prior to harvest. Heavy rainfall after the kernels have matured may cause mold growth, sprouting, discoloration, and a general weathered appearance. Inadequate soil nutrients or adverse environmental conditions may result in underdeveloped kernels. The kernels are further subjected to physical trauma and the introduction of foreign material during harvesting. Kernels may be shattered by the combine reel or cracked as the result of high cylinder speed. Foreign material will remain with the grain to varying degrees depending on crop condition and harvesting techniques.

Consequently, by the time grain sorghum enters the commercial market, its quality may vary significantly. The challenge of the marketplace is to assess the value of the grain sorghum and move it on to the ultimate users or consumers in the most efficient manner available. This is accomplished, in part, through the use of grading standards and an inspection system. This chapter addresses the relationship of quality and trading standards for grain sorghum. Only minor mention is made of millet, a grain produced in the United States but not marketed at a level requiring national standards.

MARKETING

Grain has been marketed in the United States for over 150 years. During this time, grain merchandisers continuously adjusted marketing practices to keep pace with the advancing production, harvesting, handling, and transportation technologies. As grain production and marketing became more sophisticated and widespread, a need for a more structured means of assessing and describing grain quality evolved. This need ultimately led to the passage of the United States Grain Standards Act in 1916 and the creation of the U.S. Standards for Grain.

Early on, buyers were able personally to examine the quality and condition of grain to determine its value. However, the westward expansion of the grain-producing areas created new and distant grain distribution centers, preventing buyer and seller from simultaneously examining the grain and mutually agreeing on its value. The inability of the seller to display the quality of grain being offered for sale and the inability of the buyer to examine the grain to determine its suitability for its intended use created marketing problems. To facilitate the process, a common language was needed to express the class, quality, and condition of the grain (Murphy, 1955).

Early attempts to resolve the problem through the use of standards failed because of the ambiguity and number of standards in effect. Various states and grain trade organizations created their own sets of standards and their own methods of application. Terms, such as "dry," "well-cleaned," "unsound," and "unmerchantable," were used to describe the quality and condition of grain. The use of these nondescript terms, along with the application of multiple grading standards, created intermarket problems because there were no guarantees that the quality designation in one market would be the same in another (Combs and Smith, 1940). Towards the turn of the century, the National Grain and Feed Association (formerly known as the Grain Dealers National Association) and other trade groups, discouraged by the multiple grading standards in effect, campaigned for the adoption of voluntary, national grade standards. Eventually, the inefficiencies created by the use of multiple, nondescript standards led to legislative intervention (Corbett et al, 1936). On August 11, 1916, the Grain Standards Act was signed and authorized the Secretary of Agriculture to establish U.S. standards for grain and provide for a uniform inspection and grading system that would promote the orderly and timely marketing of grain in interstate and foreign commerce.

GRAIN STANDARDS

The marketing standards that have been developed by the United States are not the only standards by which sorghum is marketed worldwide. Other countries have developed their own sets of standards with which to trade sorghum in foreign and domestic commerce, most notably, Argentina and Australia. The Argentine standards for sorghum divide quality into three numerical grades based on the percentage of damaged kernels, foreign material and nonsorghum grains, and broken kernels. Australian standards, at least for sorghum exports, establish a single set of quality parameters that includes a minimum kilogram per hectoliter level and maximum limits on moisture content, broken kernels, foreign material, and damaged kernels. International standards have also been or are being developed. An international standard for sorghum has been drafted by the International Organization for Standardization, and a worldwide standard for sorghum destined for human consumption has been established by the Codex Alimentarius Commission (standard 172-1989). The Codex standard establishes one acceptable quality level with limits set on moisture, ash, protein, fat, tannin content, and total defects which includes diseased kernels, broken kernels, foreign material, and other grains.

The U.S. standards allow producers, intermediate users, and end users to establish value and determine the benefits and risks involved in marketing by dividing sorghum quality into five grades. Sorghum may be designated grade Nos. 1, 2, 3, 4, or Sample grade, with No. 1 being of highest quality and Sample grade the lowest quality designation (Table 1). The grade is determined by the lowest quality factor. For example, if the test weight of a sample is 54.5 lb/bu, it will grade no higher than No. 3 even though it may be No. 1 on all other factors. The factors contributing to the quality determinations in sorghum include test weight per bushel, broken kernels and foreign material (BNFM), foreign material (FM), heat-damaged kernels, damaged kernels total, odor, class, special grades, and Sample grade criteria. In addition to these "grade determining" factors, each grade certificate must include the percentage of moisture content.

TABLE 1
Grades and Grade Requirements for Sorghum[a]

	Grades, U.S. Nos.[b]			
Grading Factors	**1**	**2**	**3**	**4**
Minimum pound limits				
Test weight, lbs/bu	57.0	55.0	53.0	51.0
Maximum percent limits				
Damaged kernels				
Heat (part of total)	0.2	0.5	1.0	3.0
Total	2.0	5.0	10.0	15.0
Broken kernels and foreign material				
Foreign material (part of total)	1.5	2.5	3.5	4.5
Total	4.0	7.0	10.0	13.0
Maximum count limits				
Other material				
Animal filth	9	9	9	9
Castor beans	1	1	1	1
Crotalaria seeds	2	2	2	2
Glass	1	1	1	1
Stone[c]	7	7	7	7
Unknown foreign substance	3	3	3	3
Cockleburs	7	7	7	7

U.S. Sample grade
Sorghum that
1) Does not meet the requirements for U.S. Nos. 1, 2, 3, or 4; or
2) Has a musty, sour, or commercially objectionable foreign odor (except smut odor); or
3) Is badly weathered, heating, or distinctly low quality.

[a]Source: Federal Grain Inspection Service, U.S. Department of Agriculture, 1993.
[b]Sorghum that is distinctly discolored shall not grade higher than U.S. No. 3.
[c]Aggregate weight of stones must also exceed 0.2% of the sample weight.

Although the structure and terminology may differ between standards from different industry groups or countries, they all share a common goal of providing a reliable and effective means to communicate quality. For the most part, standards divide quality into two basic categories: soundness and purity.

Soundness pertains to the general condition of the grain and includes such determinations as test weight per bushel, moisture content, and color. Soundness also refers to the condition of the kernels. Are the kernels broken or cracked? Have they been invaded by molds or insects? Has deterioration of the natural kernel structure taken place? Purity refers to material or substances other than the natural kernel. Chaff, stems, weed seeds, other grains, stones, and insects are impurities.

The composition of grain or the inherent makeup of the kernels is a quality attribute rarely addressed in marketing standards. The amount of protein, oil, fiber, fat, and starch contained in the grain are examples of composition.

Proper application of the grain standards influences market efficiency. Simply providing merchants with an understandable means of communicating grain quality does not promote efficient marketing. Improper application of a test will alter the results and create inconsistencies that may impede rather than facilitate the orderly and timely marketing of grain. In the United States, a national inspection system is used to avoid these shortcomings. The system provides the grain industry with an independent, third party to assess quality and apply the grain standards. Federal, state, and private inspectors use uniform sampling and testing procedures as well as approved equipment to grade grain.

SAMPLING

Sampling represents perhaps the most important aspect of determining sorghum quality. The precision and accuracy of any quality analysis is dependent on the adequacy of the sample tested.

Sampling Variability

When grain is handled, segregation of particle sizes occurs, which introduces a source of nonrandom variation. This is true for both static and in-motion grain. As grain comes to rest in a container, the fine particles concentrate near the center of the grain mass while the coarse particles flow to the outside. On conveyer belts, the fine material concentrates on the bottom while the coarse material tends to collect on the outer edges and on top of the grain stream. Experience shows the best sample is obtained from in-motion grain that permits a complete cross-sectioning of a falling stream of grain. When sampling grain at rest, extreme care must be taken to avoid misrepresenting factors that are influenced by particle size. Figures 1 and 2 exemplify typical differences found in the level of broken corn and foreign material in trucklots and carlots of corn. While the uneven distribution of grain may create more problems for factors that are influenced by particle size, they are not the only factors affected. Factors such as test weight, moisture, and damaged kernels are also affected.

The variations associated with the uneven distribution of grain are compounded by the unavoidable random variation that occurs whenever results are based on anything less than the entire lot. To illustrate the effects of random variation, Hill et al (1976) created three 1,000-g samples of corn, each containing a different number of colored corn kernels (four, five, and 15, respectively). Using a Boerner divider, each sample was divided into 250-g subsamples, and the number of colored kernels (which could represent damaged kernels) was counted. Subsamples were randomly selected and inspected 100 times for each of the three levels. Table 2 shows the number of colored kernels appearing in a sample and the number of times each possibility appeared.

How well a sample represents the true quality of grain depends on the number and location of subsamples taken from a given quantity of grain. Since testing an entire grain shipment (each kernel) is impractical, sampling methods and procedures are designed to optimize accuracy and cost effectiveness.

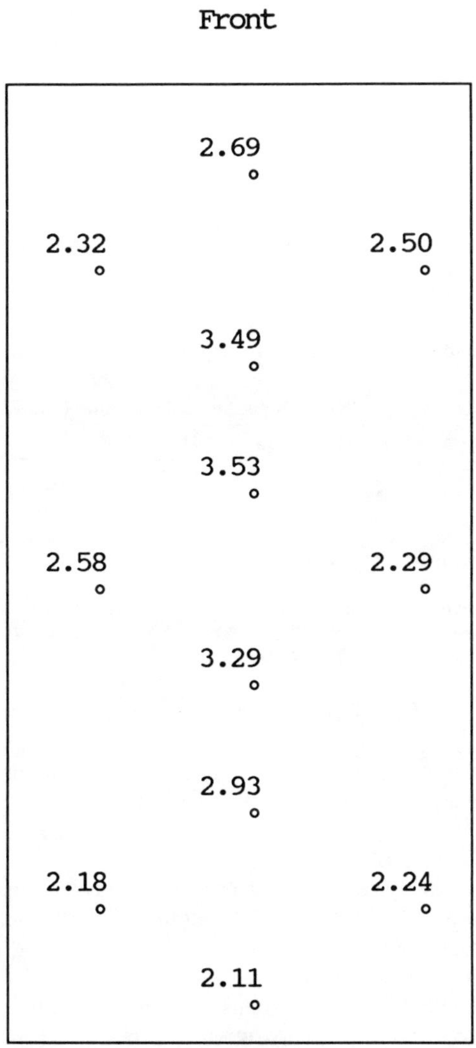

Fig. 1. Distribution of broken corn and foreign material percentages in 51 straight-truck lots of corn. Source: Hurburgh (1990); used by permission.

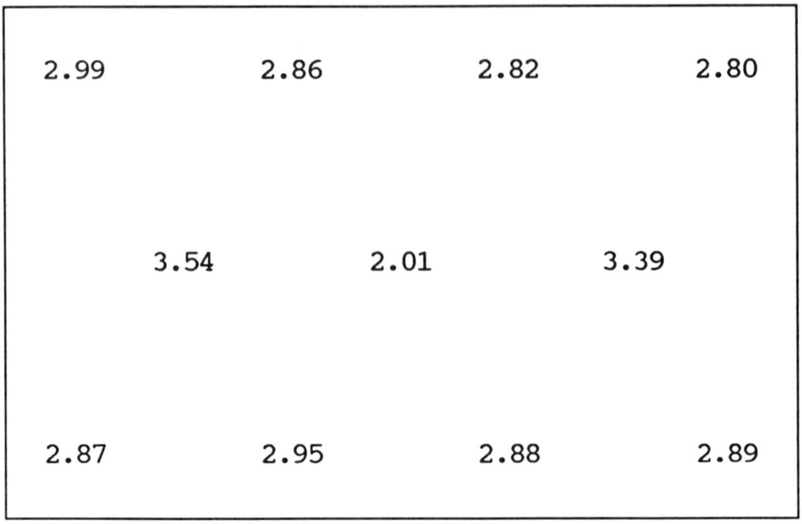

Fig. 2. Average percentages of broken corn and foreign material by probe for 50 carlots of corn. Data from the Federal Grain Inspection Service of the U.S. Department of Agriculture.

TABLE 2
Distribution of Colored Kernels in Repeated Samplings of a 1,000-g Population (100 trials)[a]

1,000-g Lot		No. of Subsamples Containing Colored Kernels[b]									
Colored Kernels	Percent of Lot	0	1	2	3	4	5	6	7	8	9–15
4	0.14	30	36	23	11	0	...	...	...	...	...
5	0.18	23	37	29	10	1	0	...	...	...	...
15	0.53	0	6	16	32	17	11	11	6	1	0

[a]Source: Hill et al (1976); used with permission.
[b]A 250-g sample was randomly selected from the 1,000-g population containing the number of colored kernels shown in column 1. The number of colored kernels appearing in a sample and the number of times each possibility appeared are shown on the right side of the table. Since the process was repeated 100 times, these numbers can also be read as percentages.

Sampling Methods and Devices

Producers, elevator operators, millers, and others in the grain industry use various methods and devices for sampling both static and in-motion grain.

SAMPLING GRAIN AT REST

Grain shipped in trucks, railcars, and barges and grain stored in bins is often sampled while static in the carrier or bin. The only effective methods of obtaining representative samples from static grain are with either a hand probe (sometimes referred to as a trier) or a mechanical probe.

Hand probes are made to varying lengths from brass or aluminum and consist of two tubes, one inside the other. Compartmented probes have an inner tube that is divided into compartments (Fig. 3). The outer tube has slots that match the compartment openings of the inner tube. When the tubes are aligned, grain flows into

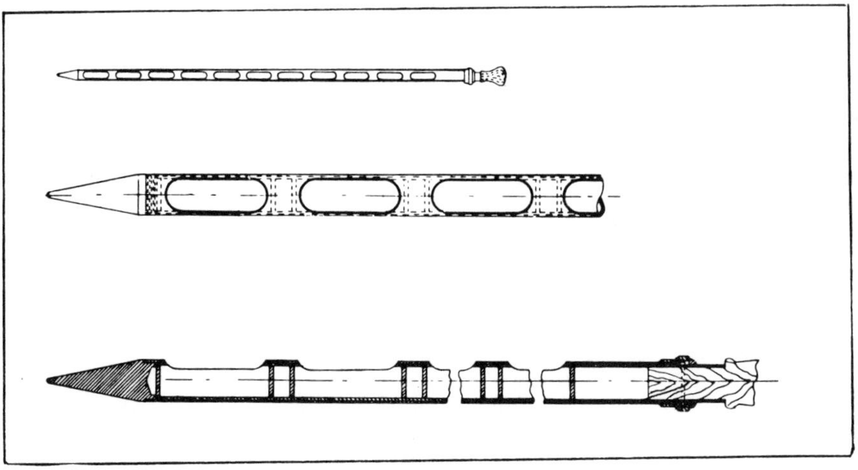

Fig. 3. Double-tube compartmented probe.

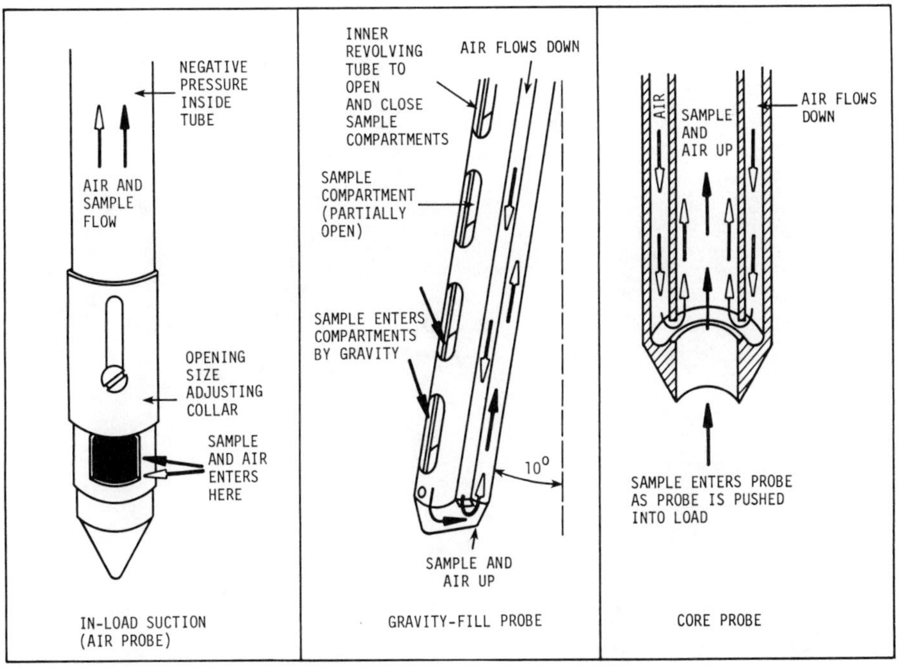

Fig. 4. The three types of mechanical probes.

the compartments. This design facilitates the sampling and examination of grain from all levels of the grain. Open-throat probes have an inner tube that is open, as is the probe handle. This feature allows the sample to be poured from the probe directly into the sample container. However, open-throat probes tend to draw more sample from the upper portions of the grain mass. Therefore, the composition of a sample obtained with an open-throat probe may differ slightly from that of a sample drawn with a compartmented probe.

Mechanical probes (Fig. 4), like hand probes, are used to sample static grain in trucks and other open-top carriers. There are three types of mechanical probes. The gravity-fill probe and the core probe work well. A third type, the in-load suction probe, tends to over sample fine material, such as chaff, stems, small weed seeds, and pieces of broken kernels. Like the manual probe, the gravity-fill probe is a double-tube, compartmented probe. The compartments are filled by gravity, and the contents are pneumatically transported to a collection box. The core probe

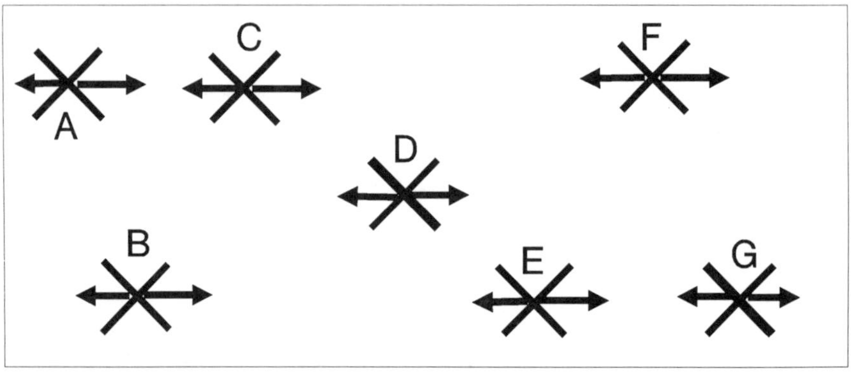

5

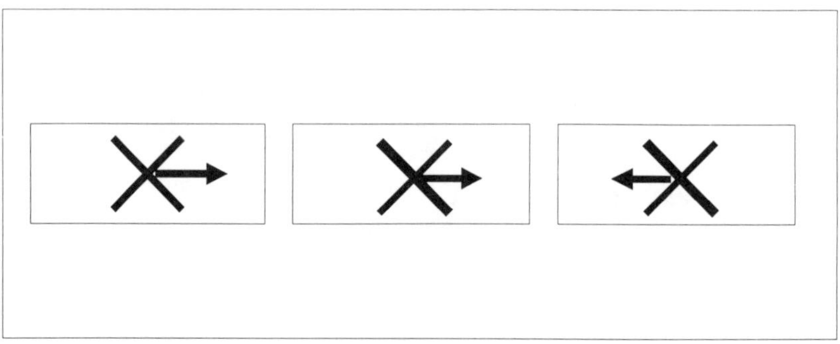

6

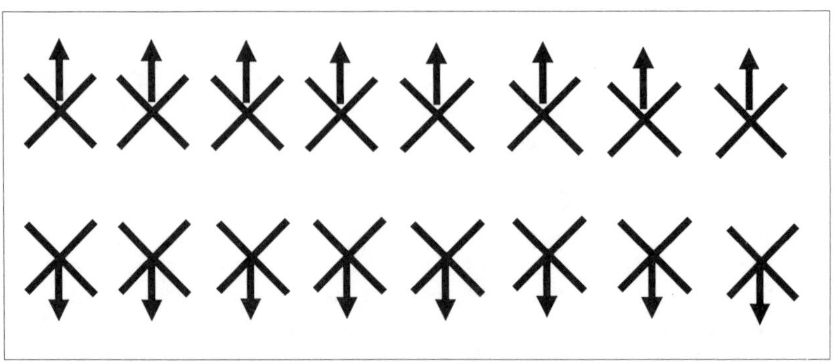

7

Figs. 5-7. Sampling pattern used by the Federal Grain Inspection Service.

consists of concentric tubes that are open at the bottom. Air flows down between the inner and outer tube and, by means of a vacuum, moves upward through the inner tube. As grain is introduced into the tip of the probe, it is pneumatically delivered to a collection box.

The number and location from which probe samples are obtained depends on 1) the design of the container or carrier being sampled, 2) the amount of sample desired, and 3) the general condition of the grain. Figures 5, 6, and 7 illustrate some of the sampling patterns used in the United States by the Federal Grain Inspection Service (FGIS). If a general observation of the grain reveals an unusual condition, such as heating, or sour or musty grain, additional probe samples of the questionable grain should be taken. The grain depth governs the length of probe used. Generally, a 6-ft probe is used to sample trucks, a 10-ft probe for hopper railcars, and a 12-ft probe for barges.

SAMPLING GRAIN IN MOTION

Diverter-Type Mechanical Samplers. Diverter-type (DT) mechanical samplers are used to sample grain moving through grain spouts or off of the end of conveyor belts (Fig. 8). DTs draw a sample by periodically moving a pelican-like device through the entire grain stream. The movement of this device is electrically timed and powered by an air cylinder or electric motor. The frequency of cross-sectioning the grain stream is governed by the grain flow-rate. If the flow-rate is 4,000 bu or less per hour, the timer is typically set to cross-section the grain once every 3 min. If the flow-rate is faster than 4,000 bu per hour, the timer is set to sample the grain every 200 bu. The DT sampler is well suited for most grain handling facilities. If

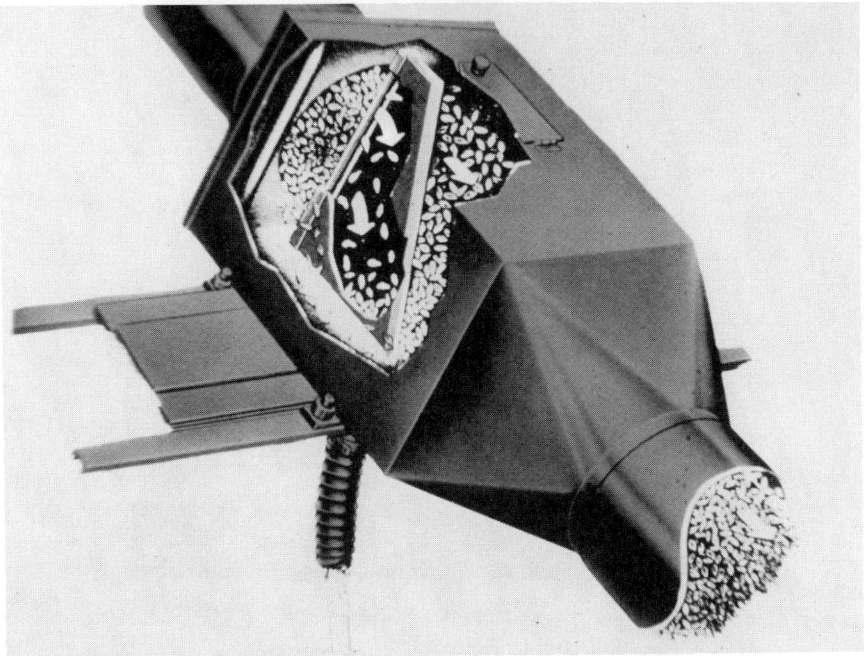

Fig. 8. Diverter-type mechanical sampler.

necessary, the sample obtained from the DT is subdivided using a mechanical secondary sampler or manual cargo divider (Fig. 9).

Pelican. The pelican sampling device, like a DT sampler, is designed to sample a falling grain stream. However, the pelican is a manual sampling method limited to use in areas where a falling grain stream is accessible (e.g., loading spout and for slow flow-rates). Typically, the pelican pouch is made of reinforced leather with an opening 18-in. long, 6-in. deep, and 2-in. wide (Fig. 10) and is attached to a long pole. A metal band along the edge of the pouch holds it open. A cross-section of the grain is obtained approximately once every 500 bu by swinging the pelican through the grain stream in one continuous motion.

Truck (Tailgate) Samplers. Truck samplers, which are simple metal containers attached to wooden handles, are used to sample grain being unloaded from trucks. A truck sampler is used in the same manner as a pelican sampler; i.e., pulled or

Fig. 9. Cargo divider.

swung through the falling stream of grain. And, like pelicans, truck samplers work best when the grain is falling at a moderate rate. The number of samples that should be drawn with the truck sampler depends on the amount that is needed and the amount of grain being discharged from the truck. Samples should be taken at random intervals by holding the truck sampler next to one side of the falling stream and pulling it through the grain stream in one continuous motion.

Ellis Cup. The Ellis cup is a manual sampling device constructed of light weight aluminum and used to sample grain moving on a conveyor belt (Fig. 11). Subsamples are obtained by inserting the heel of the Ellis cup into the grain stream and then, with the open-ended side of the cup facing the opposite direction of the grain flow, quickly lowering the opening into the grain until the bottom is parallel with the belt. A set of subsamples (center and both sides of the grain stream) should be taken approximately once every 500 bu.

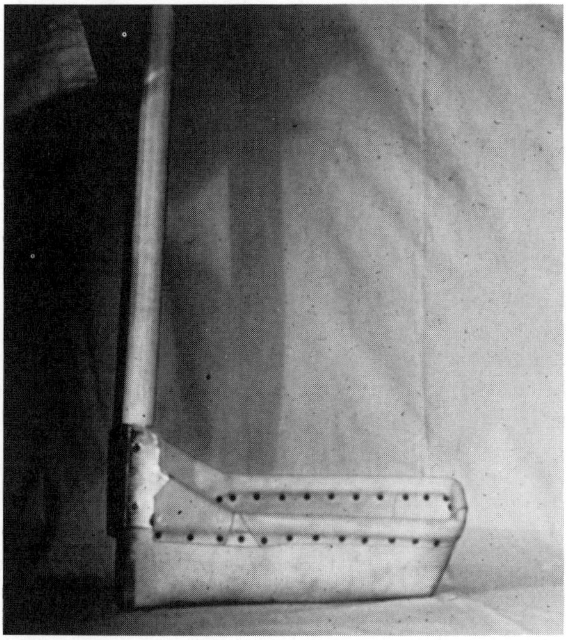

Fig. 10. Pelican.

Fig. 11. Ellis cup.

Fig. 12. Woodside sampler.

Woodside-Type Mechanical Samplers. The Woodside sampler mechanically samples grain moving on a conveyer belt (Fig. 12). Small sample buckets attached to three chains (center and both sides of grain stream) are used to collect the samples. The chains are connected to an upper and lower sprocket. A special roller, mounted approximately 10 in. from the lower sprocket, serves as the drive shaft. As grain passes over the roller, it becomes suspended in the air and falls into the sample buckets. The full buckets empty into a collection hopper.

INSPECTION PROCESS

The methods of analysis used by different countries to apply grain standards often differ. In many cases, these differences influence the very definition of quality. This is especially true for quality factors involving particle size and density separations. The method of analysis for any given quality attribute should be selected on the basis of accuracy, precision, sensitivity, and practicability, with preference given to methods which have applicability for routine use.

This section focuses on how the U.S. standards for sorghum and related inspection methods address sorghum quality. Comparisons to other country and international standards are provided for certain quality attributes.

Sorghum Classes

The U.S. standards divide sorghum (*Sorghum bicolor* (L.) Moench) into four classes: White, Sorghum, Brown, and Mixed. The distinction between classes is based on the pericarp color and, in the case of Brown sorghum, the presence of a pigmented subcoat. In most incidences, a general review of a sample is sufficient to determine the appropriate classification. However, when an analysis is necessary, 25 g of cleaned sorghum (dockage and machine-separated broken kernels and for eign material removed) is examined, and the percentage of sorghum types is determined. Differentiating between the Sorghum and Tannin sorghum on the basis of color alone is, at times, complicated by the similarities that occur in the pericarp coloring. Depending on the sorghum varieties present, it may be necessary to remove the pericarp (scrape or dissolve with a mild solvent) and determine whether or not a pigmented subcoat (testa), which is unique to Tannin sorghum varieties (Hahn et al, 1984), is present.

CLASSIFICATION

In 1993, the U.S. Department of Agriculture revised the classification terminology and corresponding definitions in the U.S. standards for sorghum to more clearly discriminate between the two classes Sorghum (formerly Yellow Sorghum) and Tannin Sorghum (formerly Brown Sorghum). The revisions seek to minimize the confusion caused by relying solely on color as a means to distinguish the types of sorghum. The different classes are defined in the U.S. standards (FGIS, 1993) in the following manner:

Sorghum. Sorghum that is low in tannin content due to the absence of a pigmented testa (subcoat) and contains less than 98.0% White sorghum and not more than 3.0% Tannin sorghum. The pericarp color of this class may appear white, yellow, pink, orange, red, or bronze.

Tannin Sorghum. Sorghum that is high in tannin content due to the presence of a pigmented testa (subcoat) and contains not more than 10.0% non-Tannin sorghum. The pericarp color of this class is usually brown but may also be white, yellow-pink, orange, red, or bronze.

White Sorghum. Sorghum that is low in tannin content due to the absence of a pigmented testa (subcoat) and contains not more than 2.0% sorghum of other classes. The pericarp color of this class is white or translucent and includes sorghum containing spots that, singly or in combination, cover 25.0% or less of the kernel.

Mixed Sorghum. Sorghum that does not meet the requirements for any of the classes Sorghum, Tannin sorghum, or White sorghum.

Physical Condition

GENERAL

Aside from a subjective observation of the condition of sorghum, the two most frequently used measurements to determine the physical quality of sorghum are test weight per bushel, or the equivalent in kilograms per hectoliter, and moisture content. Secondary to these analyses are analyses for factors such as the percentage of broken kernels and foreign material, foreign material, damaged kernels, and unusual conditions.

Under the U.S. inspection process, sorghum with an unnatural odor or an odor that is indicative of spoilage, or sorghum containing stones, animal filth, toxic substances, or other inferior conditions, is labeled Sample grade, the lowest grade designation. Special grades are used to denote special quality considerations in sorghum. If live insects injurious to grain are present, sorghum is designated the special grade Infested. Likewise, if sorghum is infested with smut spores, it is designated the special grade Smutty.

MOISTURE

The moisture content of grain is a critical factor from the time of harvest until final processing into feed or food. Moisture content governs the stability of sorghum; is used as the basis for reporting other quality attributes, such as percent protein; and influences sorghum value as the commercial market adjusts price to compensate for moisture loss, drying costs, and storage risks. Consequently, a reliable and accurate means of measuring moisture throughout the marketing process is essential.

Several different methods of measuring the moisture content of grain are employed throughout the world. These methods vary in accuracy, precision, and practicality. Multon (1991) categorized the methods as follows:

1) The "basic" or "absolute" methods, which provide a fundamental reference; these standard methods are employed in specialized laboratories to calibrate the reference methods or as ultimate measurements of moisture content.
2) The methods for "practical reference," which are easily performed and calibrated with respect to the absolute methods.
3) The so-called "rapid methods," intended for on-site use, are thermal or electrical and calibrated with respect to the practical reference methods.

In the United States and most other countries, dielectric moisture meters are most widely used in grain merchandising. Other rapid, practical methods that are being used include wide-line nuclear magnetic resonance spectroscopy, rapid automated drying ovens, and near-infrared spectroscopy (Watson, 1987).

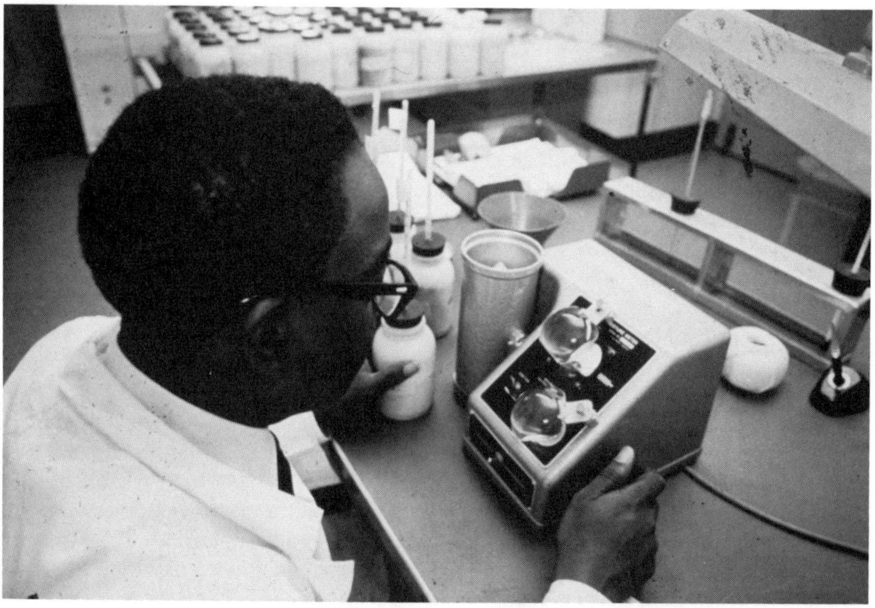

Fig. 13. Motomco moisture meter.

FGIS has approved and currently uses the Motomco moisture meter, model 919 (Fig. 13). This moisture meter is calibrated to provide equivalent results to the USDA air-oven reference method, which involves drying a 2- to 3-g ground sample of sorghum for 1 hr at 130°C in an air oven. To ensure meter accuracy, FGIS annually reverifies the calibration equations.

Moisture analysis using the Motomco moisture meter requires a representative portion of exactly 250 g of sorghum before the removal of dockage. Conversion charts, which have been standardized for a grain temperature of 77°F, have been developed to convert the meter reading to a percentage of moisture. A correction factor is applied if the grain temperature deviates from 77°F.

Market standards recognize the importance of moisture content by setting maximum limits. The limits typically reflect the environmental conditions under which the sorghum is produced, marketing trends or patterns, and final end-use requirements. The Codex standard for sorghum sets a 14.5% maximum limit, Argentina a 15% limit, and Australia a 14% limit. The U.S. market relies on a 14% limit for trading purposes, but the Official U.S Standards for Sorghum no longer include a maximum moisture limit. Prior to 1985, the U.S. standards included moisture limits for each grade. The U.S. Nos. 1, 2, 3, and 4 grades for sorghum had maximum moisture limits of 13.0, 14.0, 15.0, and 18.0%, respectively. In September 1985, these requirements were eliminated because they served no useful marketing

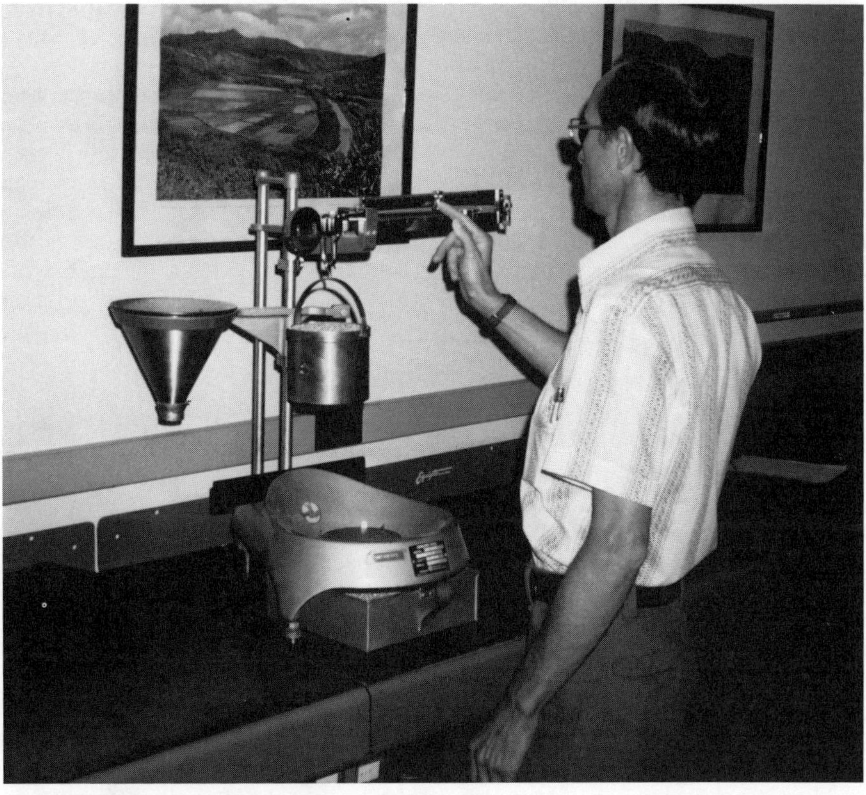

Fig. 14. Test weight per bushel apparatus.

purpose. The market uses 14% moisture as a benchmark and assesses price adjustments based on the actual moisture to account for weight loss, drying cost, and handling risks.

TEST WEIGHT

Test weight per bushel is the weight of the volume of sorghum required to fill a Winchester bushel (2,150.42 in.^{3}) to capacity. The apparatus (Fig. 14) used for this determination approximates the weight of a bushel of grain. The test is performed on a sample before the removal of dockage and requires a sample size of approximately 1,000–1,050 g. The weight of the grain filling the laboratory kettle is converted to the test weight per bushel by either placing the filled kettle on the calibrated beam as depicted in Figure 14 or by simply weighing the contents and applying a conversion formula.

Test weight is a measure of bulk density and is generally used by the industry to estimate the amount of grain that can be stored, transported, or processed (steeping-tempering bins) in a particular container. Test weight also serves as a general indicator of overall quality and end-product value. Sorghum processors use the information as a qualitative yardstick to determine the relative condition of the sorghum with respect to maturity (plumpness), physical and biological damage, and foreign material. Test weight results are negatively impacted by immature sorghum kernels or by excessive amounts of foreign material or damaged kernels. The sorghum wet-milling industry has found that sorghum with plump kernels, high test weight, and minimum damage provides optimum processing efficiency (Watson, 1970).

The relevance of test weight per bushel for sorghum used as feed has been challenged in the past. Sorghum feeding trails show only small differences between the feeding value of high test weight sorghum and low test weight sorghum. Deyoe (1975) concluded, "...test weight should not be used to indicate feeding value except under those conditions where test weights are reduced below approximately 45 pounds per bushel. Grain with less than 45 pounds per bushel may have reduced feeding value for poultry and swine due to lower energy but will have a higher quality protein. Cattle appear to utilize low test weight sorghum (under 40 pounds per bushel) efficiently."

SOUNDNESS AND PURITY

A buyer of sorghum prefers to receive only whole, mature kernels free from disease, insect infestation, fungi, and bacteria. The value of the sorghum may vary significantly depending on the type of defects present or the composition of the impurities. Consequently, the quantity and composition of defects and impurities are important quality criteria. The methods of analysis for defects and impurities are influenced by the degree of precision and separation of material desired.

Dockage. The U.S. inspection process defines dust, chaff, small weed seeds, and very small pieces of sorghum kernels as dockage. All material from a 1,000-g sample passing through a 2.5/64-in. (1-mm) round-hole sieve when operated in a Carter-Day dockage tester (Fig. 15) is considered dockage. In addition to having little nutritional value, this material creates problems for both the intermediate and end user. High concentrations of fine material reduce dryer efficiency and increase the risk of insect and mold infestation. Grain screenings and dust generally have mold spore concentrations several times higher than whole grain (Sauer et al,

1991). Fine impurities also affect the operational efficiency of steeping or tempering sorghum during processing. Consequently, processors thoroughly clean sorghum to remove any undesirable materials.

The amount of dockage does not govern the numerical grade assigned sorghum under the U.S. standards. The amount is simply reported as part of the inspection

Fig. 15. Carter dockage tester.

process, and the market typically applies a weight adjustment when dockage exceeds 1.0% since the material has little or no value. Other countries, such as Argentina and Australia, include fine material comparable to dockage in their foreign material definitions.

Broken Kernels and Foreign Material. The U.S. sorghum standards combine broken kernels and foreign material as one quality factor for grading purposes. This factor, commonly referred to as BNFM, often represents the quality responsible for the assigned numerical grade. In 1993, the standards were modified to limit the amount of foreign material included in the combined factor BNFM. The inclusion of a subfactor limit for foreign material is intended to improve market efficiency, encourage the production and delivery of cleaner sorghum of more consistent quality; and provide more complete end-use value information. The foreign material consists of large, material (including other grains) easily removed through a mechanical sieving process and material of the same size and density as sorghum, which must be separated manually. Broken kernels are defined as material passing through a 5/64-in. (1.98-mm) triangular-hole sieve and over a 2.5/64-in. (1-mm) round-hole sieve.

Using the combined total of broken kernels and foreign material as a grade-determining factor eliminates the need for precise separation of these components. This facilitates the inspection process and provides the marketplace with quality information reflective of commercial practices. Separating broken kernels from foreign material is costly and commercially impractical. Processing sorghum for food or feed requires preliminary cleaning to maintain operational efficiency and to achieve an acceptable finished product. Commercial cleaning removes dockage, BNFM, and, in many cases, some sound, mature sorghum. Commercial separation of these components is unnecessary. Although the cleaned out material has value as filler for feed or other by-product uses, the U.S. market has not demonstrated a need for the standards to describe sorghum quality by establishing limits for the separate factors.

Typically, the ratio of broken kernels to foreign material is 3:1. This may suggest that a quality distinction between broken kernels and foreign material would be of value. The Argentine and Australian standards do address broken kernels separately from foreign material, which suggests a possible market advantage for considering the nutritional and economic value of broken kernels compared to foreign material. Obviously, the relative value of broken kernels depends on the amount and intended use.

Damaged Kernels. Damage to sorghum may occur before harvest in the field, during harvest, as the result of artificial drying, while in storage, or during handling and transportation. The type and severity of damage varies depending on the cause. In the United States, the quality factor damaged kernels total (DKT) includes damage other than cracked or broken kernels. Conditions, such as kernel discoloration, mold growth, sprouting, insect damage, and germ degradation are used to define damaged kernels. The factor is used to determine the numerical grade and provides information relative to the amount of pre- and postharvest damage from unfavorable environmental conditions or poor handling and storage practices.

Disease or fungi may invade and damage sorghum in the field. Likewise, adverse environmental conditions (heavy frost, flooding, drought, etc.) may result in immature, discolored, or sprouted kernels at the time of harvest. Harvesting may break kernels or mar the pericarp, thus increasing the likelihood of postharvest degradation by mold, insects, or bacteria.

Artificial drying with excessive heat will damage the kernel composition and cause discoloration. The grading process labels this type of damage "heat damage." Improper drying may also create minor cracks in the kernel, which increases the chance of further breakage or invasion by fungi and bacteria.

Damage may also occur during storage. If the storage moisture and temperature are not adequately controlled, sorghum can quickly deteriorate by molding or heating. Growth of fungi and bacteria will break down and absorb the constituents of the kernels. This may cause discoloration of the germ and excessive respiration. Insect infestation is also of concern, causing direct physical destruction of the kernels.

The grading process relies on a physical examination of the sorghum to identify damage that adversely affects storage or processing value. The most common types of damage found in sorghum are germ, mold, and sprout. Other types of damage occasionally found include weather, heat, and insect damage.

The determination of damage, excluding germ damage, is based on a visual appraisal of approximately 15 g after the removal of dockage and mechanically cleaned BNFM. After removing the other types of damage from the subportion, the remaining kernels are placed into a sodium hypochlorite (NaOCl)/potassium hydroxide (KOH) solution to remove the pericarps for ease in determining germ damage. This method of exposing the germ is preferred over scraping to reduce analysis time and, more importantly, to reduce the risk of disturbing or dislodging the germ. The process of identifying the various types of damage relies on the judgment of an inspector to determine whether the degree of deterioration is sufficient to be considered damage under the grading standards. To promote accuracy and uniformity between individuals as they determine whether a kernel is damaged, a set of 35-mm interpretive line slides is used. These slides depict the minimum amount of discoloration or deterioration permitted for the various types of damage.

Special Grades. In addition to the numerical or Sample grade designation, special grades are assigned under the U.S. grading system to alert grain merchants of an unusual quality or condition that may influence the economic value of the grain. The sorghum standards address two such conditions: infested and smutty.

Whether the sample being examined for insects represents the lot or a component, the tolerances are the same. Sorghum will be termed "infested" when a representative sample contains two live weevils, one live weevil and five other live insects (OLI) injurious to stored grain, or 10 OLIs. Weevils are a generic term used for boring insects, such as the rice, maize, and granary weevils as well as the lesser grain borer. OLIs are those nonboring insects injurious to stored grain. These include saw-toothed grain beetles, rusty grain beetles, red flour beetles, and other such insects.

The special grade "smutty" is applied to sorghum that has kernels sufficiently covered with smut spores to give the sample a smutty appearance or that contains 20 or more smut balls full of teliospores in a 100-g sample of sorghum.

Other Grading Criteria. The numerical grading factors may not always portray sorghum quality sufficiently. Therefore, the grading process also involves a general observation of the overall appearance of the sorghum. Sorghum exhibiting a distinctly low quality will be graded Sample grade, the lowest quality designation. Sorghum is also graded Sample grade if it does not meet the numerical 1–4 factor requirements or, based on a representative sample, the sorghum contains small stones (eight or more with an aggregate weight in excess of 0.2%), small pieces of glass (two or more), crotalaria seeds (three or more), castor beans (two or more), animal filth (10 or more), and cockleburs (eight or more). Furthermore, badly discolored sorghum or sorghum with a musty, sour, or commercially objectionable foreign odor falls into the Sample grade category.

Sorghum that is partially discolored with a slight gray or blackened appearance can be graded no higher than U.S. No. 3. Inspectors use photographs to aid in the distinction between badly discolored and partially discolored.

Intrinsic Factors. The current U.S. grading standards and other major standards used throughout the world describe physical and biological characteristics of sorghum. The standards provide little direct information about the yield and quality of a final product. For example, the United States, Argentine, and Australian sorghum standards do not include measurements of protein, starch, oil, or free-fatty acid although this information is important to a processor.

The standards do not include intrinsic quality information due to technology constraints for accurate measurement meeting market needs, the diversity of end users for sorghum, and the absence of a strong market demand for such information.

Current sorghum marketing practices demonstrate minimum interest concerning the variability in the nutritional composition of sorghum. The availability of other feed grains and supplemental feed ingredients may, in part, minimize the economic value of distinguishing sorghum based on intrinsic factors for marketing purposes.

Grading Accuracy

How well inspection results reflect true sorghum quality depends on the amount of random and nonrandom variation introduced into the inspection process. Variability related to sampling was discussed earlier. Variability also occurs due to differences in equipment accuracy, test portion sizes, and the precision of the test methodology (objective vs. subjective methods).

The variables influencing accuracy and reducibility can never be eliminated but can be reduced or stabilized by standardizing the equipment and test procedures. Equipment design and performance standards are essential. Furthermore, specific operator instructions that cover test procedures and diagnostic instrument checks are needed to maintain an accurate and reliable system.

Test portion size also contributes to the stability of inspection results. In the past, FGIS illustrated the effects of portion size on corn damage by making a mixture of 5% yellow and 95% white corn. Two hundred samples, each 1,000 g, were prepared, and the 5% yellow corn was thoroughly mixed with the white corn. The mixtures were divided into 125- and 250-g portions. The level of yellow corn found in the 125-g portions ranged from 2.4 to 7.6%; whereas, the level found in the 250-g portions ranged from 3.6 to 6.2% (*unpublished data*). Standardized portion sizes improve inspection consistency.

The interpretive line slides and prints used to aid inspectors in making subjective determinations of damage have proven useful toward maintaining uniform and consistent inspection results.

MILLET

Millet, like sorghum, has the ability to adapt and respond favorably to adverse climatic conditions (i.e., drought) and is therefore grown in many arid and semiarid regions of the world, where it is an important food source in several Asiatic and African countries. While the majority of the millet produced worldwide is for human consumption, it also used as an alternative feed source and in the production of alcoholic beverages.

According to a recent Food and Agricultural Organization of the United Nations (FAO) report (FAO, 1990), approximately 90% of the millet produced worldwide is produced and utilized by developing countries in Asia and Africa: primarily, China, India, Ethiopia, Niger, and Nigeria. Little surplus is available for commercial marketing as most of the millet produced globally is grown for local consumption. Of the estimated 30 million metric tons produced annually worldwide, less than 200,000 t is traded in international markets, most of which is supplied by four countries: Argentina, United States, Australia, and China. The insignificant role millet assumes in total annual world cereal grain production (about 1.8 billion metric tons) and exports (about 200 million metric tons), clearly shows the preference of other cereal grains over millet in international and domestic markets.

The four principal species of millet produced worldwide, in order of predominance, are pearl millet, foxtail millet, proso, and finger millet. The last is not grown in the United States. Unlike most other countries that produce millet, millet is not grown for food in the United States. Instead, it is primarily grown for hay and pasture or as components of birdseed, i.e., seed for cage, aviary, and garden visitor birds, and chick feed mixtures. Pearl millet, which accounts for about half of the millet grown in the United States, is largely used for pasture and occasionally for silage. The chief commercial use of proso millet is for birdseed and chick feed. Foxtail millet, like proso, is primarily used for birdseed (Martin et al, 1976).

As in the United States, most of the millet traded in international commerce is consumed by the pet-food industry as birdseed. Due to the limited supply of millet available for commercial sale and the limited, specialized markets in which millet is traded, it is not surprising that, with the exception of Argentina, international trading standards for millet have not been developed. Internationally, millet is generally marketed on a sample basis (FAO, 1990). Codex is, however, considering the adaptation of an African regional standard for whole and decorticated pearl millet. The standard focuses on a single set of quality parameters that includes such traits as moisture content, weight of a thousand grains, protein, ash, crude fiber, degree of decortication, fat content, and impurities.

National marketing standards have not been developed for millet in the United States. The domestic U.S. market handles millet on a limited, specialized commercial basis. Due to the abundance and availability of other, more preferred feed and food grains in the United States, it is unlikely that millet will achieve the economic importance to warrant the development a national grading standard.

LITERATURE CITED

COMBS, W. B., and SMITH, F. G. 1940. Grain grading primer. U. S. Dep. Agric. Misc. Publ. 325.

CORBETT, C. L. P., PARKER, E. C., and BOERNER, E. G., 1936. Pages 69-92 in: History of grain inspection in the United States 1838–1936. U.S. Dep. Agric., Bur. Agric. Econ., Grain Div., Washington, DC.

DEYOE, C. W. 1975. Do grain standards relate to feed usage? Kansas Formula Feed Conf., Kansas State Univ., Sec. I, pp. 1-9.

FAO. 1990. Structure and characteristics of the world millet economy. Rep. Committee on Commodity Problems, Intergovernmental Group on Grains, 24th sess. Rome, Italy.

FGIS. 1993. Official United States Standards for Grain. Fed. Grain Insp. Serv., U.S. Dep. Agric., Washington, DC.

FGIS. 1979. Grain Inspection Handbook, Book I. Grain Sampling. Fed. Grain Insp. Serv., U.S. Dep. Agric., Washington, DC.

FGIS. 1990. Grain Inspection Handbook, Book II. Grain Grading Procedures. Fed. Grain Insp. Serv., U.S. Dep. Agric., Washington, DC.

HAHN, D. H., ROONEY, L. W., and EARP, C. F. 1984. Tannins and phenols of sorghum. Cereal Foods World 29:776-779.

HILL, L. D., BERMINGHAM, S. C., and SEMPER, R. 1976. Sampling and measurement problems in grain grading. Dep. Agric. Econ., Univ. Illinois, AE-4407.

HURBURGH, C. R. 1990. How to improve your sampling procedures. Grain J., pp. 22-23.

HURBURGH, C. R., and BERN, C. J. 1983. Sampling corn and soybeans. Trans. ASAE. 26(3):930-934.

MARTIN, J. H., LEONARD, W. H., and STAMP, D. L. 1976. Pages 563-576 in: Principles of Field Crop Production. Macmillan, New York.

MULTON, J. L. 1991. Basics of moisture measurement in grain. Page 36 in: Uniformity by 2000. L. D. Hill, ed. Scherer Communications, Urbana, IL.

MURPHY, E. J. 1955. The history and philosophy of grain standards in the U.S. Northwest. Miller 253(19):10-11,26-28.

SAUER, D., MERONUCK, R., and TUITE, J. 1991. Effects of fine material on mold growth in grain. In: Fine material in grain. R. L. Stroshine, ed. NC-151, Regional Res. Committee, Ohio Agric. Res. Dev. Ctr., Wooster, OH. (In press.)

WATSON, S. A. 1970. Wet-milling process and products. Page 605 in: Sorghum production and utilization. J. S. Wall and W. M. Ross, eds. AVI Publ. Co., Westport, CT.

WATSON, S. A. 1987. Measurement and maintenance of quality. Pages 140-141 in: Corn: Chemistry and Technology. S. A. Watson and P. E. Ramstad, eds. Am. Assoc. Cereal Chem., St. Paul, MN.

INDEX

Entries for tables only are designated by the letter t and figures by the letter f after the page number